MOLECULAR APPROACHES IN NATURAL RESOUR[CE]
AND MANAGEMENT

T0276859

Recent advances in molecular genetics and genomics have be[en]
natural resource conservation. Today, several major conservation and management journals are
using the "genetics" editors of this book to deal solely with the influx of manuscripts that employ
molecular data. The editors have attempted to synthesize some of the major uses of molecular
markers in natural resource management in a book targeted not only at scientists but also at
individuals actively making conservation and management decisions. To that end, the text features
contributors who are major figures in molecular ecology and evolution – many having published
books of their own. The aim is to direct and distill the thoughts of these outstanding scientists
by compiling compelling case histories in molecular ecology as they apply to natural resource
management.

J. Andrew DeWoody is Professor of Genetics and University Faculty Scholar at Purdue University.
He earned his MS in genetics at Texas A&M University and his PhD in zoology from Texas Tech
University. His recent research in genetics, evolution, and ecology has been funded by organizations
including the National Science Foundation, the U.S. Department of Agriculture (USDA) National
Research Initiative, the Great Lakes Fishery Trust, and the National Geographic Society. His research
is published in more than thirty-five journals, and he has served as Associate Editor for the *North
American Journal of Fisheries Management*, the *Journal of Wildlife Management*, and *Genetica*.

John W. Bickham is Professor in the Department of Forestry and Natural Resources (FNR) and
Director of the Center for the Environment at Purdue University. He received his MS in biology
from the University of Dayton and his PhD in zoology from Texas Tech University. He was on the
faculty of Texas A&M University's Department of Wildlife and Fisheries Sciences for thirty years.
He has published more than two hundred articles in scientific journals in evolutionary genetics,
including comparative cytogenetics, molecular systematics, molecular ecology, and ecotoxicology.

Charles H. Michler is the Fred M. van Eck Director of the Hardwood Tree Improvement and Regen-
eration Center at Purdue University and Site Director of the National Science Foundation Industry
& University Cooperative Research Program's Center for Advanced Forest Systems. He earned his
MS and PhD in horticulture, physiology, and biochemistry from The Ohio State University. He has
published more than eighty-five scholarly works and has edited nine books and proceedings. He is
Editor of *Plant Breeding Reviews* and Associate Editor of the *Journal of Forest Research*.

Krista M. Nichols is Assistant Professor, Departments of Biological Sciences and Forestry and Natural
Resources at Purdue University. She received her MS in fisheries and wildlife from Michigan State
University and her PhD in zoology from Washington State University. She was a National Research
Council postdoctoral Fellow at the National Marine Fisheries Services, Northwest Fisheries Science
Center. Dr. Nichols has published in the fields of ecotoxicology, genetics, and ecology.

Olin E. Rhodes, Jr., is Professor in the FNR and Director of the Interdisciplinary Center for Ecological
Sustainability at Purdue University. He received his MS in wildlife biology from Clemson University
and his PhD in wildlife ecology from Texas Tech University. He was named a Purdue University
Faculty Scholar in 2006. He has published more than 135 scholarly works in ecology and genetics
and has recently served as Associate Editor of the *Journal of Wildlife Management*.

Keith E. Woeste is a research molecular geneticist for the USDA Forest Service Northern Research
Station Hardwood Tree Improvement and Regeneration Center and Adjunct Assistant Professor at
Purdue University's FNR. He received an MDiv in theology from the Jesuit School of Theology at
Berkeley, an MS in horticulture from the University of California–Davis, and his PhD in genetics
from the University of California–Davis.

Molecular Approaches in Natural Resource Conservation and Management

Edited by

J. Andrew DeWoody
Purdue University

John W. Bickham
Purdue University

Charles H. Michler
Purdue University

Krista M. Nichols
Purdue University

Olin E. Rhodes, Jr.
Purdue University

Keith E. Woeste
Purdue University

CAMBRIDGE UNIVERSITY PRESS
Cambridge, New York, Melbourne, Madrid, Cape Town, Singapore,
São Paulo, Delhi, Dubai, Tokyo, Mexico City

Cambridge University Press
32 Avenue of the Americas, New York, NY 10013-2473, USA

www.cambridge.org
Information on this title: www.cambridge.org/9780521731348

First published 2010

Printed in the United States of America

A catalog record for this publication is available from the British Library.

Library of Congress Cataloging in Publication data

Molecular approaches in natural resource conservation and management /
edited by J. Andrew DeWoody . . . [et al.].
     p.   cm.
Includes bibliographical references and index.
ISBN 978-0-521-51564-1 (hardback) – ISBN 978-0-521-73134-8 (pbk.)
1. Biodiversity conservation.   2. Genetic resources conservation.
3. Molecular genetics.   4. Conservation of natural resources.
I. DeWoody, J. Andrew, 1969–   II. Title.
QH75.M647   2010
333.95′16–dc22   2009050318

ISBN 978-0-521-51564-1 Hardback
ISBN 978-0-521-73134-8 Paperback

Contents

Color plates follow page 174.

Contributors

Gerard J. Allan
Department of Biological Sciences
Environmental Genetics and
  Genomics Facility
Northern Arizona University
Flagstaff, AZ

Giridhar N. R. Athrey
Department of Biology
University of Louisiana
Lafayette, LA

John C. Avise
Ecology and Evolutionary Biology
School of Biological Sciences
University of California
Irvine, CA

Joseph K. Bailey
Department of Biological Sciences and
  the Merriam-Powell Center for
  Environmental Research
Northern Arizona University
Flagstaff, AZ

Amy B. Baird
National Museum of Natural History –
  Naturalis
Leiden, The Netherlands

Randy K. Bangert
Biological Sciences
Idaho State University
Pocatello, ID

Robert C. Barbour
Cooperative Research Centre for
  Sustainable Production Forestry
School of Plant Science
University of Tasmania
Australia

Kelly R. Barr
Department of Biology
University of Louisiana
Lafayette, LA

Anbreen Bashir
Department of Biology
St. Louis University
St. Louis, MO

Dario Beraldi
Wild Evolution Group
Institute of Evolutionary Biology
School of Biological Sciences
University of Edinburgh
Edinburgh, UK

John W. Bickham
Department of Forestry and Natural
  Resources and Center for the
  Environment
Purdue University
West Lafayette, IN

Joseph D. Busch
Microbial Genetics and Genomics
  Center
Northern Arizona University
Flagstaff, AZ

Robert H. Devlin
Fisheries and Oceans Canada
West Vancouver, BC
Canada

J. Andrew DeWoody
Departments of Forestry and
  Natural Resources and Biological
  Sciences
Purdue University
West Lafayette, IN

Alexis Ducousso
UMR Biodiversité Gènes et
  Communautés
Institut National de la Recherche
  Agronomique
Cestas, France

Yousry A. El-Kassaby
Faculty of Forestry
University of British Columbia
Vancouver, BC
Canada

Julie R. Etterson
Department of Biology
University of Minnesota Duluth
Duluth, MN

Luke M. Evans
Department of Biological Sciences
Northern Arizona University
Flagstaff, AZ

Dylan G. Fischer
The Evergreen State College
Olympia, WA

Catherine A. Gehring
Department of Biological Sciences
Northern Arizona University
Flagstaff, AZ

Jake Gratten
Department of Animal and Plant
  Sciences
University of Sheffield
Sheffield, UK

Matthew C. Hale
Department of Biological Sciences
Purdue University
West Lafayette, IN

J. L. Hamrick
Department of Plant Biology
University of Georgia
Athens, GA

Richard G. Harrison
Department of Ecology and
  Evolutionary Biology
Cornell University
Ithaca, NY

David M. Hillis
Section of Integrative Biology
University of Texas
Austin, TX

Jeffrey D. Holland
Department of Entomology
Purdue University
West Lafayette, IN

Rebecca M. Holmstrom
Department of Biology
University of Minnesota Duluth
Duluth, MN

Rodney L. Honeycutt
Natural Science Division
Pepperdine University, Seaver College
Malibu, CA

Jamie A. Ivy
Department of Collections
San Diego Zoo
San Diego, CA

Sansanee Jamjod
Faculty of Agriculture
Chiang Mai University
Chiang Mai, Thailand

Antoine Kremer
UMR Biodiversité Gènes et
    Communautés
Institut National de la Recherche
    Agronomique
Cestas, France

Robert C. Lacy
Department of Conservation Science
Chicago Zoological Society
Brookfield, IL

Richard F. Lance
Environmental Lab
U.S. Army Engineer Research and
    Development Center
Vicksburg, MS

Emily K. Latch
Department of Biological Sciences
University of Wisconsin–Milwaukee
Milwaukee, WI

Peter W. Lawson
Conservation Biology Division
NOAA Fisheries
Northwest Fisheries Science Center
Newport, OR

Valérie Le Corre
UMR Biologie et Gestion des
    Adventices
Institut National de la Recherche
    Agronomique
Dijon, France

Paul L. Leberg
Department of Biology
University of Louisiana
Lafayette, LA

Carri J. LeRoy
The Evergreen State College
Olympia, WA

Wesley J. Leverich
Department of Biology
St. Louis University
St. Louis, MO

Denise L. Lindsay
Environmental Lab
U.S. Army Engineer Research and
    Development Center
Vicksburg, MS

Chanya Maneechote
Faculty of Agriculture
Chiang Mai University
Chiang Mai, Thailand

Marjorie Matocq
Department of Natural Resources and
    Environmental Science
University of Nevada
Reno, NV

Michelle M. McClure
Conservation Biology Division
NOAA Fisheries
Northwest Fisheries Science Center
Seattle, WA

Jennifer McCreight
Department of Forestry and Natural
    Resources
Purdue University
West Lafayette, IN

Paul McElhany
Conservation Biology Division
NOAA Fisheries
Northwest Fisheries Science Center
Seattle, WA

Allan F. McRae
Genetic Epidemiology Group
Queensland Institute of Medical
 Research
Herston, Australia

Charles H. Michler
Department of Forestry and Natural
 Resources and Hardwood Tree
 Improvement and Regeneration
 Center
Purdue University
West Lafayette, IN

David B. Neale
Department of Plant Sciences
University of California at Davis
Davis, CA

Krista M. Nichols
Departments of Forestry and Natural
 Resources and Biological Sciences
Purdue University
West Lafayette, IN

Sunisa Niruntrayakul
Faculty of Agriculture
Chiang Mai University
Chiang Mai, Thailand

Stephen R. Palumbi
Department of Biology
Hopkins Marine Station
Stanford University
Pacific Grove, CA

Josephine M. Pemberton
Wild Evolution Group
Institute of Evolutionary Biology
School of Biological Sciences
University of Edinburgh
Edinburgh, UK

Rémy J. Petit
UMR Biodiversité Gènes et
 Communautés
Institut National de la Recherche
 Agronomique
Cestas, France

Dudley M. Potts
Cooperative Research Centre for
 Sustainable Production Forestry
School of Plant Science
University of Tasmania
Australia

Amena Prommin
Faculty of Agriculture
Chiang Mai University
Chiang Mai, Thailand

Adirek Punyalue
Faculty of Agriculture
Chiang Mai University
Chiang Mai, Thailand

Tonapha Pusadee
Faculty of Agriculture
Chiang Mai University
Chiang Mai, Thailand

Benjavan Rerkasem
Faculty of Agriculture
Chiang Mai University
Chiang Mai, Thailand

Olin E. Rhodes, Jr.
Department of Forestry and Natural
 Resources
Purdue University
West Lafayette, IN

Kermit Ritland
Department of Forest Sciences
University of British Columbia
Vancouver, BC
Canada

Barbara Schaal
Department of Biology
Washington University
St. Louis, MO

Jennifer A. Schweitzer
Ecology and Evolutionary Biology
University of Tennessee
Knoxville, TN

Lee R. Shugart
LR Shugart and Associates, Inc.
Oak Ridge, TN

Theerasak Sintukhiew
Faculty of Agriculture
Chiang Mai University
Chiang Mai, Thailand

Jon Slate
Department of Animal and Plant
    Sciences
University of Sheffield
Sheffield, UK

Fredrik L. Sundström
Fisheries and Oceans Canada
West Vancouver, BC
Canada

Athitya Suta
Faculty of Agriculture
Chiang Mai University
Chiang Mai, Thailand

Chris W. Theodorakis
Biology Department
Southern Illinois University at
    Edwardsville
Edwardsville, IL

Peter M. Visscher
Queensland Institute of Medical
    Research
Royal Brisbane Hospital
Queensland, Australia

Thomas C. Wainwright
Fish Ecology Division
NOAA Fisheries
Northwest Fisheries Science Center
Newport, OR

Lisette Waits
Fish and Wildlife Resources
University of Idaho
Moscow, ID

Robin S. Waples
Conservation Biology Division
NOAA Fisheries
Northwest Fisheries Science Center
Seattle, WA

Peter M. Waser
Department of Biological Sciences
Purdue University
West Lafayette, IN

David J. Weston
Environmental Sciences Division
Oak Ridge National Laboratory
Oak Ridge, TN

Thomas G. Whitham
Department of Biological Sciences and
    Merriam-Powell Center for
    Environmental Research
Northern Arizona University
Flagstaff, AZ

Alastair J. Wilson
Institute of Evolutionary Biology
School of Biological Sciences
University of Edinburgh
Edinburgh, UK

Keith E. Woeste
Department of Forestry and Natural
    Resources and Hardwood Tree
    Improvement and Regeneration
    Center
Purdue University
West Lafayette, IN

Anupong Wongtamee
Faculty of Agriculture
Chiang Mai University
Chiang Mai, Thailand

Lisa Worthen
Department of Forestry and Natural
  Resources
Purdue University
West Lafayette, IN

Stan D. Wullschleger
Environmental Sciences Division
Oak Ridge National Laboratory
Oak Ridge, TN

Kelly R. Zamudio
Department of Ecology and
  Evolutionary Biology
Cornell University
Ithaca, NY

Preface

The world would be a wonderful place if our natural resources (e.g., forests, fish, and wildlife) needed no management and conservation was not a concern. In a world with a global human population approaching 7 billion and where most developed nations overconsume these resources, however, conservation is a concern and management is necessary for sustainable use. Historically, natural resource management strategies were determined by the collection and interpretation of basic field data. Today, as challenges to the sustainability and conservation of our natural resources arise, managers often need data that cannot be acquired using conventional methods. For example, a natural resource manager might want to know the number of successful breeders in a population or if genetic variation was being depleted because of a management practice. Traditional field craft alone cannot directly address such questions, but the answers can be determined with some precision if the field work is coupled with modern molecular genetic techniques.

Molecules can enlighten us about biological attributes that are virtually impossible to observe in the field (Avise 2004). Parentage analysis is one such arena in which genetic data can inform management practices (DeWoody 2005), but there are a host of others. For example, molecular data have revealed deep evolutionary splits in stocks at one time thought to be homogeneous. This finding has concomitant management implications (Hoffman et al. 2006). Similarly, molecules can enlighten us about biologies that are virtually impossible to observe in the field, such as pollen flow (Hamrick, this volume) or the physiology of migration (Nichols et al. 2008).

Recent advances in molecular genetics and genomics have been embraced by many scientists in natural resource conservation. Today, several major conservation and management journals (e.g., *Journal of Wildlife Management, North American Journal of Fisheries Management, Plant Breeding Reviews*) are now using "genetics" editors to deal solely with the influx of manuscripts that employ molecular data. We have attempted to synthesize some of the major uses of molecular markers in natural resource management in a book targeted not only at scientists but also at individuals actively making conservation and management decisions. To that end, we have identified contributors who are major figures in molecular ecology and evolution; many have published books of their own. Our aim has been to direct and distill the thoughts of these outstanding scientists by compiling compelling case histories in molecular ecology as they apply to natural resource management.

Clearly, we hope this book will appeal to academics interested in conservation genetics, molecular ecology, and the quantitative genetics of wild organisms. We think this book could be used as an educational tool – as a text for graduate ecology/genetics courses but also, perhaps, in advanced undergraduate courses. Furthermore, we hope this book will be useful to audiences in natural resource management, education, and research by clarifying how genetic approaches can be used to answer resource-related questions.

ABOUT THE EDITORS

Our collective expertise spans from molecular population genetics in the wild to genomics and quantitative genetics of managed or cultured species. We all study the genetics of natural resources, however, and we find that similar issues arise in wildlife, forestry, and fisheries. For example, when the forest geneticists began asking how many sires contributed pollen to a nut-bearing hardwood tree, it turns out that fisheries geneticists had already studied this problem from the perspective of a male fish guarding a nest full of developing embryos, and they had created computer programs to estimate the number of parents contributing gametes to a nest (DeWoody et al. 2000). Another such intersection of research across disciplines lies in the study of genetic processes in small populations; the same conceptual and analytical approaches being used to elucidate the genetic consequences of wildlife reintroductions (Latch & Rhodes 2005) are employed to evaluate genetic diversity in hardwood tree species subjected to severe habitat fragmentation (Victory et al. 2006). Our desire to produce a book stems from our mutual interests in understanding how molecular genetics can be used to inform and improve natural resource management.

In addition to our research interests, we teach several courses that directly pertain to this book. These courses include *Conservation Genetics* (DeWoody), *Molecular Ecology and Evolution* (DeWoody), and *Evolutionary Quantitative Genetics* (Nichols). Furthermore, several of us (DeWoody, Michler, Rhodes) have served as "genetics" editors for conservation and management journals, including *Journal of Wildlife Management, North American Journal of Fisheries Management*, and *Plant Breeding Reviews*.

ACKNOWLEDGMENTS

In addition to the authors, most of whom also provided reviews on other chapters and/or boxes, we thank the following individuals for their invaluable feedback: Jean Beaulieu, Tasha Belfiore, John Burke, Dave Coltman, Tariq Ezaz, Ben Fitzpatrick, Anthony Fiumera, Mike Goodisman, Rick Howard, Irby Lovette, Bill Muir, Patty Parker, Devon Pearse, Joe Quattro, Kim Scribner, Ron Sederoff, and Rod Williams. All provided insightful comments that directly strengthened

Book contributors at an October 2008 meeting, held at the John S. Wright Forestry Center (Purdue University). Row 1: Krista Nichols, Kelly Zamudio, Charles Michler, Yousry El-Kassaby, Tom Whitham, Jamie Ivy, Emily Latch, Lisette Waits, and Marjorie Matocq. Row 2: Lee Shugart, Dave Neale, Dave Hillis, John Avise, Andrew DeWoody, Robin Waples, Rodney Honeycutt, Paul Leberg, and John Bickham. Row 3: Kermit Ritland, Antoine Kremer, Stan Wullschleger, Keith Woeste, Peter Waser, Jim Hamrick, Gene Rhodes, and John Patton. Photo credit: Caleb D. Phillips. *See Color Plate I.*

individual chapters and boxes, and we trust that this book has been enhanced by their efforts.

This volume was largely possible because of the financial and logistical support of the Department of Forestry and Natural Resources at Purdue University. In particular, the department sponsored an October 2008 meeting at Purdue where many of the book contributors congregated for three days of scientific discourse and fellowship before finalizing their respective chapters or boxes.

Our own research programs have been supported by a variety of organizations, including the National Science Foundation (DeWoody, Bickham, Michler, Nichols), the U.S. Department of Agriculture (DeWoody, Michler, Nichols, Rhodes, Woeste), the State of Indiana (DeWoody, Michler, Rhodes), the National Oceanic and Atmospheric Administration (Bickham), the Great Lakes Fishery Trust (DeWoody, Nichols), and the U.S. Forest Service (Michler, Woeste). We thank them all for investing in science.

REFERENCES

Avise JC (2004) *Molecular Markers, Natural History, and Evolution*. Sinauer, Sunderland, MA.

DeWoody JA (2005) Molecular approaches to the study of parentage, relatedness and fitness: practical applications for wild animals. *Journal of Wildlife Management*, **69**, 1400–1418.

DeWoody JA, DeWoody YD, Fiumera A, Avise JC (2000) On the number of reproductives contributing to a half-sib progeny array. *Genetical Research* (Cambridge), **75**, 95–105.

Hoffman JI, Matson CW, Amos W, Loughlin TR, Bickham JW (2006) Deep genetic subdivision within a continuously distributed and highly vagile marine mammal, the Steller's sea lion (*Eumetopias jubatus*). *Molecular Ecology*, **15**, 2821–2832.

Latch EK, Rhodes OE Jr (2005) The effects of gene flow and population isolation on the genetic structure of reintroduced wild turkey populations: are genetic signatures of source populations retained? *Conservation Genetics*, **6**, 981–997.

Nichols KM, Felip A, Wheeler P, Thorgaard GH (2008) The genetic basis of smoltification-related traits in *Oncorhynchus mykiss*. *Genetics*, **179**, 1559–1575.

Victory E, Glaubitz JC, Rhodes OE Jr, Woeste KE (2006) Genetic homogeneity in *Juglans nigra* (Juglandaceae) at nuclear microsatellites. *American Journal of Botany*, **93**, 118–126.

1 Biodiversity discovery and its importance to conservation

Rodney L. Honeycutt, David M. Hillis, and John W. Bickham

During the eighteenth, nineteenth, and early twentieth centuries, scientific inventories of biodiversity flourished as naturalists participated in expeditions throughout different geographic regions of the world (Köhler et al. 2005). These expeditions and the various journals produced by many prominent naturalists provided materials for extensive scientific collections as well as accounts of the habits and habitats of both plant and animal species. Charles Darwin and Alfred Russel Wallace were part of this tradition, and both were students of biodiversity. They chronicled their adventures in South America, the Malay Archipelago, the Galapagos Islands, New Zealand, and Australia as they discovered new species, described geology, and encountered various cultures (Darwin 1845; Wallace 1869). These adventures honed their observational skills, and their experiences culminated in their parallel proposals of the theory of biological evolution by means of natural selection. The biodiversity and natural environments encountered by Darwin and Wallace have been altered, and both habitats and species described in their journals have and are being impacted at a drastic rate. The yellow-bridled finch (*Melanodera xanthogramma*), noted by Darwin as "common" in the Falkland Islands, is now gone, and, as predicted by Darwin, the Falkland Islands fox or warrah (*Dusicyon australis*) went extinct in 1876 (Armstrong 1994). The Borneo forest harbors fewer *Mias* or orangutans, and it is unlikely that one would be allowed to collect specimens like Wallace describes (Wallace 1869). Even "pristine" regions, such as those seen by Darwin in Patagonia and the southwest Atlantic coast of Argentina, are still poorly understood, yet they are threatened by numerous human activities (Bortolus & Schwindt 2007).

Owing primarily to the fact that the probability of massive species extinction is inevitable, interest in an all-species inventory and the derivation of a Tree of Life has increased over the last two decades. In 1992, the systematics community in the United States, through funding by the National Science Foundation, organized a meeting to set an agenda for the upcoming millennium. As a consequence, Systematics Agenda 2000 (1994) established three major goals: 1) to conduct a worldwide survey and inventory of all species and the taxonomic description of new species; 2) to derive a phylogeny or Tree of Life for all species that would serve as the basis for a classification as well as a framework for other researchers in the life sciences; and 3) to develop an information retrieval system for managing data on biodiversity.

Although our knowledge of biodiversity on planet Earth has increased as a consequence of the endeavors of early naturalists and these new initiatives, we are still far from a complete census of all species, and many will go extinct before their discovery. Such an inventory is essential because it provides a baseline for understanding the stability of ecosystems and the impact of anthropogenic processes that may eventually result in our own demise.

This chapter relates specifically to the inventory of biodiversity as an important step for its conservation. The first section provides a general overview of the importance of biodiversity to society, presents a survey of its global distribution, and identifies groups and geographic regions threatened by human activities. The second section reviews our current knowledge of worldwide biodiversity in terms of its discovery and description and identifies groups that are poorly known. The third section discusses the future of inventorying biodiversity and reviews how molecular approaches and phylogenetic methods provide means for accelerating the overall processes of species discovery and the construction of the Tree of Life. Finally, we emphasize the importance of an information retrieval system that makes data on biodiversity accessible to the entire scientific community.

BIODIVERSITY

Why is biodiversity important?

Biodiversity is defined as "the variability among living organisms from all sources including, inter alia, terrestrial, marine and other aquatic ecosystems and the ecological complexes of which they are part; this includes diversity within species, between species and of ecosystems" (Glowka et al. 1994). The importance of biodiversity and the need for its conservation worldwide cannot be over-emphasized. Not only are diverse forms of organisms responsible for sustaining human populations, they also serve important roles in the maintenance of ecosystems.

Advances in human medicine have benefited directly from biodiversity (Bernstein & Ludwig 2008; Harvey 2008). For instance, species of bacteria discovered over the last forty years have helped minimize transplant rejection, provided antibiotics and antifungal drugs that help combat infections from harmful pathogens, and revolutionized molecular biology by providing thermostable DNA polymerases used in polymerase chain reaction (PCR), a procedure employed broadly in medical diagnostics. Other plant and animal species provide drugs useful for treating cancer and serve as model organisms for studying molecular processes associated with disease and neurological disorders. Approximately 50% of the most broadly used drugs were derived from natural resources (Bernstein & Ludwig 2008), and biodiversity continues to serve as an important resource for the development of clinically important pharmaceuticals and other health-related products (Harvey 2008).

Aside from medicine, humans receive both direct and indirect benefits from biodiversity through the provision of food, fuel, clean water, and fertile soil that enhances agriculture. According to Wilson (1985), more than seven thousand species of plants have been used for food, and twenty species are essential as worldwide food sources. Related species to those currently used for food are genetic reservoirs containing genes that may serve as sources of resistance to pathogens and pests as well as to potential climatic changes. In fact, enhancing both genetic diversity and crop diversity to create more complex "agroecosystems" may provide a more natural means of not only increasing production but also reducing the need for excessive use of pesticides and other chemicals (Altieri 2004). Biodiversity also provides a host of benefits and services to ecosystems. Processes associated with the recycling of nutrients, carbon and nitrogen cycling, formation of soils, climate stabilization, plant pollination, and decomposition of pollutants are influenced by biodiversity, and many of these processes are important to worldwide economies (Pimentel et al. 1997).

How is biodiversity distributed?

Biodiversity is not randomly distributed worldwide. Comparisons across biogeographic regions reveal areas that differ significantly in terms of species diversity and levels of endemism (Gaston 2000). Two major patterns associated with differences in biodiversity are considered relevant to conservation issues. First, species richness varies over a latitudinal gradient, with more species occurring in the tropics than in more temperate regions (Gaston 1996). This general pattern appears to hold for many different taxa, such as plants, mammals, and birds, yet there are exceptions. Although amphibians are more diverse in the tropics, their general pattern of diversity is not completely correlated with latitude in that they do show local exceptions, such as their high diversity in the mountains of the eastern United States relative to other areas of North America and Europe (Buckley & Jetz 2007). This latitudinal gradient associated with species richness also appears to be asymmetrical, with the gradient stronger in the northern than in the southern hemisphere (Chown et al. 2004). As indicated by several authors (Gaston 2000; Hawkins 2001; Ricklefs 2004; Buckley & Jetz 2007; Dyer et al. 2007), the mechanisms responsible for the latitudinal gradient are widely debated but probably relate to several different factors including ecological (e.g., species interactions), environmental (e.g., habitat quality, energy, and degree of stability), and historical processes (e.g., degree of isolation, rates of extinction, migration, and speciation). Second, species richness increases with size of area, and, like latitudinal gradients, species-area curves are a pattern observed for plants, animals, and bacteria (Rosenzweig 1995; Horner-Devine et al. 2004). According to Rosenzweig (1992), latitudinal gradients and species-area curves occur at different temporal and spatial scales, with the latter occurring recently and at a more local or regional scale. In terms of predicting species diversity, this area effect probably relates to habitat heterogeneity (Rosenzweig 1992). For instance, Horner-Devine and colleagues (2004) observed an increase in bacterial species diversity with an increase in area that appeared related to an increase in overall habitat heterogeneity as the distance between sites in a salt marsh increased, and Báldi (2008) found that arthropod diversity on several reserves varied in response to habitat heterogeneity rather than to area.

Regardless of the mechanisms for latitudinal gradients and species-area curves, both of these general observations have been used to establish priorities for maximizing the conservation of biodiversity through the identification of regions (termed "biodiversity hotspots") that should receive high conservation priority (Myers 1988). Two criteria are commonly used to identify biodiversity hotspots. First, the number of endemic species (i.e., species that cannot be replaced if lost from a region) is considered a more important indicator than species richness, which is potentially biased toward broadly distributed species. Second, areas with high levels of endemism and under threat of habitat loss receive the highest conservation priority. The overall establishment of biodiversity hotspots is based on an extrapolation of better-known species, especially those that are indicators of habitats. As such, plant diversity is a common means of ranking hotspots. For instance, some of the first hotspots (e.g., ten sites in tropical rainforests) were recognized based on plant diversity (Myers 1988). Similarly, Mittermeier and colleagues (1998) identified twenty-four biodiversity hotspots on the basis of plant species endemism and the degree to which vegetation cover was being removed (some as high as 98%). These original twenty-four hotspots represented approximately 2% of land surface and approximately 46% of endemic plant species (Mittermeier et al. 1998). Currently, Myers and colleagues (2000) recognize twenty-five terrestrial hotspots, encompassing 1.4% of the world's land area and representing a large percentage of plant and vertebrate species. As before, the primary indicator of these biodiversity hotspots is percentage of endemic plant species and secondarily the percentage of endemic species of mammals, birds, reptiles, and amphibians.

There is an inherent assumption that uniqueness of (primarily) plants and (secondarily) vertebrates, as indicators of hotspots, can be extrapolated to lesser-known taxa such as invertebrates. The establishment of global priorities of conservation based on this assumption is somewhat problematic. For instance, Grenyer and colleagues (2006) examined the distribution of three vertebrate groups (birds, mammals, and amphibians) and found similar species richness among regions, yet little congruence in terms of the identification of hotspots based on the distribution of rare and vulnerable species associated with each group. This finding suggests that setting global priorities on the basis of surrogate taxa may be inappropriate, especially when identifying smaller, regional areas for conservation activities (Reid 1998). The finding also implies that broader taxonomic coverage is required for the identification of hotspots that encompass the majority of rare and endemic species.

Is the extinction of biodiversity a problem?

> Those living today will either win the race against extinction or lose it, the latter for all time. They will earn either everlasting honor or everlasting contempt.
>
> (E. O. Wilson 2006, p. 99)

Extant species represent approximately 1–2% of the Earth's historical biodiversity (May et al. 1995). Therefore, *extinction*, the loss of a lineage with no replacement, is a natural process that appears nonrandom, relative to the species that go extinct, and "episodic" in the fossil record (Raup 1986, 1994). This pattern of extinction implies that the average life span for most species is short, between one and ten million years (May et al. 1995). Theoretically, the Tree of Life can withstand random and "vigorous pruning" and recover from major extinction events (Nee & May 1997). Therefore, if extinction is a natural process and the Tree of Life is capable of responding to large extinction events, why is extinction a major concern of persons and groups interested in the conservation of biodiversity? The answers are twofold. The first is from a selfish point of view: The composition of communities that will appear subsequent to such pruning is likely to be different. The loss of important ecosystem services necessary for human survival implies that *Homo sapiens* might be a casualty of rapid and random extinction processes. Even if biodiversity loss does not cause extinction of our species, it is sure to have profound negative impacts on our society. The second answer, of more immediate importance, is the estimated rate at which biodiversity is currently going extinct. Based on the fossil record, Earth has experienced five mass extinctions, each resulting in a net loss of 75–95% of species (Raup 1994). Although difficult to quantify, most evidence suggests that current rates of extinction may be approaching those experienced during mass extinctions. On the basis of annual loss from deforestation and International Union for Conservation of Nature (IUCN) listings, May and colleagues (1995) calculated a range of 200–500 years for the current life span of a species. In a separate study, the current rate of extinction was estimated to be 100–1,000 times faster than the rate estimated prior to humans (Pimm et al. 1995).

According to IUCN's Red List assessment (Baillie et al. 2004), the rate of extinction for birds, amphibians, and mammals over the last century is 50–500 times higher than background extinction. Since the 1500s, 884 extinctions (784 total extinctions and 60 extinctions in the wild) of all species assessed by IUCN have been verified (Baillie et al. 2004). The rate of extinction for amphibians, reptiles, and mammals has increased since the beginning of the twentieth century, whereas extinction of birds started increasing in the eighteenth century, especially on Oceanic islands (Nilsson 2005; Fig. 1–1).

Extinction is an ongoing process, and although many currently recognized species are not extinct, a large number are increasing in vulnerability to extinction. For instance, of the 44,838 species assessed by IUCN (2008), 38% are threatened with extinction, and, except for birds and mammals, the other vertebrate groups show an increased rate of addition to the threatened list between the years 1996 and 2008, owing primarily to an increase in the number of species assessed for these groups (Fig. 1–2). Nearly complete assessments of mammals, birds, and amphibians were performed, and a summary of results is shown in Table 1–1.

Most species of birds, mammals, and amphibians have been evaluated by IUCN, and, among these three groups of vertebrates, amphibians worldwide show the highest risk of extinction. As of January 2010, 6,603 living species have been described (AmphibiaWeb 2010). Table 1–1 shows the number of species considered by IUCN in 2008; approximately 32% are threatened with extinction (Wake & Vredenburg 2008), which represents a potential rate of extinction that may approach 45,000 times the background rate. In comparison to birds and

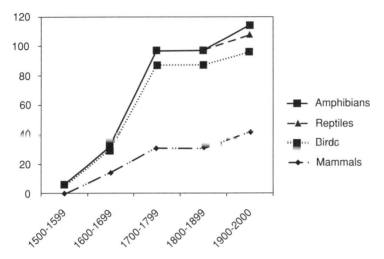

Figure 1-1: Extinction of vertebrate species between 1500 and 2000 (modified from Nilsson 2005).

mammals, twice as many species of amphibians are listed as critically endangered, and nearly one-third of the extinctions of amphibians have occurred in the last thirty years (Stuart et al. 2004). The current rate of amphibian decline has been referred to as "enigmatic" (Stuart et al. 2004) in that the processes responsible are complex, being caused by a host of potential culprits including fungal pathogens, loss of habitat, and changes in climate. The declines are not random. Amphibian communities in the neotropics, especially those in streams, are highly threatened (Dudgeon et al. 2006). In addition, of the 220 species of amphibians in Madagascar, 55 are threatened (Andreone et al. 2005).

A quarter of mammalian species (both marine and terrestrial forms) are vulnerable to extinction, and a high percentage of species show evidence of ongoing

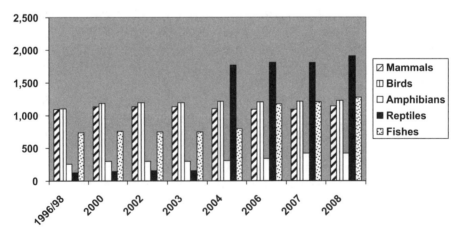

Figure 1-2: Changes in numbers of species of vertebrates in the threatened categories (critical, endangered, vulnerable) from 1996 through 2008 (derived from IUCN's Red List of Threatened Species™ 2008).

Table 1–1. Statistics on threatened species compiled from Table 1 of IUCN's Red List of Threatened Species™ (2008)

	Described species	Evaluated by IUCN 2008	Threatened	% Threatened based on number evaluated
Vertebrates	61,259	26,604	5,966	22
Mammals	5,488	5,488	1,141	21
Birds	9,990	9,990	1,222	12
Reptiles	8,734	1,385	423	31
Amphibians	6,341	6,260	1,905	30
Fish	30,700	3,481	1,275	37
Invertebrates	1,232,384	6,161	2,496	41
Insects	950,000	1,259	626	50
Molluscs	81,000	2,212	978	44
Crustaceans	40,000	1,735	1,735	35
Corals	2,175	856	235	27
Arachnids	98,000	32	18	56
Others	61,209	67	33	49
Plants	298,506	12,055	8,457	70
Lichens, Mushrooms, Brown Algae	50,040	18	9	50
Total	1,642,189	44,838	16,929	38

population decline (Schipper et al. 2008). The trend in mammalian declines is not random in that some regions and groups of species are at greater risk. For instance, nearly 80% of all species of primates from Southeast Asia are threatened (Schipper et al. 2008), and larger mammals in general are more at risk, especially those that have been or are being overexploited by humans, such as the African elephant and many carnivores. Populations of many species of great whales were reduced to near extinction by commercial whaling. Some species have recovered to some degree, but others, including the Atlantic right whale, the Spitsbergen and Okhotsk Sea stocks of bowhead whales, and the western Pacific population of gray whales, are critically endangered and have not recovered despite the cessation of commercial whaling (IWC 2007). There are multiple causes for the demise of mammalian diversity including loss of habitat such as tropical deforestation, overharvesting and bycatch, pollution, and climate change. The Yangtze River dolphin or baiji (*Lipotes vexillifer*) has been declared as extinct, and its extinction represents the first mega-vertebrate extinction in fifty years and the first human-caused extinction of a cetacean. Moreover, it is the fourth mammalian family to become extinct in 500 years (Turvey et al. 2007).

Relative to the total number of species, birds (at 14%) are the least threatened. Extinctions are better documented for birds than probably any other group, however, and the patterns and processes of avian extinctions merit discussion. Some of the more vulnerable regions for birds are islands where, historically, most avian extinctions have occurred. Today, nearly 40% of avian species listed as threatened occur on islands (Johnson & Stattersfield 2008) and have an extremely high probability of extinction relative to mainland species (Trevino et al. 2007). The causes of both extinction and increased vulnerability of island birds include loss of habitat, overexploitation by humans, and the introduction of invasive species

(Johnson & Stattersfield 2008). On some islands, bird diversity has been severely depleted as a result of one or more of these causes. For instance, on Guam, ten of the thirteen species of forest birds are now extinct as a result of the brown tree snake, a species accidentally introduced after World War II (Fritts & Rodda 1998). A large percentage of threatened birds, however, occur in forested mainland habitats, many of which are subject to deforestation and fragmentation (Brooks et al. 1999) that together are accelerating the probability of avian extinctions. One particular region that has experienced considerable loss of habitat is the Atlantic Forest of Brazil and Argentina. According to Ribon and colleagues (2003), approximately 60% of the bird species in this region are either extinct or vulnerable to extinction.

Thorough assessments of both reptiles and fishes are lacking, but both are threatened as a result of overharvesting and loss of habitat. For reptiles, the percentages of threatened chelonians (turtles and tortoises) and crocodilians are 42% and 43%, respectively (Baillie et al. 2004). Although fish diversity is poorly known relative to that of other groups of vertebrates, freshwater ecosystems in general are extremely threatened, and, according to Lundberg and colleagues (2000), approximately 40% (10,000) of all described species of fish occupy freshwater, which makes up 0.01% of the world's water. As indicated by Dudgeon and coworkers (2006), freshwater systems represent the "ultimate conservation challenge" as a result of increased use of this resource worldwide. This increased use threatens not only fish but also other vertebrates, invertebrates, and microbes that rely on freshwater habitats, and extinction rates may be five times higher than predicted for terrestrial ecosystems, reaching nearly 50% in North America (Ricciardi & Rasmussen 1999).

ENUMERATION OF BIODIVERSITY

> We need an expedition to planet Earth, where probably fewer than 10 percent of the life forms are known to science, and fewer than 1 percent of those have been studied beyond a simple anatomical description and a few notes on natural history.
>
> (Wilson 2006, p. 116)

Status of species discovery and description

Species represent the basic units by which biodiversity is measured, and, as such, the first goal of Systematics Agenda 2000 is critical. Accuracy in the estimation of extinction rates, the establishment of conservation priorities based on biodiversity hotspots, and the designation of lineages that are essential for ecosystem function and the long-term survival of biodiversity require knowledge of the approximate number of species currently inhabiting the Earth.

How far have we progressed in our discovery and description of species since Linnaeus? According to Mayr (1969), Linnaeus's *Systema Naturae* lists 4,162 species, and since Linnaeus's time, the enumeration of total species has shown progress, with the current number of discovered species being between 1.5 million and 1.9 million (May 1988, 1990, 1992). Species discovery for birds (Mayr 1946; Monroe and Sibley 1990; Peterson 1998; MacKinnon 2000), mammals (May 1988; Wilson & Reeder 1993, 2005; Patterson 2001; IUCN Red List of Threatened Species 2008), amphibians (Glaw & Köhler 1998; Köhler et al. 2005; Frost 2006; AmphibiaWeb 2010), and turtles (Bickham et al. 2007) is reasonably well documented, and all these groups show an increase in species discovery since Linnaeus, with most species of birds described early in the last century (Fig. 1–3).

The overall rate of species discovery is clearly increasing. For instance, in 2006 (State of Observed Species 2008), 16,969 new species of plants and animals were discovered, with the majority represented by vascular plants and invertebrates (Fig. 1–4). The rate of discovery of amphibian species increased by approximately 26% between 1992 and 2003, and in some geographic regions (e.g., Madagascar) the increase was 42% (Köhler et al. 2005). Mammalian species continue to be discovered. According to Patterson (2000), the rate of discovery of mammalian species in the neotropics is ten times that seen for birds. This rate of discovery is especially high for smaller mammals, such as rodents and bats (Patterson 2001), and in some cases species were rediscovered from existing collections and more recent genetic studies (Patterson 2000).

Although microbial diversity is essential to ecosystem function (Woese 1994), only approximately 5,000 species have been described (Pace 1997). In the past, this lack of species discovery was hindered by the fact that approximately 99% of microbes cannot be cultured (Amann et al. 1995). Most knowledge of bacterial species diversity comes from nucleotide sequences of ribosomal ribonucleic acid (rRNA) (Pace et al. 1986), and molecular markers are now being used to survey microbial diversity in a variety of habitats including soil (Schloss & Handelsman 2006), air (Brodie et al. 2007), marine communities (Sogin et al. 2006; Frias-Lopez et al. 2008), and extreme environments (Huber et al. 2007).

Despite the overall increase in the rate of species discovery, the tally of all species is incomplete and varies greatly across groups. Previous estimates of the potential number of species range between 5 million and 50 million, and the most current estimates are between 15.6 million and 19 million species (Erwin 1982; May 1988, 1992, 1998; Hammond 1992; Stork 1993; Ødegaard 2000; Novotny et al. 2007). Therefore, based on these numbers, our knowledge is limited to about 10% of the Earth's biodiversity (Fig. 1–5). Even for some groups of vertebrates, an all-species inventory is far from complete. For instance, with the exception of turtles, which are reasonably well known and assessed (Baillie et al. 2004; Bickham et al. 2007), the conservation status of many species of reptiles and fishes is less well known, partially as a consequence of the lack of an all-species inventory for these two groups (Table 1–1).

The most species-rich groups of organisms are even less well known than reptiles and fishes (Fig. 1–5 and Table 1–1). Approximately 59% of all described species are insects and 75% of all described species are invertebrates, yet the conservation status for most of these species has not been evaluated (IUCN 2008; Table 1–1). Even more disturbing is the fact that 80–95% of insect species have not been discovered (Stork 2007), and the number of arthropod species may range between five million and thirty million (Erwin 1982; Ødegaard 2000). On the basis of molecular markers, we are far from determining the number of microbial

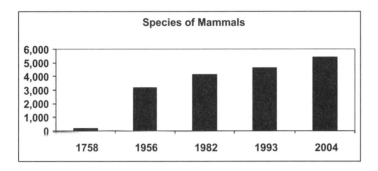

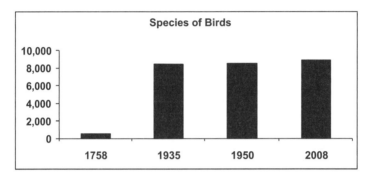

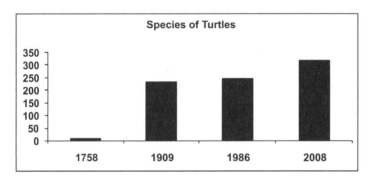

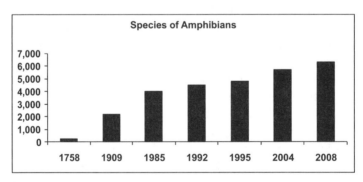

Figure 1–3: Species of vertebrates discovered since Linnaeus.

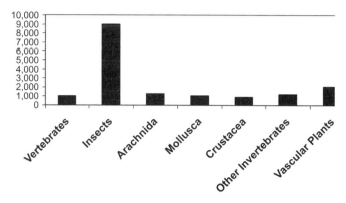

Figure 1–4: New species of plants and animals discovered in 2006 (modified from State of Observed Species 2008).

species, which may be 100 times higher than the number estimated using conventional techniques (Sogin et al. 2006).

Factors limiting the rate of species discovery

Assuming that fifteen million species are undiscovered, the rate of discovery required to establish a complete inventory by the end of the century is approximately 150,000 per year, a rate more than 22 times higher than the average of the previous 250 years (approximately 6,800 species per year since the time of Linnaeus). The rate of species discovery in recent years has been higher than the 250-year average; as shown in Fig. 1–4, the current rate is approximately 17,000 species per year. Thus, the current rate of species discovery and description would need to increase approximately ninefold to reach fifteen million described species by 2100.

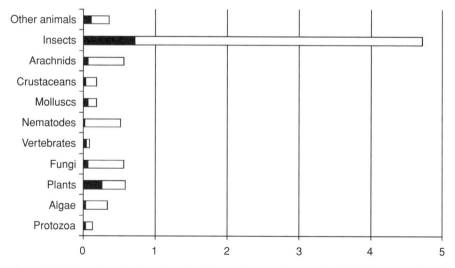

Figure 1–5: Described (black) and estimated number of extant species (white) from May (1998) with values in millions.

At this point, we have mostly described the easiest cases (large species in accessible places that can be distinguished morphologically), and much of the undiscovered biodiversity represents new challenges for systematists. Currently, several factors limit the feasibility of an all-species inventory. First, the discovery and description of new species, especially those in more diverse groups, require taxonomic expertise, and the number of experts varies greatly across groups. The slow pace of species descriptions is exacerbated by the loss of taxonomists familiar with some of the more diverse groups of taxa. This decline in the number of experts is unfortunate because studies of biodiversity rely on the accuracy of taxonomic descriptions and the establishment of a formal classification, both of which serve a "utilitarian" role for the identification and enumeration of species (Mayr 1969; Dubois 2003). As indicated by Wheeler (2004), the "infrastructure of taxonomy" needs to be reestablished through the funding and training of taxonomists.

Second, detailed surveys and inventories of geographic regions harboring large numbers of species are lacking. Decline of species diversity in these regions is increasing as a result of the alteration of natural habitats (Gibbons et al. 2000; Dudgeon et al. 2006). The process of discovery requires detailed assessments of geographic regions, some parts of which (e.g., North America, Europe, Japan, and Australia) are better known than others. Even though 32% of the described species of reptiles occur in the neotropics, less is known about the basic biology of many species (Urbina-Cardona 2008). Advances are being made to determine the status and distribution of reptiles worldwide (Cox et al. 2006), and a Global Biodiversity Assessment was started by IUCN in 2004. Regions of the world with high diversity of fishes are also poorly known, with the distribution of most species not well defined and diversity poorly surveyed (Lundberg et al. 2000; Dudgeon et al. 2006; Abell et al. 2008). As noted by Dudgeon and colleagues (2006), new species were being discovered at an average of several hundred per year between 1976 and 1994. Recently, existing information for more than 13,400 freshwater fish species was used to assign species to ecoregions characterized by level of endemism and diversity (Abell et al. 2008). More than half of these species are endemic to a specific ecoregion. Some of the highest species richness is observed in regions of Africa, the Amazonian Basin in South America, and parts of Asia. As indicated by these authors, some designated regions are "data poor." Nevertheless, this global assessment is a first step toward establishing conservation priorities for freshwater fish species.

Finally, the rate at which detailed species inventories are conducted needs to be increased, and technological advances in molecular genetics, especially those related to nucleotide sequencing, offer the best opportunity for accelerating the rate at which new taxa are discovered and placed within a phylogenetic context. The discovery of microbial species and their relationships has definitely benefited from the application of molecular techniques. For instance, phylogenies derived from rRNA sequences are the bases for a classification scheme that recognizes three major domains of life (Bacteria, Archaea, and Eukarya) as well as thirty or more major clades (Woese 1987; Pace 1997). In addition, the rate of discovery of new microbial species remains high, with considerable differences in species richness in different types of habitats (Schloss & Handelsman 2004). For instance, in thirty grams of soil from forest habitat at least 500,000 species were found (Dykhuizen 1998), and in a relatively well-known habitat like the Sargasso Sea, 148 new species or phylotypes were discovered, suggesting that the actual diversity may be as high as 47,000 species (Venter et al. 2004). Recent surveys of deep sea vents also revealed 2,700 phylotypes of Archaea and 37,000 Bacteria (Huber et al. 2007), and in a recent study of the microbial ecology of human skin, 15 undescribed species were discovered (Gao et al. 2007).

Is it possible to accomplish this goal of a relatively complete assessment of the Earth's biodiversity in the twenty-first century by following traditional approaches employed by systematics? We argue that new technology and approaches, changes in taxonomic practice and culture, and a sustained increase in funding and training are needed to reach this goal. In the following section, we will discuss some of these issues as they relate to a total biodiversity inventory.

FUTURE INVENTORY OF BIODIVERSITY

As indicated by Mayr (1969, p. 9), "The ultimate task of the systematist is not only to describe diversity of the living world but also to contribute to its understanding." The discipline of systematic biology is dedicated to the study of organic diversity, and the overall processes responsible for that diversity. The two major subdisciplines of systematics are phylogenetics and taxonomy. Taxonomy is required for the discovery, description, and identification of species, and the disciplines of taxonomy and phylogenetics merge in the creation of a classification scheme that reflects phylogenetic relationships. Formal names and recognized categories provide a mechanism for information retrieval that allows for the cataloguing and identification of worldwide biodiversity. Procedures used in classical taxonomy provide universal access to and communication about this information, and, as such, systematics in general should play a central role in the discovery and conservation of biodiversity. In fact, historical records derived from floras and faunas of various regions of the world and museum records provide the baseline for current information used to designate biodiversity hotspots and to assess the number of species threatened with extinction.

Although we feel that traditional systematics is essential to any enterprise designed to both discover and name all species on Earth, there are tools available for enhancing the efficiency and accuracy of species inventories. "DNA-based technology," improved phylogenetic methods, and databases that are Web-based and open access provide the necessary infrastructure for a concerted effort to survey and inventory all existing life forms on this planet. In the following sections, we will address some of these new technologies, including their appropriate contribution to the overall goal of the conservation of biodiversity and the completion of an all-species inventory.

Importance of phylogenetics to the delimitation and conservation of species

As buds give rise by growth to fresh buds, and these, if vigorous, branch out and overtop on all a feebler branch, so by generation I believe it has been with the

Tree of Life, which fills with its dead and broken branches the crust of the earth, and covers the surface with its ever branching and beautiful ramifications.

(Charles Darwin, 1859, Chapter 4)

Phylogenetics is sometimes poorly appreciated by many persons interested in the conservation of biodiversity and the inventorying of species. Nevertheless, Darwin's concept of descent with modification is graphically represented by a phylogeny that displays ancestor–descendent relationships and retains information about the overall pattern of biological diversification and extinction through time. As such, a phylogeny is an interpretive framework that serves several roles in conservation biology, including: 1) the delimitation of species based on the application of the phylogenetic species concept (PSC); 2) the identification of units of conservation, sometimes below the level of species; 3) the establishment of conservation priorities based on the diversity, age, and distribution of lineages; 4) the estimation of rates of speciation and extinction; and 5) the investigation of processes (e.g., climate, geology) that have influenced the historical and recent distribution and diversification of organisms.

The most obvious focal point in biodiversity science is the species because it is perceived as a real entity among biologists and is broadly appreciated by conservation biologists and the lay public. Species are also focal points for legal protection at the state, national, and international levels. Despite the significance of species in biodiversity conservation, the criteria used to delimit species are varied and sometimes contradictory, resulting in a diversity of species concepts (Mayden 1997). Studies of worldwide biodiversity require a method for delimiting species that is operational across a broad array of taxa, both sexually and asexually reproducing. According to Sites and Marshall (2003), many of the methods used in conservation biology can be subdivided into either non–tree-based or tree-based approaches. Barcoding, for instance, is a non–tree-based method that relies on magnitude of divergence as the major criterion for recognizing a species. Other non–tree-based methods emphasize either a lack of gene flow between populations or the grouping of individuals based on a set of distinguishing characteristics. In contrast, PSC criteria for tree-based methods emphasize monophyly as diagnosed on a phylogeny by shared-derived characters. This latter criterion is being broadly applied in conservation biology (Baker et al. 1995; Cracraft et al. 1998; Cracraft 2002; Wilting et al. 2007).

Application of the PSC as the major criterion for delimiting species does have consequences. In many cases, adherence to the PSC may result in higher numbers of species delimited (Agapow et al. 2004). This result is in part a consequence of recognizing subunits in a polytypic species that encompass several subspecies. Although the PSC is highly operational for delimiting species, it is stringent in its emphasis on monophyly. Part of the species problem is a consequence of the speciation process, which is a continuum with the level of divergence between lineages related to time since divergence (de Queiroz 2005). Because it is a continuum, lineages can be on different evolutionary trajectories without being monophyletic or reproductively isolated, and strict application of the PSC may result in a failure to identify unique yet recently diversifying lineages (Hey 2001). Despite this concern, however, the PSC provides a level of functionality that is useful for the discovery and delimitation of species, especially when one of the major goals of biodiversity research is to determine the number of species inhabiting our planet. Additionally, many cases of species based on phylogenetic criteria are congruent with other criteria used to designate species, and phylogenetic discontinuities provide a meaningful way to reflect units of conservation (Avise & Walker 1999).

Ryder's (1986) idea of an "evolutionary significant unit (ESU)" represents an attempt to more objectively identify units of conservation that do not rely solely on traditional taxonomic designations, especially recognized subspecies. He emphasized agreement among multiple sources of data including distribution, ecology, morphology, and genetics (see also Chapter 10 by Waples and colleagues). In contrast, Moritz (1994) emphasized the need for a more "operational" definition of ESU based on the genetic and phylogenetic distinction of particular groups. His two specific criteria for ESUs included the diagnosis of "reciprocally monophyletic" groups identified from mitochondrial gene trees and evidence for "significant divergence of allele frequencies at nuclear loci." According to Moritz (1994), this definition captures evolutionarily distinct groups that result from historical processes. A second category introduced by Moritz (1994) was the management unit (MU), which is defined as a group that fails to show reciprocal monophyly but does reveal genetic divergence at either the mitochondrial locus or nuclear loci. Presumably, reduced gene flow identified for MUs reflects more recent events. Operationally, the criteria for ESUs are similar to the PSC and would essentially have the same consequence in terms of designating units of conservation. One interesting point raised by Moritz (1994) is the use of phylogeographic concordance among several species as a means of identifying regions that should receive high conservation priorities. In effect, this approach is similar to that of the hotspot, except that it focuses on areas related to patterns of geographic variation within species.

Moritz's criteria for recognizing both ESUs and MUs have been criticized for several reasons. First, some species will not show reciprocal monophyly yet still have populations that are demographically subdivided. This situation is especially problematic for recently separated species or ESUs that have not undergone lineage sorting (Avise 2000). The paraphyletic association between brown bears and polar bears, two groups that are morphologically and ecologically considerably different, provides an example of how the concept of reciprocal monophyly depicted in a mitochondrial gene tree may result in an incorrect decision about the designation of an ESU (Paetkau 1999). Second, some species may show high levels of geographic subdivision, thus resulting in the recognition of high numbers of ESUs distinguished by reciprocal monophyly. Third, if the concept of an ESU is analogous to that of a phylogenetic species, then formal taxonomic recognition may be preferred (Cracraft et al. 1998).

DeSalle and Amato (2004) subdivide genetic methods for the recognition of ESUs into two categories, a tree-based method and a "diagnostic character-based" method. The tree-based approach is best represented by the methodology introduced by Moritz (1994), which suggests that reciprocal monophyly defined by mitochondrial DNA (mtDNA) data and evidence of restricted nuclear gene flow are objective criteria for recognizing ESUs. The approach based on diagnostic characters does not require a gene tree but rather a collective set of substitutions that are diagnostic for a particular population (e.g., a suite of nucleotide substitutions unique to a population or lineage). As indicated by DeSalle and Amato (2004), such an approach alleviates problems of gene trees failing to diagnose species trees.

Rather than setting conservation priorities based on endemism, vulnerability alone, or value based on esthetic, economic, or ecologic criteria, phylogenetic information provides a potential means of establishing priorities. A phylogeny not only depicts relationships among species but also provides estimates of amounts of divergence along lineages. In this light, branch lengths and branching patterns in a phylogeny provide a measure of the amount of evolution or genetic divergence that has occurred between species over time. Nearly all approaches designed to use phylogenies for setting conservation priorities establish criteria for ranking particular lineages. Such approaches have the potential of maximizing clade diversity and spread among clades throughout the phylogeny (Linder 1995). Vane-Wright and colleagues (1991) proposed an index that measures taxonomic distinctiveness. This particular approach combines information on both phylogeny and geographic distribution for the ranking of areas of biodiversity. In their approach, terminal taxa of a group receive less weight than more basal lineages that have little species diversity but display more evolutionary history as seen by their placement in a phylogeny. For instance, phylogenetic analyses of the freshwater fish fauna in Madagascar (a major biodiversity hotspot) reveal a large number of "basal taxa" that appear to be geographically localized and vulnerable to extinction. Therefore, the high level of endemism and large amount of evolutionary history make Madagascar a "reservoir of phylogenetic history" for freshwater fishes (Benstead et al. 2003).

Isaac and colleagues (2007) introduced the evolutionarily distinct and globally endangered (EDGE) score to identify species that should receive top conservation priority. In this approach, a species-level phylogeny is used as the interpretive framework for quantifying the overall score of a species. First, evolutionary distinctiveness (ED) is determined by calculating a value for each branch (length divided by number of species delimited by the branch) followed by the summation of all values from the base of the phylogeny to the terminal taxon of interest. Second, risk of extinction (GE) of a particular taxon is quantified based on the IUCN Red List category weight, and this value is combined with ED to provide an overall EDGE score. This particular method of setting conservation priorities was tested for a species-level phylogeny of mammals, and the results indicated that nearly half of the mammalian species with high EDGE scores did not coincide with current conservation priorities. This finding implies that if these taxa do not receive higher priority, the class Mammalia will lose a large portion of its phylogenetic diversity as estimated by EDGE scores.

Faith (1992) proposed a measure termed *phylogenetic diversity* (PD), which represents the summation of all branch lengths associated with a particular set of taxa in a phylogeny. Rather than focusing on species, this approach emphasizes the maximization of phylogenetic variance, as revealed by increasing levels of PD, and priorities of overall geographic regions or localities can be established based on the overall level of phylogenetic diversity associated with regions rather than estimates of either species richness or endemism.

Do the twenty-five currently recognized biodiversity hotspots capture a large majority of PD? Phylogenies for both primates and carnivores were used to estimate the amount of evolutionary history or PD represented by the currently designated hotspots (Sechrest et al. 2002). The measure of both "clade evolutionary history" (sum of branch lengths of groups of species occurring in a particular area) and "species evolutionary history" (branch length associated with a species back to its most recent bifurcation) is in millions of years, as estimated using a molecular clock and branch lengths derived from a species-level phylogeny. The results of this approach indicated that hotspots exclusively contain one-third of the evolutionary history of these two groups. Therefore, although the establishment of conservation priorities based on habitat and level of endemism has been criticized, estimates of PD based on phylogenies of primates and mammalian carnivores indicate that these designated hotspots capture a considerable amount of PD and evolutionary history for these two groups.

Although a phylogenetic approach for delimiting species and the establishment of a natural classification are well justified in research related to biodiversity, the establishment of conservation priorities based on phylogenies is more tenuous. One must assume that not all species can be saved from extinction, and it is likely that many will go extinct before being discovered. It is also true, however, that the ability to pick lineages with an evolutionary future is impossible. For instance, the removal of one lineage during the history of the mammalian radiations could have resulted in the elimination of our species. Who would have predicted that this lone lineage would be so successful at exploiting our planet?

Molecular taxonomy and phylogenetics of prokaryotes

In terms of species identification and discovery, microbes provide an excellent example of how molecular techniques can enhance the study of species diversity in a group that presents special difficulty with respect to culturing individual taxa. New molecular approaches have greatly expanded our knowledge of worldwide microbial diversity and have helped establish criteria for the recognition of species of bacteria (Ward 2002).

A distance-based approach represents one of the more traditional means of recognizing species of bacteria. For whole-genome comparisons, based on DNA/DNA reassociation, lineages that have 70% or greater similarity are considered strains within a species, whereas lineages less than 70% similar are considered different species (Embley & Stackebrandt 1997; Goodfellow et al. 1997; Cohan 2002; Gevers et al. 2005). Likewise, estimates of genetic divergence based on 16S rRNA sequences also consider different lineages as species if they are 3% or more divergent. Some studies have even used levels of divergence (3% to differentiate species, 5% genera, etc.) to diagnose categories in bacterial taxonomy (Wayne et al. 1987; Embley & Stackebrandt 1997; Schloss & Handelsman 2004). Although this approach has proven useful for assessing bacterial diversity, some consider such a phenetic or distance-based approach to be arbitrary. For instance, divergence based on small fragments of the 16S rRNA gene results in unstable estimates of relationships among species, and hypervariable regions in this gene show varying degrees of divergence across groups (Embley & Stackebrandt 1997; Goodfellow et al. 1997).

In contrast to a distance-based method, a phylogenic approach relies on sequencing (often the 16S rRNA gene) followed by the identification and placement of phylotypes in a phylogeny produced with the use of existing sequences (many from species that can be cultured and characterized) as well as new unknown sequences (Curtis & Sloan 2004). This approach provides a means of assigning species to functional groups, thus allowing for an evaluation of bacterial communities from different habitats and geographic locations (Whitaker et al. 2003; Venter et al. 2004)

Does the 16S rRNA gene provide enough information about the recognition of bacterial species and the derivation of a molecular phylogeny? As indicated by both Cohan (2002) and Gevers and colleagues (2005), the recognition of species based on sequences from the 16S rRNA gene alone is problematic in that some ecologically divergent lineages may have similar 16S rRNA sequences. Therefore, these authors suggest the recognition of species based on criteria that include both "genetic cohesion" and "ecological distinction." The former criterion suggests that lineages of bacteria tend to form phylogenetic clusters that can be characterized both phenotypically and ecologically. Determination of ecological distinction requires the use of multiple loci, termed the "multilocus sequence analysis" (MLSA) by Gevers and colleagues (2005). Phylogenetically distinct clusters, defined by either unique combinations of genes or patterns of gene expression (characteristics that suggest functional differences or ecological distinction), are considered different species. Such an approach is straightforward for strains that can be cultured and characterized in terms of their genome organization and ecological uniqueness. In contrast, species that cannot be cultured are difficult to characterize. In such cases, decisions as to whether an unknown sequence represents a new species or strain depend on its placement relative to well-characterized forms. Alternatively, shotgun sequencing and genome assembly, such as that performed by Venter and colleagues (2004), may provide a means of discovering new species of bacteria based on the criteria of genetic cohesion and ecological distinction.

Molecular taxonomy and phylogenetics of eukaryotes

As with microbes, molecular markers are widely used to discover species and to diagnose phylogenetic relationships in eukaryotes. "The Barcode of Life" is a relatively recent idea that is based on the use of short sequences (650 bp) of a mitochondrial gene (cytochrome oxidase I or *cox*1) as a taxonomic character for the identification and potential discovery of species across broad taxonomic categories (Hebert et al. 2003a,b; Hajibabaei et al. 2007). The basic procedure is as follows: 1) The *cox*1 fragment is PCR amplified and sequenced from DNA obtained from an unknown specimen. 2) The sequence is then compared against a database containing sequences from previously identified taxa. 3) Criteria are established for either the identification of a particular unknown species relative to an existing species or the discovery of a new species.

Like the more traditional approach used for microbes, DNA barcoding is a distance-based approach that assigns specific cutoffs for species-level differences. The approach appears most effective at species identification, and, as such, it provides valuable information for the identification of cryptic species and censuses designed to monitor invasive species. For instance, Hebert and colleagues (2004) used barcoding to distinguish among ten cryptic species of sympatric skipper butterflies, and the species identification was later confirmed with data from host plants, ecology, and color differences among caterpillars. Many invasive species gain entry as either larvae or forms at earlier stages of development, so identification can be difficult. Barcoding serves as an excellent means of identifying problematic invasive species (Savolainen et al. 2005).

Although extremely useful, the application of DNA barcoding does have some significant limitations, especially with respect to the general application for identification and discovery of species. First, *cox*1 is less appropriate as a marker for amphibian species in that intraspecific divergence can be high (7–14%) and can overlap with estimates of interspecific differences, and the primers suggested for amplification of the *cox*1 gene are not universal for amphibians (Vences et al. 2005). The latter problem, however, appears to be solved by modifying existing primers (Smith et al. 2008). For amphibians, a more effective molecular marker is the mitochondrial large subunit rRNA gene, which reveals less overlap between interspecific and intraspecific levels of divergence and is useful for diagnosing phylogenetic relationships among species (Vences et al. 2005). Likewise, different molecular markers appear more effective for not only species identification but also for the discovery of new species in plants. The database for *rbc*L gene sequences for plants is large, and this gene in combination with other loci (nuclear internal transcribed spacer [ITS] region and other chloroplast genes) provides an effective means of establishing phylogenetic relationships among taxa (Chase et al. 2005).

Second, the success of accurately identifying an existing species or discovering an undescribed species depends on the extensiveness of the existing database (Ekrem et al. 2007). Such databases are being assembled at GenBank and at the European Molecular Biology Laboratory (EMBL), and both organizations have established identifiers for searches of the barcode database. Nevertheless, these databases are limited by the existing numbers of species entries. Therefore, one problem with current searches of existing databases for the identification of either known or new species is that perfect matches may not occur (this is more likely when the databases are incomplete). There needs to be a concerted effort to increase sequence databases, especially for genes that are already being used for a broad number of species. For instance, the mitochondrial cytochrome *b* gene has been extensively examined for mammals (Bradley & Baker 2001). Therefore, mammalogists should make a concerted effort to enhance this database.

Third, barcoding currently lacks the ability to accurately place unknown specimens in a phylogenetic context. The derivation of an accurate phylogeny and the placement of unknown species in that phylogeny require the diagnosis of relationships among species and higher categories using an approach that emphasizes multiple genes and their products (DeSalle et al. 2005). The distance-based, single-gene approach used by barcoding can result in mistakes in the assignment of unknown specimens to particular groups (e.g., species complexes or genera) and can fail to identify the proper phylogenetic placement. This latter point is extremely problematic when one relies on data from a single mitochondrial gene rather than on independent data sources and overall congruence, especially if the goal is to both discover new species and to accurately determine phylogenetic relationships among lineages. As more effective and accurate phylogenetic approaches and increased databases containing multiple gene sequences for species are developed, the overall accuracy of phylogenetic placement should be improved (Munch et al. 2008).

Finally, because mtDNA can be transferred between species through introgressive hybridization, the sole reliance on a mitochondrial marker may make it difficult to differentiate some taxa. MtDNA tracks maternal lineages, which in such cases do not reflect species lineages. One example is the North American deer of the genus *Odocoileus*, in which historical hybridization between mule deer (*O. hemionus*) and white-tailed deer (*O. virginianus*) has resulted in the establishment of a white-tailed deer mtDNA lineage within the mule deer. Nuclear markers, including Y-chromosomal sequences and allozymes, show a sister relationship between mule deer and black-tailed deer (both are *O. hemionus*), whereas mtDNA shows a sister relationship between mule deer and white-tailed deer (Carr et al. 1986; Cathey et al. 1998). In cases such as the North American deer complex, multiple genetic markers are required to resolve phylogenetic reticulations, and the use of genetic markers that track the four genetic transmission systems of mammals is an effective way to solve such evolutionary complexities (Lim et al. 2008; Trujillo et al. 2009).

Molecular approaches to the discovery of cryptic species

Similar to some species of bacteria, the phenotypic characteristics of which are unknown as a result of their inability to be cultured, many groups of eukaryotes have species complexes that contain a number of cryptic species that are indistinguishable (or difficult to distinguish) at the phenotypic level. Like research on microbes, the advent of PCR and nucleotide sequencing has enhanced the ability to identify cryptic species, many of which are physiologically, behaviorally, or otherwise distinct, despite their morphological similarity. According to two recent surveys, articles dealing with the discovery of cryptic species based on molecular data are increasing exponentially, with between 2,235 and 3,500 articles reporting cryptic species published over the last two to three decades (Bickford et al. 2006; Pfenninger & Schwenk 2007). For the most part, if one corrects for differences in species richness, the discovery of cryptic species appears to be evenly distributed in terms of taxonomic groups and geographic distribution, with examples being found in a diversity of metazoan phyla (Fig. 1–6).

In many cases, broadly distributed species that are morphologically homogeneous throughout their range actually consist of several cryptic species. For example, bonefishes of the genus *Albula* have a pantropical distribution and traditionally have been considered a single species, *Albula vulpes*. Based on a detailed phylogenetic study of bonefishes throughout most of their range, as many as eight divergent lineages can be identified with the use of mitochondrial sequences. Many of these divergent lineages occur in areas of sympatry yet demonstrate no morphological distinction (Colborn et al. 2001) (see Box 1). The cosmopolitan species of moss *Grimmia laevigata* is morphologically similar throughout its broad distribution yet, based on amplified fragment length polymorphism (AFLP) data,

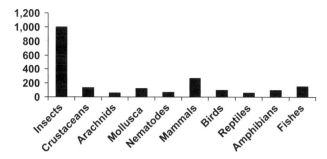

Figure 1–6: Number of reports of cryptic species between 1978 and 2006 (compiled from Pfenninger and Schwenk 2007).

the species in California consists of two cryptic species (Fernandez et al. 2006). Similarly, the sea star, *Parvulastra exigua*, a species broadly distributed in the southern hemisphere, consists of several cryptic species as defined on the basis of mtDNA divergence (Hart et al. 2006).

Amphibian diversity has been severely underestimated, partially due to the lack of morphological distinction among some forms and partially because so many species are narrowly endemic to small geographical areas. In Southeast Asia, the two broadly distributed species of frogs (*Odorrana livida* and *Rana chalconota*) actually represent as many as fourteen cryptic species, many of which are sympatric (Stuart et al. 2006). Given the rate at which habitat is being destroyed in this region of the world, such information is necessary for the proper identification of regions of endemism and the establishment of conservation priorities. Underestimates of amphibian biodiversity are not limited to Southeast Asia. Fouquet and colleagues (2007) used data from the mitochondrial 16S rRNA gene to examine frog diversity in the neotropics. On the basis of these molecular data, they identified twice as many candidate species (129) as the number of named species examined. Likewise, the frog *Eleutherodactylus ockendeni* in Ecuador probably represents at least three genetically distinct species (Elmer et al. 2007).

The problem of morphologically cryptic species has hindered research on some model organisms. For many decades, a single species of leopard frog, *Rana pipiens*, was thought to be distributed from Canada to Panama, throughout North and Middle America (Moore 1944). For much of the twentieth century, *Rana pipiens* was used extensively in research, especially in studies of physiology and endocrinology (Hillis 1988). As source populations for experimental animals changed, however, laboratory biologists who found different populations showed markedly different physiological responses. Studies of behavior (e.g., Littlejohn and Oldham 1968; Mecham 1971), reproductive timing (e.g., Hillis 1981; Frost & Platz 1983), and genetic compatibility (e.g., Moore 1975; Frost & Bagnara 1977) all indicated the existence of many species of biologically distinct, mostly cryptic species of leopard frogs throughout North and Middle America. Sorting out the cryptic species required extensive and careful analyses of behavioral, morphological, and genetic data, although the various species eventually proved readily distinguishable using analyses of proteins (Hillis et al. 1983) or DNA sequences (Hillis & Wilcox 2005). Although molecular studies of the phylogeny of the *Rana*

BOX 1: GENETIC IDENTIFICATION OF CRYPTIC SPECIES: AN EXAMPLE IN *RHOGEESSA*

Amy B. Baird

Problem

Understanding and describing the diversity of life on Earth is a daunting task. This problem is made especially difficult when species cannot be distinguished from one another based on traditional means. Cryptic species occur that are morphologically indistinguishable yet are genetically, behaviorally, or otherwise quite divergent. Biologists must take these differences into account when determining taxonomic status, as well as when planning conservation and management issues for the species of interest.

Case Study

Cryptic species among mammals are relatively rare in some groups yet common in others (Baker & Bradley 2007). An example of a group of mammals that illustrates this is the bat genus *Rhogeessa*. Within the *R. tumida* complex, there are multiple species that are morphologically indistinguishable but were elevated to species status based on genetic differences (Box Fig. 1–1, a).

Historically, *Rhogeessa tumida* has been found from northern Mexico to northern South America. Chromosome banding studies performed within the last 30 years, however, have shown a high degree of karyotypic variation throughout this range (Bickham & Baker 1977). Allozyme analyses have confirmed that these various chromosome races were, in fact, genetically distinct groups (Baker et al. 1985), and they were later described as unique species based on these differences. Those species found to be karyotypically distinct were *R. aeneus*, *R. genowaysi*, and *R. io*.

Recent advances in DNA sequencing technology have allowed researchers to more accurately test the hypotheses of taxonomic status and degree of gene flow between members of the *R. tumida* species complex. By sequencing markers from mtDNA, Y-chromosomal DNA, and nuclear autosomal loci, researchers were able to confirm the taxonomic status of previously described members of the *R. tumida* complex (Baird et al. 2008, 2009). They showed that these species are genetically well differentiated and the molecular phylogenies are consistent with them being unique species (i.e., they are well-supported monophyletic groups; Box Fig. 1–1, b). These data also showed that with one possible exception (an ancient hybridization event between *R. tumida* and *R. aeneus*), the species in the *R. tumida* complex have been genetically isolated for a long period of time.

DNA data can often detect more subtle differences among populations than can karyotypic analyses. One surprising result of the molecular studies of the *R. tumida* complex was the finding of additional variation that did not correspond with karyotypic changes. These genetically distinct populations represent an additional two new species of *Rhogeessa* that are karyotypically identical to *R. tumida* (Baird et al. 2009). They also showed that a population of *Rhogeessa* in

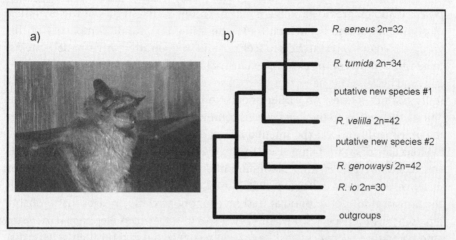

Box Figure 1–1: (a) Image of one of the putative new species of *Rhogeessa*. (b) Phylogenetic relationships of members of the *R. tumida* species complex based on mtDNA sequences, modified from Baird and colleagues (2008). Note that branch lengths are not drawn to scale.

Ecuador, although karyotypically identical to *R. genowaysi*, was phylogenetically distinct based on mtDNA sequences and should be considered a separate species (named *R. velilla*). These results are significant because they were not predicted based on karyotypic or morphological analyses.

The case study of *Rhogeessa* highlights several important lessons for biodiversity studies. First, to understand the diversity of life on Earth, it is necessary to collect large amounts of DNA sequence data and analyze them in both phylogeographic and phylogenetic contexts. Second, efforts to conserve biodiversity should include an understanding of genetic variation so as to account for unknown cryptic species that might occur. This lesson is well illustrated by *Rhogeessa* because one of the cryptic species, *R. genowaysi*, is listed on the 2008 IUCN Red List as an endangered species due to habitat fragmentation and decline. This species is known only from a highly restricted range along the Pacific coast of Chiapas, Mexico, where the forests have been largely cleared for agriculture. Without genetic analyses, this species would never have been recognized; sadly, it might already be extinct.

REFERENCES

Baker RJ, Bickham JW, Arnold ML (1985) Chromosomal evolution in *Rhogeessa (Chiroptera: Vespertilionidae)*: possible speciation by centric fusions. *Evolution*, **39**, 233–243.

Baker RJ, Bradley RD (2007) Speciation in mammals and the genetic species concept. *Journal of Mammalogy*, **87**, 643–662.

Baird AB, Hillis DM, Patton JC, Bickham JW (2008) Evolutionary history of the genus *Rhogeessa (Chiroptera: Vespertilionidae)* as revealed by mitochondrial DNA sequences. *Journal of Mammalogy*, **89**, 744–754.

Baird AB, Hillis DM, Patton JC, Bickham JW (2009) Speciation by monobrachial centric fusions: a test of the model using nuclear DNA sequences from the bat genus *Rhogeessa*. *Molecular Phylogenetics and Evolution*, **50**, 256–267.

Bickham JW, Baker RJ (1977) Implications of chromosomal variation in *Rhogeessa (Chiroptera: Vespertilionidae)*. *Journal of Mammalogy*, **58**, 448–453.

pipiens complex provide a rich context for comparative studies of the evolution of physiological and behavioral traits, the initial taxonomic complexity of this group forced many researchers to seek out and develop alternative model systems. In addition, much of the existing extensive literature on *Rana pipiens* is difficult to interpret because this name was used for so many decades to refer to many biologically distinct species. An earlier understanding of the species and relationships among the species in the *Rana pipiens* complex would have greatly facilitated the use of leopard frogs as experimental model organisms.

Molecular data are being used to discover new cryptic species of mammals and to modify the existing taxonomy of some well-known forms (Box 1). For instance, Brown and colleagues (2007) used genetic data to examine variation in the African giraffe, and found at least seven monophyletic lineages that probably represent distinct species. Recently, two species of African elephant (*Loxodonta africana* and *Loxodonta cyclotis*) have been recognized based on their genetic distinction at several nuclear gene loci (Roca et al. 2001). In the case of the elephants, the two species are not cryptic in that they do show some morphological differences. In other cases, molecular phylogenies derived primarily from mitochondrial sequences have been used to modify the existing taxonomy of mammals, either by relegating subspecies to specific-level status or rearranging existing subspecies boundaries. One case involves the Sumatran tiger (*Panthera tigris sumatrae*), which was designated as a species on the basis of its unique phylogenetic position relative to mainland forms (Cracraft et al. 1998). In this case, the authors used a PSC to justify this taxonomic change. Modification of subspecies boundaries of the common chimpanzee, *Pan troglodytes*, was recommended based on mitochondrial data that subdivided the three recognized subspecies into two monophyletic groups (Gonder et al. 2006).

The discovery of cryptic species is important to biodiversity research as well as to other areas of science. Identification of morphologically similar, yet genetically distinct, species is important to conservation efforts, especially if the establishment of conservation priorities is based on the uniqueness of particular lineages. Distinguishing cryptic species may result in partitioning patterns of endemism into finer spatial scales that are more conducive to conservation efforts, such as seen in Australian freshwater systems (Cook et al. 2008). Cryptic species also have implications for evolutionary biology in terms of understanding morphological stasis, speciation, ecological overlap, species recognition, host/race speciation, and many other topics. Finally, the recognition of cryptic species has applications in both medicine and agriculture, especially as it relates to the identification of human pathogens and plant pests and pathogens. Therefore, any detailed assessment of worldwide biodiversity will benefit from the use of genetic markers for identification and discovery of species. Without such an approach, our overall species count might be a severe underestimate.

ENHANCING RATE OF SPECIES DISCOVERY

Taxonomic practice reveals that not all taxonomic characters are equally useful. Some are powerful indicators of relationship, others are not. The usefulness of a

> character depends on its information content, that is, on its correlation with the natural groupings of taxa produced by evolution.
>
> (Mayr 1969, p. 123)

Analyzing molecular characters in a phylogenetic framework offers a means of accelerating the rate of species discovery and identification, especially in groups containing either cryptic species or large numbers of species. Although molecular-based approaches are important for studying biodiversity, the application of traditional taxonomy is essential if our information databases are to be biologically sound and meaningful (Wheeler 2004).

Taxonomic databases derived from molecular markers exist for microbes, and newly developed molecular approaches have greatly increased the rate of species discovery and identification of microbial diversity. Application of these methods, combined with genomics, methods of sequence assembly, robotics, and the use of informational databases, has greatly increased overall estimates of microbial diversity worldwide. All of these approaches emphasize the acceleration of species identification and discovery with the use of high-throughput methods.

Some of these high-throughput methods used for studies of microbes have greatly accelerated the identification of microbial species, thus allowing for detailed studies of microbial diversity in different regions as well as the assessment of changes in diversity in response to environmental perturbations. For instance, microbial communities respond quickly to changes in the environment, yet assessing community response is hindered by the quantification of microbial diversity in both terrestrial and aquatic ecosystems. In marine ecosystems, the more traditional means of quantifying phytoplankton diversity, especially in terms of identifying species and genera, require microscopy. As indicated by Ellison and Burton (2005), identification via microscopy is more qualitative than quantitative, and even flow cytometric quantification is limited in the number of taxa that can be identified by photopigmentation. These authors have developed a method that uses DNA hybridization and bead-array technology for both the identification and quantification of species. This particular approach bypasses PCR amplification and instead assesses species diversity directly from whole DNA isolated from water samples. Taxon-specific probes containing different fluorescent tags attached to beads are hybridized to specific components of the total DNA. Flow cytometric techniques are then used for species identification and quantification, and the procedure accommodates screening on ninety-six well plates. Therefore, the method allows for rapid assessment of species diversity in different marine environments.

High-throughput methods are available for rapid assessment of bacterial diversity. Many of these methods are PCR-based and rely on assays of variation in the rRNA genes. Terminal restriction fragment length polymorphism (T-RFLP) is used to produce species-specific DNA fingerprints that can be analyzed on an automated sequencer (Schütte et al. 2008). The method uses total DNA extracted from a substrate (e.g., soil, water), fluorescently labeled primers, PCR amplification of the 16S rRNA gene, and digestion of the PCR product with specific restriction endonucleases. Existing databases of fingerprint profiles for particular species can be used to select restriction endonucleases and to identify species by comparison to known taxa in the database (Marsh et al. 2000). Serial analysis of ribosomal sequence tags (SARST) and parallel pyrosequencing provide another high-throughput method for the rapid identification of species (Neufeld et al. 2004; Ashby et al. 2007; Huse et al. 2008). These methods amplify small hypervariable regions (17–55 base pairs) of the 16S rRNA gene from total DNA, and clones as many as twenty sequence tags in a single plasmid. Microarrays are used to sequence multiple plasmids, and particular sequence tags are used to identify taxa of microbes. This technique is cost effective and allows for high-throughput and rapid identification of components of a bacterial community. Although the size of sequenced fragments is limited, making them less reliable in a detailed phylogenetic study, these markers do allow for an assessment of bacterial communities as well as the discovery of rare components of the community.

A phylogenetic approach for species discovery of microbes relies on sequencing cloned amplicons (PCR amplification products) from 16S rRNA fragments amplified from total DNA extractions (Cottrell et al. 2005; Green & Keller 2006). This tree-based approach has been the "gold standard" for studying microbial biodiversity, and new high-throughput methods of DNA isolation, PCR amplification, cloning, and sequencing allow for hundreds of samples to be processed in a short period of time. This particular approach is essential because most species of microbes are known only from nucleotide sequences (Amann et al. 1995). Another more recent approach involves shotgun sequencing of sheared DNA cloned into specific vectors (Castiglioni et al. 2004; Venter et al. 2004; Schloss & Handelsman 2005; Tringe et al. 2005). Randomly obtained sequences are assembled into contigs and scaffolds. This "metagenomic approach" provides an effective means of species discovery in a large number of habitats.

In addition to the high-throughput methods for discovering microbial diversity, new technological advancements are making it possible to accelerate the discovery of eukaryotic species by several orders of magnitude. These molecular-based methods offer the ability to produce data in a format for rapid species identification and phylogenetic placement of unknown taxa. Most of these new devices use nanotechnology that provides platforms for rapid PCR amplification and sequencing. For instance, Blazej and colleagues (2008) describe a nanoliter-scale bioprocessor capable of all the steps necessary for sequencing including PCR, purification of PCR products, and capillary electrophoresis. This "lab-on-a-chip" device uses low amounts of DNA template and provides a means of sequencing more than 550 bp at high accuracy and low cost. Another instrument based on chip technology provides a means of PCR amplification and capillary analysis (Govind et al. 2003). Drmanac and colleagues (1998) also present a high-throughput technique, termed *sequencing by hybridization* (SBH), that uses labeled oligonucleotide probes (of known sequence) in replicate arrays that are hybridized to template DNA. As a result of ongoing technological advancements, it is not far-fetched to imagine a relatively inexpensive, handheld device that can isolate DNA, PCR amplify specific DNA fragments, rapidly sequence PCR products, and organize sequence data for immediate phylogenetic analysis and the screening of existing databases. Such a device could be carried into the field by biologists or other interested individuals and used to quickly identify unknown species and to discover species that have never been previously identified by biologists. Such technology is needed if we have any hope of achieving a reasonably complete understanding of the biodiversity of the Earth in this century.

PHYLOGENETIC DATABASES

> Imagine an electronic page for each species of organism on Earth, available everywhere by single access on command.
>
> (Edward O. Wilson, 2003, p. 77)

Identifying the name of an organism is of little utility by itself. The value in identifying an organism is that the name ties the organism to the scientific literature and other information about that species. Thus, the final and perhaps most important link between systematics and conservation is the establishment of effective, useful, and comprehensive databases on the diversity of life. Given that biological taxonomy is based on phylogenetic relationships, such databases need to be organized and searchable using phylogenetic information. In other words, when a biologist identifies an unknown as linked to a particular part of the Tree of Life (using, e.g., the methods described in the previous sections), he or she needs to be able to connect that organism with all the information on that species. If the unknown is a new species that has never before been studied, then the best information available will be the information on phylogenetically related species. This comparative framework is essential to the use and understanding of biodiversity resources.

There have been many recent efforts to develop effective systematic databases. Many of these are quite limited and amount to little more than lists of names, perhaps linked to bibliographic information on the original description. A more effective approach is to link all of the world's species with all of the information on those species. This is the idea behind the Encyclopedia of Life project (Wilson 2003; see http://www.eol.org). After the database is created, a biologist will be able to identify an unknown organism by placing it within the Tree of Life; this placement would automatically identify the species within the framework of biological taxonomy and immediately link the organism to the information on that (and related) species. Imagine the many and varied uses of such information, from human health applications to conservation biology, to bioprospecting for new useful compounds, to basic biological research. Suddenly, systematic biology would be a critical and necessary component of almost every interaction between people and the living world.

How will automated identification and phylogenetic databases make a difference to conservation biology? Our current ability to protect and understand biodiversity on Earth is severely hampered by our ignorance of what we are trying to preserve and study. If we only know about a small fraction of life on Earth, how can we possibly understand the function of ecosystems? At present, conservation biologists are like car mechanics, who are working to keep a car running, but who only have fragmentary knowledge about the function of 10% of the engine parts. The other 90% of the parts are falling off the engine faster than they can be discovered, and it is unclear how much longer the car will keep running. In the case of the living world, systematics will help us identify and understand the various components of biodiversity, but only if biologists are willing to adopt new technologies and strategies to tackle the enormous undertaking that lies before us.

REFERENCES

Abell R, Thieme ML, Revenga C et al. (2008) Freshwater ecoregions of the world: a new map of biogeographic units for freshwater biodiversity conservation. *BioScience*, **58**, 403–414.

Agapow P-M, Bininda-Emonds ORP, Crandall KA et al. (2004) The impact of species concept on biodiversity studies. *Quarterly Review of Biology*, **79**, 161–179.

Altieri, MA (2004) Linking ecologists and traditional farmers in the search for sustainable agriculture. *Frontiers in Ecology and the Environment*, **2**, 35–42.

Amann RI, Ludwig W, Schleifer KH (1995) Phylogenetic identification and in situ detection of individual microbial cells without cultivation. *Microbiology Review*, **59**, 143–169.

AmphibiaWeb (2010) Information on amphibian biology and conservation. Berkeley, California, http://amphibiaweb.org.

Andreone F, Cadle JE, Cox N et al. (2005) Species review of amphibian extinction risks in Madagascar: conclusions from the global amphibian assessment. *Conservation Biology*, **19**, 1790–1802.

Armstrong PH (1994) Human impact on the Falkland Islands environment. *The Environmentalist*, **14**, 215–231.

Ashby MN, Rine J, Mongodin EF, Nelson KE, Dimster-Denk D (2007) Serial analysis of rRNA genes and the unexpected dominance of rare members of microbial communities. *Applied Environmental Microbiology*, **73**, 4532–4542.

Avise JC (2000) *Phylogeography: The History and Formation of Species*. Harvard University Press, Cambridge, MA.

Avise JC, Walker D (1999) Species realities and numbers in sexual vertebrates: perspectives from an asexually transmitted genome. *Proceedings of the National Academy of Sciences USA*, **96**, 992–995.

Baillie JEM, Hilton-Taylor C, Stuart SN, editors (2004) *2004 IUCN Red List of Threatened Species. A Global Species Assessment*. IUCN, Gland, Switzerland, and Cambridge, UK.

Baker AJ, Daugherty CH, Colborne R, Mclennani JL (1995) Flightless brown kiwis of New Zealand possess extremely subdivided population structure and cryptic species like small mammals. *Proceedings of the National Academy of Sciences USA*, **92**, 8254–8258.

Báldi A. (2008) Habitat heterogeneity overrides the species-area relationship. *Journal of Biogeography* **35**, 675–681.

Benstead JP, De Rham PH, Gattolliat J-L et al. (2003) Conserving Madagascar's freshwater biodiversity. *BioScience*, **53**, 1101–1111.

Bernstein AS, Ludwig DS (2008) The importance of biodiversity to medicine. *Journal of the American Medical Association*, **300**, 2297–2299.

Bickford D, Lohman DJ, Sodhi NS et al. (2006) Cryptic species as a window on diversity and conservation. *Trends in Ecology and Evolution*, **22**, 148–155.

Bickham JW, Parham JF, Philippen H-D et al. (2007) An annotated list of modern turtle terminal taxa with comments on areas of taxonomic instability and recent change. *Chelonian Research Monographs*, **4**, 173–199.

Blazej RG, Kumaresan P, Mathies RA (2008) Microfabricated bioprocessor for integrated nanoliter-scale Sanger DNA sequencing. *Proceedings of the National Academy of Sciences USA*, **103**, 7240–7245.

Bortolus A, Schwindt E (2007) What would Darwin have written now? *Biodiversity and Conservation*, **16**, 337–345.

Bradley RD, Baker RJ (2001) A test of the genetic species concept: cytochrome-b sequences and mammals. *Journal of Mammalogy*, **82**, 960–973.

Brodie EL, DeSantis TZ, Parker JPM et al. (2007) Urban aerosols harbor diverse and dynamic bacterial populations. *Proceedings of the National Academy of Sciences USA*, **104**, 299–304.

Brooks T, Tobias J, Balmford A (1999) Deforestation and bird extinctions in the Atlantic forest. *Animal Conservation*, **2**, 211–222.

Brown DM, Brenneman RA, Koepfli K-P et al. (2007) Extensive population genetic structure in the giraffe. *BMC Biology*, **5**, 57.

Buckley LB, Jetz W (2007) Environmental and historical constraints on global patterns of amphibian richness. *Proceedings of the Royal Society B*, **274**, 1167–1173.

Carr SM, Ballinger SW, Derr JN, Blankenship LH, Bickham JW (1986) Mitochondrial DNA analysis of hybridization between sympatric white-tailed and mule deer in west Texas. *Proceedings of the National Academy of Sciences USA*, **83**, 9576–9580.

Castiglioni B, Rizzi E, Frosini A et al. (2004) Development of a universal microarray based on the ligation detection reaction and 16S rRNA gene polymorphism to target diversity of cyanobacteria. *Applied Environmental Microbiology*, **70**, 7161–7172.

Cathey JC, Bickham JW, Patton JC (1998) Introgressive hybridization and nonconcordant evolutionary history of maternal and paternal lineages in North American deer. *Evolution*, **52**, 1224–1229.

Chase MW, Salamin N, Wilkinson M et al. (2005) Land plants and DNA barcodes: short-term and long-term goals. *Philosophical Transactions of the Royal Society B*, **360**, 1889–1895.

Chown SL, Sinclair BJ, Leinaas HP, Gaston KJ (2004) Hemispheric asymmetries in biodiversity – a serious matter for ecology. *PLoS Biology*, **2**, 1701–1707.

Cohan FM (2002) What are bacterial species? *Annual Review of Microbiology*, **56**, 457–487.

Colborn J, Crabtree RE, Shaklee JB, Pfeiler E, Bowen BW (2001) The evolutionary enigma of bonefishes (Albula spp.): cryptic species and ancient separations in a globally distributed shorefish. *Evolution*, **55**, 807–820.

Cook BD, Page TJ, Hughes JM (2008) Importance of cryptic species for identifying 'representative' units of biodiversity for freshwater conservation. *Biological Conservation*, **141**, 2821–2831.

Cottrell MT, Waidner LA, Yu L, Kirchman DL (2005) Bacterial diversity of metagenomic and PCR libraries from the Delaware River. *Environmental Microbiology*, **7**, 1883–1895.

Cox N, Chanson J, Stuart S (2006) *The Status and Distribution of Reptiles and Amphibians of the Mediterranean Basin*. The World Conservation Union (IUCN), Gland, Switzerland, and Cambridge, UK.

Cracraft J (2002) The seven great questions of systematic biology: an essential foundation for conservation and the sustainable use of biodiversity. *Annals of the Missouri Botanical Garden*, **89**, 127–144.

Cracraft J, Feinstein J, Vaughn J, Helm-Bychowski K (1998) Sorting out tigers (*Panthera tigris*): mitochondrial sequences, nuclear inserts, systematics, and conservation genetics. *Animal Conservation*, **1**, 139–150.

Curtis TP, Sloan WT (2004) Prokaryotic diversity and its limits: microbial community structure in nature and implications for microbial ecology. *Current Opinion in Microbiology*, **7**, 221–226.

Darwin CR (1845) *The Voyage of the Beagle*, 2nd edn. John Murray, London.

Darwin CR (1859) *On the Origin of Species by Means of Natural Selection, or the Preservation of Favored Races in the Struggle for Life*. John Murray, London.

DeSalle R, Amato G (2004) The expansion of conservation genetics. *Nature Reviews*, **5**, 702–712.

DeSalle R, Egan MG, Siddall M (2005) The unholy trinity: taxonomy, species delimitation and DNA barcoding. *Philosophical Transactions of the Royal Society B*, **360**, 1905–1916.

de Queiroz K (2005) Ernst Mayr and the modern concept of species. *Proceedings of the National Academy of Sciences USA*, **102**, 6600–6607.

Drmanac S, Kita D, Labat I et al. (1998) Accurate sequencing by hybridization for DNA diagnostics and individual genomes. *Nature Biotechnology*, **16**, 54–58.

Dubois A (2003) The relationships between taxonomy and conservation biology in the century of extinctions. *Comptes Rendus Biologies*, **326**, S9–S21.

Dudgeon D, Arthington AH, Gessner MO et al. (2006) Freshwater biodiversity: importance, threats, status and conservation challenges. *Biological Review*, **81**, 163–182.

Dyer LA, Singer MS, Lill JT et al. (2007) Host specificity of Lepidoptera in tropical and temperate forests. *Nature*, **448**, 696–700.

Dykhuizen DE (1998) Santa Rosalia revisited: why are there so many species of bacteria? *Antonie van Leeuwenhoek*, **73**, 25–33.

Ekrem T, Willassen E, Stur E (2007) A comprehensive DNA sequence library is essential for identification with DNA barcodes. *Molecular Phylogenetics and Evolution*, **43**, 530–542.

Ellison CK, Burton RS (2005) Application of bead array technology to community dynamics of marine phytoplankton. *Marine Ecology Progress Series*, **288**, 75–85.

Elmer KR, Dávila JA, Lougheed SC (2007) Cryptic diversity and deep divergence in an upper Amazonian leaflitter frog, *Eleutherodactylus ockendeni*. *BMC Evolutionary Biology*, **7**, 247.

Embley TM, Stackebrandt E (1997) Species in practice: explaining uncultured prokaryotic diversity in natural samples. In: *Species: The Units of Biodiversity* (eds. Claridge MF, Dawah HA, Wilson MR), pp. 61–81. Chapman & Hall, London.

Erwin TL (1982) Tropical forests: their richness in Coleoptera and other arthropod species. *Coleoptera Bulletin*, **36**, 74–75.

Faith DP (1992) Conservation evaluation and phylogenetic diversity. *Biological Conservation*, **61**, 1–10.

Fernandez CC, Shevock JR, Glazer AN, Thompson JN (2006) Cryptic species within the cosmopolitan desiccation-tolerant moss *Grimmia laevigata*. *Proceedings of the National Academy of Sciences USA*, **103**, 637–642.

Fouquet A, Gilles A, Vences M et al. (2007) Underestimation of species richness in neotropical frogs revealed by mtDNA analyses. *PLoS ONE*, **10**, e1109.

Frias-Lopez J, Shi Y, Tyson GW et al. (2008) Microbial community gene expression in ocean surface waters. *Proceedings of the National Academy of Sciences USA*, **105**, 3805–3810.

Fritts TH, Rodda GH (1998) The role of introduced species in the degradation of island ecosystems: a case history of Guam. *Annual Review of Ecology and Systematics*, **29**, 113–140.

Frost DR (2006) *Amphibian Species of the World, an Online Reference* (American Museum of Natural History, New York), Version 4. Available at http://research.amnh.org/herpetology/amphibia/index.php.

Frost JS, Bagnara JT (1977) An analysis of reproductive isolation between *Rana magnaocularis* and *Rana berlandieri forreri* (*Rana pipiens* complex). *Journal of Experimental Zoology* **202**, 291–306.

Frost JS, Platz JE (1983) Comparative assessment of modes of reproductive isolation among four species of leopard frogs (*Rana pipiens* complex). *Evolution*, **37**, 66–78.

Gao Z, Tseng C, Pei Z, Blaser MJ (2007) Molecular analysis of human forearm superficial skin bacterial biota. *Proceedings of the National Academy of Sciences USA*, **104**, 2927–2932.

Gaston KJ (1996) Biodiversity – latitudinal gradients. *Progress in Physical Geography*, **20**, 466–476.

Gaston KJ (2000) Global patterns in biodiversity. *Nature*, **405**, 220–227.

Gevers D, Cohan FM, Lawrence JG et al. (2005) Re-evaluating prokaryotic species. *Nature Reviews in Microbiology*, **3**, 733–739.

Gibbons JW, Scott DE, Ryan TJ et al. (2000) The global decline of reptiles, déjà vu amphibians. *BioScience*, **50**, 653–666.

Glaw F, Köhler J (1998) Amphibian species diversity exceeds that of mammals. *Herpetological Review*, **29**, 11–12.

Glowka L, Burhenne-Guilmin F, Synge H (1994) *A Guide to the Convention on Biological Diversity*. International Union for Conservation of Nature and Natural Resources (IUCN), Gland, Switzerland, and Cambridge, UK.

Gonder MK, Disotell TR, Oates JF (2006) New genetic evidence on the evolution of chimpanzee populations and implications for taxonomy. *International Journal of Primatology*, **27**, 1103–1127.

Goodfellow M, Manfio GP, Chun J (1997) Towards a practical species concept for cultivable bacteria. In: *Species: The Units of Biodiversity* (eds. Claridge MF, Dawah HA, Wilson MR), pp. 25–60. Chapman & Hall, London.

Govind VK, Hoang VN, Stickel A et al. (2003) An inexpensive and portable microchip-based platform for integrated RT-PCR and capillary electrophoresis. *Analyst*, **133**, 331–338.

Green BD, Keller M (2006) Capturing the uncultivated majority. *Current Opinion in Biotechnology*, **17**, 236–240.

Grenyer R, Orme CDL, Jackson SF et al. (2006) Global distribution and conservation of rare and threatened vertebrates. *Nature*, **444**, 93–96.

Hajibabaei M, Singer GAC, Hebert PDN, Hickey DA (2007) DNA barcoding: how it complements taxonomy, molecular phylogenetics, and population genetics. *Trends in Genetics*, **23**, 167–172.

Hammond PM (1992) Species inventory. In: *Global Diversity: Status of the Earth's Living Resources* (ed. Groombridge B), pp. 17–39. Chapman & Hall, London.

Hart MW, Keever CC, Dartnall AJ, Byrne M (2006) Morphological and genetic variation indicate cryptic species within Lamarck's little sea star, *Parvulastra (=Patiriella) exigua*. *Biological Bulletin*, **210**, 158–167.

Harvey AL (2008) Natural products in drug discovery. *Drug Discovery Today*, **13**, 894–901.

Hawkins BA (2001) Ecology's oldest pattern? *Trends in Ecology & Evolution*, **16**, 470.

Hebert PDN, Cywinska A, Ball SL, De Waard JR (2003a) Biological identification through DNA barcodes. *Proceedings of the Royal Society London B*, **270**, 313–321.

Hebert PDN, Ratnasingham S, De Waard JR (2003b) Barcoding animal life: cytochrome c oxidase subunit I divergences among closely related species. *Proceedings of the Royal Society London B*, **270**, S96–S99.

Hebert PDN, Penton EH, Burns JM, Janzen DH, Hallwachs W (2004) Ten species in one: DNA barcoding reveals cryptic species in the neotropical skipper butterfly Astraptes fulgerator. *Proceedings of the National Academy of Sciences USA*, **101**, 14812–14817.

Hey J (2001) The mind of the species problem. *Trends in Ecology & Evolution*, **16**, 326–329.

Hillis DM (1981) Premating isolating mechanisms among three species of the *Rana pipiens* complex in Texas and southern Oklahoma. *Copeia*, 1981, 312–319.

Hillis DM (1988) Systematics of the Rana pipiens complex: puzzle and paradigm. *Annual Review of Ecology and Systematics*, **19**, 39–63.

Hillis DM, Frost JS, Wright D (1983) Phylogeny and biogeography of the *Rana pipiens* complex: a biochemical evaluation. *Systematic Zoology*, **32**, 132–143.

Hillis DM, Wilcox TS (2005) Phylogeny of the New World true frogs (*Rana*). *Molecular Phylogenetics and Evolution*, **34**, 299–314.

Horner-Devine MC, Lage M, Hughes JB, Bohannan BJM (2004) A taxa-area relationship for bacteria. *Nature*, **432**, 750–753.

Huber JA, Mark Welch DB, Morrison HG et al. (2007) Microbial population structures in the deep marine biosphere. *Science*, **318**, 97–100.

Huse SM, Dethlefsen L, Huber JA et al. (2008) Exploring microbial diversity and taxonomy using SSU rRNA hypervariable tag sequencing. *PLoS Genetics*, **4**, 1–10.

Isaac NJB, Turvey ST, Collen B, Waterman C, Baillie JEM (2007) Mammals on the EDGE: conservation priorities based on threat and phylogeny. *PLoS ONE*, **3**, e296.

IUCN Red List of Threatened Species™ (2008) *State of the World's Species*. Available at http://www.iucn.org/redlist.

IWC (2007) Report of the Scientific Committee. *Journal of Cetacean Research and Management*, 10 (Suppl.).

Johnson TH, Stattersfield AJ (2008) A global review of island endemic birds. *Ibis*, **132**, 167–180.

Köhler J, Vieites DR, Bonett RM et al. (2005) New amphibians and global conservation: a boost in species discoveries in a highly endangered vertebrate group. *BioScience*, **55**, 693–696.

Lim BK, Engstrom, MD, Bickham JW, Patton JC (2008) Molecular phylogeny of New World sheath-tailed bats (*Emballonuridae: Diclidurini*) based on loci from the four genetic transmission systems in mammals. *Biological Journal of the Linnean Society*, **93**, 189–209.

Linder HP (1995) Setting conservation priorities: the importance of endemism and phylogeny in the southern African orchid genus Herschelia. *Conservation Biology*, **9**, 585–595.

Littlejohn MJ, Oldham RS (1968) *Rana pipiens* complex: mating call structure and taxonomy. *Science*, **162**, 1003–1005.

Lundberg JG, Kottelat M, Smith GR, Stiassny MLJ, Gill AC (2000) So many fishes, so little time: an overview of recent ichthyological discovery in continental waters. *Annals of the Missouri Botanical Garden*, **87**, 26–62.

MacKinnon J (2000) New mammals in the 21st century? *Annals of the Missouri Botanical Garden*, **87**, 63–66.

Marsh TL, Saxman P, Cole J, Tiedje J (2000) Terminal restriction fragment length polymorphism analysis program, a web-based research tool for microbial community analysis. *Applied and Environmental Microbiology*, **66**, 3616–3620.

May RM (1988) How many species are there on Earth? *Science*, **241**, 1441–1449.

May RM (1990) How many species? *Philosophical Transactions Royal Society London B*, **330**, 293–304.

May RM (1992) How many species inhabit the Earth? *Scientific American* (October), 18–24.

May RM (1998) The dimensions of life on Earth. In: *Nature and Human Society: The Quest for a Sustainable World* (ed. Raven PH), pp. 30–45. National Academy Press, Washington, DC.

May RM, Lawton JH, Stork NE (1995) Assessing extinction rates. In: *Extinction Rates* (eds. Lawton JH, May RM), pp. 1–25. Oxford University Press, Oxford.

Mayden RL (1997) A hierarchy of species concepts: the denouement in the saga of the species problem. In *Species: The Units of Biodiversity* (eds. Claridge MF et al.), pp. 381–424, Chapman & Hall, London.

Mayr E (1946) The number of species of birds. *The Auk*, **63**, 64–69.

Mayr E (1969) *Principles of Systematic Zoology*. McGraw-Hill, Inc., New York.

Mecham JS (1971) Vocalizations of the leopard frog, *Rana pipiens*, and three related Mexican species. *Copeia*, 1971, 505–516.

Mittermeier RA, Myers N, Thomsen JB, da Fonseca GAB, Olivieri S (1998) Biodiversity hotspots and major tropical wilderness areas: approaches to setting conservation priorities. *Conservation Biology*, **12**, 516–520.

Monroe BL Jr, Sibley CG (1990) *Distribution and Taxonomy of Birds of the World*. Yale University Press, New Haven, CT.

Moore JA (1944) Geographic variation in *Rana pipiens Schreber* of eastern North America. *Bulletin of the American Museum of Natural History*, **82**, 345–370.

Moore JA (1975) *Rana pipiens* – the changing paradigm. *American Zoologist*, **15**, 837–849.

Moritz C (1994) Defining 'evolutionary significant units' for conservation. *Trends in Ecology and Evolution*, **9**, 373–375.

Munch K, Boomsma W, Willerslev E, Nielsen R (2008) Fast phylogenetic DNA barcoding. *Philosophical Transactions of the Royal Society B*, **263**, 3997–4002.

Myers N (1988) Threatened biotas: "hotspots" in tropical forests. *Environmentalist*, **8**, 187–208.

Myers N, Mittermeier RA, Mittermeier CG, da Fonseca GAB, Kent J (2000) Biodiversity hotspots for conservation priorities. *Nature*, **403**, 853–858.

Nee S, May RM (1997) Extinction and the loss of evolutionary history. *Science*, **278**, 692–694.

Neufeld JD, Yu Z, Lam W, Mohn WW (2004) Serial analysis of ribosomal sequence tags (SARST): a high-throughput method for profiling complex microbial communities. *Environmental Microbiology*, **6**, 131–144.

Nilsson G (2005) *Endangered Species Handbook*. Animal Welfare Institute, Washington, DC.

Novotny V, Miller SE, Hulcr J et al. (2007) Low beta diversity of herbivorous insects in tropical forests. *Nature*, **448**, 692–697.

Ødegaard F (2000) How many species of arthropods? Erwin's estimate revised. *Biological Journal of the Linnean Society*, **71**, 583–597.

Pace NR (1997) A molecular view of microbial diversity and the biosphere. *Science*, **276**, 734–740.

Pace NR, Stahl DA, Lane DL, Olsen GJ (1986) The analysis of natural microbial populations by rRNA sequences. *Advances in Microbial Ecology*, **9**, 1–55.

Paetkau D (1999) Using genetics to identify intraspecific conservation units: a critique of current methods. *Conservation Biology*, **13**, 1507–1509.

Patterson BD (2000) Patterns and trends in the discovery of new neotropical mammals. *Diversity and Distributions*, **6**, 145–151.

Patterson BD (2001) Fathoming tropical biodiversity: the continuing discovery of neotropical mammals. *Diversity and Distributions*, **7**, 191–196.

Peterson AT (1998) New species and new species limits in birds. *The Auk*, **115**, 555–558.

Pfenninger M, Schwenk K (2007) Cryptic animal species are homogeneously distributed among taxa and biogeographical regions. *BMC Evolutionary Biology*, **7**, 121.

Pimentel D, Wilson C, McCullum C et al. (1997) Economic and environmental benefits of biodiversity. *BioScience*, **47**, 747–757.

Pimm SL, Russell GJ, Gittleman JL, Brooks TM (1995) The future of biodiversity. *Science*, **269**, 347–350.

Raup DM (1986) Biological extinction in Earth history. *Science*, **231**, 1528–1533.

Raup DM (1994) The role of extinction in evolution. *Proceedings of the National Academy of Sciences USA*, **91**, 6758–6763.

Reid WV (1998) Biodiversity hotspots. *Trends in Ecology & Evolution*, **13**, 275–280.

Ribon R, Simon JE, de Mattos GT (2003) Bird extinctions in Atlantic forest fragments of the Viçosa region, southeastern Brazil. *Conservation Biology*, **17**, 1827–1839.

Ricciardi A, Rasmussen JB (1999) Extinction rates of North American freshwater fauna. *Conservation Biology*, **13**, 1220–1222.

Ricklefs RE (2004) A comprehensive framework for global patterns in biodiversity. *Ecology Letters*, **7**, 1–15.

Roca AL, Georgiadis N, Pecon-Slattery J, O'Brien SJ (2001) Genetic evidence for two species of elephant in Africa. *Science*, **293**, 1473–1477.

Rosenzweig ML (1992) Species diversity gradients: we know more and less than we thought. *Journal of Mammalogy*, **73**, 715–730.

Rosenzweig ML (1995) *Species Diversity in Space and Time*. Cambridge University Press, Cambridge, UK.

Ryder OA (1986) Species conservation and systematics: the dilemma of subspecies. *Trends in Ecology and Evolution*, **1**, 9–10.

Savolainen V, Cowan RS, Vogler AP, Roderick GK, Lane R (2005) Towards writing the encyclopedia of life: an introduction to DNA barcoding. *Philosophical Transactions of the Royal Society B*, **360**, 1805–1811.

Schipper J, Chanson JS, Chiozza F et al. (2008) The status of the world's land and marine mammals: diversity, threat, and knowledge. *Science*, **322**, 225–230.

Schloss PD, Handelsman J (2004) Status of the microbial census. *Microbiology and Molecular Biology Reviews*, **68**, 686–691.

Schloss PD, Handelsman J (2005) Metagenomics for studying unculturable microorganisms: cutting the Gordian knot. *Genome Biology*, **6**, 229.

Schloss PD, Handelsman J (2006) Toward a census of bacteria in soil. *PLoS Computational Biology*, **2**, e92.

Schütte UME, Abdo Z, Bent SJ et al. (2008) Advances in the use of terminal restriction fragment length polymorphism (T-RFLP) analysis of 16S rRNA genes to characterize microbial communities. *Applied Microbiology and Biotechnology*, **80**, 365–380.

Sechrest W, Brooks TM, da Fonseca GAB et al. (2002) Hotspots and the conservation of evolutionary history. *Proceedings of the National Academy of Sciences USA*, **99**, 2067–2071.

Sites JW Jr, Marshall JC (2003) Delimiting species: a Renaissance issue in systematic biology. *Trends in Ecology and Evolution*, **18**, 462–470.

Smith MA, Poyarkov NA Jr, Hebert PDN (2008) CO1 DNA barcoding amphibians: take the chance, meet the challenge. *Molecular Ecology Resources*, **8**, 235–246.

Sogin ML, Morrison HG, Huber JA, Welch DM, Huse SM et al. (2006) Microbial diversity in the deep sea and the underexplored "rare biosphere." *Proceedings of the National Academy of Sciences USA*, **103**, 12115–12120.

State of Observed Species (2008) International Institute for Species Exploration, Arizona State University, http://species.asu.edu/index.

Stork NE (1993) How many species are there? *Biodiversity and Conservation*, **2**, 215–232.

Stork NE (2007) World of insects. *Nature*, **448**, 657–658.

Stuart BL, Inger RF, Voris HK (2006) High level of cryptic species diversity revealed by sympatric lineages of Southeast Asian forest frogs. *Biological Letters*, **2**, 470–474.

Stuart SN, Chanson JS, Cox NA et al. (2004) Status and trends of amphibian declines and extinctions worldwide. *Science*, **306**, 1783–1786.

Systematics Agenda 2000 (1994) *Systematics Agenda 2000: Charting the Biosphere*. New York: Society of Systematic Biologists, American Society of Plant Taxonomists, Willi Hennig Society, Association of Systematics Collections.

Trevino HS, Skibiel AL, Karels TJ, Dobson FS (2007) Threats to avifauna on oceanic islands. *Conservation Biology*, **21**, 125–132.

Tringe SG, von Mering C, Kobayashi A et al. (2005) Comparative metagenomics of microbial communities. *Science*, **308**, 554–557.

Trujillo RG, Patton JC, Schlitter DA, Bickham JW (2009) Molecular phylogenetics of the bat genus *Eptesicus* (Chiroptera: Vespertilionidae): perspectives from paternally and maternally inherited genomes. *Journal of Mammalogy* **90**, 548–560.

Turvey ST, Pitman RL, Taylor BL et al. (2007) First human-caused extinction of a cetacean species? *Biological Letters*, **3**, 537–540.

Urbina-Cardona JN (2008) Conservation of neotropical herpetofauna: research trends and challenges. *Tropical Conservation Sciences*, **1**, 359–375.

Vane-Wright RI, Humphries CJ, Williams PH (1991) What to protect? – Systematics and the agony of choice. *Biological Conservation*, **55**, 235–254.

Vences M, Thomas M, Van Der Meijden A, Chiari Y, Vietes DR (2005) Comparative performance of the 16S rRNA gene in DNA barcoding of amphibians. *Frontiers in Zoology*, **2**, 5.

Venter J, Remmington K, Heidelberg J et al. (2004) Environmental genome shotgun sequencing of the Sargasso Sea. *Science*, **304**, 66–74.

Wake DB, Vredenburg VT (2008) Are we in the midst of the sixth mass extinction? A view from the world of amphibians. *Proceedings of the National Academy of Sciences USA*, **105**, 11466–11473.

Wallace AR (1869) *The Malay Archipelago: The Land of the Orang-utan, and the Bird of Paradise. A Narrative of Travel, with Studies of Man and Nature*. Harper & Brothers, New York.

Ward BB (2002) How many species of prokaryotes are there? *Proceedings of the National Academy of Sciences USA*, **99**, 10234–10236.

Wayne LG, Brenner DJ, Colwell RR et al. (1987) Report of the ad hoc committee on reconciliation of approaches to bacterial systematics. *International Journal of Systematic Bacteriology*, **37**, 463–464.

Wheeler QD (2004) Taxonomic triage and the poverty of phylogeny. *Philosophical Transactions of the Royal Society London B*, **359**, 571–583.

Whitaker RJ, Grogan DW, Taylor JW (2003) Geographic barriers isolate endemic populations of hyperthermophilic Archaea. *Science*, **301**, 976–978.

Wilson DE, Reeder DM, editors (1993) *Mammal Species of the World: A Taxonomic and Geographic Reference*, 2nd edn. Smithsonian Institution Press, Washington.

Wilson DE, Reeder DM, editors (2005) *Mammal Species of the World: A Taxonomic and Geographic Reference*, 3rd edn. Johns Hopkins University Press, Baltimore, Maryland.

Wilson EO (1985) The biological diversity crisis. *BioScience*, **35**, 700–706.

Wilson, EO (2003) The encyclopedia of life. *Trends in Ecology and Evolution*, **18**, 77–80.

Wilson EO (2006) *The Creation: An Appeal to Save Life on Earth*. W. W. Norton and Co., Inc., New York.

Wilting A, Buckley-Beason VA, Feldhaar H et al. (2007) Clouded leopard phylogeny revisited: support for species recognition and population division between Borneo and Sumatra. *Frontiers in Zoology*, **4**, 15.

Woese CR (1987) Bacterial evolution. *Microbiology Review*, **51**, 221–271.

Woese CR (1994) There must be a prokaryote somewhere: microbiology's search for itself. *Microbiological Reviews*, **58**, 1–9.

2 Gene flow, biodiversity, and genetically modified crops: Weedy rice in Thailand

Barbara Schaal, Wesley J. Leverich, Sansanee Jamjod, Chanya Maneechote, Anbreen Bashir, Amena Prommin, Adirek Punyalue, Athitya Suta, Theerasak Sintukhiew, Anupong Wongtamee, Tonapha Pusadee, Sunisa Niruntrayakul, and Benjavan Rerkasem

The domestication of plants and animals and the development of agriculture some 10,000 years ago has led to profound changes in the environment and to biodiversity (Diamond 1997; Smith 1998). As natural communities were replaced by pastures and fields, native species were displaced or their habitats fragmented (Heywood 1995; Millennium Ecosystem Assessment 2005). The extirpation of native species began with the earliest agricultural communities in the Middle East and Asia and continues today. Every major advance in agriculture, from the development of new crops to mechanized farming, has environmental consequence. The most recent change in agricultural practice is the planting of genetically modified (GM) crops. First developed and legalized in the 1990s, today the majority of crops in the United States are GM, with approximately 90% of the U.S. soybean crop GM for herbicide tolerance (U.S. Department of Agriculture 2008).

GM crops are varieties that have been transformed by using a biological or physical method to insert specific genes into a genome (Chrispeels & Sadava 2003). The inserted genes, *transgenes*, can come from another species or from the same species. In contrast, most varieties of nontransgenic crops are produced by traditional and modern methods of crop improvement and selective breeding (Chrispeels & Sadava 2003). Other descriptions for such GM crops are recombinant or genetically engineered crops. The specific methods of genetic manipulation used to produce a GM crop are not thought to have any serious consequences (National Research Council 2002), but rather the consideration of most concern is the specific nature of the introduced transgene. Because the method of crop improvement has little effect, some researchers have argued that the distinction between GM crops and non-GM crops is artificial; crops produced by traditional means of plant breeding are also GM (Federoff & Brown 2004). This point is important: The issues and concerns that have been raised about recently developed GM crops are also of concern regarding traditional crop varieties. In fact, here we present a case study of rice in a modern, non-GM agricultural setting where traditional practices have led to adverse effects and which can be used to infer some of the potential consequences of introducing GM crops.

This work was supported by the McKnight Foundation, USAID grant GDG–G00–01–00012–00, and the Thailand Research Fund.

There has been an active, and sometimes acrimonious, debate about the potential risks and benefits of GM crops on human health and the environment, as well as other concerns related to ethics or the economy (for an example in salmon, see Box 2). Although there are potential positive effects of GM agriculture on biodiversity (e.g., reduction in agrochemical use), more attention has been paid to potential negative environmental or ecological effects. The National Research Council (2002) lists four categories of environmental hazard from GM crops: 1) the movement of transgenes and their expression in other plants; 2) harmful effects from the transgenic crop itself; 3) negative effects on nontarget organisms; and 4) the development of resistance in GM crops. Here, we are concerned only with the first category. Specifically, what are the potential effects on biodiversity of gene flow from GM crops? Gene flow is a topic that has long concerned evolutionary biologists, yet, in the past, assessing the impact of gene flow has been exceedingly difficult because of the lack of suitable genetic markers. The development of *deoxyribonucleic acid* (DNA)-based markers such as microsatellites, the markers used in DNA fingerprinting, has led to many studies of gene flow, hybridization, and introgression of native plants. We now have a much clearer picture of gene migration in nature; gene flow in many cases is higher than expected (Ellstrand 2003). Also, these genetic markers now allow for the detailed assessment of the consequences for biodiversity of gene flow in agricultural systems. Here, we first consider the issues surrounding gene flow and biodiversity in general. Next, we discuss an example of gene flow, hybridization, and the formation of weedy rice in a nontransgenic agricultural system in Thailand and then infer consequences for biodiversity of gene flow in this system.

POTENTIAL EFFECTS OF GENE FLOW ON BIODIVERSITY

Gene flow is defined as the transfer of genes from one population into another. Gene flow has two necessary processes: the movement of genes between populations and the subsequent incorporation of those genes into the gene pool of the recipient population. Gene flow may occur between different populations within the same species, between GM crops and non-GM crops, between crop and wild relative of the same species, or between species. Gene flow is a natural and nearly universal process (Ellstrand 2003). In plants, gene flow can be extensive among populations as well as frequent between species. Because gene flow is nearly ubiquitous in plant species, it is appropriate to consider the potential affect of gene flow from GM crops on biodiversity (see Chapter 3 by Whitham and colleagues). Here, we will limit our consideration of gene flow to between GM crops and their wild (noncultivated) relatives. The wild relative can be a weedy form of the crop, the wild ancestor of the crop, or some other closely related species that is reproductively compatible. Another aspect of gene flow, adventitious presence of a transgene in the growth, transportation, and processing of crops, is discussed elsewhere (Council for Agricultural Science and Technology 2007). Here, we consider only escape into "natural" systems that are not under agricultural control and the potential effect on biodiversity.

BOX 2: ENVIRONMENTAL RISK ASSESSMENT OF GENETICALLY ENGINEERED SALMON

Robert H. Devlin and Fredrik L. Sundström

Problem

Genetically engineered fish strains have been developed over the past 25 years, many with the objective of altering traits to enhance aquaculture production efficiency. Although transgenic fish have not been released and have not escaped to nature to date, significant concern exists regarding potential ecological effects if such strains entered natural environments. Effects of a transgenic strain to ecosystems are anticipated to highly depend on phenotype alterations that 1) influence survival and reproductive success relative to conspecifics, and 2) affect the interaction of the strain with other species in a community that could influence their survival. In addition, many transgenic fish strains possess complex pleiotropic physiological and behavioral changes (Devlin et al. 2006) that could influence their fitness and consequences in nature. A major issue facing risk assessors, however, is whether experimental information on transgenic fish, necessarily generated in bioconfinement laboratory facilities, can consistently provide reliable predictions of impacts to ecosystems in nature.

Case Study

Transgenic coho salmon with enhanced growth have been engineered for risk assessment studies using growth hormone gene constructs, resulting in strains with remarkably enhanced daily growth rates and age-specific sizes. Gene constructs are designed with novel molecular structures (e.g., fusing new promoters to complementary DNA [cDNA] or full-length gene-coding regions). When these integrate into host chromosomes, they are organized in strain-specific fashions (e.g., integrated transgenes can possess single inserts or multiple direct-tandem duplications of gene constructs, as well as insertions and deletions of host DNA at the chromosomal site of integration; see, e.g., Uh et al. 2006). Such structural information allows development of specific molecular (polymerase chain reaction [PCR], quantitative PCR [Q-PCR], or Southern blot) diagnostics for the presence or absence of the transgene, its structural stability (copy number and organization), and whether it is present in hemizygous or homozygous form. These assays are critical for assessing transgene effects on phenotype for laboratory-based ecological risk assessments. Laboratory studies have shown a plethora of phenotypic transformations in growth hormone (GH) transgenic coho salmon, including increased feeding motivation, enhanced feed intake and conversion efficiency, altered prey-selection characteristics, reduced predator avoidance, enhanced dispersal tendencies, reduced swimming ability, altered metabolic rates, impaired disease resistance, precocious development, early sexual maturation, and reduced spawning ability. The molecular and endocrine bases of these phenotypic changes are now being revealed in part through the use of genomic and other molecular techniques (e.g., microarray and Q-PCR analyses of gene expression;

see, e.g., Rise et al. 2006; Raven et al. 2008;Devlin et al. 2009). Measurements
of survival relative to wild type in laboratory (tank) studies have revealed strain-
specific effects, but such simple environments generate much higher survival
than in nature, indicating that critical fitness-altering parameters are missing.
Thus, extrapolating such laboratory data could be misleading, and this is particularly so if antagonistic pleiotropic effects are operating. For example, in rich
simple environments, with abundant feed and no predation, elevated feeding
motivation of transgenic fish enhances their fitness (survival) relative to wild
type, whereas in naturalized environments with predators, enhanced feeding
motivation causes transgenic fish to suffer increased predation (Devlin et al.
2004; Sundström et al. 2004). How to combine such opposite effects accurately
to reflect net fitness in nature is complex. In addition, the rearing environments
of transgenic and wild fish prior to experimentation are also critical due to their
remarkable phenotypic plasticity (Box Fig. 2–1). For example, GH transgenic
coho salmon raised in culture conditions in tank facilities display high growth
rates and consequent high predation activity. In contrast, transgenic salmon
reared in naturalized environments (e.g., complex habitat, natural food only,
predators present) show much reduced growth rate, and significantly fewer consequences to natural prey, relative to wild-type fish from the same rearing background (Sundström et al. 2007). Because phenotypes of transgenic and wild-type
fish can overlap when they are reared in seminatural environments, transgene-
specific molecular markers capable of distinguishing these genotypes are essential.
In addition, phenotypic effects of GH transgenes can be influenced by genetic
background (Devlin et al. 2001), suggesting that a transgene entering naturally
occurring populations and encountering a range of allelic variation could cause
multiple phenotypic changes. Such differences would be subject to natural selection and would be anticipated to alter the fitness consequences of a transgene
in a population over time, requiring risk assessments to understand the evolution of phenotype caused by the transgene in population and genomic contexts.
Coupling phenotypic assessments with genomic analyses identifying molecular markers for quantitative trait loci (QTLs) affecting growth rate will be critical to provide a full understanding of the interaction between anthropogenic
(i.e., GH transgene) and natural variation in populations. Thus, identification of
physiological, behavioral, and ecological mechanisms influencing the fitness and
environmental consequences of transgenic fish may be identified in laboratory
studies. If they are consistently observed across environments (i.e., displaying
parallel reaction norms) or are of an obviously severe nature, such data may be
useful predictors of effects in wild ecosystems. In other cases (particularly for

Box Figure 2–1: Phenotypic plasticity of growth rates between wild-type and growth-hormone
transgenic coho salmon under different environmental conditions.

species from large complex ecosystems), accurately estimating phenotypic effects, their interactions, and magnitudes and rates of evolutionary change under a suitably broad range of naturalized experimental conditions (to allow meaningful predictions of harms to natural ecosystems) remains a complex task. If transgenic fish should ever enter ecosystems, tracking the persistence of transgenes in populations using specific molecular PCR tests will be an essential component of any monitoring or extirpation program.

REFERENCES

Devlin RH, Biagi CA, Yesaki TY, Smailus DE, Byatt JC (2001) Growth of domesticated transgenic fish. *Nature*, **409**, 781–782.

Devlin RH, D'Andrade M, Uh M, Biagi CA (2004) Population effects of GH transgenic salmon are dependant upon food availability and genotype by environment interactions. *Proceedings of the National Academy of Sciences USA*, **101**, 9303–9308.

Devlin RH, Sakhrani D, Tymchuk WE, Rise ML, Goh B (2009) Domestication and growth hormone transgenesis cause similar changes in gene expression profiles in salmon. *Proceedings of the National Academy of Sciences USA*, **106**, 3047–3052.

Devlin RH, Sundström LF, Muir WF (2006) Interface of biotechnology and ecology for environmental risk assessments of transgenic fish. *Trends in Biotechnology*, **24**, 89–97.

Raven PA, Uh M, Sakhrani D et al. (2008) Endocrine effects of growth hormone over-expression in transgenic coho salmon. *General and Comparative Endocrinology*, **159**, 26–37.

Rise ML, Douglas SE, Sakhrani D et al. (2006) Multiple microarray platforms utilized for hepatic gene expression profiling of GH transgenic coho salmon with and without ration restriction. *Journal of Molecular Endocrinology*, **37**, 259–282.

Sundström LF, Lõhmus M, Johnsson JI, Devlin RH (2004) Growth hormone transgenic salmon pay for growth potential with increased predation mortality. *Proceedings of the Royal Society B*, **271**, S350–S352.

Sundström LF, Lõhmus M, Tymchuk WE, Devlin RH (2007) Gene–environment interactions influence ecological consequences of transgenic animals. *Proceedings of the National Academy of Sciences USA*, **104**: 3889–3894.

Uh M, Khattra J, Devlin RH (2006) Transgene constructs in coho salmon (*Oncorhynchus kisutch*) are repeated in a head-to-tail fashion and can be integrated adjacent to horizontally-transmitted parasite DNA. *Transgenic Research*, **15**, 711–727.

Initially, it was thought that gene flow was not a significant risk for GM crops (Day 1987). In part, this assumption was due to a paucity of gene-flow studies. The development of molecular markers to trace gene flow in crops has led to a reevaluation of the likelihood of transgene escape and has implicated gene flow as a potential risk for biodiversity (McRae & Beier 2007; Pasquet et al. 2008). In a number of species, hybridization or the potential for hybridization between crop and wild relative has been observed. This includes work on *Sorghum* (Tesso et al. 2008), flax (Jhala et al. 2008), *Phaseolus* (Martinez-Castillo et al. 2007), *Brassica* (Wilkinson et al. 2003; Ceddia et al. 2007), and *Agrostis* (Watrud et al. 2004). Recent studies include assessment of hybrid fitness and the first documentation of multiple herbicide resistance in *Brassica* (Knispel et al. 2008; Warwick et al. 2008). Documenting gene flow, however, does not adequately address potential environmental risks from gene flow; one needs to determine the fate of the introduced genes.

The fate of introduced genes is influenced by the population genetics of gene flow and selection in crop systems (Chapman & Burke 2006). Gene flow typically occurs by the transfer of pollen and the subsequent formation of a hybrid seed between GM crop and wild relative. Although somewhat less likely, gene flow can also occur if crop seeds are moved into a wild population and the resulting plants cross and produce seed with the wild relative. Such gene flow is likely if the crop has a seed pool, where seeds are stored over long periods of time in the soil (Nielson et al. 2009). The movement of genes is not in itself harmful. To determine any negative consequences, several factors need to be considered (Pilson & Prendeville 2004, Clark 2006). The wild population and the GM crop must be compatible; that is, they can produce seeds by crossing. Then the relative fitness of the hybrids needs to be determined. Will the hybrids have superior, inferior, or the same fitness as the members of the wild population? The ultimate fate of the transgene in the wild population and the potential for an effect on biodiversity rest on the answer to this question.

If the fitness of plants containing the transgene is less than that of the native population, we expect the gene to decrease in frequency and ultimately be eliminated by selection. This process in itself may have negative consequences for the native population. Genetic diversity can be lost, and, if there is strong selection, population size may decrease as individuals are selectively eliminated (Levin et al. 1996; Andow & Zwahlen 2006). If the transgene has no effect on fitness, one expects the gene frequency, on average, to equal the migration rate, and the fate of the gene is determined by genetic drift. The transgene most commonly should be eliminated by genetic drift unless migration rate is high and new genes are continually being reintroduced into the population. If there is continual migration from the source into the recipient population, it is possible for many crop alleles to be assimilated into the wild population (Haygood et al. 2003), and these, in turn, can reduce fitness and cause potential local extinction (Snow 2002; Ellstrand 2003).

Of most interest is the case in which the transgene can convey a selective advantage. Whether the transgene will be positively selected depends on the nature of the gene. Genes for pharmaceutical production or fruit traits might have neutral to slightly negative affects. Genes for herbivore, pathogen, or drought resistance may be positively selected. In the case of herbicide resistance, one might expect neutral (Zhang et al. 2003) to negative selection, unless there is an application of herbicide to the environment that would then convey strong positive selection. Assessing fitness of hybrids is difficult and tedious, and far fewer studies have been conducted to date (e.g., Allainguillaume et al. 2006). A pioneering study of transgenic insect-resistant *Helianthus annuus* documented gene flow, hybridization, and natural selection for the transgene and predicted its frequency to increase in native sunflower populations (Snow et al. 2003; Pilson et al. 2004).

Even more difficult than determining hybrid fitness is assessing any community-wide changes that might occur because of the transgene. The escape of transgenes in high frequency might affect seed production, population size, or habitat use by the native species. In turn, these changes may affect native herbivores, pathogens, competitors, and symbionts, which then could potentially cascade through an ecological community. These possible effects are difficult to study, and there is an overall lack of data to evaluate such indirect effects. Finally, such change might potentially result in the invasion of previous unsuitable habitats and in the evolution of weediness.

If hybrids have higher fitness than wild parents, the hybrids may be invasive and evolution of weediness may occur. Gene flow from crops to wild relatives is implicated in the evolution of weediness in more than half of the world's most significant crops (Ellstrand 2003). In addition, crop weeds cause huge losses globally, resulting in economic losses and reducing the food security of the developing world. Assessing the risk of evolving weediness is exceptionally difficult because we as yet cannot specify exactly what makes a plant weedy. Purrington and Bergelson (1995) list fourteen traits associated with weediness, but many weeds seem to defy any such generalities. The cause of weediness in specific cases is quite clear, however. In rice, for example, weediness is the result of seed shattering (dropping from the plant when ripe), perennial habit, and dispersal via awns and hairs.

Much of the concern about gene flow from GM crops centers on the formation of aggressive weeds. Direct studies of weed formation by transgene escape are, in general, avoided because any experimental plants may escape and spread. Because the population genetics of transgenes, their dispersion, and their incorporation is not different than those of other crop genes, an avenue for assessing risks to biodiversity is to use a conventional crop–weed system (Hancock 2003) or modeling (Muir & Howard 2002). In the next section, we discuss our work on the origin of weedy rice in Thailand in a nontransgenic system. We use this information to infer the implications for GM rice by using Thai weedy rice as a model system.

THE ORIGIN OF WEEDY RICE IN THAILAND

Rice is grown on more than 9 million ha in Thailand each year (Office of Agriculture Economics [OAE] 2008). It is not only the major source of food for the Thai people; it is also an important economic export. In the past decade, rice production in Thailand has been challenged by the appearance of a weedy form of rice that is responsible for the precipitous decline in harvest in some areas. Weedy rice is a common pest of cultivated rice fields in most rice-growing parts of the world, including the United States. The origin of weedy rice is debated, with hypotheses ranging from de-domestication of cultivated rice, to hybridization among cultivated rice varieties, to formation by gene flow and hybridization between cultivated and wild rice. In our work described here, we evaluate the hybrid origin hypothesis by using field, crossing, and genetic studies.

Thailand has diverse rice varieties and cultivation practices. Modern semi-dwarf photoperiod-insensitive varieties (often called high-yielding varieties [HYVs]) are grown on land with year-round irrigation. High-quality varieties (e.g., aromatic Thai Jasmine) derived from local germplasm are grown in rain-fed fields. Also, genetically diverse traditional local varieties, or landraces, are grown on the remaining one-fifth of the land, including upland villages of indigenous peoples and in flood-prone areas, where more than 100 distinct local varieties are recognized (Sommut 2003). As with many other crops, varieties of GM rice have been or are currently being developed with such transgenes as beta-carotene, herbicide resistance, and enhanced nutrition. Although Thailand currently bans GM rice, there will be increasing pressure both internally and externally for its adoption.

Wild rice (*Oryza rufipogon*), the wild ancestor of domesticated rice, *Oryza sativa* (Londo et al. 2006), is found in most of the rice landscapes in the country, from the intensive, irrigated areas in the Central Plain and Lower North to the rain-fed area of the Northeast. Perennial populations are found in year-round wet places, on the shallow edges of large lakes, and on the banks of waterways and irrigation canals in close proximity to cultivated rice fields. Annual populations are found in seasonally wet places, roadside ditches, farm ditches, and depressions where water collects during the May-to-October monsoons. *O. rufipogon* generally flowers at the end of the monsoon in October or November. Some populations can also flower again in April and May, however (Punyalue 2006; Wongtamee 2008). As cultivation practices intensify, there is increasing overlap between the flowering time of wild rice and that of various cultivars of rice. Wild rice represents an important reservoir of genetic biodiversity for cultivated rice. *O. rufipogon* populations are highly diverse for morphology and for genetic markers, and they often segregate for agronomically important traits such as pathogen or insect resistance. Wild rice populations have been repeatedly used for rice improvement. Green revolution rice varieties were developed in part from alleles found in native *O. rufipogon* populations.

The first documented observation of invasive weedy rice was in 2001 approximately 150 km north of Bangkok (Maneechote et al. 2004). Weedy rice has since spread almost exponentially and now is found in all provinces where the HYVs are grown, from the Central Plain to the Lower North. It covered a total area of more than 32,000 ha by 2007 (Maneechote 2007). A close monitoring of the spread of weedy rice in the village of Khao Sam Sib Harb in Kanchanaburi, where HYV rice is double-cropped, found that beginning with a few plants in a field, weedy rice can take over an entire village after just five to six crops. Yield loss increases with increasing level of infestation: 1.1% of yield declines for every 1% increase in infestation (Maneechote et al. 2004), which has resulted in rapid declines of rice production and, in some cases, farmers abandoning their fields. Weedy rice displaces the crop plants and appears to be a better competitor for light. Several aspects of weedy rice in Thailand are noteworthy: its recent origin, its rapid spread, and the mixture of wild ancestor and cultivated rice throughout the rural landscape.

Weedy rice has a distinct morphology that is easy to recognize. It has many of the morphological traits of rice's wild ancestor, *O. rufipogon*, including awns; branched panicles; red pericarp color; shattering, multiple tillers; and tall stature. Within an infested field, weedy rice shows a great deal of morphological variation for these traits, with a nearly complete spectrum from the morphology of cultivated rice to that of *O. rufipogon*. The observations that weedy rice had many of the traits of *O. rufipogon* and that *O. rufipogon* often grows at the margins of cultivated rice fields led to the hypothesis that weedy rice is of hybrid origin. That is, weedy rice is the result of gene flow between cultivated and wild rice.

Table 2–1. Cross-fertilization between cultivated (female parent) and wild rice (male parent). All parents germinated 100%, all seedlings were normal, except 10% of the Lampoon wild rice seedlings, which were abnormal. (From Sintukhiew 2004)

Female parent (cultivated)	Panicles pollinated	Seed set (% of spikelets pollinated)	% F1 seed germinated	Seedlings that grew normally (% of seed)
With Lampoon wild rice as male parent				
KDML105	4	36.9 ± 13.1	65	20
Kum Doi Saket	12	31.1 ± 16.5	50	50
Niaw Sanpatong	5	15.6 ± 14.2	75	75
RD6	3	22.7 ± 20.1	40	40
RD10	4	35.8 ± 10.4	80	55
With Acc. 18883 wild rice as male parent				
Sew Machan	15	35.4 ± 25.2	85	70
CNT1	8	31.9 ± 27.7	75	15

To begin testing the hypothesis of gene flow and a hybrid origin of weedy rice, crossing studies between varieties of cultivated rice and wild rice were undertaken to determine if gene flow and crossing could occur. Five cultivated rice varieties were crossed with plants from two geographically distinct *O. rufipogon* populations. Seeds produced from manual pollinations varied from 15% up to 36.9%, indicating that wild and cultivated rice are compatible with the resulting hybrid seed (Table 2–1). The hybrid seeds were viable and germinated with germination ranging from 50 to 80%. The percentage of resulting viable seedlings with normal growth ranged from 15 to 75%. These results are interesting for several reasons. First, clearly wild and cultivated rice in Thailand are capable of gene flow; viable hybrid seeds and seedlings are produced. Next, there is a great deal of variability in the cultivated–wild rice system. The ability to produce hybrids and their viability vary by the specific type of both the cultivated rice and the wild rice populations; reciprocal crosses set seed but often with different frequency. In a more extensive study, some crosses between wild and cultivated rice were incompatible, whereas others had exceedingly high yields (Sintukhiew 2004). Finally, because cultivated rice is self-fertile and wild rice is outcrossed, gene flow will tend to be from cultivated crops into wild populations.

Another set of crosses examined the morphology and fitness of the F2 hybrids (Prommin 2007). F_2 seed were produced from selfed F_1 plants (250 seeds per cross), and the germination percentage was recorded. The F_2 seeds germinated at the same levels (74–82%) as the cultivated rice parents and the F_1 hybrids, indicating no loss of fitness associated with germination. As expected, the F_2 plants segregated for morphology and allowed for some inferences about mode of inheritance. Plant type, stigma color, apiculus color, and awn color were consistent with single-gene mode of inheritance, whereas spikelet awns appeared to be controlled by at least three genes. The F_2 plants showed transgressive segregation in all crosses for number of tillers per plant, culm length, and panicle length, all traits associated with greater reproduction. A normal distribution was found for number of primary branches per panicle. For spikelets per panicle, most F_2 plants fell into the range closer to wild rice parent, while for seed fertility, most F_2 plants fell into the range closer to that of cultivated rice. Of particular interest is that

Table 2–2. Presence of foreign alleles in farmers' rice seeds detected with microsatellite markers (RM1, RM341, RM444, and RM586[a]) (From Suta, 2008)

| | | Number of individuals | | |
| | | With at least 1 foreign allele, of | | |
Seed variety	Total	Wild rice	Other variety	% Individuals with weedy rice contamination[b]
SPR1	270	5	3	1.8
PSL2	60	8	14	13.3
CNT80	20	0	8	0
Total	350	13	25	2.86

[a] For SPR1 and PSL2 only.
[b] With alleles from wild rice.

the morphology of the F_2 population was similar to the range of morphologies found in weedy rice of Thailand, reinforcing the hypothesis of gene flow and hybrid origin of weedy rice.

Next, we asked if there is genetic evidence of gene flow in the Thai weedy rice system. Cultivated rice seeds were collected from several farmers' fields. A set of four microsatellite markers was chosen to distinguish cultivated varieties of rice from wild rice (RM1, RM341, RM444, and RM586; Londo & Schaal 2007). These markers were used to assay the genotype of weedy rice plants to determine if alleles from wild rice were present. In a survey of weedy rice plants found at the Khao Sam Sib Harb village, wild rice alleles were found in 5.7% of weedy individuals (Niruntrayakul, in preparation). These data clearly indicate that weedy rice contains alleles from wild rice and they provide strong genetic support for gene flow. These results are consistent with a recent study of weedy rice in the United States that detected wild rice alleles within weedy rice across the U.S. rice-growing region (Londo & Schaal 2007).

One unusual aspect of the Thai system is the rapid spread of weedy rice among fields. The increase of weedy rice within a single field can be explained by rapid population growth assisted by seed shattering. It is unclear, however, why the weedy rice spreads so quickly among fields. Farmers often save seed from one year to the next for planting, and they often share seed with each other. If weedy rice seed has the same appearance as cultivated rice seed, farmers may be inadvertently planting and dispersing the weed. A molecular analysis was conducted on rice seeds collected from these fields (Table 2–2). The seeds all had the appropriate appearance for each variety – that is, there was no morphological evidence for weedy rice in the seed lots. The frequency of contamination by wild rice alleles within a variety varied from 0.0 in a new, recently released variety to 13.3% in PSL2, a frequently planted variety. When multiple seed samples were analyzed from different fields, wild rice contamination was found in up to 50% of the PSL2 fields (Table 2–3). These results suggest that gene flow has not only resulted in weedy rice but has also introduced alleles into cultivated rice. The reverse study, detecting cultivated alleles in wild rice populations, is currently underway. Preventing gene flow in this system is difficult. Not only are genes dispersed by wind via pollen, but seed dispersal also can move genotypes across

Table 2–3. Farmers' rice seed samples with wild rice alleles (From Suta, 2008)

	Number of samples			
		With at least 1 foreign allele, of		% Samples with weedy rice contamination
Seed variety	Total	Wild rice	Contamination[a]	
SPR1	27	2	2	7.4
PSL2	6	3	2	50.0
CNT80	2	0	1	0
Total	35	5	5	14.3

[a] Number of alleles from wild rice.

fields and villages. In these agricultural systems, it is difficult to effectively isolate any variety of rice.

Perhaps the most important information for predicting the potential effects on biodiversity of gene flow from GM rice is a thorough determination of the fitness of hybrids. The relative fitness of hybrids, compared to wild and GM rice, will determine the ultimate fate of the transgene. How can GM hybrids be studied? GM rice has been developed, but it is not possible to conduct field trials of gene flow, introgression, and fitness because of both government regulation and the potential negative environmental consequences should transgenic hybrids escape. For this reason, greenhouse studies of the relative fitness of GM rice, wild rice, and their F_1 hybrids were used to evaluate the likelihood of transgene escape and the formation of weedy rice. The GM rice used was a variety that had been transformed with the late embryogenesis-abundant protein gene, HVA1, that conveys drought resistance (Xu et al. 1996). Seeds were germinated of each parental variety and the hybrid. Plants were compared for growth and reproduction (Bashir 2006) in both a standard greenhouse environment and also under drought stress. Results of this comparison are shown in Table 2–4. There was little evidence for a fitness advantage of the hybrid plants in either standard or stress environments. In fact, the hybrid had lower seed production than either the GM rice or wild rice in both environments, suggesting that the plants may be at a reproductive disadvantage. These data taken alone would suggest that the transgene would be eliminated.

Although these results are interesting, they do not mimic conditions in a rice field where competition for space can be intense. The relative fitness of hybrids was studied under competition by a deWitt replacement experiment, where the

Table 2–4. Relative fitness of GM crop, *O. rufipogon* and hybrid plants (From Bashir 2006)

Fitness trait	Nonstress environment	Drought-stress environment
# leaves	*O. rufi* > hybrid > GM	Similar
# tillers	*O. rufi* > hybrid > GM	Similar
Flowering time	GM > hybrid > *O. rufi*	Hybrid > GM > *O. rufi*
Final biomass	Hybrid > *O. rufi* > GM	*O. rufi* > GM > Hybrid
Total # seeds	GM > *O. rufi* > hybrid	GM > *O. rufi* > Hybrid

Table 2–5. Fitness and competition. Competition of GM × O. *rufipogon* hybrids with GM rice and wild rice, O. *rufipogon* in a deWitt replacement experiment

Mean per plant		0 GM 8 hybrid	2 GM 6 hybrid	4 GM 4 hybrid	6 GM 2 hybrid	8 GM 0 hybrid
Total seeds	GM	–	51.0	34.0	50.4	42.5
	hybrid	329.7	346.7	360.8	357.4	–
		0 O. rufi 8 hybrid	2 O. rufi 6 hybrid	4 O. rufi 4 hybrid	6 O. rufi 2 hybrid	8 O. rufi 0 hybrid
Total seeds	O. rufi	–	236.7	245.7	181.4	313.3
	hybrid	340.0	291.8	365.0	379.1	–

hybrid plants compete against either the O. *rufipogon* parent or the GM rice parent in varying frequencies (Bashir 2006). Table 2–5 shows the design and results of this experiment. In every case, under competition, the hybrids have greater reproduction (and biomass; data not shown) than either the wild rice parent or the GM variety. Differences in fitness can be extreme. When 6 GM plants were grown with 2 hybrids (intense competition), the GM plants produced, on average, 50 seeds, whereas the hybrid plants produced 357 seeds. Likewise, when O. *rufipogon* was grown in competition with hybrid plants, O. *rufipogon* produced 181 seeds, whereas hybrid plants produced 379 seeds, on average. These results indicate a high fitness advantage for GM × wild rice hybrids under competition. Most likely, this advantage has little to do with the inserted transgene because plants in the competition experiments were grown with ample water. More likely, the advantage may be due to heterosis or some other general genetic feature of the hybrids. Regardless of the mechanism, these results suggest that not only will weedy rice be formed with GM rice, but that, under some conditions, the newly produced weeds also will have a selective advantage and spread.

CONCLUSIONS

These results suggest that the formation of weedy intermediate forms of rice by gene flow between GM rice and wild rice will be a potential problem in areas where wild rice is endemic. In particular, the movement of alleles from crop to wild rice and vice versa via intermediate weedy rice can potentially threaten the biodiversity of native O. *rufipogon*. At this point, we do not know how frequently alleles from cultivated rice are found in natural O. *rufipogon* populations, and we do not know if these alleles would alter the selection regimes or levels of diversity in wild rice. The answers to these important questions are essential for fully inferring the consequences of gene flow for biodiversity. Such work will require new genetic information, including genomic studies, which are currently being undertaken. It is clear, however, that gene flow between wild and cultivated rice results in the formation of an invasive weedy intermediate that both reduces agricultural yield and potentially can displace the native species growing within marshy regions of Thailand. Moreover, when we consider possible effects of GM rice, the specific transgenes may have a strong effect, particularly if they enhance the fitness of plants that contain them. For example, if transgenic rice were to introduce herbicide resistance into weedy populations, the consequences for native biodiversity could be large if weedy rice invaded natural areas.

Can this study be generalized to other crop systems? Is gene flow always a concern for biodiversity? For many crops that are grown outside the range of their wild relatives, gene flow will not affect biodiversity because there is no chance for genes to escape into the environment. (Adventitious presence, contamination of non-GM crops, remains a concern for farmers, however.) If a crop is grown within the native range of its reproductively compatible relatives, however, the potential for gene flow exists. If native species are rare and reproductively isolated, and if they grow far removed from agricultural lands, the likelihood of environmental effects is low. In contrast, for crops like rice or sunflowers that are reproductively compatible and are grown in close proximity to large populations of wild relatives, assessment of gene flow and its consequences is necessary.

Molecular markers are essential tools for this assessment of potential effects of gene flow on biodiversity. Because microsatellite markers are highly variable with many alleles per locus, they provide precise information on parentage and, hence, gene flow. In the case of potential hybrid weed formation between crop and native species, molecular markers easily identify hybrids, as seen in the case of Thai rice. Also, markers will be useful for addressing additional issues. At this point, we do not know how much total gene flow has occurred between crops and their wild relatives in the past. Is this a frequent occurrence? Are native populations already "contaminated" by gene flow (and thus has biodiversity already been lost)? Or, are native species still genetically unchanged? Which crop systems and which species are affected? Are levels of gene flow changing in response to new agricultural practices or climate change? Molecular markers provide a powerful means to address these questions and to fully understand the impact of agriculture and GM crops on biodiversity.

REFERENCES

Allainguillaume J, Alexander M, Bullock JM et al. (2006) Fitness of hybrids between rapeseed (*Brassica napus*) and wild *Brassica rapa* in natural habitats. *Molecular Ecology*, **15**, 1175–1184.

Andow DA, Zwahlen C (2006) Assessing environmental risks of transgenic plants. *Ecology Letters*, **9**, 196–214.

Bashir, A (2006) *Hybrid Fitness and the Potential for Gene Flow Between Cultivated Rice, Oryza sativa and its Wild Relative, Oryza rufipogon*. Ph.D Dissertation, St. Louis University, St. Louis, Missouri.

Ceddia MG, Bartlett M, Perrings C (2007) Landscape gene flow, coexistence and threshold effect: the case of genetically modified herbicide tolerant oilseed rape (*Brassica napus*). *Ecological Modeling*, **205**, 169–180.

Chapman MA, Burke JM (2006) Letting the gene out of the bottle: the population genetics of GM crops. *New Phytologist*, **170**, 429–443.

Chrispeels MJ, Sadava DE (2003) *Plants, Genes, and Crop Biotechnology*. Jones and Bartlett Publishers, Boston.

Clark EA (2006) Environmental risk of genetic engineering. *Euphytica*, **148**, 47–60.

Council for Agricultural Science and Technology (2007) Implications of gene flow in the scale-up and commercial use of biotechnology-derived crops: economic and policy considerations. *Issue Paper*, **37**, 1–24.

Day PR (1987) Concluding remarks. *Regulatory Considerations: Genetically-Engineered Plants*. Center for Science Information, San Francisco.

Diamond J (1997) *Guns, Germs and Steel: The Fates of Human Societies*. WW Norton, New York.

Ellstrand NC (2003) *Dangerous Liaisons? When Cultivated Plants Mate with Their Wild Relatives*. Johns Hopkins University Press, Baltimore.

Federoff N, Brown NM (2004) *Mendel in the Kitchen: A Scientist's View of Genetically Modified Foods*. Joseph Henry Press, Washington, DC.

Hancock JF (2003) A framework for assessing the risk of transgenic crops. *BioScience*, **53**, 512–519.

Haygood R, Ives AR, Andow DA (2003) Consequences of recurrent gene flow from crops to wild relatives. *Proceedings of the Royal Society B: Biological Sciences*, **279**, 1879–1886.

Heywood VH (1995) *Global Biodiversity Assessment*. Cambridge University Press, Cambridge, UK.

Jhala AJ, Hall LM, Hall JC (2008) Potential hybridization of flax with weedy and wild relatives: an avenue for movement of engineered genes. *Crop Science*, **48**, 825–840.

Knispel AL, McLachlan SM, Van Acker RC et al. (2008) Gene flow and multiple herbicide resistance in escaped canola populations. *Weed Science*, **56**, 72–80.

Levin DA, Francisco-Ortega J, Jansen RK (1996) Hybridization and the extinction of rare plant species. *Conservation Biology*, **10**, 10–16.

Londo J, Schaal B (2007) Origins and population genetics of weedy rice in the United States. *Molecular Ecology*, **16**, 4522–4535.

Londo JP, Chiang YC, Hung KH, Chiang TY, Schaal BA (2006) Phylogeography of Asian wild rice, *Oryza rufipogon*, reveals multiple independent domestications of cultivated rice *Oryza sativa*. *Proceedings of the National Academy of Sciences, USA*, **103**, 9578–9583.

Maneechote C (2007) *Weedy Rice: Problem and Control*. Department of Agriculture, Thailand Research Fund and Chiang Mai University, Chiang Mai, Thailand.

Maneechote C, Jamjod S, Rerkasem B (2004) Invasion of weedy rice in rice fields in Thailand: problems and management. *International Rice Research Notes*, **29**, 20–22.

Martinez-Castillo J, Zizumbo-Villarreal D, Gepts P et al. (2007) Gene flow and genetic structure in the wild-weedy-domesticated complex of Phaseolus lunatus L. in its Mesoamerican center of domestication and diversity. *Crop Science*, **47**, 58–66.

McRae BH, Beier P (2007) Circuit theory predicts gene flow in plant and animal populations. *Proceedings of the National Academy of Sciences*, **104**, 19885–19890.

Millennium Ecosystem Assessment (2005) *Ecosystems and Well-Being*. Island Press, Washington, DC.

Muir WM, Howard RD (2002) Assessment of possible ecological risks and hazards of transgenic fish with implications for other sexually reproducing organisms. *Transgenic Research*, **11**, 101–114.

National Research Council (2002) *Environmental Effects of Transgenic Plants: The Scope and Adequacy of Regulation*. National Academy Press, Washington, DC.

Nielson RL, McPherson MA, O'Donovan JT et al. (2009) Seed-mediated gene flow in wheat: seed bank longevity in western Canada. *Weed Science*, **57**, 124–132.

Office of Agriculture Economics (OAE) 2008. *Thailand Agriculture Year Book 2550*. OAE, Ministry of Agriculture and Co-operatives, Bangkok, Thailand.

Pasquet RS, Peltier A, Hufford MB et al. (2008) Long-distance pollen flow assessment through evaluation of pollinator foraging range suggests transgene escape distances. *Proceedings of the National Academy of Sciences USA*, **105**, 13456–13461.

Pilson D, Prendeville HR (2004) Ecological effects of transgenic crops and the escape of transgenes into wild populations. *Annual Review of Ecology, Evolution and Systematics*, **35**, 149–174.

Pilson D, Snow AA, Rieseberg LH et al. (2004) A protocol for evaluating the ecological risks associated with gene flow from transgenic crops into their wild relatives: the case of cultivated sunflower and wild Helianthus annuus. In: *Introgression from Genetically Modified Plants into Wild Relatives* (eds. den Nijs HCM, Bartsch D, Sweek J), pp. 219–233. CAB International, Oxfordshire, UK.

Prommin A (2007) *Segregation in F2 Generation of Crosses between Common Wild Rice (Oryza rufipogon Griff.) and Cultivated Rice (O. sativa L.).* MSc Thesis, Chiang Mai University, Thailand.

Punyalue A (2006) *Characterization of Common Wild Rice Populations from Main Rice Growing Regions of Thailand.* MSc Thesis, Chiang Mai University, Thailand.

Purrington CB, Bergelson J (1995) Assessing weediness of transgenic crops: industry plays plant ecologist. *Trends in Ecology and Evolution,* **10**, 340–342.

Sintukhiew T (2004) *Interspecific Hybridization between Cultivated Rice (Oryza sativa L.) and Wild Rice (O. rufipogon).* MSc Thesis, Chiang Mai University, Thailand.

Smith, B (1998) *The Emergence of Agriculture.* Scientific American Library, New York.

Snow AA (2002) Transgenic crops – why gene flow matters. *Nature Biotechnology,* **20**, 542.

Snow AA, Pilson D, Rieseberg LH et al. (2003) A Bt transgene reduces herbivory and enhances fecundity in wild sunflowers. *Ecological Applications,* **13**, 279–286.

Sommut W (2003) *Changes in Flood-Prone Rice Ecosystems in Thailand, Crop Year 2000–2001.* Department of Agriculture. Bangkok, Thailand.

Suta A. 2008. *Weedy Rice Contamination in Farmers' Seed.* MSc Thesis, Chiang Mai University, Thailand.

Tesso T, Kapran I, Grenier C et al. (2008) The potential for crop-to-wild gene flow in sorghum in Ethiopia and Niger: a geographic survey. *Crop Science,* **48**, 1425–1431.

U.S. Department of Agriculture, National Agricultural Statistics Service (2008) *Adoption of Genetically Engineered Crops in the U.S.* Available at www.ers.usda.gov/Data/ BiotechCrops.

Warwick SI, Legere A, Simard MJ et al. (2008) Do escaped transgenes persist in nature? The case of an herbicide resistance transgene in a weedy *Brassica rapa* population. *Molecular Ecology,* **17**, 1387–1395.

Watrud LS, Lee EH, Fairbrother A et al. (2004) Evidence for landscape level, pollen-mediated gene flow from genetically modified creeping bent-grass with CP4 EPSPS as a marker. *Proceedings of the National Academy of Sciences USA,* **101**, 14533–14538.

Wilkinson MJ, Elliott LJ, Allainguillaume J et al. (2003) Hybridization between *Brassica napus* and *B. rapa* on a national scale in the United Kingdom. *Science,* **302**, 457–459.

Wongtamee A (2008) *Evaluation of Exotic Genes Contamination from Crop Rice* (Oryza sativa L.) *in Common Wild Rice* (Oryza rufipogon Griff.) *Populations.* MSc Thesis, Chiang Mai University, Thailand.

Xu D, Duan X, Wang B et al. (1996) Expression of a late embryogenesis abundant protein gene, HVA1, from barley confers tolerance to water deficit and salt stress in transgenic rice. *Plant Physiology,* **110**, 249–257.

Zhang NY, Linscombe S, Oard J (2003) Outcrossing frequency and genetic analysis of hybrids between transgenic glufosinate herbicide resistant rice and the weed, red rice. *Euphytica,* **130**, 35–45.

3 A community and ecosystem genetics approach to conservation biology and management

Thomas G. Whitham, Catherine A. Gehring, Luke M. Evans, Carri J. LeRoy,
Randy K. Bangert, Jennifer A. Schweitzer, Gerard J. Allan, Robert C. Barbour,
Dylan G. Fischer, Bradley M. Potts, and Joseph K. Bailey

INTRODUCTION

The emerging field of community and ecosystem genetics has so far focused on
how the genetic variation in one species can influence the composition of associated communities and ecosystem processes such as decomposition (see definitions in Table 3–1; reviews by Whitham et al. 2003, 2006; Johnson & Stinchcombe
2007; Hughes et al. 2008). A key component of this approach has been an emphasis on understanding how the genetics of foundation plant species influence a
much larger community. It is reasoned that because foundation species structure their ecosystems by creating locally stable conditions and provide specific
resources for diverse organisms (Dayton 1972; Ellison et al. 2005), the genetics
of these species as "community drivers" are most important to understand and
most likely to have cascading ecological and evolutionary effects throughout
an ecosystem (Whitham et al. 2006). For example, when a foundation species'
genotype influences the relative fitness of other species, it constitutes an indirect genetic interaction (Shuster et al. 2006), and when these interactions change
species composition and abundance among individual tree genotypes, they result
in individual genotypes having distinct community and ecosystem phenotypes.
Thus, in addition to an individual genotype having the "traditional" phenotype that population geneticists typically consider as the expression of a trait at
the individual and population level, community geneticists must also consider
higher-level phenotypes at the community and ecosystem level. The predictabil-
ity of phenotypes at levels higher than the population can be quantified as community heritability (i.e., the tendency for related individuals to support similar
communities of organisms and ecosystem processes; Whitham et al. 2003, 2006;
Shuster et al. 2006).

We thank the National Science Foundation (NSF) for a Frontiers of Integrative Biological
Research grant to bring together diverse scientists to help integrate and develop the emerging field of community and ecosystem genetics. NSF Integrative Graduate Education and
Research Traineeship (IGERT) and Science Foundation Arizona grants supported student
research; The Australian Research Council (grants DP0451533 and DP0773686) supported
parallel studies in Australia, and NSF grant DEB-0816675 supported climate change research
in the United States. The Ogden Nature Center, the Utah Department of Natural Resources
and Reclamation, provided lands for common gardens, restoration, and public outreach.

Table 3–1. Definitions of community genetics terms

Community and ecosystem genetics − The study of the genetic interactions that occur among species and their abiotic environment in complex communities

Community and ecosystem phenotypes − The effects of genes at levels higher than the population. These phenotypes result from interspecific indirect genetic effects, which can be summarized as a univariate trait for statistical analyses.

Community heritability − The tendency for related individuals to support similar communities of organisms and ecosystem processes

Foundation species − Species that structure a community by creating locally stable conditions for other species, and by modulating and stabilizing fundamental ecosystem processes

Minimum viable population (MVP) − The size of a population that, with a given probability, will ensure the existence of the population for a stated period of time

Minimum viable interacting population (MVIP) − The size of a population that is required to maintain genetic diversity at levels required by *other* dependent and interacting species in a community context

---

Numerous studies in diverse systems demonstrate that community and ecosystem phenotypes exist (e.g., cottonwoods – Dickson & Whitham 1996; Whitham et al. 2003; Bailey et al. 2004, 2006; Fischer et al. 2004, 2006, 2007; Schweitzer et al. 2004, 2005a,b, 2008; Wimp et al. 2005, 2007; LeRoy et al. 2006, 2007; Shuster et al. 2006; aspen – Madritch et al. 2006, 2007; willows – Fritz & Price 1988; Hochwender & Fritz 2004; metrosideros – Treseder & Vitousek 2001; oaks – Madritch & Hunter 2002, 2004; Tovar-Sanchez & Oyama 2006; aspen – Madritch et al. 2006, 2007; Donaldson & Lindroth 2007; seagrass – Hughes & Stachowicz 2004; goldenrod – Maddox & Root 1987; Crutsinger et al. 2006; Crawford et al. 2007; eucalypts – Dungey et al. 2000; Barbour et al. 2009; evening primrose – Johnson & Agrawal 2005; Johnson et al. 2006; Johnson 2008; common reed – Araki et al. 2005; birch – Silfver et al. 2007; alder – Lecerf & Chauvet 2008; grape – English Loeb & Norton 2006; *Baccharis* – Rudgers & Whitney 2006). Although most of the species examined in these studies are considered to be foundation species in their respective ecosystems, the evening primrose is a notable exception. Thus, it would appear that a community genetics approach need not be limited to just foundation species.

Having demonstrated that different plant genotypes support different communities, new emphasis must be placed on understanding the implications of these findings. For example, relatively few studies have quantified how the genetic variation in foundation species might also affect the evolution of dependent community members (e.g., review by Mopper 1996 and more recent studies on oaks – Mopper et al. 2000; goldenrod – Craig et al. 2007; cottonwood – Evans et al. 2008), how the genetic variation in a foundation species in one community might affect another community (e.g., terrestrial affects aquatic – Driebe & Whitham 2000; LeRoy et al. 2006, 2007; Lecerf & Chauvet 2008), and how a community might feed back to affect the survival and performance of the genotype expressing the trait (e.g., soil microbial community impacts on a host tree; Fischer et al. 2006; Schweitzer et al. 2008). These frontiers of research remain important for a community genetics approach to reach its full conceptual potential. Based on our current knowledge, however, it seems clear that this approach can offer productive insights into conservation biology and management. The remainder of this chapter considers the conservation implications of the existence of heritable community and ecosystem phenotypes.

CONSERVATION IN A COMMUNITY CONTEXT

There are two main reasons to support a community genetics approach for conservation and management. First, because all species have evolved in a community context of many interacting species and their abiotic environments, it is unrealistic to study any one species in isolation as if it had evolved in a vacuum (e.g., Thompson 2005; Wade 2007). Thus, the community context of species evolution is far more realistic as it considers more of the variables that are likely to have influenced the evolution of any one species.

Second, and perhaps more important from a manager's perspective, a community approach is less likely to result in management errors. For example, as a two-way interaction, mistletoes are clearly parasites of flowering plants. Few interactions are just two-way interactions, however. In their studies of one-seed juniper (*Juniperus monosperma*) and its associated mistletoe (*Phoradendron juniperinum*), van Ommeren and Whitham (2002) found that the berries produced by both the juniper and the mistletoe were primarily dispersed by birds such as Townsend's Solitaires (*Myadestes townsendi*). If the seeds of mistletoe attract more birds that end up dispersing more juniper berries, then as a three-way interaction, mistletoe might be considered a mutualist rather than a parasite. In support of this hypothesis, they found that juniper stands with mistletoe attracted up to 3 times more birds than did stands with little or no mistletoe. Most important, the number of juniper seedlings was more than twofold greater in stands with high mistletoe density compared with stands that had little or no mistletoe. A manager would be justified in the case of a two-way interaction in removing mistletoe to promote tree performance. In the case of a three-way interaction, however, the opposite strategy might be warranted because the mistletoe benefits juniper recruitment by promoting avian seed dispersal. Thus, just by adding one additional species to the analysis, the fundamental relationships of juniper and mistletoe can "switch or reverse" from negative to positive, and management decisions would also shift accordingly. Because the species composing a community interact, the sign of the ecological interaction and the appropriate management decision is dependent on the number of interacting players and their combined interactions.

Although the previously mentioned study demonstrates how a community analysis versus a population analysis can result in a qualitative change in interpretation that affects management decisions, it is important to know whether such reversals in interpretation are common enough to warrant concern. Bailey and Whitham (2007) reviewed 85 studies and 417 tests that varied in the number of factors examined in scientific studies, which included numbers of species, geographic scale (e.g., local to global), and number of years of the study (Fig. 3–1). When the study included only one factor, there was no chance of detecting either a significant interaction or a reversal in outcome as it is beyond the ability of the statistical approach. When two factors were included in the study, however, there was a 55% probability of detecting a significant interaction

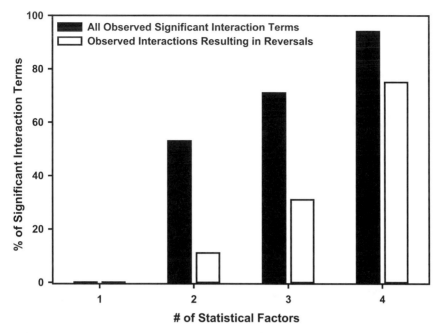

Figure 3–1: A survey of the literature shows that interaction effects and pattern reversals are common, which argues for a community approach to avoid potentially costly errors in conservation and management decisions. Solid bars show that as the number of factors in a study increases, the observed percentage of significant interaction effects also increases (i.e., both quantitative and qualitative interactions). Open bars show that as the number of factors in a study increases, the observed percentage of reversals also increases (i.e., a qualitative change in the sign of an interaction). Figure from Bailey and Whitham (2007).

term (solid bars), and this probability increased to nearly 95% when four factors were included in the analysis. Note that such an interaction term could be just a "quantitative" change in which the sign of an interaction would still remain the same (i.e., negative remains negative and positive remains positive). More important, they found that as the number of factors increased, the number of reversals also increased such that when the analysis included four factors, there was a nearly 80% probability that the sign of the interaction of one or more factors reversed from negative to positive or from positive to negative. These "qualitative" changes in interpretation are the most vexing, as they have a high probability of resulting in fundamental management errors. This study concluded that to avoid such errors, conservation biologists and managers need to adopt a long-term, community approach that covers a large geographic area.

Because species interact with one another (i.e., it is a multivariate world), only by including the major interacting species can one really understand the sign of the interactions. This is why a community approach is desirable and is less likely to result in management errors that are both costly and produce unintended effects.

HERITABLE COMMUNITY AND ECOSYSTEM PHENOTYPES

To justify a community genetics approach, it is essential to demonstrate that community and ecosystem phenotypes exist and that they are heritable. Although

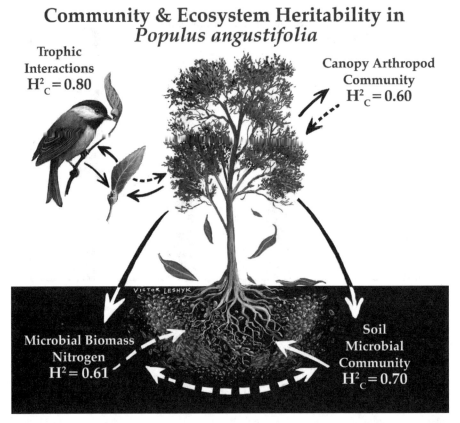

Community & Ecosystem Heritability in
Populus angustifolia

Trophic
Interactions
$H^2_C = 0.80$

Canopy Arthropod
Community
$H^2_C = 0.60$

Microbial Biomass
Nitrogen
$H^2 = 0.61$

Soil
Microbial
Community
$H^2_C = 0.70$

Figure 3–2: Community and ecosystem phenotypes of individual tree genotypes of *P. angustifolia* show broad-sense heritability (H^2). Significant heritability of the canopy arthropod community, soil microbial community, trophic interactions between birds and insects, and soil nutrient pools demonstrate how trait variation in a foundation tree can structure communities and ecosystem processes (H^2_C for community traits; H^2 for single ecosystem trait). Because soil microorganisms mediate many ecosystem processes, including litter decomposition and rates of nutrient mineralization, the formation of these communities may feed back to affect plant fitness. Solid lines indicate known interactions; dashed lines indicate possible interactions. Quantitative genetic patterns such as these argue that genomic approaches will enhance our understanding of how interacting community members influence ecosystem processes. Figure from Whitham and coworkers (2008). *See Color Plate III.*

numerous studies in diverse systems have shown that different genotypes support different community members and ecosystem processes (see references in the introduction), only a few studies have quantified the heritability of community and ecosystem phenotypes (e.g., arthropod diversity on evening primrose – Johnson & Agrawal 2005; Johnson et al. 2006; arthropod community composition, soil microbial community composition, and trophic interactions on narrowleaf cottonwood – Bailey et al. 2006; Shuster et al. 2006; Schweitzer et al. 2008; respectively). For example, Fig. 3–2 shows that the canopy arthropod community on narrowleaf cottonwood (*Populus angustifolia*) was under strong genetic control such that regardless of where replicate clones of the same tree genotype were planted within a common garden, they tended to accumulate and support the same community of arthropods (Shuster et al. 2006). Because Shuster and coworkers used replicate clones of individual tree genotypes, they calculated broad-sense community heritability (H^2_c), which estimates that genetic factors account for approximately 60% of the variation in the canopy arthropod community ($H^2_c = 0.60$).

In another common garden study of the soil microbial community, Schweitzer et al. (2008) assessed the microbial community composition in bulk soil beneath tree canopies by using phospholipid fatty acid (PLFA) biomarkers (Fig. 3–2). PLFAs are unique to major taxonomic groups (e.g., Gram-positive and -negative bacteria, and fungi), and thirty different groups were identified. Their analysis found a strong tree genetic component to microbial community composition ($H^2_c = 0.70$). Schweitzer et al. (2008) also found that soil microbial biomass nitrogen exhibited high broad-sense heritability ($H^2 = 0.61$), indicating a genetic basis for soil nitrogen pools stored in microorganisms. This latter finding is important because soil microorganisms mediate many ecosystem processes, including litter decomposition, rates of nutrient mineralization, and nutrient pools. Because these nutrient pools are genotype dependent, the formation of different microbial communities beneath different tree genotypes may feed back to affect plant fitness. If so, such feedbacks would establish a crucial link between the microbial community and plant fitness that could pave the way for ecosystem science to be placed within an evolutionary framework (Bever et al. 1997; Hooper et al. 2000; Whitham et al. 2003; Bartelt-Ryser et al. 2005).

The potential for similar feedbacks on plant performance also exists aboveground in the interactions of insect herbivores and their avian predators (Fig. 3–2). For example, the presence or absence of the leaf-galling aphid (*Pemphigus betae*) is determined by susceptible or resistant tree genotypes. This interaction predictably affects other trophic levels and alters the composition of a diverse community of fungi, insects, spiders, and avian predators (Dickson & Whitham 1996; Bailey et al. 2006). In combination, these and other studies described previously suggest that indirect genetic interactions among relatively few foundation species (e.g., tree, herbivore, predator, mutualist, and/or pathogen) and their environment may structure the composition and abundance of a much larger community of organisms, which could then feed back to affect the performance of individual tree genotypes.

An important goal for understanding the ecology, evolution, conservation, and management of any ecosystem is to define how many foundation species (plants, animals, microbes) define a given ecosystem and their extended effects on associated species. Understanding the genetics (genotypic or genetic variation therein) of foundation species in a community context is likely to be more informative than the genetics of most other species (e.g., rare species) and should result in fewer management errors than would single-factor or population approaches. So far, our findings suggest that the study of relatively few foundation species in a community context can tell us much about a larger system involving many species. If correct, then the conservation and management of whole ecosystems may be more tractable than previously thought as all species are not equally important in "driving" the system. Focusing on foundation species does not imply that other organisms such as rare species are not important or deserving of preservation. It simply argues that the interactions of relatively few species structure an ecosystem and that many species respond to rather than create the conditions required by other species for their survival. If concentrating on the community genetics of foundation species proves generally correct, it would greatly simplify the dilemma of having to understand the complex interactions of all species to make appropriate management decisions, which is a daunting if not impossible task.

CONSERVATION CONSEQUENCES OF HERITABLE PHENOTYPES

If different genotypes of a foundation tree species support different community members and ecosystem processes as the previously mentioned heritability studies demonstrate, several key predictions can be made, which are of conservation and management value.

First, if different tree genotypes support different communities, we would expect that the genotypes that are genetically most similar will support the most similar communities. At both local and regional levels, Bangert and colleagues (2006a,b) found that genetically similar cottonwoods supported similar arthropod communities, whereas genetically dissimilar plants supported more different communities (i.e., a genetic similarity rule hypothesis). For example, by using a Mantel test to compare the molecular marker and arthropod community matrices of individual trees, we found that trees that were the most genetically similar were also the most similar in their arthropod communities (Fig. 3–3). This pattern holds for trees growing in a common garden, for trees growing in the wild in a single river (Fig. 3–3; Bangert et al. 2006a), and for trees growing throughout the American West in multiple river systems covering 720,000 km^2 (Bangert et al. 2006b). Studies with the forest tree (*Eucalyptus globulus*) in Australia showed a similar pattern in which the most genetically similar trees and geographic races also supported more similar canopy communities (Barbour et al. 2009). In both studies, the correlation in divergence in molecular and community variation was mediated through correlated divergence in quantitative traits such as leaf morphology and chemistry. Although it is well known that many specialist insects may show corresponding divergence that is correlated with the divergence of their host plant species (e.g., Abbot & Withgott 2004), it appears that the same pattern holds for less specialized community members. From a conservation perspective, the findings of Figs. 3–2 and 3–3 argue that genetic diversity in a foundation tree species matters for the dependent community of microbes, arthropods, and vertebrates.

Second, if different tree genotypes support different communities, then greater genetic diversity in a foundation species should be directly correlated with greater community diversity. Based on this prediction, Wimp and colleagues (2004) used amplified fragment length polymorphism (AFLP) molecular markers to quantify genetic diversity in eleven different stands of cottonwoods growing in a hybrid zone along the Weber River in Utah. The arthropod community was then recorded for the same trees, and the researchers found that stand genetic diversity accounted for approximately 60% of the variation in the diversity of an arthropod community composed of 207 species (Fig. 3–4). Similar findings were obtained

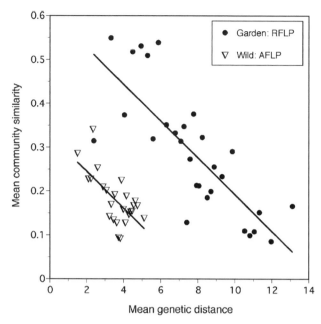

Figure 3–3: Because genetically similar trees support similar communities of organisms (i.e., the Genetic Similarity Rule), genetic diversity in a foundation plant species is likely to support greater biodiversity in the associated community. On average, Mantel tests show that arthropod communities become less similar as plant genetic distance increases (i.e., small genetic distance indicates more closely related trees). Genetic distances of the common garden trees (solid circles) were based on restriction fragment length polymorphism (RFLP) markers. The genetic distances of the naturally growing trees in the wild (open triangles) were based on AFLP marker data and include only the F_1 and backcross hybrids, thus removing the leverage of the two pure parental species from the analysis. Figure from Bangert and colleagues (2006a).

with oaks (*Quercus* spp.; Tovar-Sanchez & Oyama 2006). Experimental studies that manipulated genetic diversity with evening primrose (*Oenothera biennis*) and goldenrod (*Solidago altissima*) also showed that genetically diverse patches supported significantly greater species richness than did monocultures (Johnson et al. 2006; Crutsinger et al. 2006; respectively).

Third, if individual plant genotypes support different communities, they also may drive genetic differentiation and speciation in dependent community members. Several studies have found that populations of herbivores are adapted to individual host genotypes (Karban 1989; Mopper et al. 1995, 2000; Mopper 1996; Evans et al. 2008). Because speciation may follow from locally adapted, differentiated populations of arthropods (Via 2001; Dres & Mallet 2002), conserving genotypic diversity in foundation species can conserve locally adapted populations of arthropods. Conserving genotypic diversity in foundation species to conserve locally adapted populations of dependent species may extend to multiple trophic levels. For example, multiple herbivores (Stireman et al. 2005) and their parasitoids (Stireman et al. 2006) are genetically differentiated on two closely related sympatric species of *Solidago*, indicating that host plants can affect species' evolution at multiple tropic levels. A major conclusion from such studies is that genetic diversity is far more than just an endangered species issue; it is also

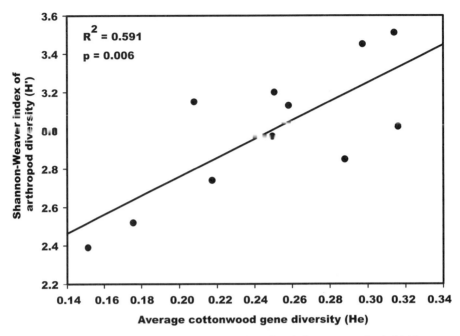

Figure 3–4: At the stand level, genetic diversity in a foundation tree species is highly corre-lated with the diversity of an arthropod community of 207 species. Cottonwood genetic diversity predicted arthropod diversity at a stand level. Each point in the scatterplot represents the Shannon–Weaver index of arthropod diversity (H') as a function of cottonwood gene diversity (He) in the same stand; a total of eleven stands were analyzed. The diagonal line represents the line of best fit that describes the relationship between increasing arthropod diversity as a result of increasing cottonwood genetic diversity. Figure from Wimp and colleagues (2004).

important for common foundation species in which different plant genotypes support different community members.

Fourth, when different genotypes of a foundation species support different communities, management practices based on species-area relationships may need to shift to genetic diversity-area relationships. In other words, is the species-area curve really driven by area, is it driven by the genetic diversity of the founda-tion species, or is it driven by both? To answer this question, Bangert et al. (2008) quantified the arthropod species accumulation curve for broadleaf cottonwoods (Fig. 3–5, a). This curve shows that as the arthropod communities of more trees are sampled (i.e., area), arthropod species richness increases predictably and begins to plateau. This pattern is common and is a primary justification for the con-servation practice of preserving large areas to capture the species pool. When molecular marker diversity for the same trees was analyzed to produce a molec-ular marker richness accumulation curve, the same pattern emerged. This curve shows that as the molecular markers of more trees are sampled (i.e., area), marker richness also increases predictably and begins to plateau. The two curves are remarkably similar. Furthermore, arthropod species richness is highly correlated with molecular marker richness ($r^2 = 0.92$; Fig. 3–5, b). This strong correlation raises an issue: Is the species-area curve driving biodiversity, is it the genetic diversity of the foundation species that drives biodiversity, or is there some com-bination of the two? Although area and genetic diversity of foundation species

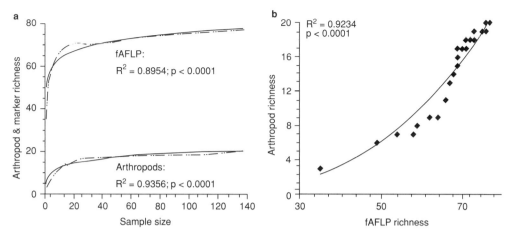

Figure 3–5: The species-area curve is also a species-genetic diversity curve, which may affect management decisions to preserve communities based on area, genetic diversity of foundation species, or both. (a) Scaling in which both arthropods and molecular markers scale similarly as a function of the number of broadleaf cottonwoods sampled (sample size or number is also a proxy for area). At their asymptotes, adequate sampling captures most of the arthropod community and genetic diversity, respectively. (b) Consistent with the above similarity in scaling, there is a strong positive relationship between arthropod species richness and marker richness. Because trees were selected at random, no geographic structure is reflected in these patterns. Figure from Bangert and colleagues (2008).

are undoubtedly autocorrelated, from a manager's perspective it is important to know whether a landscape should be managed from an area perspective or from the genetic diversity of a foundation species' perspective. Furthermore, if future conservation efforts try to establish living repositories for individual species, it is important to define the genetic diversity of these repositories that will be required to support their dependent communities.

A COMMUNITY GENETICS PERSPECTIVE ON MINIMUM VIABLE POPULATIONS

Historically, an important concept in conservation biology is the minimum viable population (MVP), which is the size of a population that, with a given probability, will ensure the existence of the population for a stated period of time (Shaffer 1981; Gilpin 1987). This concept is especially important for rare and endangered species whose populations are often small, highly inbred, and lacking in genetic diversity. Although this concept attempts to understand what population sizes are required to save a single species, it falls short of understanding what is necessary to save dependent community members and especially species that require a single host species or a subset of the host's genotypes for their survival.

An example of how the distribution, population growth rate, and evolution of one species can be dependent on genotypic differences of another species is that of the gall-forming mite (*Aceria parapopuli*) on cottonwoods. McIntyre and Whitham (2003) found that 99% of the mite's population was concentrated on naturally occurring F_1 hybrids between narrowleaf and Fremont cottonwood

(*P. angustifolia* and *P. fremontii*, respectively). Because suitable F_1 hybrid hosts are rare in the wild, the "actual" host population for these mites is a small subset of the larger cottonwood population. Furthermore, Evans et al. (2008) showed that the mites are even further differentiated such that the mites on the hybrids and parental hosts are different cryptic species. Transfer experiments showed that even among F_1 hybrids, the mites can be adapted to individual tree genotypes. Thus, even though cottonwoods are common trees in riparian forests, a dependent species can still be limited by the number of suitable host genotypes.

Even generalist species can be genetically differentiated and highly specialized at the local level (Thompson 1994). The eastern tiger swallowtail (*Papilio glaucus*) has a large list of host species, but at a local scale, it can be host specific (Scriber 1986). Other examples are provided by Feder and colleagues (1988), Wood and Keese (1990), Roinen and coworkers (1993), and Thompson (1994). Microorganisms can exhibit even greater specificity. Gene-specific interactions between *Rhizobium leguminosarum* bv. *viciae* and native Afghani pea plants (*Pisum sativum*) regulate symbiosis. The loss of a single gene will disrupt the symbiotic interaction and prevent nitrogen fixation (Vijn et al. 1993). Specificity in gene-for-gene interactions in molecular control points between plants and microorganisms in the rhizosphere is thought to be widespread in many foundation species (Hirsch et al. 2003; Phillips et al. 2003).

Such specificity suggests that MVP sizes (Shaffer 1981) in one species may be too small to support important interactions with other species (see Thompson 1994 as well as Chapter 10 by Waples and colleagues). Other researchers have recognized the weaknesses associated with the conservation of individual species rather than communities or ecosystems (e.g., Rohlf 1991; Simberloff 1998; Lindenmayer et al. 2007). A community genetics approach enhances our understanding of why the genetics of foundation species are so important, provides genetically based mechanisms to support these criticisms, and has the potential to merge different points of view as it spans genes to ecosystems.

Because of these issues, Whitham and colleagues (2003) proposed the minimum viable interacting population (MVIP) size that represents the size of a population needed to maintain genetic diversity at levels required by *other* dependent and interacting species in a community context. For example, MVP sizes for conserving a foundation tree species are probably much smaller than the population size required to conserve other dependent community members that require a subset of the tree's genotypes for their survival (MVIP). Thus, MVP represents the lower end of the population size that is required to conserve the foundation species, and MVIP represents the upper end of the population size of the foundation species that is required to conserve the dependent community and their interactions. Fig. 3–6 provides an initial guide as to where the differences between MVP for a foundation species and MVIP for a foundation species and its dependent community would be expected to be highest and lowest. High community and ecosystem heritability (H^2_c) suggests that many community members and the ecosystem processes they drive are dependent on the individual genotypes of a foundation species, whereas a low or nonsignificant H^2_c suggests that the distributions of most community members are not genotype dependent. Thus, as community heritability increases for any given foundation species, we would expect that the difference between MVP and MVIP would also increase.

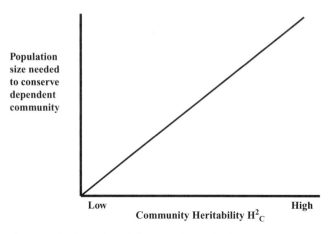

Figure 3–6: Illustration of the hypothesized relationship between high community heritability (i.e., different genotypes of a foundation species support different communities of organisms and ecosystem processes) and the population size needed to preserve the dependent community. As community heritability increases, the population size of the foundation species must increase to support its dependent community and the ecosystem processes they promote. The MVIP takes into account the genetics-based interactions among species in a community context. See text for contrast of MVP and MVIP.

MVP and MVIP represent end points on the population-to-community continuum. They provide a conceptual mechanism to explain why conservation efforts may flounder if they do not consider the consequences of genetic variation in foundation species on the dependent community (see Chapter 9 by Leberg and colleagues). Numerous studies show that temperate and boreal trees exhibit moderate to strong clines in ecologically important traits, suggesting that populations are genetically differentiated and locally adapted (Aitken et al. 2008). Associated with this genetic structuring, common garden studies by Barbour et al. (2009) showed that different geographic races of the same eucalypt species supported different communities. Although current conservation strategies target rare species, it may be even more important to conserve the genetic variation in foundation species with broad distributions because individual genotypes and races have different community and ecosystem phenotypes.

COMMUNITY PHENOTYPES OF GENETICALLY ENGINEERED ORGANISMS

Although the incorporation of novel genes into an organism through recombinant deoxyribonucleic acid (DNA) technology has enormous potential benefits, including increasing crop yields and reducing reliance on chemical pesticides (Avise 2003), the genes introduced into genetically engineered organisms (GEOs) may have unintended community and ecosystem phenotypes that are important to evaluate. Several issues need to be considered. First, GEOs now have been introduced over large areas in both industrialized and developing economies. For example, the global area planted with transgenic crops exceeded 110 million hectares in 2007, a 12% increase over the previous year (Lawrence 2008). Because many of these crops are grown as monocultures with little genetic variation, they are, in effect, both a foundation species and a foundation genotype.

Second, there is increasing evidence that transgenic organisms can move beyond their intended destinations where they have the potential to persist and to hybridize with native species (see Chapter 2 by Schaal and colleagues). Maximal gene-flow distances of 21 km were observed in herbicide (glyphosate)-resistant creeping bent grass (*Agrostis stolonifera* L.), a wind-pollinated, out-crossing, perennial turf grass planted in test fields in Oregon (Watrud et al. 2004). Three years after the test fields were taken out of production, 62% of plants tested were glyphosate resistant (Zapiola et al. 2008). There is evidence that transgenic canola (*Brassica napus*) and sunflower (*Helianthus annuus*) hybridize with wild relatives, although the fitness of these hybrids is highly variable and frequently not greater than wild genotypes (reviewed in Chapman & Burke 2006). Recent research also suggests that the risk of environmental impact from horizontal gene transfer of bacterial genes from crops to surrounding soil microorganisms is much higher than initially supposed (Heinemann & Traavik 2004). In their review, Marvier and Van Acker (2005) concluded that movement of transgenes beyond their intended destinations is a "virtual certainty."

A third reason that transgenics may have unintended ecological consequences for communities and ecosystems is that transgenes are designed to have significant fitness consequences (e.g., reducing herbivory, disease, or competition) thereby increasing the likelihood that transgenics will interact significantly with members of the communities in which they are introduced. Furthermore, crops with stacked traits (i.e., two or more transgenes) are now more widely planted than those containing only insect-resistance traits (Lawrence 2008).

Transgenic plants have been shown to have nontarget community and ecosystem phenotypes. For example, Bt corn byproducts can enter headwater streams where they are stored, consumed, and transported downstream to other bodies of water. Feeding trials in the laboratory indicate that Bt corn residues negatively affect nontarget stream insects (Rosi-Marshall et al. 2007). Recent meta-analyses, however, highlight the need for comparison of the effects of transgenes to those of traditional agricultural practices. An analysis of forty-two experiments showed that Bt cotton and corn altered the abundance of some taxa of nontarget arthropods, but the effects of the transgenes were much smaller than those of traditional pesticides (Marvier et al. 2007). Likewise, consistent effects of Bt corn, cotton, and potatoes on particular functional guilds of nontarget arthropods were not observed (Wolfenbarger et al. 2008). A recent review of Bt plant effects on the soil biota also found few significant impacts (Icoz & Stotzky 2008). Note that most studies of nontarget effects of transgenic organisms are relatively short-term and thus may ignore effects that accumulate slowly or are triggered by a change in environmental conditions.

A community genetics analysis provides a method for quantifying the community and ecosystem phenotypes of transgenes and their heritability. Importantly, in many ecological studies, the direct effects of a trait on an organism's phenotype (e.g., increased resistance to pest attack; see Box 3) may be less important than the indirect effects on the community phenotype (e.g., increased resistance to pest attack alters competitive interactions with other species to alter

BOX 3: LANDSCAPE GENETICS OF AN AMERICAN CHESTNUT BORER

Jeffrey D. Holland

Molecular identification tools have the potential to optimize reintroduction efforts by offering new ways to parameterize models of complex interactions between the introduced species and pest species. The American chestnut tree [*Castanea dentata* (Marsh.) Borkh.] was an important source of valuable wood, wildlife habitat, and food across the eastern United States and the southern tip of Ontario until the early to mid-twentieth century (Merkle et al. 2007). Although reintroduction efforts must focus on resistance to the chestnut blight [*Cryphonectria parasitica* (Murrill) Barr] that almost exterminated the American chestnut (Chapter 12 Box), success will also hinge on dealing with native and exotic insect threats.

Native species of wood-boring insects were a source of American chestnut mortality before the arrival of *C. parasitica*. Felt (1905) lists nine wood-boring insects that attack the tree. At least three of the long-horned beetles (Coleoptera: Cerambycidae) listed by Linsley and Chemsak (1997) as developing within the American chestnut may do so in living wood. These three borers also have oak as a host tree, and this alternate host has allowed these beetles to persist despite the rarity of mature American chestnut. The two-lined chestnut borer [*Agrilus bilineatus* (Weber)] is a metallic wood-boring beetle (Coleoptera: Buprestidae) of particular concern. This congener of the emerald ash borer [*A. planipennis* (Fairmaire)] and the oak splendor beetle [*A. biguttatus* (Fabricius)] was the main insect pest of the American chestnut (Haack & Acciavatti 1992) and has remained extant on oak (Côté & Allen 1980).

Agrilus bilineatus usually causes mortality in trees already under stress (Dunn et al. 1987). American chestnut, however, is susceptible to gypsy moth caterpillars [*Lymantria dispar* (L.)] (Montgomery 1991; Liebhold et al. 1995), and this defoliation of trees can increase susceptibility to *A. bilineatus* (Muzika et al. 2000). With *L. dispar* covering approximately half the former range of American chestnut and advancing (Box Fig. 3–1; U.S. Forest Service [USFS] 2008), it becomes important to optimize reintroduction so that trees have an opportunity to become established before facing such potentially lethal insect interactions.

Molecular scatology can be used to evaluate potential reintroduction sites for wood-borers of concern as the abrasive diet of these insects results in much genetic material from their guts being passed into the frass. Shukle and Holland have positively identified DNA from wood-borer frass and have dated the tree damage using dendrochronology. The insect damage was dated as seven years old, and DNA in frass taken from the damaged area was matched to the suspected borer species (Box Fig. 3–2). The presence of a specific borer species at a site therefore can be determined using adults, larvae, eggs, or frass. *Agrilus bilineatus* has been recognized as containing at least two different subspecies with one subspecies not attacking the American chestnut (Bright 1987). With the near disappearance of the American chestnut, it is possible that unrecognized subspecies that fed on this tree are now either extinct or largely constrained to small areas. Determining the current status of such subspecies and the potential

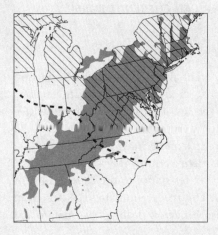

Box Figure 3–1: Range of gypsy moth as of 2006 (USFS, 2008; hatching) and former range of American chestnut (USGS, 1999; gray shading). Heavy broken line indicates estimated 2025 gypsy moth range. Geographical range of *A. bilineatus* covers entire range of American chestnut and is not shown. All data to create the map were taken from government Web sites and are free of copyright. The map was created by Jeffrey Holland.

of other, still wide-ranging subspecies of *A. bilineatus* to attack the American chestnut using molecular identification should be a research priority. Identification among species of *Agrilus* can be difficult, and mitochondrial DNA (mtDNA) tools to distinguish among subspecies of *A. bilineatus* could be important in this effort.

If parentage can be determined from DNA in frass, it may be possible to determine how far individuals move. The dispersal kernel, or probability-density function of movement distance, for a pest species then would be ideally formed from the distances between the larval host trees of successive generations. These movement data could then be used to accurately parameterize models of pest-insect diffusion and spread. Alternatively, the spatial scale of extant populations of genetically related individuals and the spread of the same could be used to model likely spread as the pest species expand to track the introduced tree species. The combination of spatial ecology and genetics offers several tools that can help determine the occurrence of pest insects and understand their movement patterns to optimize reintroduction strategies.

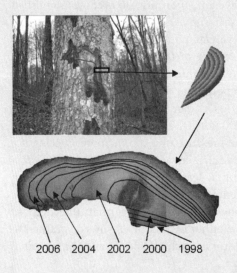

Box Figure 3–2: Dating wood-borer occurrence. Tree was not killed, so growth rings continued around wound. This section of the wound occurred seven years previously, in 2000. Wood slice is from a different actual tree. Dark lines superimposed on tree rings for clarity. Photos were taken by Jeffrey Holland. Wood slice image was created by the author using POV-Ray ray-tracing software.

2006   2004   2002   2000   1998

Acknowledgments

Laura M. Blackburn provided recent geographic information system data and metadata on gypsy moth range.

REFERENCES

Bright DE (1987) *The Metallic Wood-Boring Beetles of Canada and Alaska. The Insects and Arachnids of Canada, Part 15*. Agriculture Canada, Research Branch, Canadian Government Publishing Centre, Ottawa.

Côté WA, Allen DC (1980) Biology of the two-lined chestnut borer, *Agrilus bilineatus*, in Pennsylvania and New York. *Annals of the Entomological Society of America*, **73**, 409–413.

Dunn JP, Kimmerer TW, Potter DA (1987) Winter starch reserves of white oak as a predictor of attack by the twolined chestnut borer, *Agrilus bilineatus* (Weber) (Coleoptera: Buprestidae). *Oecologia*, **74**, 352–355.

Felt EP (1905) *Insects Affecting Park and Woodland Trees*. New York State Museum Memoir 8. New York State Education Department, Albany.

Haack RA, Acciavatti RE (1992) Twolined chestnut borer. *Forest Insect & Disease Leaflet 168*. U.S. Department of Agriculture, Forest Service, Washington, DC.

Liebhold AM, Gottschalk KW, Muzika RM et al. (1995) Suitability of North American tree species to gypsy moth: a summary of field and laboratory tests. General Technical Report NE-211. U.S. Department of Agriculture, Forest Service, Radnor, Pennsylvania.

Linsley EG, Chemsak JA (1997) *The Cerambycidae of North America, Part VIII: Bibliography, Index and Host Plant Index*. University of California Publications in Entomology Vol. 117. University of California Press, Los Angeles.

Merkle SA, Andrade GM, Nairn CJ, Powell WA, Maynard CA (2007) Restoration of threatened species: a noble cause for transgenic trees. *Tree Genetics & Genomes*, **3**, 111–118.

Montgomery ME (1991) Variation in the suitability of tree species for gypsy moth. In: *Proceedings of the U.S. Department of Agriculture Interagency Gypsy Moth Research Review 1990* (eds. Gottschalk KW, Twery MJ, Smith SI). Radnor, Pennsylvania, January 22–25. U.S. Department of Agriculture Forest Service General Technical Report NE-146.

Muzika RM, Liebhold AM, Twery MJ (2000) Dynamics of twolined chestnut borer *Agrilus bilineatus* as influenced by defoliation and selection thinning. *Agricultural and Forest Entomology*, **2**, 283–289.

U.S. Forest Service (USFS) (2008) *Alien Forest Pest Distribution* (Geographic Information Systems data). U.S. Department of Agriculture Forest Service, Morgantown, West Virginia.

U.S. Geological Service (USGS) (1999) Digital representation of "Atlas of United States Trees" by Elbert L. Little, Jr. (Geographic Information Systems data). USGS, Denver, Colorado.

community structure). The movement of transgenes into wild populations, via hybridization and introgression, could alter interspecific interactions, potentially resulting in changes in the distribution of the wild population and its dependent community. Few studies of GEOs examine the consequences of transgenes on interactions, particularly indirect interactions, among organisms. In addition, subtle, initial changes in interactions between transgenic and wild populations may not be detected in short-term studies that are frequently limited in spatial scale as well. The community and ecosystem consequences of transgenes are also important to define because of the growing view that transgenes will escape from their intended destination and that they cannot be retracted after they have escaped (Marvier & Van Acker 2005). When released, some transgenes may have irreversible conservation implications for communities and ecosystems. These possible negative consequences of GEOs should be considered along with their potential benefits, such as promoting agricultural efficiency and reducing pesticide use.

COMMUNITY GENETICS AND CLIMATE CHANGE

Climate change is a "wild card" in conservation and restoration. Because of concerns about restoring only with locally adapted genotypes, the policies of most agencies require the use of local stock in restoration. Whereas this is a valid view in an unchanging world, in a dynamic world facing major vegetation shifts in response to climate change, the "local" environment may well shift so fast or so much that local genotypes will no longer be locally adapted (see also Aitken et al. 2008). In the American Southwest, the effects of climate change are pronounced and affect the distributions of pinyon pine (*Pinus edulis*), cottonwood (*Populus fremontii*), ponderosa pine (*Pinus ponderosa*), quaking aspen (*Populus tremuloides*), and pointleaf manzanita (*Arctostaphylos pungens*), which are foundation species of their respective communities from low elevation deserts to mountain forests (Breshears et al. 2005; Mueller et al. 2005; Gitlin et al. 2006). For example, climate-change models of Rehfeldt and colleagues (2006) predict that by 2090, the climate envelopes for both pinyon pine (*P. edulis*) and saguaro cactus (*Carnegiea gigantea*) will shift out of Arizona, where they are currently foundation species that define their respective communities (Fig. 3–7). Such dramatic shifts in the distributions of these species are likely to represent both ecological and evolutionary events and major conservation challenges. If these shifts come about, Saguaro National Monument will no longer contain saguaros, and for saguaros to physically reach their new climate envelope in eastern Nevada, human-assisted migration will likely be required (McLachlan et al. 2007; Aitken et al. 2008).

One factor that can potentially alleviate such dramatic shifts in distribution is the potential for selection to favor more drought-adapted genotypes that allow species to persist in areas where they might otherwise go locally extinct. For example, Franks and colleagues (2007) found rapid evolution in flowering time of field mustard (*Brassica rapa*) in response to altered growing seasons over just seven years. Postdrought descendants of this annual bloomed earlier than did predrought ancestors, advancing first flowering by as much as nine days. The intermediate flowering time of Ancestor × Descendant hybrids supports an additive genetic basis for this divergence. Even vertebrate species appear capable of rapid evolutionary responses to climate change. Over a thirty-year period, Grant and Grant (2002) observed rapid evolution in the shape and size of beaks of finches in association with climate change–driven changes in seed availability.

So, what is the community genetics connection to climate change? If climate change acts as an agent of selection to alter the frequency distribution of genotypes of foundation species, then their dependent communities will also be affected. For example, Whitham and coworkers (2003) showed that approximately 1,000 species of organisms from microbes to vertebrates were differentially associated with insect-resistant and -susceptible pinyon pine (*P. edulis*). Recent studies by Sthultz and colleagues (2009) showed that insect-resistant trees were

Carnegiea gigantea

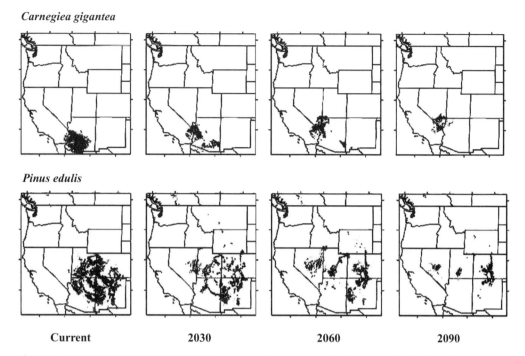

Pinus edulis

| Current | 2030 | 2060 | 2090 |

Figure 3–7: Climate change is predicted to have rapid and major effects on the distribution of foundation plant species across the American West. Modeled climate profiles for saguaro and pinyon pine are shown for the contemporary climate and for the decades beginning in 2030, 2060, and 2090 (left to right). Figure from Rehfeldt and colleagues (2006).

3 times more likely to die than susceptible trees were during a record drought and that the change in the distribution of resistant and susceptible phenotypes on the landscape had changed from resistance-dominated stands to near parity. Thus, even if a foundation species survives in a given habitat, the change in its genetic structure is likely to affect dependent community members whose distributions are also dependent on the genotypes of the foundation plant species (see previous sections on heritable phenotypes and minimum viable populations).

Rehfeldt and colleagues (2006) used climate envelopes to predict the future distributions of many tree species of the western United States, including *P. edulis*. They concluded that *P. edulis* would be extirpated from Arizona within eighty years. This prediction, however, did not include potentially drought-adapted genotypes of *P. edulis* or reflect the genetic variation among populations that might exist over its large geographic distribution. In contrast, a few other models have incorporated intraspecific genetic variation and demonstrated its potential influence on climate-change projections. For example, based on provenance trials involving 140 populations tested at 62 sites throughout most of the range of lodgepole pine (*P. contorta*), O'Neill and coworkers (2008) included into their models genetic differences in the performance of these populations. Because populations tend to be locally adapted, these models predicted that different populations would vary greatly in their responses to climate change and that many populations would be maladapted to their new local environment. This provenance-trial approach also provides an experimental method for identifying which populations or genotypes might best survive predicted climate changes. Thus, for species for which such provenance trials already exist, the use of this powerful experimental approach could greatly enhance the precision of climate-change models.

From these examples, it would appear that climate change creates a conundrum for conservation biologists and managers. If our management policies require restoration with local genotypes, in a climate-change scenario are we then planting the very genotypes that are doomed to fail? Few studies with foundation species are in place to provide clear information for managers to make appropriate decisions (but, see O'Neill et al. 2008). We need a national network of common gardens for foundation species in which genotypes from the whole range of a species' distribution are planted reciprocally within their current range and their predicted new ranges. Only with such empirical information can we identify the specific genotypes and source populations that are best suited to current and predicted future local conditions.

MANAGEMENT CONCLUSIONS

A community and ecosystem genetics approach is based on the fundamental finding that individual genotypes or populations of foundation species have heritable community and ecosystem phenotypes (Fig. 3–2). Because this finding has emerged from diverse systems on different continents, our review explores five management implications: 1) Management practices should be based on conserving genetic diversity in foundation species. Because different genotypes support different communities (Figs. 3–2 and 3–3), greater biodiversity will be achieved by preserving the greatest genetic diversity in the foundation species (e.g., Fig. 3–4). 2) It is important for managers to determine if they should be managing for area, the genetic diversity of the foundation species, or some combination of both. This is due to the fact that the species-area curve (i.e., the current basis of the management practice of setting aside large geographical areas to capture the maximum number of species in preserves) is also highly correlated with the genetic diversity of the foundation species (Fig. 3–5). 3) Management practices for conserving biodiversity should be based on the MVIP, which is likely to be much larger than the MVP. Because the distribution of many species can be dependent on the genotypes of foundation species, the population size of the foundation species required to support all of its dependent community members is much larger than that needed to save the foundation species itself (Fig. 3–6). 4) We need management protocols that specifically quantify the unintended ecosystem effects of genetically engineered organisms (i.e., their community and ecosystem phenotypes) as part of the evaluation process for commercial release. 5) Managing for climate change by establishing provenance trials of foundation species throughout their current and predicted range is needed to identify the genotypes and source populations that are most likely to survive, reproduce, and support their dependent communities in a new climate. These are a few of the management issues that arise when we apply a community and ecosystem genetics approach rather than just a population genetics approach.

As an initial step to achieve these goals, it would be especially helpful and cost effective if managers tagged and mapped the donors used for propagation (both for seeds and cuttings) and then tagged the individuals used in restoration plantings. Although such "bookkeeping" would represent a small fraction of the overall cost of a restoration project, the value added could be great and result in cost savings in future restoration projects. If multiple-source populations and genotypes were used, managers could identify which genotypes and source populations were best adapted to a specific planting site. Long-term data are especially important because those genotypes that perform best under current climatic conditions may not be the best with future climatic conditions (e.g., Sthultz et al. 2009). Performance data combined with climate change models could be used to identify which source populations and genotypes might best suit a predicted future climate. Because many plants used in restoration are long-lived, after tagged restoration sites are established, they also can be used to identify the communities of organisms that individual plant genotypes support and identify source populations that will perform best in the restoration project. Because species evolve in a community context, then we must ultimately conserve and manage species in a community context.

REFERENCES

Abbot P, Withgott JH (2004) Phylogenetic and molecular evidence for allochronic speciation in gall-forming aphids (Pemphigus). *Evolution*, **58**, 539–553.
Aitken SN, Yeaman S, Holliday JA et al. (2008) Adaptation, migration or extirpation: climate change outcomes for tree populations. *Evolutionary Applications*, **1**, 95–111.
Araki R, Mori M, Mori M et al. (2005) Genetic differences in nitrate uptake in two clones of the common reed, *Phragmites australis*. *Breeding Science*, **55**, 297–302.
Avise JC (2003) The best and the worst of times for evolutionary biology. *BioScience*, **53**, 247–255.
Bailey JK, Schweitzer JA, Rehill BJ et al. (2004) Beavers as molecular geneticists: a genetic basis to the foraging of an ecosystem engineer. *Ecology*, **85**, 603–608.
Bailey JK, Whitham TG (2007) Biodiversity is related to indirect interactions among species of large effect. In: *Ecological Communities: Plant Mediation in Indirect Interaction Webs* (eds. Ohgushi T, Craig TP, Price PW), pp. 306–328. Cambridge University Press, Cambridge, UK.
Bailey JK, Wooley SC, Lindroth RL et al. (2006) Importance of species interactions to community heritability: a genetic basis to trophic-level interactions. *Ecology Letters*, **9**, 78–85.
Bangert RK, Lonsdorf EV, Wimp GM et al. (2008) Genetic structure of a foundation species: scaling community phenotypes from the individual to the region. *Heredity*, **100**, 121–131.
Bangert RK, Turek RJ, Rehill B et al. (2006a) A genetic similarity rule determines arthropod community structure. *Molecular Ecology*, **15**, 1379–1392.
Bangert, RK, Allan, GJ, Turek, RJ et al. (2006b). From genes to geography: a genetic similarity rule for arthropod community structure at multiple geographic scales. *Molecular Ecology*, **15**, 4215–4228.
Barbour RC, O'Reilly-Wapstra JM, De Little DW et al. (2009) A geographic mosaic of genetic variation within a foundation tree species and its community-level consequences. *Ecology*, **90**, 1762–1772.
Bartelt-Ryser J, Joshi J, Schmid B et al. (2005) Soil feedbacks of plant diversity on soil microbial communities and subsequent plant growth. *Perspectives in Plant Ecology, Evolution and Systematics*, **7**, 27–49.

Bever JD, Westover KM, Antonovics J (1997) Incorporating the soil community into plant population dynamics: the utility of the feedback approach. *Journal of Ecology*, **85**, 561–573.

Breshears DB, Cobb NS, Rich PM et al. (2005) Regional vegetation die-off in response to global-change-type drought. *Proceedings of the National Academy of Sciences USA*, **102**, 15144–15148.

Chapman MA, Burke JM (2006) Letting the gene out of the bottle: the population genetics of genetically modified crops. *New Phytologist*, **170**, 429–443.

Craig TP, Itami JK, Craig JV (2007) Host plant genotype influences survival of hybrids between *Eurosta solidaginis* host races. *Evolution*, **61**, 2607–2613.

Crawford KM, Crutsinger GM, Sanders NJ (2007) Host-plant genotypic diversity mediates the distribution of an ecosystem engineer. *Ecology*, **88**, 2114–2120.

Crutsinger GM, Collins MD, Fordyce JA et al. (2006) Plant genotypic diversity predicts community structure and governs an ecosystem process. *Science*, **313**, 966–968.

Dayton PK (1972) Toward an understanding of community resilience and the potential effects of enrichments to the benthos at McMurdo Sound, Antarctica. In: *Proceedings of the Colloquium on Conservation Problems* (ed. Parker BC), pp. 81–96. Allen Press, Lawrence, Kansas.

Dickson LL, Whitham TG (1996) Genetically-based plant resistance traits affect arthropods, fungi, and birds. *Oecologia*, **106**, 400–406.

Donaldson JR, Lindroth RL (2007) Genetics, environment, and their interaction determine efficacy of chemical defense in trembling aspen. *Ecology*, **88**, 729–739.

Dres M, Mallet J (2002) Host races in plant-feeding insects and their importance in sympatric speciation. *Philosophical Transactions of the Royal Society of London B*, **357**, 471–492.

Driebe EM, Whitham TG (2000) Cottonwood hybridization affects tannin and nitrogen content of leaf litter and alters decomposition. *Oecologia*, **123**, 99–107.

Dungey HS, Potts BM, Whitham TG et al. (2000) Plant genetics affects arthropod community richness and composition: evidence from a synthetic eucalypt hybrid population. *Evolution*, **54**, 1938–1946.

Ellison AM, Bank MS, Clinton BD et al. (2005) Loss of a foundation species: consequences for the structure and dynamics of forested ecosystems. *Frontiers in Ecology and the Environment*, **3**, 479–486.

English-Loeb G, Norton A (2006) Lack of trade-off between direct and indirect defence against grape powdery mildew in riverbank grape. *Ecological Entomology*, **31**, 415–422.

Evans LM, Allan GJ, Shuster SM et al. (2008) Tree hybridization and genotypic variation drive cryptic speciation of a specialist mite herbivore. *Evolution*, **62**, 3027–3040.

Feder JL, Chilcote CA, Bush GL (1988) Genetic differentiation between sympatric host races of the apple maggot fly *Rhagoletis pomonella*. *Nature*, **336**, 61–64.

Fischer DG, Hart SC, LeRoy CJ et al. (2007) Decoupled fine root production and total belowground carbon allocation in *Populus* forests. *New Phytologist*, **176**, 415–425.

Fischer DG, Hart SC, Rehill BJ et al. (2006) Do high-tannin leaves require more roots? *Oecologia*, **149**, 668–675.

Fischer DG, Hart SC, Whitham TG et al. (2004) Ecosystem implications of genetic variation in water-use of a dominant riparian tree. *Oecologia*, **139**, 188–197.

Fischer DG, Hart SC, Whitham TG et al. (2006) Hidden belowground responses to defense: do highly defended leaves require more roots? *Oecologia*, **149**, 668–675.

Franks SJ, Sim S, Weis AE (2007) Rapid evolution of flowering time by an annual plant in response to a climate fluctuation. *Proceedings of the National Academy of Sciences*, **104**, 1278–1282.

Fritz RS, Price PW (1988) Genetic variation among plants and insect community structure: willows and sawflies. *Ecology*, **69**, 845–856.

Gilpin ME (1987) Minimum viable populations: a restoration ecology perspective. In: *Restoration Ecology: A Synthetic Approach to Ecological Research* (eds. Jorden WR III, Gilpin ME, Aber JD), pp. 301–305. Cambridge University Press, Cambridge, UK.

Gitlin AR, Sthultz CM, Bowker MA et al. (2006) Mortality gradients within and among dominant plant populations as barometers of ecosystem change during extreme drought. *Conservation Biology*, **20**, 1477–1486.

Grant PR, Grant BR (2002) Unpredictable evolution in a 30-year study of Darwin's finches. *Science*, **296**, 707–711.

Heinemann JA, Traavik T (2004) Problems in monitoring horizontal gene transfer in field trials of transgenic plants. *Nature Biotechnology*, **23**, 488.

Hirsch AM, Bauer WD, Bird DM et al. (2003) Molecular signals and receptors: controlling rhizosphere interactions between plants and other organisms. *Ecology*, **84**, 858–868.

Hochwender CG, Fritz RS (2004) Plant genetic differences influence herbivore community structure: evidence from a hybrid willow system. *Oecologia*, **138**, 547–557.

Hooper DU, Bignell DE, Brown VK et al. (2000) Interaction between aboveground and belowground biodiversity in terrestrial ecosystems: patterns, mechanisms and feedbacks. *BioScience*, **50**, 1049–1061.

Hughes AR, Inouye BD, Johnson MTJ et al. (2008) Ecological consequences of genetic diversity. *Ecology Letters*, **11**, 609–623.

Hughes AR, Stachowicz JJ (2004) Genetic diversity enhances the resistance of a seagrass ecosystem to disturbance. *Proceedings of the National Academy of Sciences USA*, **101**, 8998–9002.

Icoz I, Stotzky G (2008) Fate and effects of insect-resistant Bt crops in soil ecosystems. *Soil Biology and Biochemistry*, **40**, 559–586.

Johnson MTJ (2008) Bottom-up effects of plant genotype on aphids, ants and predators. *Ecology*, **89**, 145–154.

Johnson MTJ, Agrawal AA (2005) Plant genotype and the environment interact to shape a diverse arthropod community on evening primrose (*Oenothera biennsis*). *Ecology*, **18**, 549–551.

Johnson MTJ, Lajeunesse MJ, Agrawal AA (2006) Additive and interactive effects of plant genotypic diversity on arthropod communities and plant fitness. *Ecology Letters*, **9**, 24–34.

Johnson MTJ, Stinchcombe JR (2007) An emerging synthesis between community ecology and evolutionary biology. *Trends in Ecology and Evolution*, **22**, 250–257.

Karban R (1989) Fine-scale adaptation of herbivorous thrips to individual host plants. *Nature*, **340**, 60–61.

Lawrence S (2008) Brazil surpasses US in new transgenic crop plantings. *Nature Biotechnology*, **26**, 260.

Lecerf A, Chauvet E (2008) Intraspecific variability in leaf traits strongly affects alder leaf decomposition in a stream. *Basic and Applied Ecology*, **9**, 598–605.

LeRoy CJ, Whitham TG, Keim P et al. (2006) Plant genes link forests and streams. *Ecology*, **87**, 255–261.

LeRoy CJ, Whitham TG, Wooley SC et al. (2007) Within-species variation in foliar chemistry influences leaf-litter decomposition in a Utah river. *Journal of the North American Benthological Society*, **26**, 426–438.

Lindenmayer DB, Fischer J, Felton A et al. (2007) The complementarity of single-species and ecosystem-oriented research in conservation research. *Oikos*, **116**, 1220–1226.

Maddox GD, Root RB (1987) Resistance to 16 diverse species of herbivorous insects within a population of goldenrod, *Solidago altissima*: genetic variation and heritability. *Oecologia*, **72**, 8–14.

Madritch MD, Donaldson JR, Lindroth RL (2006) Genetic identity of *Populus tremuloides* litter influences decomposition and nutrient release in a mixed forest stand. *Ecosystems*, **9**, 528–537.

Madritch MD, Donaldson JR, Lindroth RL (2007) Canopy herbivory can mediate the influence of plant genotype on soil processes through frass deposition. *Soil Biology and Biochemistry*, **39**, 1192–1201.

Madritch MD, Hunter MD (2002) Phenotypic diversity influences ecosystem functioning in an oak sandhills community. *Ecology*, **83**, 2084–2090.

Madritch MD, Hunter MD (2004) Phenotypic diversity and litter chemistry affect nutrient dynamics during litter decomposition in a two species mix. *Oikos*, **105**, 125–131.

Marvier M, McCreedy C, Regetz J et al. (2007) A meta-analysis of effects of Bt cotton and maize on non-target invertebrates. *Science*, **316**, 1475–1477.

Marvier M, Van Acker R (2005) Can crop transgenes be kept on a leash? *Frontiers in Ecology and the Environment*, **3**, 99–106.

McIntyre PJ, Whitham TG (2003) Plant genotype affects long-term herbivore population dynamics and extinction: conservation implications. *Ecology*, **84**, 311–322.

McLachlan JS, Hellman JJ, Schwartz MW (2007) A framework for debate of assisted migration in an era of climate change. *Conservation Biology*, **21**, 297–302.

Mopper S (1996) Adaptive genetic structure in phytophagous insect populations. *Trends in Ecology & Evolution*, **11**, 235–238.

Mopper S, Beck M, Simberloff D et al. (1995) Local adaptation and agents of selection in a mobile insect. *Evolution*, **49**, 810–815.

Mopper S, Stiling P, Landau K et al. (2000) Spatiotemporal variation in leafminer population structure and adaptation to individual oak trees. *Ecology*, **81**, 1577–1587.

Mueller RC, Scudder CM, Porter ME et al. (2005) Differential mortality in response to severe drought: implications for long-term vegetation shifts. *Journal of Ecology*, **93**, 1085–1093.

O'Neill GA, Hamann A, Wang T (2008) Accounting for population variation improves estimates of the impact of climate change on species' growth and distribution. *Journal of Applied Ecology*, **45**, 1040–1049.

Phillips DA, Ferris H, Cook DR et al. (2003) Molecular control points in rhizosphere food webs. *Ecology*, **84**, 816–826.

Rehfeldt GE, Crookston NL, Warwell MV et al. (2006) Empirical analyses of plant-climate relationships for the western United States. *International Journal of Plant Sciences*, **167**, 1123–1150.

Rohlf DJ (1991) Six biological reasons why the Endangered Species Act doesn't work – and what to do about it. *Conservation Biology*, **5**, 273–282.

Roinen H, Vuorinen J, Tahvanainen J et al. (1993) Host preference and allozyme differentiation in shoot galling sawfly, *Euura atra*. *Evolution*, **47**, 300–308.

Rosi-Marshall EJ, Tank JL, Royer TV et al. (2007) Toxins in transgenic crop byproducts may affect headwater stream ecosystems. *Proceedings of the National Academy of Sciences USA*, **104**, 16204–16208.

Rudgers JA, Whitney KD (2006) Interactions between insect herbivores and a plant architectural dimorphism. *Journal of Ecology*, **94**, 1249–1260.

Schweitzer JA, Bailey JK, Fischer DG et al. (2008) Plant-soil microorganism interactions: a heritable relationship between plant genotype and associated soil microorganisms. *Ecology*, **89**, 773–781.

Schweitzer JA, Bailey JK, Hart SC et al. (2005a) Nonadditive effects of mixing cottonwood genotypes on litter decomposition and nutrient dynamics. *Ecology*, **86**, 2834–2840.

Schweitzer JA, Bailey JK, Hart SC et al. (2005b) The interaction of plant genotype and herbivory decelerate leaf litter decomposition and alter nutrient dynamics. *Oikos*, **110**, 133–145.

Schweitzer JA, Bailey JK, Rehill BJ et al. (2004) Genetically based trait in a dominant tree affects ecosystem processes. *Ecology Letters*, **7**, 127–134.

Scriber JM (1986) Origins of the regional feeding abilities in the tiger swallowtail butterfly: ecological monophagy and the *Papilio glaucus australis* subspecies in Florida. *Oecologia*, **71**, 94–103.

Shaffer ML (1981) Minimum population sizes for species conservation. *BioScience*, **31**, 131–134.

Shuster SM, Lonsdorf EV, Wimp GM et al. (2006) Community heritability measures the evolutionary consequences of indirect genetic effects on community structure. *Evolution*, **60**, 991–1003.

Silfver T, Mikola J, Rousi M et al. (2007) Leaf litter decomposition differs among genotypes in a local *Betula pendula* population. *Oecologia*, **152**, 707–714.

Simberloff D (1998) Flagships, umbrellas, and keystones: is single-species management passé in the landscape era? *Biological Conservation*, **83**, 247–257.

Sthultz CM, Gehring CA, Whitham TG (2009) Deadly combination of genes and drought: increased mortality of herbivore-resistant trees in a foundation species. *Global Change Biology* **15**, 1949–1961.

Stireman JO III, Nason JD, Heard S (2005) Host-associated genetic differentiation in phytophagous insects: general phenomenon or isolated exceptions? Evidence from a goldenrod-insect community. *Evolution*, **59**, 2573–2587.

Stireman JO III, Nason JD, Heard S et al. (2006) Cascading host-associated genetic differentiation in parasitoids of phytophagous insects. *Proceedings of the Royal Society of London B*, **273**, 523–530.

Thompson JN (1994) *The Coevolutionary Process*. University of Chicago Press, Chicago.

Thompson JN (2005) *The Geographic Mosaic of Coevolution*. University of Chicago Press, Chicago.

Tovar-Sanchez E, Oyama K (2006) Effect of hybridization of the *Quercus crassifolia* x *Quercus crassipes* complex on the community structure of endophagous insects. *Oecologia*, **147**, 702–713.

Treseder KK, Vitousek PM (2001) Potential ecosystem-level effects of genetic variation among populations of *Metrosideros polymorpha* from a soil fertility gradient in Hawaii. *Oecologia*, **126**, 266–275.

van Ommeren RJ, Whitham TG (2002) Changes in interactions between juniper and mistletoe mediated by shared avian frugivores: parasitism to potential mutualism. *Oecologia*, **130**, 281–288.

Via S (2001) Sympatric speciation in animals: the ugly duckling grows up. *Trends in Ecology & Evolution*, **16**, 381–390.

Vijn I, das Neves L, van Kammen A et al. (1993) Nod factors and nodulation in plants. *Science*, **260**, 1764–1765.

Wade MJ (2007) The co-evolutionary genetics of ecological communities. *Nature Reviews Genetics*, **8**, 185–195.

Watrud LS, Lee HE, Fairbrother A et al. (2004) Evidence for landscape-level, pollen-mediated gene flow from genetically modified creeping bentgrass with CP4 EPSPS as a marker. *Proceedings of the National Academy of Science USA*, **101**, 14533–14538.

Whitham TG, Bailey JK, Schweitzer JA et al. (2006) A framework for community and ecosystem genetics: from genes to ecosystems. *Nature Reviews Genetics*, **7**, 510–523.

Whitham TG, DiFazio SP, Schweitzer JA et al. (2008) Extending genomics to natural communities and ecosystems. *Science*, **320**, 492–495.

Whitham TG, Young WP, Martinsen GD et al. (2003) Community and ecosystem genetics: a consequence of the extended phenotype. *Ecology*, **84**, 559–573.

Wimp GM, Martinsen GD, Floate KD et al. (2005) Plant genetic determinants of arthropod community structure and diversity. *Evolution*, **59**, 61–69.

Wimp GM, Wooley S, Bangert RK et al. (2007) Plant genetics predicts intra-annual variation in phytochemistry and arthropod community structure. *Molecular Ecology*, **16**, 5057–5069.

Wimp GM, Young WP, Woolbright SA et al. (2004) Conserving plant genetic diversity for dependent animal communities. *Ecology Letters*, **7**, 776–780.

Wolfenbarger LL, Naranjo SE, Lundgren JG et al. (2008) Bt crop effects on functional guilds of non-target arthropods: a meta-analysis. *PLoS ONE*, **3**, e2118. doi:10.1371/journal.pone.0002118.

Wood TK, Keese MC (1990) Host plant-induced assortative mating in Enchenopa tree hoppers. *Evolution*, **44**, 619–628.

Zapiola M, Campbell C, Butler M et al. (2008) Escape and establishment of transgenic glyphosate-resistant creeping bentgrass (*Agrostis stolonifera*) in Oregon, USA: a 4-year study. *Journal of Applied Ecology*, **45**, 486–494.

4 Vertebrate sex-determining genes and their potential utility in conservation, with particular emphasis on fishes

J. Andrew DeWoody, Matthew C. Hale, and John C. Avise

GENETIC MARKERS IN WILDLIFE CONSERVATION

Individual identification

Often, animals leave clues that can provide some information about their individual identities. These may be conventional fingerprints, which are extremely useful in courts of law but can be physically altered or removed and are restricted to humans. In contrast, deoxyribonucleic acid (DNA) fingerprinting (Avise 2004) is important not only in human forensics and paternity analysis but also in conservation biology and resource management (Table 4–1). When monitoring DNA fingerprints, biologists are capitalizing on the permanent genetic tags by which nature has labeled each individual. In principle, DNA fingerprints (e.g., from leaves, root tips, blood, hair, or feathers) can be traced over space and time, thereby yielding insights into organismal behavior, population structure, and population demography.

Sexing assays and conservation

In concert with individual identification via DNA fingerprinting, molecular sexing has proven valuable in conservation and management (see Box 4 by Lisette Waits). Molecular assays that distinguish males from females can be informative in many ways. For example, the sexes of most dioecious plants are indistinguishable prior to sexual maturity, yet molecular assays have revealed that sex ratios change as seeds develop into reproductively mature plants (Korpelainen 2002; Korpelainen & Kostamo 2008). For vertebrates, most efforts in molecular sexing have focused on mammals and birds in part because of the special interest in these animals, but also because the mode of sex determination

We thank Bob Devlin, Tariq Ezaz, and members of the DeWoody group for their comments on an earlier version of the manuscript. The lake sturgeon research described herein was funded by grants to J. Andrew DeWoody from the Great Lakes Fishery Trust and the Indiana Department of Natural Resources through a State Wildlife Grant (T07R05).

Table 4–1. DNA fingerprinting plays a major role in the modern management of our natural resources

Organism	Insights	References
Polar bear (*Ursus maritimus*)	Long annual migrations, but little gene flow	Paetkau et al. 1995
Atlantic salmon (*Salmo salar*)	Geographic origin of fraudulent fish	Primmer et al. 2000
Painted turtle (*Chrysemys picta*)	Genetic mark-recapture estimates of population size	Pearse et al. 2001
Passerine birds	Genetic promiscuity despite social monogamy	Griffith et al. 2002
Gray (*Halichoerus grypus*) and harbor (*Phoca vitulina*) seals	Dietary preferences via scatology	Reed et al. 1997
Imperial eagles (*Aquila heliaca*)	Monogamy, turnover, population size	Rudnick et al. 2005, 2008

is known and conserved within each of these two taxonomic groups. Thus, reliable molecular-sexing assays have been relatively straightforward to develop and also to transfer across species within each group.

To date, DNA-based sexing has been most valuable in the conservation and management of mammals. For instance, Zhan and colleagues (2007) discovered via molecular sexing that dispersal in giant pandas (*Ailuropoda melanoleuca*) is female-biased, driven largely by female competition for birthing dens. In another mammalian example, Blejwas and coworkers (2006) used DNA from the saliva of attack wounds to determine that male coyotes (*Canis latrans*) are most often responsible for sheep kills. Finally, Bradley and colleagues (2008) used fecal DNA to assess the demography of gorillas (*Gorilla gorilla*) at nesting sites and showed that individual age and/or sex determinations based on dung size alone may be misleading.

With regard to such studies in birds, Millar and coworkers (1997) sexed monomorphic individuals of the highly endangered New Zealand black stilt (*Himantopus novaezelandiae*), thus allowing captive individuals to be paired on the basis of gender. Clout and colleagues (2002) showed by molecular sexing that supplemental feeding of critically endangered kakapos (*Strigops habroptilus*) in the wild led to male-biased clutches because of the ability of female birds to manipulate the sex ratio of their offspring. Finally, Rudnick and coworkers (2005) coupled DNA fingerprinting with sexing assays to infer genetic parentage, demographic turnover, and variance in chick sex ratios over six years in a population of imperial eagles (*Aquila heliaca*).

The utility of DNA-based sexing assays among mammals and birds is due largely to the conserved mode of sex determination within each of these taxonomic groups, thus allowing sex assays developed in particular species to be applied with only slight modifications to many taxa. The evolution of sex-determining mechanisms has been addressed by Bull (1983) and several others (e.g., Mank et al. 2006). As we will see, genetic sex determination (GSD) is nearly universal

BOX 4: SEX IDENTIFICATION AND POPULATION SIZE OF GRIZZLY BEARS BY USING NONINVASIVE GENETIC SAMPLING

Lisette Waits

Problem

Globally, the numbers of large carnivores have declined due to human persecution and loss of habitat. As a result, there are many vulnerable and endangered populations of conservation concern. To manage these populations, researchers need to collect basic data on population size, population trend, and sex ratios. Historically, this type of data was collected by capturing and marking animals. Capture rates are generally low for large carnivores, however, and capture techniques pose a risk for humans and the study organism. The application of molecular tools in the form of noninvasive genetic sampling of hair and fecal samples has provided an alternative approach for counting individual animals and determining sex.

Case Study

Grizzly bears (*Ursus arctos horribilis*) were listed as threatened under the U.S. Endangered Species Act in 1975, and six recovery zones were identified. The grizzly bear population of the Northern Continental Divide Ecosystem (NCDE) Recovery Zone in northwestern Montana, including Glacier National Park, is particularly important to the long-term viability of grizzly bears in the lower forty-eight states due to its connection to grizzly bear populations in Canada.

In 1998, bear managers launched a large-scale noninvasive genetic sampling effort to estimate the number of bears in a 7,933 km² area in the northern one-third of the NCDE (Boulanger et al. 2008; Kendall et al. 2008). In 1998 and 2000, almost 15,000 hair samples were collected using barbed wire hair traps (with a scent lure attractant) systematically distributed using a 125-unit grid of 8 × 8 km cells and naturally occurring bear rub trees distributed along 1,185 km of trails. Grizzly bear samples were separated from black bear samples by amplifying a section of the mitochondrial DNA control region with a diagnostic length variant. Individual bears were identified by genotyping six microsatellite loci. Sex identification was initially conducted on all unique individuals using the SRY marker and was verified using a size polymorphism in the amelogenin gene. Both methods produce a single PCR fragment for females and two PCR fragments for males. These data were used to estimate population size and sex ratio (Boulanger et al. 2008; Kendall et al. 2008).

Capture probability varied by sex and by sampling method. Males were captured more frequently at rub trees, and females were captured more frequently at hair traps (Box Table 4–1). Sex ratio estimates varied depending on the sampling method, but results from hair trapping were closest to the estimates obtained from a mark-recapture population estimate using the full data set (Box Table 4–1).

Box Table 4–1. Minimum (Min) count of male (M) and female (F) grizzly bears, sex ratio, and mark-recapture population estimate determined using DNA analysis of hair samples collected using hair traps and rub trees in the Greater Glacier Ecosystem (from Kendall et al. 2008)

Year	Min count hair traps			Min count rub trees			Population estimate		
	F	M	F:M	F	M	F:M	F	M	F:M
1998	91	56	1.63	25	37	0.68	125.4	115.3	1.09
2000	85	70	1.21	54	81	0.67	146.0	94.6	1.54

Molecular sex identification has become an important tool for the conservation and management of wildlife species. This study demonstrates the value of noninvasive sampling and molecular sex identification to the conservation and management of grizzly bears and highlights the importance of considering potential biases introduced by sampling methods when using noninvasive genetic approaches to estimate sex ratios.

REFERENCES

Boulanger JB, Kendall KC, Stetz JB et al. (2008) Use of multiple data sources to improve DNA-based mark-recapture population estimates of grizzly bears. *Ecological Applications*, **18**, 557–589.

Kendall KC, Stetz JB, Roon DA et al. (2008) Grizzly bear density in Glacier National Park. *Journal of Wildlife Management*, **72**, 1693–1705.

in mammals and birds but occurs far more sporadically in reptiles, amphibians, and fishes (Table 4–2).

Sex determination is chromosomal (with males being XY and females XX) in most of the 5,000 extant therian mammals. Exceptions include some voles in which both sexes are XX or XO (Just et al. 1995, 2007) and the monotremes (platypus and echidnas) in which sex determination includes five X and five Y chromosomes; the Xs are homologous to the bird Z chromosome (Grützner et al. 2004; Veyrunes et al. 2008). Sex determination is probably chromosomal in most if not all of the 10,000 extant species of birds (Mank & Ellegren 2007), but males are the homogametic sex (ZZ) and females are heterogametic (ZW).

SEX-DETERMINING GENES

Mammals and birds

The molecular mechanism of sex determination is also conserved across most mammals. The mammalian sex-determining gene *Sry* encodes a transcription factor (SRY) that initiates the male cascade of sexual differentiation (Table 4–3; Sinclair et al. 1990; Ferguson-Smith 2007). Known exceptions (species that do not have the *Sry* gene or that do not express SRY) include the five monotremes

Table 4–2. Vertebrate sex-determining mechanisms are diverse

Taxa	Sex determination	Examples	Citations
Mammals	XY	Nearly all therians	Schafer & Goodfellow 1996
	XO in both sexes	*Ellobius lutescens*	Just et al. 1995
	XX in both sexes	*Ellobius tancrei*	Just et al. 1995
	XXXXX/YYYYY	Monotremes (platypus and echidnas)	Grützner et al. 2004
Birds	ZW	Presumably all	Smith & Sinclair 2004
Reptiles	XY	*Bassiana duperreyi*[1]	Radder et al. 2007
	ZW	*Pogona vitticeps*[1]	Ezaz et al. 2005
	ESD (TSD)	*Amphibolurus muricatus*	Warner & Shine 2008
	GSD+ESD	*Gekko japonicus*	Valenzuela et al. 2003
Amphibians	XY	*Rana rugosa*[2]	Ogata et al. 2008
	ZW	*Rana rugosa*[2]	Ogata et al. 2008
	XY (with reversals[3])	*Rana temporaria*	Matsuba et al. 2008
Fishes	XY	*Takifugu rubripes*	Kikuchi et al. 2007
	ZW	*Characidium* cf. *gomesi*	Vicari et al. 2008
	$X_1X_1X_2X_2/X_1X_2Y$	*Lutjanus quinquelineatus*	Ueno & Takai 2008
	XX/XY_1Y_2	*Ancistrus* sp. 1 Balbina	Ribeiro de Oliveira et al. 2008
	$Z_1Z_1Z_2Z_2/Z_1Z_2W_1W_2$	*Ancistrus* sp. 2 Barcelos	Ribeiro de Oliveira et al. 2008
	Autosomal modifiers[4]	*Oreochromis niloticus*	Mair et al. 1991
	GSD+ESD	*Menidia menida*	Conover & Kynard 1981

[1] Environmental factors can also influence sex determination in these species, so they might also be categorized as GSD+ESD.
[2] Some populations of *R. rugosa* have XY sex determination, whereas others are ZW.
[3] Reversals may be caused by environmental factors.
[4] Sex determined chromosomally but can be subsequently reversed depending on the autosomal gene complement.

(Graves 2002), the mole vole (*Ellobius lutescens*; Just et al. 2007), and the spiny rat (*Tokudaia osimensis*; Arakawa et al. 2002).

Molecular sexing assays exploit the presence or absence of *Sry* as revealed by polymerase chain reaction (PCR) and electrophoresis (e.g., Pomp et al. 1995). They yield definitive results, at least in principle, because *Sry* normally determines maleness. Furthermore, recombination typically is suppressed near the master sex-determining gene, so *Sry* is nearly always in strong linkage disequilibrium with nearby loci that also may be sound indicators of gender. For example, length differences between zinc finger genes (*Zfx* and *Zfy*), often in conjunction with *Sry*, can be used to determine mammalian sex (Pomp et al. 1995; Cathey et al. 1998). For loci more distant from *Sry*, the categorical error rate in molecular sexing should increase proportionally with the recombination rate (i.e., vary as a function of the history of recombination).

Table 4–3. Some of the primary gene products involved in mammalian sex determination

Gene	Role
SRY	Primary testis-determining factor
SOX9	Main target of SRY
DAX1	Antagonist to SRY
AMH	Activated by SOX9
WNT4, WNT7	Required by AMH
DMRT1	Required by Sertoli and germ cells
ATRX	Required for testis development
LHX9, M33	Required for gonad development
TDA1,2,3	Promote ovarian development

Many vertebrates share the same cascade of genes for sexual differentiation, but the primary switch that initiates the cascade (e.g., *SRY* in mammals) varies. Modified from Ferguson-Smith (2007).

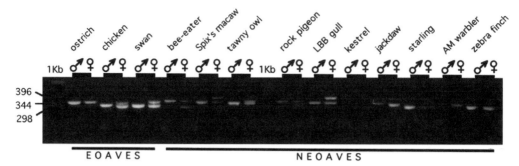

Figure 4–1: A sex-determining assay based on the avian *Chd* gene. This test can be used to sex most nonratite birds (Griffiths et al. 1998). Such broad assays are of great general utility, as evidenced by the more than 800 citations to Griffiths et al. (1998) in the last decade. Reprinted from Griffiths et al. 1998.

In birds, it has long been unclear how genes on the Z or W chromosome cause sexual differentiation (Ferguson-Smith 2007). Some evidence suggested that Z-linked genes such as *Dmrt1* may masculinize chicks, but there is also evidence for feminization by W genes like *Asw* and/or *Fet1* (Smith & Sinclair 2004). Recently, Smith and colleagues (2009) proposed that a Z-linked *DMRT1* gene is the master sex-determinant in the chicken (*Gallus gallus*), such that two doses of the DMRT1 protein produces roosters (ZZ) and one dose gives rise to hens (ZW). It seems very likely that *DMRT1* is indeed the master sex-determining gene in chickens, but there is still a possibility that *DRMT1* is a downstream subordinate of a yet-unidentified master switch (Kuroiwa, 2009; Smith et al. 2009).

Regardless of whether *DMRT1* is the master sex-determining gene, *Chd* (the avian chromo-helicase-DNA-binding gene) seems to be tightly linked to the master control gene in most birds and thus can be used to differentiate the sexes. When amplified via PCR, distinct amplicons emerge for males and females because the introns differ in size between the Z and W chromosomes (Fig. 4–1). Various recent conservation studies on birds have used this or similar sexing assays (Clout et al. 2002; Rudnick et al. 2005; Jarvi & Farias 2006).

Reptiles and amphibians

As we will see, sex determination is more complicated in reptiles and amphibians than in mammals and birds. That said, some species have systems of chromosomal sex determination (XY or ZW) that are similar to those seen in mammals or birds. For instance, snake sex is determined by ZW chromosomes that, depending on the species, may be morphologically similar to or quite differentiated from the autosomes (Matsubara et al. 2006). The Australian central bearded dragon (*Pogona vitticeps*) also has a ZW chromosome system (Ezaz et al. 2005). Unlike mammals, in which sex is determined by the dominance of *Sry*, sex in *P. vitticeps* is determined by the dosage of gene(s) on the Z chromosome (one dose gives rise to a ZW female, two doses produce a ZZ male) (Quinn et al. 2007).

Many reptiles have environmental sex determination (ESD) rather than GSD (Shine 1999). ESD is thought to occur in all crocodilians, many turtles, and some lizards. ESD in reptiles is often manifested as temperature-dependent sex determi-nation (TSD) whereby eggs incubated at high temperatures yield mostly females and those incubated at low temperatures produce mostly males (Bull 1981; Valen-zuela & Lance 2004). Conversely, in many turtles, high incubation temperatures tend to produce females and low incubation temperatures bias toward males. Sex determination is not always straightforward, however (Valenzuela et al. 2003). For example, genotypic male (ZZ) lizards reared at high temperatures reverse to phenotypic females, so in some cases, ESD in effect overrides what normally is GSD (Quinn et al. 2007).

The evolution of GSD from TSD has been studied in Testudines (Janzen & Krenz 2004; Valenzuela et al. 2006). The loss of TSD in GSD species is thought to be caused by downstream genes that override the signal from the TSD switch and thus prevent phenotypic response to temperature (Valenzuela 2008). Temperature-sensitive genes in *Chrysemys picta* (painted turtle) include the Wilms' tumor-suppressor gene *Wt1* and the steroidogenic factor 1 gene *Sf1* (Valenzuela 2008). In mammals, *Wt1* encodes a transcription factor necessary for the maintenance of Sertoli cells and the development of testes, and *Sf1* is a nuclear receptor that controls the expression of *Wt1* (Goa et al. 2006). In the GSD species *Apalone mutica* (smooth softshell turtle), both *Dax1* (which targets downstream *Sf1*) and *Wt1* are differentially expressed according to temperature, but this temperature sensitivity does not seem to skew the sex ratio (Valenzuela 2008). The broader role (if any) of *Sf1* in reptilian sex determination needs further investigation.

Among amphibians, GSD is the rule. Although extreme temperatures can skew sex ratios, they seldom seem to do so in nature (Nakamura 2009; see also Valenzuela et al. 2003). Anthropogenic agents including exogenous steroids (Hayes 1998; Nakamura 2009) and some herbicides (Hayes et al. 2002, 2010) can also induce alterations (including intersexuality) in many amphibians, but here we will restrict our discussion to natural means of sex determination.

Amphibians generally lack morphologically distinguishable sex chromosomes, but sex reversal and breeding experiments indicate that the mechanism of GSD in various species may be XY, ZW, or OO/OW (Eggert 2004; Ezaz et al. 2006). This evolutionary plasticity sometimes is reflected within a single species. For instance, some local populations of the Japanese wrinkled frog *Rana rugosa* are XY (heterogametic males) whereas others are ZW (heterogametic females) (Ogata et al. 2008). Presumably, mutations in the primary sex-determining gene could cause phenotypic sex reversals in frogs. A similar situation occurs in medaka (*Oryzias latipes*), a fish normally displaying an XY system of sex determination but in which XX males and XY females occur infrequently (Shinomiya et al. 2004; Otake et al. 2008).

The master sex-determining gene in amphibians has proven elusive, but several genes involved in sexual differentiation have been cloned. These genes include *Sf1, Dmrt1*, and various *Sox* genes (reviewed in Nakamura 2009), all of which appear to be downstream loci controlled by a master switch. In the African clawed frog *Xenopus laevis*, the master switch may be the *DM-W* gene (Yoshimoto et al. 2008). Because the mode of GSD is evolutionarily labile in amphibians, it is likely that – as in fishes – different master genes have evolved in different lineages.

Table 4–4. Sex-determining systems across more than twenty taxonomic orders of teleost fishes

Order	GSD			Hermaphroditism				
	XY	ZW	Protandry	Protogyny	Simul.	Unisexuality	ESD
Anguilliformes	X	X		X	X		
Atheriniformes						X	X
Aulopiformes		X			X		
Beloniformes	X						X
Beryciformes	X						
Characiformes	X	X					
Clupeiformes	X		X				
Cypriniformes	X	X		X		X	X
Cyprinodontiformes	X	X		X	X	X	X
Gasterosteiformes	X	X					
Gymnotiformes	X						
Myctophiformes	X						
Osteoglossiformes	X						
Perciformes	X	X	X	X	X		X
Pleuronectiformes	X	X					X
Salmoniformes	X						X
Scorpaeniformes	X						
Siluriformes	X	X	X				X
Stephanoberyciformes	X						
Stomiiformes	X		X				
Synbranchiformes	X			X			
Tetraodontiformes	X						
Zeiformes	X						

After Mank et al. (2006); some of these orders, most notably Perciformes, are polyphyletic.

SEX DETERMINATION IN FISHES

Sex determination in fishes is complicated, and interested readers are urged to consult Devlin and Nagahama (2002) for more detail than can be provided here. Their 173-page monograph covers gonad development, differentiation, sex-determining systems, environmental effects (EE), and marker development. Sex-determination modes in fishes include male-heterogametic (XY-like) and female-heterogametic (ZW-like) systems, various forms of hermaphroditism (protandry, protogyny, and simultaneous hermaphroditism; Avise & Mank 2010), unisexuality (in which clonal or hemiclonal taxa consist solely of females; Avise 2008), and various forms of ESD (Table 4–4). Alternative mechanisms often are found even in closely related fish taxa (e.g., confamilial species, congeners, and sometimes even conspecifics), thus indicating a high level of evolutionary lability in how sex is determined (Mank et al. 2006; Mank & Avise 2009). Here, we briefly summarize several of these mechanisms.

ESD as determined by behavior

In many fishes, sex determination is environmental (Penman & Piferrer 2008). For example, individuals in species that are sequentially hermaphroditic may switch sex in response to changes in the behavior or social status of conspecifics.

The change of sex can be from female to male (protogynous) or from male to female (protandrous), and the process of sex reversal often involves complete reorganization of the gonads, hormones, and behaviors (Devlin & Nagahama 2002). Such phenotypic changes can be completed in a few days or may take several years, depending on the species (Hattori 1991; Godwin 1994). In fishes that exhibit this mode of ESD, the social structure normally consists of many individuals of one sex that are submissive to one or a few members of the opposite sex. When the dominant individual departs or dies, the individual that ranks highest in the new hierarchy may switch sex and fill the vacated role. Most species that exhibit behaviorally induced ESD live in small social groups or relatively confined spaces (e.g., reef fishes).

Behaviorally induced ESD is exhibited by some gobies (family Gobiidae), in which both protogyny and protandry occur. In protogynous species, the removal of a dominant male can induce sex reversal of the largest female, but that individual retains both ovarian and testicular tissues and can revert back to a female state if the social settings again change (e.g., if a larger male immigrates from a nearby population; Sunobe & Nakazono 1993). Some gobies possess both ovarian and testicular tissues simultaneously. In the "female phase," the ovaries are well developed and functional whereas the testes are shrunken and inactivated; the reverse applies to individuals in the "male phase" (Kobayashi et al. 2005). In aquarium experiments involving pairs of female Okinawa rubble gobies (*Trimma okinawae*), the larger female always changed to a functional male within five days; when males were paired, the smaller male changed to a functional female within ten days (Kobayashi et al. 2005).

Wrasses provide another example of behaviorally induced ESD. In *Thalassoma bifasciatum*, all individuals first develop a rudimentary female gonad that subsequently differentiates into ovaries or testes. Fish that differentiate as males remain male for life and are called primary males (Warner & Swearer 1991). Fish that differentiate as females either remain female *or* subsequently become male; the latter are known as secondary males. The gonads of primary males show no evidence of ovarian tissue, whereas those of secondary males do (Munday et al. 2006). Similarly, in the bluestreak cleaner wrasse (*Labroides dimidiatus*), sexual differentiation is under behavioral control. Bluestreak cleaner wrasses are protogynous; each social group consists of about six to eight individuals, including one adult male plus females and juveniles. If the male dies, the dominant female changes sex. Coral reef fish seem particularly prone to behavior-induced ESD (Munday et al. 2006).

ESD as determined by temperature

In homeothermic mammals and birds, a cascade of genes normally dictates an embryo's sex. Fish, by contrast, are poikilothermic, so their embryos are exposed to a wide range of environmental conditions, including temperature. TSD has been studied closely in reptiles, where a temperature change as little as $2°C$ can alter the sex ratio in a clutch from 100% female to 100% male (Bull & Vogt 1979). These effects arise in part through the increased production of aromatase and subsequent estradiol synthesis at elevated temperatures (Crews 1996). Estradiol secretion in the common carp can range 20-fold over a $5°C$ change in temperature.

Temperature also affects steroid production in trout and carp (Manning & Kimme 1985). TSD appears to be widespread in fishes, having been documented provisionally in at least nine taxonomic families (Devlin & Nagahama 2002). Valenzuela and colleagues (2003) argue, however, that some of these purported cases may not be true TSD but rather instances of GSD with EE.

One such well-documented case involves the Atlantic silverside (*Menidia menidia*). In some populations, early-season offspring (which are reared in lower temperatures) are mostly female, whereas late-season offspring (reared in warmer temperatures) are mostly male (Conover & Kynard 1981). Presumably, embryos exposed to colder environments have the additional time that is needed to develop ovarian tissues (Conover 1984). TSD is not uniform across the range of *M. menidia*, however. For example, fish in some Canadian waters do not respond to fluctuations in temperature, whereas populations from South Carolina do (Lagomarsino & Conover 1993). Sex determination in *M. menidia* also appears to have a genetic component because progeny from different females respond differently to the same temperature fluctuations and different sires seem to have a strong effect on the sensitivity of their offspring to temperature (Conover & Kynard 1981).

In another such case, sex determination in various tilapias normally has a genetic basis (involving either XY or ZW chromosomal systems depending on the species), yet temperature can influence sex ratios as well. In the Nile tilapia (*Oreochromis niloticus*), for example, high temperature induces masculinization that can override genetic influences (in an XY system) on sexual development (Baras et al. 2001). In Mozambique tilapia (*O. mossambicus*), crosses between sex-reversed males (genotype XX) and normal females (also XX) should produce 100% female progeny, but approximately 90% of embryos reared at low temperatures develop as males (Mair et al. 1990). Temperature also influences sex ratio in *O. aureus* (Desperz & Melard 1998). The differences between the sex chromosomes in tilapias are minor and may reflect the recent evolution of heteromorphic sex chromosomes from a TSD ancestor. Baras and coworkers (2001) suggested that remnant temperature sensitivity might be retained if it provides a selective advantage, such as a greater capacity for dispersal via males.

Sex determination can range from pure GSD to GSD+EE to pure TSD (Valenzuela et al. 2003; Penman & Piferrer 2008), so it can be difficult to categorize sex-determining mechanisms using published data from fishes. For example, some species (e.g., *Poeciliopsis lucida* and *Ictalurus punctatus*) have heteromorphic sex chromosomes – often considered a defining feature of GSD – but temperature nevertheless seems to influence progeny sex ratios in at least some cases (Schultz 1993; Patino et al. 1996). Overall, sex determination in fishes is extremely complex and evolutionarily labile.

Genetic sex determination

Despite the prevalence of ESD in fishes, many species exhibit strict GSD (Penman & Piferrer 2008). Furthermore, several genes involved in sex determination have been uncovered in various fish species (Table 4–5). Unfortunately, several primary piscine genetic models – the zebrafish (*Danio rerio*) and two pufferfish (*Takifugu rubripes* and *Tetraodon nigroviridis*) – lack known sex chromosomes and sex-determining markers (although see Cui et al. 2006).

Table 4–5. Loci involved in sex determination of fishes

Species	Locus/Gene	Comments	Citations
Gasterosteus aculeatus (stickleback)	LG19	XX–XY system, master gene near Idh	Peichel et al. 2004
Oreochromis spp. (tilapia)	LG23; Amh	Master gene?	Shirak et al. 2006
Oreochromis spp. (tilapia)	LG23; Dmrta2	Master gene?	Shirak et al. 2006
Oreochromis spp. (tilapia)	SOX14	Minor effect gene	Cnaani et al. 2007
Oreochromis aureus (tilapia)	LG1 and LG3	Both loci sex associated	Cnaani et al. 2008
Oreochromis karongae (tilapia)	LG3	WZ-ZZ system	Cnaani et al. 2008
Oreochromis mossambicus (tilapia)	LG1 and LG3	Both loci sex associated	Cnaani et al. 2008
Oreochromis niloticus (tilapia)	LG1	XX–XY system	Cnaani et al. 2008
Oryzias dancena (medaka)	LG10	XY system, but master gene is not DMY	Takehana et al. 2007a
Oryzias hubbsi (medaka)	LG5	ZW system, unlike *O. latipes* and *O. dancena*	Takehana et al. 2007b
Oryzias latipes (medaka)	DMY (dmrt1bY)	Master gene, derived from DMRT1	Matsuda et al. 2002
Takifugu rubripes (pufferfish)	LG19	amhrII or inhbb master genes?	Kikuchi et al. 2007
Tilapia mariae (tilapia)	LG3	WZ-ZZ system	Cnaani et al. 2008
Tilapia zillii (tilapia)	LG1	XX–XY system	Cnaani et al. 2008
Xiphophorus maculatus (platyfish)	LG24	X, Y, Z system	Volff et al. 2007

Where known, the candidate master gene is indicated.

The best-characterized example of a sex-determining gene in fish involves the medaka (*Oryzias latipes*). In that species, a protein variously designated as DMRT1bY or DMY – encoded by a gene (*DMY/DMRT1bY*) housed on the sex-determining portion of the Y chromosome – is expressed specifically in Sertoli cells of the testis (Matsuda et al. 2002; Nanda et al. 2002). The gene shows 90% nucleotide sequence similarity to *DMRT1*, which is autosomal but is not itself a master sex-determining gene (Brunner et al. 2001). In mammals and birds, a comparable gene is expressed differentially in males and females. It produces a protein with a DNA-binding domain – also called the DM domain – that is involved in testis determination and differentiation (Volff et al. 2003). Similarly, in the invertebrates *Caenorhabditis elegans* and *Drosophila melanogaster*, homologous DM domain proteins are integral components of the male sex-determining cascade (Raymond et al. 1998). Clearly, proteins with DM domains play an evolutionarily conserved role in the testis-forming pathway in a wide variety of animals.

The medaka was the first nonmammalian vertebrate in which the primary sex-determining gene has been characterized (Volff & Schartl 2002; Matsuda 2005; Volff et al. 2007). *DMY/DMRT1bY* is not the master sex-determining gene in all fishes, however. The duplication that gave rise to the nascent *DMY/DMRT1bY* gene was recent (~10 million years ago; Kondo et al. 2004) as *DMY/DMRT1bY* is absent in more distantly related members of the genus (Kondo et al. 2004), and the sex-determining region in medaka corresponds to an autosomal linkage group (LG) in *Oryzias dancena* (Takehana et al. 2007a). In the Asian swamp eel (*Monopterus albus*), some of the variants of four alternatively spliced transcripts of *DMRT1* are expressed more in one sex than the other (Huang et al. 2005).

Alternative splicing and differential expression of DM domain proteins also have been reported in the zebrafish (Guo et al. 2005), honeycomb grouper (Alam et al. 2008), Nile tilapia (Wang & Tsai 2006), channel catfish (Liu et al. 2007), and lake sturgeon (Hale et al. 2010). In all cases, *DMRT1* (or its alternative transcripts) is expressed in both sexes but substantially more so in males than in females. These results again suggest a key role for this gene in testis differentiation. On its initial discovery, *DMRT1* was believed to be the master sex-determining gene in both birds and fish, analogous to *SRY* in mammals. Current thought is that although DMRT1 may not be the master switch in all fishes, it probably plays a key role in the sex-determining pathway (Volff et al. 2003; Ferguson-Smith 2007).

Many other fish species exhibit GSD. For example, the platyfish (*Xiphophorus maculatus*) has three sex chromosomes; males can be either XY or YY, and females can be XX, XW, or YW. One model to explain this pattern proposes that male-determining genes are present on all three chromosomes, but only the Y chromosome allele is active (and perhaps the YW females arise because the W chromosome carries a suppressor for the master sex-determining gene on the Y chromosome) (Kallman 1984). A different model proposes that sex determination depends on gene dosage; a male-determining gene is present in different copy numbers on the Y (two copies), X (one copy), and W (no copies) chromosomes (Volff & Shartl 2001). The higher the copy number, the higher the gene dosage, which in turn leads to the male phenotype (Volff et al. 2007). The sex chromosomes in platyfish are barely distinguishable from each other or from autosomes, and recombination occurs along most of their length. The sex-determining region has been located, but no sex-determining genes have been documented to date (Volff & Schartl 2001; Schultheis et al. 2006).

Some other fishes with GSD have simpler sex-determining systems. For example, the three-spined stickleback (*Gasterosteus aculeatus*) is male heterogametic (Peichel et al. 2004), with the sex chromosomes being cytogenetically indistinguishable from autosomes, and is thus probably young evolutionarily. As expected, recombination rates are reduced on the sex chromosomes, particularly in the region suspected to contain the master sex-determination gene. Various salmonids including rainbow trout (*Oncorhynchus mykiss*), arctic charr (*Salvelinus alpines*), brown trout (*Salmo trutta*), and Atlantic salmon (*Salmo salar*) also are XY male heterogametic (Woram et al. 2003). Mapping studies have shown that the sex-determining region of the sex chromosomes is short and is located on different LGs in each salmonid species (in contrast to the extensive synteny exhibited across most of their genomes). These data suggest that novel Y chromosomes have evolved in each species (Phillips et al. 2001; Woram et al. 2003). Despite extensive efforts, a master sex-determining gene has not yet been discovered in either the sticklebacks or the salmonids (although see Griffiths et al. 2000 for the description of a stickleback sexing marker).

Sex determination in tilapia has been studied in detail. Both genetic and linkage maps are available for *Oreochromis niloticus*, among others (Katagiri et al. 2005; Cnaani et al. 2008). As previously mentioned, *O. niloticus* has an XY system with influences also from autosomal loci and rearing temperature (Mair et al. 1991; Griffin et al. 2002). Sex chromosomes are indistinguishable, and YY males are viable, suggesting recent and limited divergence between the X and Y chromsomes (Cnaani et al. 2008). Mapping studies have localized the sex-determining region to LG1, an area with reduced recombination in males. Thirteen candidate genes known to be involved in the sex-determining pathway in other species were mapped, but none occurred in LG1 (Lee & Kocher 2007), so the master sex-determining gene remains unknown. GSD in two other tilapias (*T. mariae* and *O. karongae*) entails female heterogamety. Markers on LG3 appear to be associated with gender in that specific alleles at several loci are seen only in the maternal line, suggesting the presence of female heterogamety (Cnaani et al. 2008). These results indicate that at least two major loci are involved in sex determination in tilapiine fish.

In some fish species, sex determination is "autosomal" and occurs in the absence of well-differentiated sex chromosomes (Mank et al. 2006). The number and identity of genes involved are unknown. Polygenic sex determination, however, is thought to be evolutionarily unstable and easily invaded by single-locus sex determination (Rice 1986). Thus, autosomal mechanisms of sex determination seem unlikely to persist over evolutionary timescales, and this may explain their relative scarcity in fishes compared to GSD involving differentiated sex chromosomes.

Each particular sex chromosome characterized to date in extant fish lineages seems evolutionarily young (perhaps arising no more than approximately 10 million years ago; Kondo et al. 2004; Peichel et al. 2004), especially compared to the ancient X and Y chromosomal systems (and the *Sry* gene) of mammals. Furthermore, fishes with GSD have evolved a region of suppressed recombination around the sex-determining region, a direct result of heteromorphic sex-chromosome evolution (Charlesworth et al. 2005). The suppression of recombination permits the joint inheritance of genes that are advantageous for one sex. It also avoids the transfer of genes from one sex chromosome to the other, which could have detrimental effects (Bull 1983; Rice 1987). Another consequence of the suppression of recombination is the genetic degeneration of the hemizygous chromosome (Charlesworth et al. 2005). For homeothermic vertebrates, this degeneration is demonstrated by the Y chromosome of mammals and the W chromosome of birds, which are both much smaller and contain far fewer genes than their X and Z counterparts. The small size and reduced gene content of the heteromorphic sex chromosome mean that YY and WW genotypes are lethal in mammals and birds (a situation that contrasts with the viability and often fertility of such genotypes in fishes).

In summary, an emerging theme with regard to GSD fishes is a repeated, independent evolution of sex chromosomes in different lineages, most likely accompanied or followed by the emergence of different sex-determining genes (Mank et al. 2006; Volff et al. 2007). This greatly complicates the scientific search for master sex-determining loci in this largest vertebrate class.

ISOLATION OF SEXING MARKERS FOR FISH MANAGEMENT

We envision many potential uses of sexing assays in fish conservation and management. A few of these uses have already been achieved, mostly in salmonids. For example, molecular sexing assays have latent utility in aquaculture. Devlin and colleagues have developed a culture methodology, based on a molecular sexing assay, for the production of monosex salmon populations. After removing immature juvenile XY males from the population, females produce substantially more roe than do mixed-sex cultures (Devlin et al. 1994; Henry et al. 2004). Monosex cultures are not only of interest to salmon culturists but probably also to caviar producers who would prefer to rear only females if the fish somehow could be sexed during early-life-history stages (e.g., from fin clips). Likewise, tilapia mature and breed before they reach marketable size, so aquaculturists often wish to rear monosex cultures so that fish expend less energy on reproduction (Lee et al. 2004).

Molecular sexing markers could also inform studies of population structure, which can differ between the sexes because of sex-biased dispersal (e.g., in cichlids and brook trout; Knight et al. 1999; Hutchings & Gerber 2002). Such markers also could facilitate estimates of the operational sex ratio (OSR). [The OSR influences effective population size (N_e) so that as the ratio of N_f:N_m departs from unity, N_e decreases according to the expression: $N_e = (4N_{ef} N_{em})/(N_{ef} + N_{em})$.] In many fishes, spawning aggregations are markedly male-biased, so N_e may be much smaller than expected if sex ratios were near unity. For instance, in red drum (*Sciaenops ocellatus*), N_e is only 1/1,000 of the adult census size (Turner et al. 2002) and, in theory, this could be partly due to the OSR.

There are few empirical examples that illustrate the potential utility of fish-sexing markers better than the OSR. In some Pacific salmonids, the OSR is skewed toward males. Biologists wondered if this skewing was because of sex reversals or differential survivorship between the sexes. This question has since been answered in large part because of the fortuitous discovery of a Y-linked pseudogene (GH-Ψ) that contains a restriction site not found in the functional gene (Du et al. 1993). The presence of GH-Ψ was used to evaluate sex-specific survivorship in chinook salmon (*Oncorhynchus tshawytscha*) and in coho salmon (*O. kisutch*). In both species, freshwater survivorship was similar between juvenile males and females, but male survivorship was substantially greater than female survivorship in the marine realm (Spidle et al. 1998; Olsen et al. 2006). This is a fundamental biological insight that was realized only when biologists applied molecular sexing assays to juvenile salmon that could not be phenotypically sexed. Unfortunately, the phylogenetic distribution of GH-Ψ appears to be constrained to a few Pacific salmonids (Devlin et al. 2001), and thus its utility is limited.

Biased OSRs might be due to differential survivorship, but they might also be due to sex reversals (Olsen et al. 2006). For example, the sex ratio in a young GSD species might be near 1:1, but if one sex predominates among adults, then we must consider environmental factors, such as contaminants. Williamson & May (2002) employed a Y-specific sexing marker to determine that sex reversals are likely in chinook salmon, perhaps due to endocrine-disrupting chemicals. In contrast, Fernandez and colleagues (2007) found no such evidence for sex reversals in a different population.

In principle, the molecular identification of sexing markers like GH-Ψ is straightforward in species with GSD; one need only isolate and exploit the genomic region found in one sex but not the other. Occasionally, such markers are identified fortuitously (Du et al. 1993). One example is the discovery that *Idh* is sex-linked in the three-spined stickleback (Withler et al. 1986; Peichel et al. 2004). Subsequent efforts with amplified fragment length polymorphisms (AFLPs) produced a PCR-based sexing assay in that species (Griffiths et al. 2000). Sometimes sexing markers can be isolated by subtractive hybridization, whereby male and female portions of the genome are compared and the similar portions are discarded (Devlin et al. 1991; Wallner et al. 2004). In theory, this should reveal the region in linkage disequilibrium with the master sex-determining gene. Alternatively, the candidate gene approach might seem to be an effective means of characterizing novel markers of gender (as in the use of medaka *DMY/DMRT1bY* as a probe for another fish species). Master sex-determining genes evolve so rapidly in fishes, however, that this approach is likely to work only for closely related taxa.

Next, we focus on yet another possible approach for the isolation of sexing markers: comprehensive transcriptome analyses involving the characterization of genes expressed in the gonads of males and females. We are now using next-generation sequencing technologies to evaluate pools of complementary DNA (cDNA) synthesized from messenger ribonucleic acid (RNA) isolated from fish gonads, in anticipation that comparisons of genes transcribed in testes versus those transcribed in ovaries will help to elucidate the sex-determining cascade in a primitive fish, the lake sturgeon.

CASE STUDY: THE IDENTIFICATION OF SEX-DETERMINING GENES IN LAKE STURGEON

Sturgeons comprise an ancient lineage of benthic fishes characterized by cartilaginous skeletons and bony plates called scutes. The lake sturgeon (*Acipenser fulvescens*) is a freshwater North American species whose historic distribution ranged from Hudson Bay drainages down through the lower reaches of the Mississippi River, but whose range today is much more restricted (Peterson et al. 2007). Mature individuals are large bottom feeders; specimens can grow to more than 100 kg and more than 2 m in length.

Since the arrival of Europeans, lake sturgeon populations have declined precipitously because of habitat degradation and overfishing (Peterson et al. 2002, 2007). Dams now prevent natural spawning migrations, and pollution has impacted many watersheds. Other sturgeon species are famous for their eggs (caviar), but lake sturgeon also were harvested for isinglass to clarify wine and beer, for fish oil that purportedly was used to fuel steamboats, and for animal (pig) feed. According to the U.S. Fish and Wildlife Service (2008), "In Lake Michigan, commercial harvest was closed in 1929 after the catch declined to only 2000 pounds compared to 3.8 million pounds harvested in 1879." Similarly, the Mississippi River harvest declined from 113,000 kg in 1894 to 3,100 kg in 1922 to 0 kg in 1931 (Peterson et al. 2007).

Today, lake sturgeon in Great Lakes watersheds are of significant conservation concern (Jackson et al. 2002; Peterson et al. 2002, 2007). The restoration of lake sturgeon populations is complicated by their biology, including delayed sexual maturity, infrequent spawning, and sexual monomorphism. Lake sturgeon are long-lived, as evidenced by wild centenarians that live to be older than

150 years (Peterson et al. 2007). Males become sexually mature at approximately twelve to twenty years of age and females at eighteen to thirty years. Males may spawn every second or third year, but most females probably spawn on a four- to nine-year cycle (Peterson et al. 2007). Males arrive on the spawning grounds before females, and a half-dozen or so males flank spawning females and simultaneously attempt to fertilize her eggs. There is no parental care, as females depart soon after spawning and males wait for the next available female.

The sex of a nonbreeding lake sturgeon cannot be determined reliably without invasive surgical examination or ultrasound (Vecsei et al. 2003). Furthermore, even for individuals that are sexually mature but nonspawning, current hormonal assays require anesthetization and a substantial blood sample (Feist et al. 2004). Thus, a general conservation and management problem is that gender cannot be determined easily or consistently in most specimens.

Studies on sex reversal and artificial gynogenesis strongly suggest that at least some sturgeon have GSD (Van Eenennaam et al. 1999; Hett & Ludwig 2005). Furthermore, cytogenetic analyses indicate that female white sturgeon have two different sex chromosomes (Z and W) that are similar in appearance, whereas males have two identical sex chromosomes (Z and Z). Thus, the female genome has all of the components of the male genome, plus some components that are found only on the W chromosome (Van Eenennaam et al. 1999). The gene(s) that differentiate females from males presumably are found somewhere on the W chromosome. Thus, consistent DNA-based differences should exist in the chromosomal makeup of males and females.

Sturgeon species are polyploid, and chromosome numbers (including ~250 microchromosomes) range from 120 to approximately 500. Most lake sturgeon microsatellites are inherited in a tetrasomic (4n) fashion, but some loci are disomic (2n) and others are octasomic (8n); thus, the overall ploidy level is unclear (Ludwig et al. 2001). (Our unpublished single-nucleotide polymorphism [SNP] data generally support the microsatellite data [Hale et al. forthcoming].) The DNA content per cell (c-value) is 4.45 pg in lake sturgeon, considerably larger than the 3.50 pg in humans (Gregory et al. 2006). The large size and apparent complexity of sturgeon genomes have conspired to make the search for sex-determining genes extremely challenging. As summarized later and in McCormick and coworkers (2008), we have used a variety of molecular approaches in attempts to isolate sex-specific markers in lake sturgeon because breeding studies designed to evaluate the possibility of ESD would be difficult in this long-lived species with protracted spawning periodicity.

Candidate genes

We first considered genes known to be involved in the sex-determination pathway of other fishes. We started with the medaka *DMY* gene, which failed to amplify in lake sturgeon. We then designed locus-specific primers for the high-mobility group domain of *Sox2, Sox3, Sox4, Sox11, Sox17, Sox19*, and *Sox21* genes as *Sox* genes are often involved in sex determination (Hett & Ludwig 2005). Of these seven genes, four (*Sox2, Sox4, Sox17*, and *Sox21*) amplified in lake sturgeon but none was sex-specific.

Subtractive hybridization

Representational difference analysis (RDA) is a technique often employed to characterize the differences between two genomes (Lisitsyn et al. 1993) and has been used to clone sex-specific markers in mammals (Perez-Perez & Barragan 1998; Wallner et al. 2004). We performed RDA in lake sturgeon first by using known female DNA as the "tester" and male DNA as the "driver" and then by reversing the sex of the tester and the driver. We cloned DNA fragments using this approach, then sequenced them and designed PCR primers for each putative sex-specific fragment, and then screened them against male and female DNA samples. One hundred fifty-nine clones were sequenced; several genes and many previously uncharacterized DNA sequences were isolated using this procedure, but no clone proved to be sex-specific.

In addition to RDA, we performed subtractive hybridization between male and female lake sturgeon genomes using streptavidin-coated magnetic beads. This procedure was carried out twice using various restriction enzymes; after nine rounds of subtractive hybridization, the enriched fragments were cloned and sequenced. Three hundred sixteen clones, representing approximately thirty unique DNA fragments, were analyzed, but no primer set developed from these clones yielded a sex-specific marker.

Random markers

We also attempted to generate a sex-specific lake sturgeon marker using random primers. Randomly amplified polymorphic DNA (RAPD) fragments were generated using a commercial kit. We first iteratively assessed fifty RAPD primers singly before proceeding to evaluations of primer pairs. This procedure was used first on genomic DNA, then on ovary and testis cDNA samples, and finally on the RDA and subtractive hybridization products. More than 120 unique PCRs were performed and thousands of discrete amplicons were generated. Random primers produced no sex-specific DNA fragments, however.

Alternatives to genomic DNA?

De novo characterization of a sturgeon sexing marker has proven difficult for other research groups as well. Hett and Ludwig (2005) evaluated candidate genes in Atlantic sturgeon (*Acipenser sturio*), Wuertz and coworkers (2006) used RAPDs, AFLPs, and inter-sample sequence repeat markers in four *Acipenser* species, and Keyvanshokooh and colleagues (2007) tried RAPDs in beluga sturgeon (*Huso huso*). All of these attempts relied primarily on genomic DNA, and all failed to identify sex-specific markers. Might such searches be enhanced if they focused on candidate genes expressed in gonads? In principle, a focus on expressed genes (i.e., cDNA) might simplify otherwise fruitless searches for sex-linked markers in these extraordinarily complex sturgeon genomes.

Pyrosequencing of transcriptomes

To narrow our search to expressed genes, we sampled gonads from lake sturgeon collected in Lake Oneida, New York, in June 2007 and May 2008. Based on stocking records and established length-at-age relationships (Jackson et al. 2002), lake sturgeon collected in 2007 were most likely twelve years old and those in 2008 were thirteen years old. The males expressed milt during handling and were apparently ready to mate, whereas gametes from the two females were not fully developed (Colombo et al. 2007). From these fish, we surgically collected gonad biopsies and froze them immediately in liquid nitrogen for subsequent RNA extraction and cDNA synthesis (Hale et al. 2009, 2010).

These cDNA fragments were sequenced using the Roche 454 platform (Margulies et al. 2005). The Roche 454 platform does not use conventional dideoxy sequencing, but instead uses a massive sequencing-by-synthesis approach in which photons of light are emitted during elongation (i.e., pyrosequencing). Briefly, DNA is fragmented using air pressure, adapters are ligated to both ends of fragments, and single-stranded DNA is attached to microbeads (one unique fragment per bead). All beads with a unique genomic DNA fragment are contained in a single reaction tube but are subsequently separated into hydrophobic bubbles for emulsion PCR (ePCR). Each unique DNA fragment is amplified via ePCR to coat each bead with a single sequencing template. The four nucleotides (ACGT) are sequentially washed over the entire plate for chain extension, and then CCD cameras detect photon emissions from each of 1.6×10^6 wells on a plate the size of a microscope slide.

Pyrosequencing generates a vast amount of data. We have generated more than 500,000 sequences from lake sturgeon gonads, and they span some 125 MB (Fig. 4–2; Hale et al. 2010). These lake sturgeon sequences include (among others) genes involved in protein binding, DNA repair, translation, and apoptosis. One major advantage to pyrosequencing is that unlike subtractive hybridization or random priming, useful data are recovered regardless of whether the gene or genes of interest are identified. For example, our 454 surveys of lake sturgeon gonads have identified hundreds of single nucleotide polymorphisms (SNPs) in functional genes as well as in genes that are differentially expressed between the sexes (Hale et al. 2009, 2010). Furthermore, our transcriptome surveys have uncovered a number of reproductive proteins that are apparently expressed on the surface of gametes. A number of these reproductive proteins bear signatures of strong selection, suggesting that genes encoding these proteins may have played a key role in the diversification (speciation) of sturgeons. Finally, our surveys of genes expressed in lake sturgeon gonads suggest that exogenous parasites (schistosomes) and pathogens (trichomonads) apparently infect lake sturgeon. The schistosomes horizontally transferred their genes to the lake sturgeon genome, perhaps through the ingestion of snail hosts, whereas the trichomonad sequences were probably derived from a contemporary infection (Hale et al. 2010).

With regard to the search for candidate sex-determining genes in lake sturgeon, we have used a multipronged approach (Fig. 4–3; Hale et al. 2010). By comparing the sequences of known sex-determining genes in other organisms (e.g., *DMY/DMRT1bY, SOX9, Sry*) to the lake sturgeon gene sequences, we generated a list of candidate genes that were then evaluated as sexing markers using PCR. We also compared sequences derived from lake sturgeon testes to those derived from ovaries (and vice versa). Those genes expressed in only one sex were considered as candidate sex-determining genes and were again evaluated as sexing markers with PCR. Presumably, this set of genes includes those that are

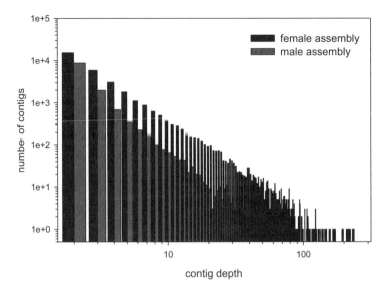

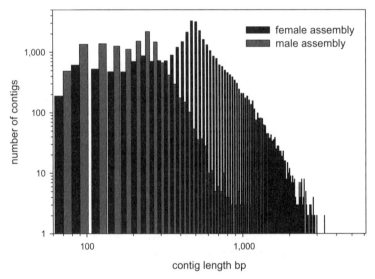

Figure 4–2: The data from 454 runs can be used to generate thousands of long contigs that each consists of many individual sequencing reads. From Hale et al. (2010). *See Color Plate IV.*

expressed in one sex but not the other, as should occur if the genes exist in both sexes but are differentially regulated.

The 454 sequence data indicate that at least two lake sturgeon genes (*DMRT1* and *Tra-1*) are differentially expressed between the sexes (Hale et al. 2010). These expression differences are interesting from an evolutionary perspective as they suggest that gene dosage might play a role in lake sturgeon sex determination. Our *DMRT1* data, considered in light of similar data from chickens, medaka, and lizards, further point to *DMRT1* as a key determinant of vertebrate sex (Marshall Graves 2009). It remains to be determined if sex in lake sturgeon is determined simply by gene dosage (e.g., *DMRT1*; Hale et al. 2010), but at this point, it seems likely. Unfortunately, gene expression assays based on dosage have little practical

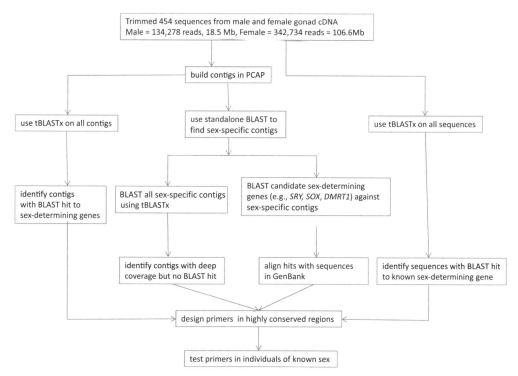

Figure 4–3: Methods employed to search for candidate sex determining. From Hale et al. (2010).

utility for conservation efforts because they require gonad tissue. We uncovered no evidence for a master gene that determines sex by its presence or absence, but we remain optimistic that the evaluation of gonad transcriptomes will generally prove to be a profitable approach for such endeavors in fishes and in various other animals.

Of course, the possibility still remains that lake sturgeon sex determination is not genetic (Hale et al. 2010). An absence of GSD in this species would be consistent with other recent molecular work (Hett & Ludwig 2005; Wuertz et al. 2006; Keyvanshokooh et al. 2007) but apparently inconsistent with cytogenetic evidence from a related species (Van Eenennaam et al. 1999) and with unpublished observations that lake sturgeon hatcheries produce sex ratios near parity.

Lessons from lake sturgeon

The development of sexing assays in reptiles, amphibians, and fishes is not trivial. First and foremost, one needs to know that sex determination is genetic. For the reasons outlined in this chapter, GSD is suspected in lake sturgeon but not confirmed because of their delayed sexual maturity, tiny homomorphic chromosomes, and polyploidy. Second, we suspect that the development of robust sexing assays might be easier in species with small simple genomes than in polyploids. Finally, the rapid evolution of fish sex chromosomes and of master sex-determining genes means that candidate loci will typically be useful only if they are assessed in species closely related to the prototype species from which the gene originally was described. Thus, the approaches described herein need to be refined if the scientific community is to more efficiently identify and characterize various master sex-determining genes among the diversity of fishes.

Prospective

In species with GSD, molecular markers can assign sex to unconventional tissue samples such as hair, feathers, blood, or fin snips. The utility of molecular sexing assays is yet to be realized in fishes because sex-determining genes and chromosomes evolve rapidly in this taxonomic group. When such assays *are* developed – and, in species with pure GSD, there is no reason to think they will not be developed – they will likely be of greatest interest and utility in long-lived species of management or conservation concern, such as sturgeon, tunas, billfishes, paddlefish, and rockfishes. As the taxonomic distribution of complete genome sequences increases, speeded in part by new technologies such as pyrosequencing, our ability to use comparative genomics to help elucidate mechanisms of sex determination in various vertebrate lineages, including fishes, will be greatly enhanced.

REFERENCES

Alam MA, Kobayashi Y, Horiguchi R et al. (2008) Molecular cloning and quantitative expression of sexually dimorphic markers Dmrt1 and Foxl2 during female-to-male sex change in *Epinephelus merra*. *General and Comparative Endocrinology*, **157**, 75–85.

Arakawa Y, Nishida-Umehara C, Matsuda Y et al. (2002) X-chromosomal localization of mammalian Y-linked genes in two XO species of the Ryukyu spiny rat. *Cytogenetic and Genome Research*, **99**, 303–309.

Avise JC (2004) *Molecular Markers, Natural History, and Evolution*. Sinauer Press, Sunderland, Massachusetts.

Avise JC (2008) *Clonality: The Genetics, Ecology, and Evolution of Sexual Abstinence in Vertebrate Animals*. Oxford University Press, New York.

Avise JC, Mank JE (2010) Evolutionary perspectives on hermaphroditism in fishes. *Sexual Development*, **3**, 152–163.

Baras E, Jacobs B, Melard C (2001) Effects of water temperature on survival, growth and phenotypic sex of mixed (XX–XY) progenies of Nile tilapia *Oreochromis niloticus*. *Aquaculture*, **192**, 187–199.

Blejwas KM, Williams CL, Shin GT et al. (2006) Salivary DNA evidence convicts breeding male coyotes of killing sheep. *Journal of Wildlife Management*, **70**, 1087–1093.

Bradley BJ, Doran-Sheehy, DM, Vigilant L (2008) Genetic identification of elusive animals: re-evaluating tracking and nesting data for wild western gorillas. *Journal of Zoology*, **275**, 333–340.

Brunner B, Hornung U, Shan Z et al. (2001) Genomic organization and expression of the doublesex-related gene cluster in vertebrates and detection of putative regulatory regions for DMRT1. *Genomics*, **77**, 8–17.

Bull JJ (1981) Evolution of environmental sex determination from genotypic sex. *Heredity*, **47**, 173–184.

Bull JJ (1983) *Evolution of Sex-Determining Mechanisms*. Benjamin/Cummings, Menlo Park, California.

Bull JJ, Vogt RC (1979) Temperature dependent sex-determination in turtles. *Science*, **206**, 1186–1188.

Cathey JC, Bickham JW, Patton JC (1998) Introgressive hybridization and nonconcordant evolutionary history of maternal and paternal lineages in North American deer. *Evolution*, **52**, 1224–1229.

Charlesworth D, Charlesworth B, Marais G (2005) Steps in the evolution of heteromorphic sex chromosomes. *Heredity*, **95**, 118–128.

Clout MN, Elliott GP, Robertson BC (2002) Effects of supplementary feeding on the off-spring sex ratio of kakapo: a dilemma for the conservation of a polygynous parrot. *Biological Conservation*, **107**, 13–18.

Cnaani A, Lee BY, Ozouf-Costaz C et al. (2007) Mapping of sox2 and sox14 in Tilapia (Oreochromis spp.). *Sexual Development*, **1**, 207–210.

Cnaani A, Lee BY, Zilberman N et al. (2008) Genetics of sex determination in tilapiine species. *Sexual Development*, **2**, 43–54.

Colombo RE, Garvey JE, Wills PS (2007) Gonadal development and sex-specific demographics of the shovelnose sturgeon in the Middle Mississippi River. *Journal of Applied Ichthyology*, **23**, 420–427.

Conover DO (1984) Adaptive significance of temperature-dependent sex-determination in a fish. *American Naturalist*, **123**, 297–313.

Conover DO, Kynard BE (1981) Environmental sex determination: interaction of temperature and genotype in a fish. *Science*, **250**, 577–579.

Crews D (1996) Temperature-dependent sex determination: the interplay of steroid hormones and temperature. *Zoological Science*, **13**, 1–13.

Cui JZ, Shen XY, Gong QL et al. (2006) Identification of sex markers by cDNA-AFLP in Takifugu rubripes. *Aquaculture*, **257**, 30–36.

Desperz D, Melard C (1998) Effect of ambient water temperature on sex determination in the blue tilapia Oreochromis aureus. *Aquaculture*, **162**, 1–2.

Devlin RH, Biagi CA, Smailus DE (2001) Genetic mapping of Y-chromosomal DNA markers in Pacific salmon. *Genetica*, **111**, 43–58.

Devlin RH, McNeil BK, Groves TDD, Donaldson EM (1991) Isolation of a Y-chromosomal DNA probe capable of determining genetic sex in chinook salmon (*Oncorhynchus tshawytscha*). *Canadian Journal of Fisheries and Aquatic Sciences*, **48**, 1606–1612.

Devlin RH, McNeil BK, Solar II, Donaldson EM (1994) A rapid PCR-based test for Y-chromosomal DNA allows simple production of all-female strains of chinook salmon. *Aquaculture*, **128**, 211–220.

Devlin RH, Nagahama Y (2002) Sex determination and sex differentiation in fish: an overview of genetic, physiological, and environmental influences. *Aquaculture*, **208**, 191–364.

Du SJ, Devlin RH, Hew CL (1993) Genomic structure of growth hormone genes in chinook salmon (*Oncorhynchus tshawytscha*): presence of two functional genes, GH-I and GH-II, and a male-specific pseudogene, GH-Ψ. *DNA and Cell Biology*, **12**, 739–751.

Eggert C (2004) Sex determination: the amphibian models. *Reproduction Nutrition Development*, **44**, 539–549.

Ezaz T, Quin AE, Miura I et al. (2005) The dragon lizard *Pogona vitticeps* has ZZ/ZW micro-sex chromosomes. *Chromosome Research*, **13**, 763–776.

Ezaz T, Stiglec R, Veyrunes F et al. (2006) Relationships between vertebrate ZW and XY sex chromosome systems. *Current Biology*, **16**, R736–R743.

Feist G, Van Eenennaam JP, Doroshov SI et al. (2004) Early identification of sex in cultured white sturgeon, Acipenser transmontanus, using plasma steroid levels. *Aquaculture*, **232**, 581–590.

Ferguson-Smith M (2007) The evolution of sex chromosomes and sex determination in vertebrates and the key role of DMRT1. *Sexual Development*, **1**, 2–11.

Fernandez MP, Campbell PM, Ikonomou MG, Devlin RH (2007) Assessment of environmental estrogens and the intersex/sex reversal capacity for chinook salmon in primary and final municipal wastewater effluents. *Environment International*, **33**, 391–396.

Goa F, Maiti S, Alam N et al. (2006) The Wilm's tumor gene, Wt1, is required for Sox9 expression and the maintenance of tubular architecture in the developing testis. *Proceedings of the National Academy of Sciences USA*, **103**, 11987–11992.

Godwin J (1994) Behavioral aspects of protandrous sex change in the anemonefish *Amphirion melanopus* and endocrine correlates. *Animal Behavior*, **48**, 551–567.

Graves JAM (2002) Evolution of the testis-determining gene – the rise and fall of SRY. *Genetics and Biology of Sex Determination*, **244**, 86–101.

Gregory TR, Nicol JA, Tamm H et al. (2006) Eukaryotic genome size databases. *Nucleic Acids Research*, **35** (Database issue), D332–D338; doi:10.1093/nar/gkl828.

Griffin DK, Harvey SC, Campos-Ramos R et al. (2002) Early origins of the X and Y chromosomes: lessons from tilapia. *Cytogenetic and Genome Research*, **99**, 157–163.

Griffith SC, Owens IPF, Thuman KA (2002) Extra pair paternity in birds: a review of interspecific variation and adaptive function. *Molecular Ecology*, **11**, 2195–2212.

Griffiths R, Double MC, Orr K, Dawson RJ (1998) A DNA test to sex most birds. *Molecular Ecology*, **7**, 1071–1075.

Griffiths R, Orr KJ, Adam A, Barber I (2000) DNA sex identification in the three-spined stickleback. *Journal of Fish Biology*, **57**, 1331–1334

Grützner F, Rens W, Tsend-Ayush E et al. (2004) In the platypus a meiotic chain of ten sex chromosomes shares genes with the bird Z and mammal X chromosomes. *Nature*, **432**, 913–917.

Guo YQ, Cheng HH, Huang X et al. (2005) Gene structure, multiple alternative splicing, and expression in gonads of zebrafish DMRT1. *Biochemical and Biophysical Research Communications*, **330**, 950–957.

Hale MC, Jackson JR, DeWoody JA (2010) Discovery and evaluation of candidate sex-determining genes and xenobiotics in the gonads of lake sturgeon (*Acipenser fulvescens*), *Genetica* DOI 10.1007/s10709-010-9455-y.

Hale MC, McCormick CR, Jackson JR, DeWoody JA (2009) Next-generation pyrosequencing of gonad transcriptomes in the polyploid lake sturgeon (*Acipenser fulvescens*): the relative merits of normalization and rarefaction in gene discovery. *BMC Genomics*, **10**, 203.

Hattori A (1991) Socially controlled growth and size-dependent sex-change in the anemonefish *Amphiprion frenatus* in Okinawa, Japan. *Japanese Journal of Ichthyology*, **38**, 165–177.

Hayes T, Haston K, Tsui M et al. (2002) Herbicides: feminization of male frogs in the wild. *Nature*, **419**, 895–896.

Hayes TB, Khoury V, Narayan A, Nazir M, Park A, Brown T, Adame L, Chan E, Buchholz D, Stueve T, Gallipeau S (2010) Atrazine induces complete feminization and chemical castration in male African clawed frogs (*Xenopus laevis*). *Proceedings of the National Academy of Sciences USA*, **107**, 4612–4617.

Hayes TB (1998) Sex determination and primary sex differentiation in amphibians: genetic and developmental mechanisms. *Journal of Experimental Zoology*, **281**, 373–399.

Henry J, Sakhrani D, Devlin RH (2004) Production of all-female populations of coho salmon, *Oncorhynchus kisutch*, using Y-chromosomal DNA markers. *Bulletin of the Aquaculture Association of Canada*, **104-2**, 34–38.

Hett AK, Ludwig A (2005) SRY-related (Sox) genes in the genome of European Atlantic sturgeon (*Acipenser sturio*). *Genome*, **48**, 181–186.

Huang X, Hong CS, O'Donnell M et al. (2005) The doublesex-related gene XDmrt4 is required for neurogenesis in the olfactory system. *Proceedings of the National Academy of Sciences USA*, **102**, 11349–11354.

Hutchings JA, Gerber L (2002) Sex-biased dispersal in a salmonid fish. *Proceedings of the Royal Society of London B*, **269**, 2487–2493.

Jackson JR, VanDeValk AJ, Brooking TE et al. (2002) Growth and feeding dynamics of lake sturgeon, *Acipenser fulvescens*, in Oneida Lake, New York: results from the first five years of a restoration program. *Journal of Applied Ichthyology*, **18**, 439–443.

Janzen FJ, Krenz JG (2004) Phylogenetics: which was first, TSD or GSD? In: *Temperature Dependent Sex Determination in Vertebrates* (eds. Valenzuela N, Lance VA), pp. 121–130. Smithsonian Books, Washington, DC.

Jarvi SI, Farias ME (2006) Molecular sexing and sources of CHD1-Z/W sequence variation in Hawaiian birds. *Molecular Ecology Notes*, **6**, 1003–1005.

Just W, Baumstark A, Suss A et al. (2007) *Ellobius lutescens*: sex determination and sex chromosome. *Sexual Development*, **1**, 211–221.

Just W, Rau W, Vogel W et al. (1995) Absence of SRY in species of the vole *Ellobius*. *Nature Genetics*, **11**, 117–118.

Kallman KD (1984) Sex ratio and the genetics of sex determination in swordtails, *Xiphophorus Poeciliidae*. *Genetics*, **107**, S54.

Katagiri T, Kidd C, Tomasino E et al. (2005) A BAC-based physical map in the Nile tilapia genome. *BMC Genomics*, **6**, 89.

Keyvanshokooh S, Pourkazemi M, Kalbassi MR (2007) The RAPD technique failed to identify sex-specific sequences in beluga (*Huso huso*). *Journal of Applied Ichthyology*, **23**, 1–2.

Kikuchi K, Kai W, Hosokawa A et al. (2007) The sex-determining locus in the tiger pufferfish, *Takifugu rubripes*. *Genetics*, **175**, 2039–2042.

Knight ME, Oppen MJ, Smith HL et al. (1999) Evidence for male-biased dispersal in Lake Malawi cichlids from microsatellites. *Molecular Ecology*, **8**, 1521–1527.

Kobayashi Y, Sunobe T, Kobayashi T et al. (2005) Gonadal structure of the serial-sex changing gobiid fish Trimma okinawae. *Developmental Growth and Differentiation*, **47**, 7–13.

Kondo M, Nanda I, Hornung U et al. (2004) Evolutionary origin of the medaka Y chromosome. *Current Biology*, **14**, 1664–1669.

Korpelainen H (2002) A genetic method to resolve gender complements investigations on sex ratios in *Rumex acetosa*. *Molecular Ecology*, **11**, 2151–2156.

Korpelainen H, Kostamo K (2008) Excessive variation in Y chromosomal DNA in *Rumex acetosa (Polygonaceae)*. *Plant Biology*, **9**, 383–389.

Kuroiwa A (2009) No final answers yet on sex determination in birds. *Nature*, **462**, 34.

Lagomarsino IV, Conover DO (1993) Variation in environmental and genotypic sex-determining mechanisms across a latitudinal gradient in the fish, *Menidia menidia*. *Evolution*, **47**, 487–494.

Lee BY, Hulata G, Kocher TD (2004) Two unlinked loci controlling the sex of blue tilapia (*Oreochromis aureus*). *Heredity*, **92**, 543–549.

Lee BY, Kocher TD (2007) Exclusion of Wilm's tumour (Wt1b) and ovarian cytochrome P450 aromatase (CYP19A1) as candidates for sex determination genes in Nile tilapia (*Oreochromis niloticus*). *Animal Genetics*, **38**, 85–86.

Lisitsyn, N, Lisitsyn N, Wigler M (1993) Cloning the differences between two complex genomes. *Science*, **259**, 946–951.

Liu ZH, Wul FR, Jiao BW et al. (2007) Molecular cloning of doublesex and mab-3 related transcription factor1, forkhead transcription factor gene 2, and two types of cytochrome P450 aromatase in Southern catfish and their possible roles in sex differentiation. *Journal of Endocrinology*, **194**, 223–241.

Ludwig A, Belfiore NM, Pitra C et al. (2001) Genome duplication events and functional reduction of ploidy levels in sturgeon (*Acipenser, Huso*, and *Scaphirhynchus*). *Genetics*, **158**, 1203–1215.

Mair GC, Beardmore JA, Skibinski DOF (1990). Experimental evidence for environmental sex determination in *Oreochromis* species. In: Proceedings of the Second Asian Fisheries Forum (eds. Hirano R, Hanyu I), pp. 555–558. Asian Fisheries Society, Manila, Philippines.

Mair GC, Scott AG, Penman DJ et al. (1991) Sex determination in the genus *Oreochromis* 1. Sex reversal, gynogenesis and triploidy in *Oreochromis niloticus*. *Theoretical and Applied Genetics*, **82**, 144–152.

Mank JE, Avise JC (2009) Evolutionary diversity and turn-over of sex determination in teleost fishes. *Sexual Development*, **3**, 60–67.

Mank JE, Ellegren H (2007) Parallel divergence and degradation of the avian W sex chromosome. *Trends in Ecology and Evolution*, **22**, 389–391.

Mank JE, Promislow DEL, Avise JC (2006) Evolution of alternative sex-determining mechanisms in teleost fishes. *Biological Journal of the Linnean Society*, **87**, 83–93.

Manning NJ, Kimme DE (1985) The effects of temperature on testicular steroid production in the rainbow trout, *Salmo gairdneri*, in vitro and in vivo. *Genetics and Comparative Endocrinology*, **57**, 377–382.

Margulies MM, Egholm W, Altman E et al. (2005) Genome sequencing in microfabricated high-density picolitre reactors. *Nature*, **437**, 376–380.

Marshall Graves JA (2009) Sex determination: birds do it with a Z gene. *Nature*, **461**, 177–178.

Matsuba C, Miura I, Merilä J (2008) Disentangling genetic vs. environmental causes of sex determination in the common frog, *Rana temporaria*. *BMC Genetics*, **9**, 3.

Matsubara K, Tarui H, Toriba M et al. (2006) Evidence for different origins of sex chromosomes in snakes, birds, and mammals and step-wise differentiation of snake sex chromosomes. *Proceedings of the National Academy of Sciences USA*, **103**, 18190–18195.

Matsuda M (2005) Sex determination in the teleost medaka, *Oryzias latipes*. *Annual Review of Genetics*, **39**, 293–307.

Matsuda M, Nagahama Y, Schinomiya A et al. (2002) DMT is a Y-specific DM-domain gene required for male development in the medaka fish. *Nature*, **417**, 559–563.

McCormick CR, Bos DH, DeWoody JA (2008) Multiple molecular approaches yield no evidence of sex-determining genes in lake sturgeon (*Acipenser fulvescens*). *Journal of Applied Ichthyology*, **24**, 643–645.

Millar CD, Reed CEM, Halverson JL et al. (1997) Captive management and molecular sexing of endangered avian species: an application to the black stilt *Himantopus novaezelandiae* and hybrids. *Biological Conservation*, **82**, 81–86.

Munday PL, Buston PM, Warner RR (2006) Diversity and flexibility of sex-change strategies in animals. *Trends in Ecology and Evolution*, **21**, 89–95.

Nakamura M (2009). Sex determination in amphibians *Seminars in Cell and Developmental Biology*, **20**, 271–282.

Nanda I, Kondo M, Hornung U et al. (2002) A duplicated copy of DMRT1 in the sex-determining region of the Y chromosome of the medaka, *Oryzias latipes*. *Proceedings of the National Academy of Sciences USA*, **99**, 11778–11783.

Ogata M, Hasegawa Y, Ohtani H et al. (2008) The ZZ/ZW sex-determining mechanism originated twice and independently during evolution of the frog, *Rana rugosa*. *Heredity*, **100**, 92–99.

Olsen JB, Miller SJ, Harper K, Nagler JJ, Wenburg JK (2006) Contrasting sex ratios in juvenile and adult chinook salmon *Oncorhynchus tshawytscha* (Walbaum) from southwest Alaska: sex reversal or differential survival? *Journal of Fish Biology*, **69**, 140–144.

Otake H, Shinomiya A, Kawaguchi A et al. (2008) The medaka sex-determining gene DMY acquired a novel temporal expression pattern after duplication of DMRT1. *Genesis*, **46**, 719–723.

Paetkau D, Calvert W, Stirling I, Strobeck C (1995) Microsatellite analysis of population structure in Canadian polar bears. *Molecular Ecology*, **4**, 347–354.

Patino R, David KB, Schoore JE et al. (1996) Sex differentiation of channel catfish gonads: normal development and effects of temperature. *Journal of Experimental Zoology*, **276**, 209–218.

Pearse DE, Janzen FJ, Avise JC (2001) Genetic markers substantiate long-term storage and utilization of sperm by female painted turtles. *Heredity*, **86**, 378–384.

Peichel CL, Ross JA, Matson CK et al. (2004) The master sex-determination locus in three-spine sticklebacks is on a nascent Y chromosome. *Current Biology*, **14**, 1416–1424.

Penman DJ, Piferrer F (2008) Fish gonadogenesis. Part I: Genetic and environmental mechanisms of sex determination. *Reviews in Fisheries Science*, **16 (Suppl. 1)**, 14–32.

Perez-Perez J, Barragan C (1998) Isolation of four pig male-specific DNA fragments by RDA. *Animal Genetics*, **29**, 157–158.

Peterson DL, Gunderman B, Vecsei P (2002) Lake sturgeon of the Manistee River: a current assessment of spawning stock size, age and growth. *American Fisheries Society Symposium*, **28**, 175–182.

Peterson DL, Vecsei P, Jennings CA (2007) Ecology and biology of the lake sturgeon: a synthesis of current knowledge of a threatened North American *Acipenseridae*. *Reviews of Fish Biology and Fisheries*, **17**, 59–76.

Phillips RB, Konkol NR, Reed KM et al. (2001) Chromosome painting supports lack of homology among sex chromosomes in *Oncorhynchus*, *Salmo* and *Salvelinus* (*Salmonidae*). *Genetica*, **111**, 119–123.

Pomp D, Good BA, Geisert RD, Corbin CJ, Conley AJ (1995) Sex identification in mammals with polymerase chain reaction and its use to examine sex effects on diameter of day-10 or -11 pig embryos. *Journal of Animal Science*, **73**, 1408–1415.

Primmer CR, Koskinen MT, Piironen J (2000) The one that did not get away: individual assignment using microsatellite data detects a case of fishing competition fraud. *Proceedings of the Royal Society of London B*, **267**, 1699–1704.

Quinn AE, Georges A, Sarre SD et al. (2007) Temperature sex reversal implies sex gene dosage in a reptile. *Science*, **316**, 411.

Radder RS, Quinn AE, Georges A et al. (2007) Genetic evidence for co-occurrence of chromosomal and thermal sex-determining systems in a lizard. *Biology Letters*, **4**, 176–178.

Raymond CS, Shamu CE, Shen MM et al. (1998) Evidence for evolutionary conservation of sex-determining genes. *Nature*, **391**, 691–695.

Reed JZ, Tollit DJ, Thompson PM et al. (1997) Molecular scatology: the use of molecular genetic analysis to assign species, sex and individual identity to seal faeces. *Molecular Ecology*, **6**, 225–234.

Ribeiro de Oliveira R, Souza IL, Venere PC (2008) Karyotype description of three species of *Loricariidae* (Siluriformes) and occurrence of the ZZ/ZW sexual system in *Hemiancistrus spilomma*. *Neotropical Ichthyology*, **4**, 93–97.

Rice WR (1986) On the instability of polygeneic sex determination – the effects of sex specific selection. *Evolution*, **40**, 633–639.

Rice WR (1987) The accumulation of sexually antagonistic genes as a selective agent promoting the evolution of reduced recombination between primate sex-chromosomes. *Evolution*, **41**, 911–914.

Rudnick JA, Katzner TE, Bragin EA et al. (2005) Using naturally shed feathers for individual identification, genetic parentage analyses, and population monitoring in an endangered Eastern imperial eagle (*Aquila heliaca*) population from Kazakhstan. *Molecular Ecology*, **14**, 2959–2967.

Rudnick JA, Katzner TE, Bragin EA et al. (2008) A non-invasive genetic evaluation of population size, natal philopatry, and roosting behavior of non-breeding eastern imperial eagles (*Aquila heliaca*) in central Asia. *Conservation Genetics*, **9**, 667–676.

Schafer AJ, Goodfellow PN (1996) Sex determination in humans. *Bioessays*, **18**, 955–963.

Schultheis C, Zhou Q, Froschauer A et al. (2006) Molecular analysis of the sex-determining region of the platyfish *Xiphophorus maculatus*. *Zebrafish*, **3**, 299–309.

Schultz RJ (1993) Genetic regulation of temperature-mediated sex ratios on the livebearing fish *Poeciliopsis lucida*. *Copeia*, **1993(4)**, 1148–1151.

Shine R (1999) Why is sex determined by nest temperature in many reptiles? *Trends in Ecology and Evolution*, **14**, 186–189.

Shinomiya A, Otake H, Togashi K et al. (2004) Field survey of sex-reversals in the medaka, *Oryzias latipes*: genotypic sexing of wild populations. *Zoological Science*, **21**, 613–619.

Shirak A, Seroussi E, Cnaani A et al. (2006) Amh and Dmrta2 genes map to tilapia (*Oreochromis spp.*) linkage group 23 within quantitative trait locus regions for sex determination. *Genetics*, **174**, 1573–1581.

Sinclair AH, Berta P, Palmer MS et al. (1990) A gene from the human sex-determining region encodes a protein with homology to a conserved DNA-binding motif. *Nature*, **346**, 240–244.

Smith CA, Sinclair AH (2004) Sex determination: insights from the chicken. *Bioessays*, **26**, 120–132.

Smith CA, Roeszler KN, Ohnesorg T, Cummins DM, Farlie PG, Doran TJ, Sinclair AH (2009) The avian Z-linked gene *DMRT1* is required for male sex determination in the chicken. *Nature*, **461**, 267–271.

Spidle AP, Quinn TP, Bentzen P (1998) Sex-biased marine survival and growth in a population of coho salmon. *Journal of Fish Biology*, **52**, 907–915.

Sunobe T, Nakazono A (1993) Sex-change in both directions but alternation of social dominance in *Trimma okinawae* (Pisces, Gobiidae). *Ethology*, **94**, 339–345.

Takehana Y, Demiyah D, Naruse K et al. (2007a) Evolution of different Y chromosomes in two medaka species, *Oryzias dancena* and *O. latipes*. *Genetics*, **175**, 1335–1340.

Takehana Y, Naruse K, Hamaguchi S et al. (2007b) Evolution of ZZ/ZW and XX/XY sex-determination systems in the closely related medaka species, *Oryzias hubbsi* and *O. dancena*. *Chromosoma*, **116**, 463–470.

Turner TF, Richardson LR, Gold JR (2002) Temporal genetic variation of mitochondrial DNA and the female effective population size of red drum (*Sciaenops ocellatus*) in the northern Gulf of Mexico. *Molecular Ecology*, **8**, 1223–1229.

Ueno K, Takai A (2008) Multiple sex chromosome system of X1X1X2X2/X1X2Y type in lutjanid fish, *Lutjanus quinquelineatus* (Perciformes). *Genetica*, **132**, 35–41.

U.S. Fish and Wildlife Service (2008) Lake sturgeon biology and population history in the Great Lakes. Available at http://www.fws.gov/midwest/sturgeon/biology.htm. Accessed 3 February 2010.

Valenzuela N (2008) Relic thermosensitive gene expression in a turtle with genotypic sex determination. *Evolution*, **62**, 234–240.

Valenzuela N, Adams DC, Janzen FJ (2003) Pattern does not equal process: exactly when is sex environmentally determined? *American Naturalist*, **161**, 676–683.

Valenzuela N, Lance VA, editors (2004) *Temperature Dependent Sex Determination in Vertebrates*. Smithsonian Books, Washington, DC.

Valenzuela N, LeClere A, Shikano T (2006) Comparative gene expression of steroidogenic factor 1 in *Chrysemys picta* and *Apalone mutica* turtles with temperature dependent and genotypic sex determination. *Evolution and Development*, **8**, 424–432.

Van Eenennaam AL, Van Eenennaam JP, Medrano JF et al. (1999) Evidence of female heterogametic genetic sex determination in white sturgeon. *Journal of Heredity*, **90**, 231–233.

Vecsei P, Litvak MK, Noakes DLG, Rein T, Hochleithner M (2003) A noninvasive technique for determining sex of live adult North American sturgeons. *Environmental Biology of Fishes*, **68**, 333–338.

Veyrunes F, Waters PD, Miethke P et al. (2008) Bird-like sex chromosomes of platypus imply recent origin of mammal sex chromosomes. *Genome Research*, **18**, 965–973.

Vicari MR, Artoni RF, Moreira O et al. (2008) Diversification of a ZZ/ZW sex chromosome system in Characidium fish (Crenuchidae, Characiformes). *Genetica*, **134**, 311–317.

Volff JN, Nanda I, Schmid M (2007) Governing sex determination in fish: regulatory putsches and ephemeral dictators. *Sexual Development*, **1**, 85–99.

Volff JN, Shartl M (2001) Variability of genetic sex determination in poeciliid fishes. *Genetics*, **111**, 101–110.

Volff JN, Schartl M (2002) Sex determination and sex chromosome evolution in the medaka *Oryzias latipes*, and the platyfish, *Xiphophorus maculatus*. *Cytogenetic and Genome Research*, **99**, 170–177.

Volff JN, Zarkower D, Bardwell VJ et al. (2003) Evolutionary dynamics of the DM domain gene family in metazoans. *Journal of Molecular Evolution*, **57**, 241–249.

Wallner B, Piumi F, Brem G et al. (2004) Isolation of Y chromosome-specific microsatellites in the horse and cross-species amplification in the genus *Equus*. *Journal of Heredity*, **95**, 158–164.

Wang LH, Tsai CL (2006) Influence of temperature and gonadal steroids on the ontogenetic expression of brain serotonin 1A and 1D receptors during the critical period of sexual differentiation in tilapia, *Oreochromis mossambicus*. *Comparative Biochemistry and Physiology B-Biochemistry and Molecular Biology*, **143**, 116–125.

Warner DA, Shine R (2008) Maternal nest-site choice in a lizard with temperature-dependent sex determination. *Animal Behaviour*, **75**, 861–870.

Warner RR, Swearer SE (1991) Social-control of sex change in the bluehead wrasse, *Thalassoma bifasciatum* (Pisces, Labridae). *Biological Bulletin*, **181**, 199–204.

Williamson KS, May B (2002) Incidence of phenotypic female chinook salmon positive for the male Y-chromosome-specific marker OtY1 in the Central Valley, California. *Journal of Aquatic Animal Health*, **14**, 176–183.

Withler RE, MacPhail JD, Devlin RH (1986) Electrophoretic polymorphism and sexual dimorphism in the freshwater and anadromous threespine sticklebacks (*Gasterosteus aculeatus*) of the Little Campbell River, British Columbia. *Biochemical Genetics*, **24**, 701–713.

Woram RA, Gharbi K, Sakamoto T et al. (2003) Comparative genome analysis of the primary sex-determining locus in salmonid fishes. *Genome Research*, **13**, 272–280.

Wuertz S, Gaillard S, Barbisan F et al. (2006) Extensive screening of sturgeon genomes by random screening techniques revealed no sex-specific marker. *Aquaculture*, **258**, 685–688.

Yoshimoto S, Okada E, Umemoto H et al. (2008) A W-linked DM-domain gene, DM-W, participates in primary ovary development in *Xenopus laevis*. *Proceedings of the National Academy of Sciences USA*, **105**, 2469–2474.

Zhan XJ, Zhang ZJ, Wu H et al. (2007) Molecular analysis of dispersal in giant pandas. *Molecular Ecology*, **16**, 3792–3800.

5 Historical and contemporary dynamics of adaptive differentiation in European oaks

Antoine Kremer, Valérie Le Corre, Rémy J. Petit, and Alexis Ducousso

INTRODUCTION

There is growing interest in estimating rates of evolutionary change, motivated by the ongoing environmental change (Gingerich 2001; Stockwell et al. 2003; Carroll et al. 2007). Particular concerns have been raised about forest trees, which are thought to be less able to adapt to these rapid changes due to their long generation time (Reich & Oleksyn 2008). Other authors have suggested that large standing genetic variation in trees may enable rapid adaptive responses, at a pace matching that of ongoing climate change (Kremer 2007; Aitken et al. 2008). An elegant approach to get some insights on the evolutionary responses to global warming is to reconstruct past genetic changes and processes that occurred during the postglacial periods, when temperatures were steadily increasing (Petit et al. 2008). The timing and direction of spread of wind-pollinated trees following the last ice age can be reconstructed from their pollen remains in sediments (e.g., Cheddadi et al. 2005). Palynological data have now been compared to phylogeographic approaches based on range-wide surveys of genetic fingerprints of maternally inherited organelle genomes (Petit et al. 2002b for oaks, Magri et al. 2006 for beech, Cheddadi et al. 2006 for Scots pine). This combination has elucidated genetic and demographic processes associated with tree responses to environmental change. A second approach for predicting adaptive responses to environmental changes is based on theory and simulations (Bürger & Lynch 1995; Bürger & Krall 2004; Sato & Waxman 2008). These studies have been limited so far to single populations and have focused on the amount of environmental change that a population can tolerate, given its genetic and demographic properties.

We will use both approaches to elucidate evolutionary changes associated with environmental changes in trees. First, we will concentrate on adaptive differentiation (Q_{ST}) among populations as an indicator of evolutionary change. Adaptive divergence measures differences between extant populations, and inferences about evolutionary change drawn from divergence are not straightforward unless evolutionary trajectories to ancestor populations are known (Hendry & Kinnison 1999). Second, we will use temperate European white oaks [mainly sessile oak, *Quercus petraea* (Matt.) Liebl.] as models. The postglacial history of temperate oaks in Europe has been reconstructed in detail by combining genetic and historical approaches (Kremer 2002). Moreover, oaks share many genetic, demographic, and ecological attributes with other tree species (long-lived and out-crossing species, large population size, and asymmetry of seed-to-pollen gene flow). Finally, there is a vast body of data concerning population differentiation at various levels in oaks (organelle genomes, nuclear genes, and adaptive traits). To study genetic variation, forest geneticists have traditionally relied on common garden experiments – called provenance tests – by growing populations from various geographic origins in field trials at a common site (see König 2005 for a review on European trees). In the case of European oaks, provenance tests were first established in 1877 (Kleinschmit 1993), and new tests continue to be established today (Jensen & Hansen 2008).

Combining historical and theoretical approaches can overcome in part the obvious limitations of experimental studies of evolutionary change in long-lived species. The historical perspective over the past 15,000 years may disclose long-term trends that can be combined with the findings of the theoretical approach. Both methods may ascertain time frames of past evolution and help identify key drivers of current adaptive responses of trees to climate.

HISTORICAL DYNAMICS OF ADAPTIVE DIFFERENTIATION

How much evolutionary change has accumulated since the Last Glacial Maximum (LGM), 18,000 years ago? Ideally, assessments of change would compare past source populations with extant populations. Evolutionary change can hardly be monitored at the deoxyribonucleic acid (DNA) level unless DNA could be extracted and analyzed from macrofossil remains of oaks. Such assessments are impossible for traits of adaptive significance. We will instead proceed indirectly, by subdividing differentiation of extant populations, assessed in provenance tests, into historical components (e.g., ancient and recent sources of population divergence). Inferring various historical sources is possible due to the detailed knowledge of colonization dynamics provided by the combined pollen and chloroplast DNA (cpDNA) analysis in the case of sessile oak.

Ancient differentiation

The distribution of temperate white oaks in Europe has shifted recurrently from Mediterranean to boreal regions during interglacial and glacial periods (Cheddadi et al. 2005). At the end of the LGM, oak forests were largely restricted to the Iberian Peninsula, Italy, and the Balkan Peninsula (Greece and the western coast of the Black Sea) (Fig. 5–1). A pan-European survey of the pollen fossil remains (Brewer et al. 2002) showed that all refugial sites were located in mountainous areas (e.g., Sierra Nevada in Spain, the Southern Apennine chain in Italy, and the Pindos Mountains in Greece). During the postglacial period, between 13,000 and 10,000 before present (BP), oaks increased in abundance in mountainous areas (Pyrénées, Southeastern Alps, and Carpathian). The cooling of temperatures during 11,000 BP to 10,000 BP stopped this expansion and resulted in reductions of existing populations. After 10,000 BP, oaks spread throughout Europe and reached their extant distribution at approximately 6,000 BP (Fig. 5–1). The expansion was more

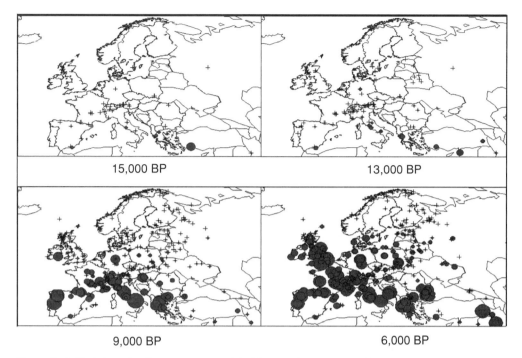

15,000 BP   13,000 BP

9,000 BP   6,000 BP

Figure 5–1: Variation of pollen percentages of deciduous oaks as extracted from the European Pollen Database (according to Brewer et al. 2002, Fig. 2). Crosses indicate locations of sites; sizes of the red circles indicate percentages varying from 1% to 50% (1%, 10%, 25%, and 50%). *See Color Plate V.*

rapid in the west and was reduced in the center and east due to the Alps and the Carpathian Mountains.

Major refugial populations were likely to be genetically differentiated at the end of the last glacial period. Refugial populations were genetically isolated. Pollen fossil and genetic data indicate that the western (Iberian Peninsula) and central (Italy) regions where oaks persisted were physically separated during the LGM. Indeed, despite the lowering of the Mediterranean Sea level by more than 100 meters during the LGM (Rabineau et al. 2006), the Iberian and Italian Peninsulas remained separated (Thiede 1978). Evidence for physical separation of the Italian and Balkan refugial population is not supported by geography or genetic data, however. The Adriatic Sea was partially filled in its northern part (Eckerle et al. 1996), favoring connections between the peninsulas. The duration of geographic separation between the refugial zones lasted for more than 100,000 years, during the last glacial period (Cheddadi et al. 2005). Shorter, milder periods did occur during the glacial period, but the migration north of the Pyrénées and connection of oak populations between Italy and Spain appear unlikely (Cheddadi et al. 2005). Finally, isolation and ancient differentiation are also suggested by genetic data (Fig. 5–2). Approximately forty-eight cpDNA haplotypes, identified in a broadscale survey of cpDNA diversity across Europe, cluster in six different maternal lineages (Petit et al. 2002b). The lineages spread as parallel longitudinal strips from the Atlantic coast to the Ural Mountains. Each lineage extends from a refugial zone to the extreme northern latitude of the current distribution,

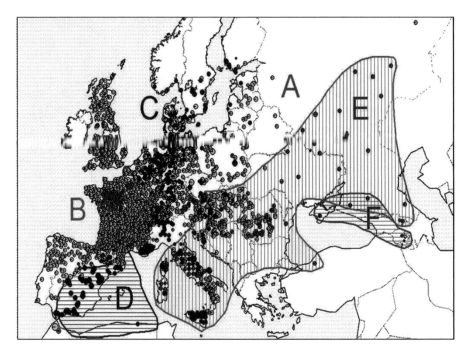

Figure 5–2: Map of cpDNA lineages in deciduous oaks (according to Petit et al. 2002b, Fig. 3). Forty-eight different haplotypes were identified that cluster in six major lineages indicated by colors and letters. *See Color Plate VI.*

with one exception. Lineage D (Fig. 5–2) was indeed restricted to Spain and did not extend across the Pyrénées. CpDNA lineages are not shared between Spain and Italy (Fig. 5–2). Italy and the Balkans, however, share three lineages, suggesting that refugial populations inhabiting these regions were not isolated. The lineages most likely date back to the Pleistocene if not to the Tertiary epoch. The spread of the lineages from extreme southern to northern latitudes allows retrospective identification of the maternal origin of modern populations (Petit et al. 2002a). Given the large sizes of oak populations, the extant distribution of cpDNA variation must reflect the original genetic structure established during colonization. Therefore, cpDNA divergence between refugial populations of Spain, Italy, and the Balkans was likely present at the end of the glacial period. There can be no such retrospective reasoning for nuclear genomes, as pollen flow may have blurred the ancient differentiation. Not only were refugia isolated from each other, they were also undergoing divergent selection pressures due to the different ecological conditions prevailing within the refugia. These speculations suggest that differentiation was widespread not only for cpDNA but also for nuclear DNA and for phenotypic traits.

Assuming that the Spanish and Italian–Balkan refugial populations were genetically separated for more than 100,000 years, we can estimate the minimum genetic differentiation accumulated by neutral factors (drift and mutation) during this period. The expected between-population genetic variance would have amounted to $2tV_m$ (Lynch 1990), where t is the number of generations separating the eastern and western refugial populations from their common ancestor, and V_m is the mutational variance. We assumed that the same traits in the two populations had the same heritability and that the mutational variance represents 10^{-3} of the environmental variance (Houle et al. 1996). Under these assumptions, genetic differentiation of traits (Q_{ST}) between the refugial populations would range from 0.75 to 0.92 for a trait whose heritability varies from 0.2 to 0.5. For adaptive traits, diversifying selection (or convergent selection) due to different (or similar) ecological conditions prevailing in the refugial areas may have further increased (or reduced) Q_{ST}. These figures provide a rough estimate of the minimal genetic differentiation created during the last glacial period due to genetic separation of the refugial areas.

Transient differentiation during colonization

The velocity of oak migration during the postglacial recolonization period (between 15,000 and 6,000 BP) averaged 500 meters/year (Huntley and Birks 1983; Brewer et al. 2002), reaching in some cases up to 1,000 meters/year (Brewer et al. 2005). These figures are much larger than predicted by diffusion models of the advancing wave of the species. If rare long-distance dispersal (LDD) (Nathan 2006) events are included in the diffusion models, however, then the overall expansion results from the combined effects of the advancing wave (diffusion) and the aggregation of the many populations that were founded by the LDD events. Rates of LDD as low as 10^{-4} (in probability) may be sufficient to account for the rapidity of the expansion as deduced from fossil pollen (Le Corre et al. 1997b; Bialozyt et al. 2006). Considering the high fecundity of oaks, LDD may have occurred repeatedly even if their frequency was low. It is interesting that the occurrence of LDD did not only increase the velocity of the colonization, it also contributed to the maintaining of the overall diversity of the species (Bialozyt et al. 2006). The immediate consequence of LDD is the creation of new populations by a limited number of colonizers generating strong founder effects. Footprints of founder effects are easily recognizable in the patchy geographic structure of cpDNA haplotypes (Petit et al. 1997). Ancient differentiation was most likely increased by founder events due to LDD. According to theory, recently founded populations at the leading edge of the distribution should increase the overall differentiation among populations (Le Corre & Kremer 1998) while colonization is proceeding. Given that colonization lasted more than 7,000 years, transient differentiation was maintained and affected organelle and nuclear genes and traits. As the range of a species expands, transient differentiation should decrease (Le Corre & Kremer 1998). After the full size of the range had been reached (6,000 BP), gene flow resulted in the decay of differentiation.

Erasing ancient differentiation

As migration proceeded northward from the different source populations, the colonization routes merged in central Europe and admixture of the different populations resulted in genetic homogenization. Ancient differentiation was almost totally erased after full admixture of the migration fronts and extensive pollen flow among populations originating from different refugia. Large-scale analysis of nuclear gene diversity across Europe has consistently found low levels of genetic differentiation among modern populations regardless of the markers used

Table 5–1. Genetic differentiation between *Quercus petraea* populations assessed with cytoplasmic and nuclear genetic markers

Marker	Number of loci	Number of pops	Geographic distribution	F_{ST}	Reference
cpDNA	1 (cpDNA molecule)	650	Range-wide	0.835	Petit et al. 2002b
Isozymes	13	7	Range-wide	0.032	Zanetto et al. 1994
Isozymes	13	81	Range-wide	0.025	Zanetto & Kremer 1995
RAPDs	31	21	Western part of natural range	0.021	Le Corre et al. 1997a
AFLPs	107	7	Range-wide	0.044	Mariette et al. 2002
Microsatellites	6	7	Range-wide	0.023	Mariette et al. 2002

(Zanetto & Kremer 1995; Le Corre et al. 1997a; Mariette et al. 2002), which we attribute to gene flow (Fig. 5–3 and Table 5–1). Extensive pollen flow has been demonstrated by parentage analyses conducted in *Q. petraea* or *Q. robur* stands (Streiff et al. 1999; Valbuena-Carabana et al. 2005). More than half of the pollen contributing to the next generation came from outside the study stands,

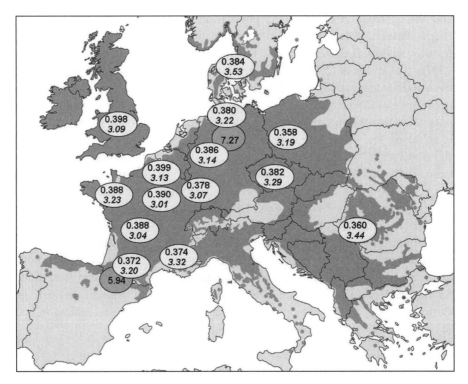

Figure 5–3: Geographic range of *Quercus petraea* and distribution of genetic diversity based on nuclear data (according to Zanetto & Kremer 1995, figs. 2 and 3, and unpublished data). Data in yellow circles correspond to thirteen isoenzymatic loci and represent observed heterozygosity (in bold characters) and allelic richness (in italic). Each circle is the mean value obtained from two to eleven populations. Data in pink circles correspond to nucleotide diversity (π values) obtained from nine candidate genes of bud burst, from fifty sequences in North Germany, and fifty sequences in the Pyrénées. Distribution map was obtained from Euforgen (http://www.bioversityinternational.org/networks/euforgen/Euf_Distribution_Maps.asp). *See Color Plate II.*

and pollen-dispersion curves are characterized by long tails. As in the case of acorns for the colonization of new territories, LDD of pollen may have played an important role in gene flow. Most investigations based on parentage analysis are limited at this point to Local Neighborhood Diffusion (LND) (Hengeveld 1989), whereas LDD has received less attention. LDD may be caused by uplifting air movements into the upper layer of the atmosphere, where the pollen can be transported long distances but is also exposed to high ultraviolet (UV) radiation. A recent study based on an atmospheric model of air movements and taking into account loss of pollen viability due to UV radiation showed that oak pollen can be transported and maintain viability over 100 kilometers (Schueler et al. 2005; Schueler & Schlünzen 2006).

The persistence of ancient differentiation among the modern populations was tested by comparing populations belonging to different cpDNA lineages, as each lineage witnesses the glacial origin of the population. A meta-analysis was conducted across all provenance tests that were established throughout Europe (Kremer et al. 2002) and for all phenotypic traits that were assessed so far. The analysis compiled data from sixty-two trait–test combinations from sixteen provenance tests in France, Germany, and England. Each test comprised between twenty and ninety-four provenances. Phenotypic traits that were assessed were related to growth, phenology, dendrometry, and morphology, most of which would affect trees' fitness. Each provenance was fingerprinted for its cpDNA and was assigned to one of the maternal lineages. The test for persistence of ancient differentiation was made in two different ways: 1) by comparing lineage and population components of variance in a hierarchical analysis of variance; and 2) by comparing genetic distances between the haplotypes assigned to populations and their phenotypic divergence using multiple Mantel tests. Among the sixty-two trait–test comparisons, only two exhibited significant associations with maternal lineage. These results strongly suggest that the ancient differentiation, if it existed, did not persist and was entirely erased by gene flow that followed colonization, by new selection pressures that occurred since colonization, or by their joint actions.

Recent differentiation

The meta-analysis revealed that despite the lack of variation of phenotypic traits among maternal lineages, there is much variation among modern populations for all traits assessed, accounting for "recent" differentiation that has built up since population establishment. It is interesting that the geographic patterns of variation observed were consistent across the different maternal lineages. This consistency is best illustrated by the bud-flushing variation in Q. petraea (Fig. 5–4) that can be observed in today's provenance test. Populations show a clear latitudinal trend of variation, southern populations flushing earlier than northern populations. The clinal pattern is consistent among the three major maternal lineages (A, B, and C), with minor differences in the regression slopes between bud-burst scores and latitude. Again, if ancient differentiation would have persisted, one would have expected different patterns of variation of bud burst in the different lineages. Depending on the part of the natural range that was prospected for

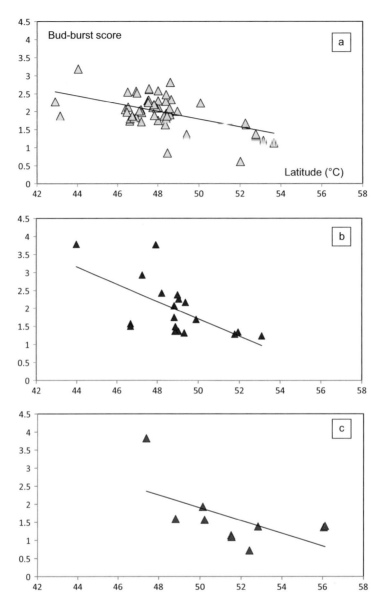

Figure 5–4: Variation of apical bud phenology of different populations of *Quercus petraea* in provenance tests. The analysis was done separately for each different maternal lineage to which modern populations belong. (a) yellow (B lineage) according to Fig. 5–2; (b) blue (A lineage) according to Fig. 5–2; (c) red (C lineage) according to Fig. 5–2. Bud development was recorded according to bud-burst scores varying from 0 (dormant) to 5 (elongating leaves). *See Color Plate VII.*

collecting populations to be installed in provenance tests, Q_{ST} values vary from 0.15 to 0.7 (Table 5–2). Many earlier studies emphasize the strong population variation observed in the provenance tests of sessile oak but did not provide data of Q_{ST} as this parameter has been used only in the past decade (Kleinschmit 1993; Liepe 1993; Ducousso et al. 1996; Jensen 2000). As extensive pollen medi-ated gene flow has homogenized populations across Europe, the differentiation now observed in provenance tests for adaptive traits was most likely created by

Table 5–2. Genetic differentiation between *Quercus petraea* populations assessed for phenotypic traits

Trait	Number of populations	Geographic distribution	Q_{ST}	Reference
Bud-burst score	21	Western part of natural range	0.503	Kremer et al. 1997
Bud-burst score	107	Range-wide	0.55	Ducousso et al. 2005
Bud-burst score	23[a]	Northern part of natural range	0.15 to 0.27	Jensen & Hansen 2008
Leaf coloration in the fall	23[a]	Northern part of natural range	0.15 to 0.18	Jensen & Hansen 2008
Leaf maintenance in the winter	107	Range-wide	0.54	Ducousso et al. 2005
Height at age 6	21	Western part of natural range	0.32	Kremer et al. 1997
Height at age 10	107	Range-wide	0.70	Ducousso et al. 2005
Length of growing season	23[a]	Northern part of natural range	0.41	Jensen & Hansen 2008
Number of branches	107	Range-wide	0.36	Ducousso et al. 2005
Number of forks	107	Range-wide	0.10	Ducousso et al. 2005
Stem form	107	Range-wide	0.68	Ducousso et al. 2005
Crown form	107	Range wide	0.71	Ducousso et al. 2005

[a] This study comprises mostly *Q. robur* populations.

diversifying selection. Temperate oaks are widely distributed in Europe and occupy diverse ecological sites, thus stimulating diversifying selection.

Summing up the conclusions of the historical and genetic survey, the pace of population differentiation during the past 15,000 years can be sketched in four phases:

(1) The glacial period ended with deciduous oaks distributed in three major refugia that were probably genetically differentiated at chloroplast markers, nuclear genes, and adaptive traits.

(2) As the climate warmed, oaks migrated northward as an advancing wave disrupted by LDD events, resulting in local founder events. Transient differentiation was generated by these founder events both for cpDNA and (probably to a lesser extent) for phenotypic traits.

(3) As oaks progressively occupied the central and northern part of Europe, pollen flow established communication between stands originating from eastern, central, and western refugia. Pollen migration increased as stands from different origins merged and reduced the ancient and transient differentiation for traits.

(4) Finally, local selection pressures acting on the recently established populations resulted again in differentiation, which constantly increased over time. New patterns of differentiation appeared, distinct from those existing at colonization.

CONTEMPORARY DYNAMICS OF DIFFERENTIATION

This historical overview indicates that the differentiation of extant populations is large and has been established recently, but it does not provide any time frame for its dynamics. Was evolutionary change generated continuously over the warming period or did it take place faster? How much change can be generated during short periods (i.e., fewer than thirty generations)? The historical overview also showed that despite extensive gene flow, differentiation of traits could accumulate, finally exceeding by far differentiation at neutral markers (Tables 5–1 and 5–2). To learn about the dynamics during shorter time frames and to investigate the apparent contradiction between differentiation and gene flow, we first subdivided complex traits into their elementary components (e.g., genes) to compare differentiation at the same level for markers and traits. In a second step, the dynamics of the components of Q_{ST} were monitored under different evolutionary scenarios mimicking the oak situation.

Components of adaptive differentiation (Q_{ST})

Using standard quantitative genetics reasoning (e.g., a trait controlled by a set of loci with additive actions of their alleles) and further assuming that each locus was biallelic (i.e., each allele having the same effect but of the opposing sign), we obtained the following relationship between population differentiation measured at the level of genes (F_{ST}, mean value of all genes contributing to the trait) and of the trait (Q_{ST}) (Le Corre & Kremer 2003).

$$Q_{ST} = (1 + \theta_B)F_{ST}/[(\theta_B - \theta_W)F_{ST} + 1 + \theta_W] \qquad (1)$$

As expected, there is a positive relationship between both measures. The relationship is affected, however, by two other parameters (θ_B and θ_W), which represent the relative importance of covariances of additive effects of alleles present at the different loci at the between- (θ_B) and within- (θ_W) population level. If there are n genes contributing to the trait, then the genetic variance (V_W) of the trait within a population can be decomposed into the variance due to allelic effects at each locus i (σ_{wi}^2) and into covariances among allelic effects between loci Cov_{wij}:

$$V_W = \sum_i \sigma_{wi}^2 + \sum_i \sum_{j=i} Cov_{wij}, \qquad (2)$$

and θ_W is the relative contribution of covariance terms to the variance terms:

$$\theta_W = \sum_i \sum_{j \neq i} Cov_{wij} / \sum_i \sigma_{wi}^2. \qquad (3)$$

Similar reasoning can be followed for the between-population genetic variance to obtain θ_B. Equation (1) provides a formal framework to understand the relationships between differentiation of a trait and its elementary components; it extends earlier approaches by decomposing the between- and within-population trait variances into covariances and variances of allelic effects (Latta, 1998, 2003; McKay and Latta, 2002).

The relationship between F_{ST} and Q_{ST} derived in the simplified two-allele model holds in a large range of multiallelic cases, as was empirically tested by Le Corre and Kremer (2003). The relationship also holds at any point in time, regardless of equilibrium or non-equilibrium conditions among selection, migration, and drift. The two θ parameters represent the amount of covariances among alleles contributing to the additive value of a trait weighted by the genic variances. Covariances themselves depend on the disequilibria among alleles and the products of allelic effects, meaning that the θ values are generated by nonrandom associations of alleles at the between- and within-population level. Positive θ_B values indicate that populations tend to assemble alleles with effects of the same sign regardless of individuals that bear those alleles. Conversely, negative θ_B values indicate that alleles with opposite signs are associated in the same populations, whereas a θ_B value close to 0 would mean that alleles are randomly associated among populations. Similarly, negative θ_W values indicate that alleles of opposite signs are associated within individuals.

From Equation (1), it can be shown that when θ_B is greater than θ_W, then Q_{ST} will always be greater than F_{ST}. Conversely, when θ_B is less than θ_W, then Q_{ST} is less than F_{ST}. Hence, the comparison of Q_{ST} and F_{ST} depends on the comparison between θ_B and θ_W. These analytical predictions were confirmed by simulations of a wide range of evolutionary scenarios with various levels of gene flow and strength of selection (Le Corre & Kremer 2003). We summarize here the main conclusions regarding the equilibrium values of the components of Q_{ST} (e.g., θ_B, θ_W, and F_{ST}) in the case of an outcrossing species with large population sizes, as is usually the case in tree species.

θ_W is negligible in most scenarios tested, except when the strength of intrapopulation selection is strong, in which case slightly negative values can be reached ($\theta_W < 0.3$). Limited θ_W is expected for large outcrossing populations, where recombination will decrease disequilibria at a constant rate over generations.

θ_B is positive when gene flow is large [Nm (= number of migrant genes per population) > 1] and when diversifying selection occurs (e.g., populations are driven by selection to different optimal values for the trait investigated), due to ecological heterogeneity across the landscape, which is the case for forest trees. It is interesting that θ_B increases as gene flow increases and as diversifying selection is more pronounced. Strength of within-population selection will, however, decrease θ_B, becoming negative in extreme cases. Overall, θ_B exceeds 1 when migration is high ($Nm > 10$), diversifying selection is important, and the strength of within-population selection is weak (Le Corre & Kremer 2003).

F_{ST} increases as diversifying selection increases, the strength of within-population selection gets stronger, and gene flow decreases. Under weak selection intensity, the F_{ST} value of genes contributing to the trait undergoing selection will remain at the level of that of neutral markers.

These opposing trends for θ_B and F_{ST} in outcrossing species comprising large populations and with extensive gene flow among populations are further inflated when the number of loci increases, to the extent that allelic frequencies of genes may reach the level observed at neutral markers (data not shown).

To sum up, in oaks, θ_B is the main driver of Q_{ST}, especially for traits controlled by a large number of loci. These results suggest that adaptive divergence is mainly caused by covariances among allelic effects and not by variances of allelic frequencies (Latta 2003). They also provide an interpretation of the persistence of large, adaptive divergence in the context of high rates of gene flow (see Chapter 8 by Hamrick et al). Intuitively speaking, large gene flow causes constant import of genes, offering new opportunities for associations with resident genes and, hence, increasing θ_B values.

These conclusions, based on theoretical developments and simulations, were recently supported by first results where differentiation was assessed at neutral markers, candidate genes putatively controlling an adaptive trait, and the trait itself, on the same set of populations. In Norway spruce (Heuertz et al. 2006) and in European aspen (Hall et al. 2007), large adaptive divergence was observed for phenological and growth traits, whereas differentiation for candidate genes amounted to levels similar to those found at neutral markers.

Dynamics of adaptive differentiation

How fast does adaptive differentiation (Q_{ST}) build up? How do the components contribute to this trend? To answer these questions, we monitored Q_{ST} in silico using the Metapop simulator (Le Corre et al. 1997b; Le Corre & Kremer 2003), available at request of the author. The simulations were designed to mimic the evolution of a set of oak populations after their installation following colonization. Simulations consisted in creating from a large source population a set of twenty-five populations exchanging pollen and seed according to the island model and undergoing different selection scenarios. We considered a trait controlled by ten genes located on different chromosomes. We considered large population sizes ($N = 500$), high gene flow ($Nm = 10$), and uniform mutation rates (10^{-5} for all loci). These input parameters were chosen to mimic realistic situations for oak forests (large population sizes, extensive gene flow, and moderate-to-low heritability) and were kept constant across all simulations. Contrasting values of parameters, mimicking the effects of natural selection, were selected to test the impact of selection on θ_B and θ_W, Q_{ST}, and F_{ST}. Selection is modeled in two different ways: 1) Stabilizing selection is considered to occur within populations, and the strength of selection is driven by ω^2 (Turelli 1984) (with ω^2 varying between 5 [strong selection] to 50 [weak selection]), and 2) diversifying selection is driven by *VarZopt*, the variance of optimal values of the trait that were assigned to each population (*VarZopt* = 1, weak selection to *VarZopt* = 5, strong diversifying selection). Populations were distributed in a two-dimensional array, and optimal values were assigned to populations along a one-dimensional gradient to reproduce clinal patterns of variation. In these scenarios, the twenty-five populations are at equilibrium at generation 0 (θ_B and $\theta_W = 0$, Q_{ST} and $F_{ST} = 0$). Population differentiation was then monitored over generations for a phenotypic trait submitted to selection, for its underlying genes, and for neutral markers.

The dynamics of θ_B and θ_W show contrasting patterns among generations (Figs. 5-5 and 5-6). These figures depict the evolution of Q_{ST}, F_{ST}, θ_B, and θ_W during the first 100 generations, whereas simulations where conducted up to 3,000 generations to obtain equilibrium values. As expected, θ_W is always extremely low. Disequilibria between alleles at different genes are disrupted at each generation by random mating in large populations. They build up only under strong within-population stabilizing selection, where θ_W values become slightly negative during the first generations when stabilizing selection is strong (Fig. 5-5, a and c). As

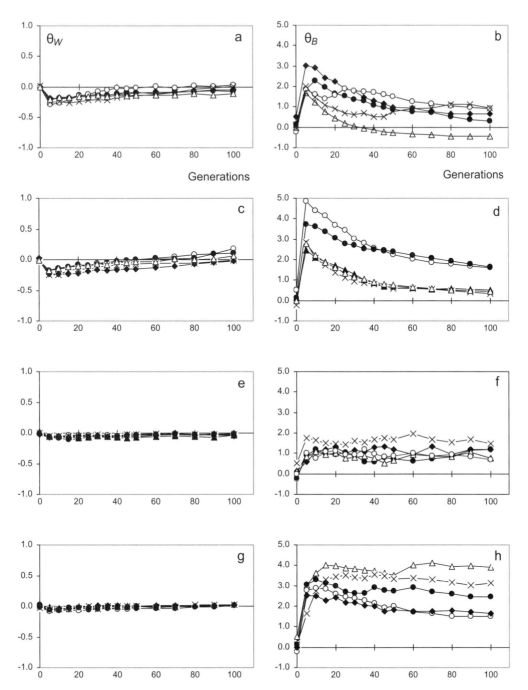

Figure 5–5: Dynamics of θ_W and θ_B along generations in different evolutionary scenarios. a, c, e, and g correspond to the variation of θ_W; b, d, f, and h correspond to the variation of θ_B. a and b: strong stabilizing selection ($\omega^2 = 5$) and weak diversifying selection (*VarZopt* = 1); c and d: strong stabilizing selection ($\omega^2 = 5$) and strong diversifying selection (*VarZopt* = 5); e and f: weak stabilizing selection ($\omega^2 = 50$) and weak diversifying selection (*VarZopt* = 1); g and h: weak stabilizing selection ($\omega^2 = 50$) and strong diversifying selection (*VarZopt* = 5). The five curves correspond to five independent repetitions of the evolutionary scenarios with the same genetic input parameters.

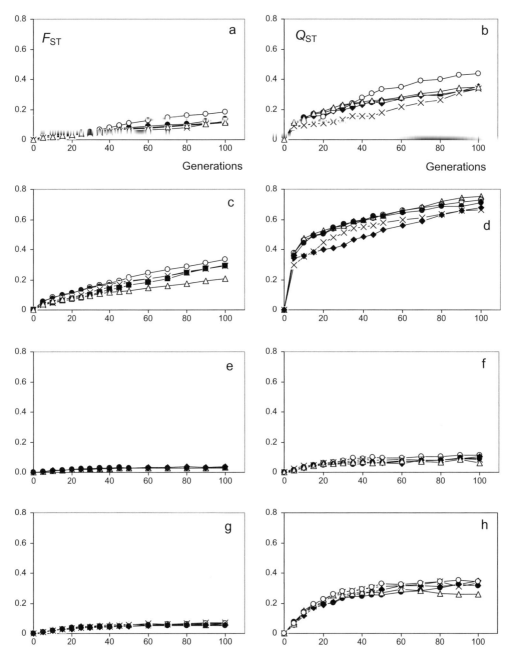

Figure 5–6: Dynamics of F_{ST} and Q_{ST} along generations in different evolutionary scenarios. a, c, e, and g correspond to the variation of F_{ST}; b, d, f, and h correspond to the variation of Q_{ST}; a and b: strong stabilizing selection ($\omega^2 = 5$) and weak diversifying selection (*VarZopt* = 1); c and d: strong stabilizing selection ($\omega^2 = 5$) and strong diversifying selection (*VarZopt* = 5); d and f: weak stabilizing selection ($\omega^2 = 50$) and weak diversifying selection (*VarZopt* = 1); g and h: weak stabilizing selection ($\omega^2 = 50$) and strong diversifying selection (*VarZopt* = 5). The five curves correspond to five independent repetitions of the evolutionary scenarios with the same genetic input parameters.

expected from earlier investigations in single populations, negative correlations between additive values at different genes built up at the early phase of selection (corresponding to the so-called Bulmer effect [Bulmer 1980]). θ_B values vary quite differently: They are in most cases positive and larger than θ_W. When diversifying selection is important (Fig. 5–5, d and h), the covariances among allelic effects are up to 5 times larger than the variances of allelic effects. The most striking feature, however, is the rapid increase of Q_{ST} during the early stages. Indeed, Q_{ST} approaches asymptotic values in fewer than thirty generations, when stabilizing selection is weak (Fig. 5–6, f and h). The increase is more progressive when stabilizing selection is stronger (Fig. 5–6, b and d), although 40–60% of the asymptotic values of Q_{ST} are reached after thirty generations. The rapid increase of differentiation during the first generations is due to the variation of θ_B. Peak values *of* θ_B are reached in fewer than fifteen generations regardless of the selection model chosen (Fig. 5–5, b, d, f, and h). After reaching their maximum, θ_B values slightly decrease along generations when stabilizing selection is strong (Fig. 5–5, b and d) but remain close to their maximum when stabilizing selection is weak (Figure 5-5, f and h). In contrast, F_{ST} steadily increases over generations. The rate of increase is low when stabilizing selection is weak (Fig. 5–6, e and g); in this case, asymptotic values of F_{ST} of genes are close to those of F_{ST} of neutral markers. The rate is higher and continuous when stabilizing selection is strong (Fig. 5–6, a and c). This finding suggests that diversifying selection generates strong associations between alleles at different loci at the between-population level. To sum up, the dynamics of adaptive differentiation can be decomposed in two stages: 1) the early stage (first thirty generations) when Q_{ST} is increasing very rapidly (this pattern is generated by the rapid increase of θ_B, regardless of the selection scenario); and 2) the subsequent stage when Q_{ST} continues to increase but at lower rates. In the case of strong stabilizing selection, F_{ST} will be the main driver of the increase of Q_{ST}. Under weak stabilizing selection, the increase of Q_{ST} is lower and is brought about equally by F_{ST} and θ_B.

Coexistence of strong adaptive differentiation and high gene flow

The following conclusions regarding the tempo of differentiation can be drawn from the comparative analysis of the dynamics of Q_{ST}, F_{ST}, and their components (θ_B and θ_W) along different evolutionary scenarios:

(1) In European oaks, extant genetic differentiation at phenotypic traits is large, in contrast with differentiation at molecular markers, and involves most traits. This differentiation is neither a result of ancient differentiation existing prior to the warming period nor the consequence of founder events at the time of colonization because pollen flow and population admixture have been extensive. It has instead accumulated recently by diversifying selection caused by the ecological heterogeneity across the European continent.

(2) Differentiation (Q_{ST}) of adaptive traits does not solely depend on differences in allelic frequencies (F_{ST}) of loci contributing to the traits. Due to their multilocus structure, covariances among alleles at different loci

$(\theta_B$ and $\theta_W)$ are additional components, whose contribution to Q_{ST} exceeds by far the contributions of individual loci (F_{ST}).

(3) This trend (e.g., the predominant contribution of θ_B to Q_{ST}, as opposed to a contribution of individual loci to F_{ST}) increases as the number of loci contributing to the trait increases. Hence, fitness-related traits that are usually more composite than others will exhibit extremely contrasted F_{ST} and Q_{ST}.

(4) The important contribution of θ_B to Q_{ST} solves the apparent contradiction of coexistence of large differentiation in the context of extensive gene flow in trees. θ_B is increasing as gene flow imports new genes in populations, offering new opportunities to increase covariances between alleles.

(5) When selection (stabilizing or diversifying) starts to operate, the main and immediate driver of Q_{ST} is θ_B. For multilocus traits, diversifying selection is capturing first beneficial allelic associations (θ_B) distributed among the loci contributing to the trait, prior to modifying allele frequencies. During the early phases of selection and even later under certain evolutionary scenarios, allelic frequencies may even be insensitive to selection and behave as neutral markers.

(6) Trees are prone to respond rapidly to selection induced by environmental change, as they possess attributes that inflate θ_B, such as high gene flow and high genetic diversity. Furthermore, the low F_{ST} values for loci controlling traits suggest that θ_B can build instantaneously as adaptive alleles are evenly distributed in natural populations.

CONCLUSION

In this review, the historical and contemporary dynamics of differentiation has focused mainly on European oaks, where fossil and genetic data have been assembled across their entire continental distribution. The conclusions can, however, be extended to many other widespread tree species as they share similar biological features such as high genetic diversity and gene flow (Petit & Hampe 2006; Savolainen et al. 2007). Another example of population differentiation in the case of extensive gene flow is illustrated by the acorn barnacle (Box 5 by S. Palumbi 2010). Assembling lessons from phylogeography, paleobotany, and simulations, we conclude that long-lived tree species have responded quite rapidly to environmental change, despite their low evolutionary rate at the gene level (Smith & Donoghue 2008). Indeed, and as suggested by theoretical investigations, the tempo of differentiation is driven by the standing level of genetic diversity and gene flow. In comparison to earlier studies that highlighted these features of trees (Hamrick 2004), our review shows the mechanisms by which the interaction between gene flow and local selection pressures accelerates the rate of differentiation, by building up favorable allelic associations among genes contributing to fitness-related traits. Intergenic allelic associations are the cause of the rapid differentiation, and they explain the coexistence of strong phenotypic differentiation in trees (as assessed in provenance tests) and extensive gene flow.

It is likely that these mechanisms were actually favored during the repeated interglacial–glacial periods allowing widespread tree species to colonize and adapt

BOX 5: ADAPTIVE SHIFTS IN NATURAL POPULATIONS OF HIGH DISPERSING SPECIES

Stephen R. Palumbi

Problem

Geographic differences in the phenotypes of individuals of a widely distributed species can signal the impact of selection (Kremer et al., this volume). How isolated from one another do populations have to be to accumulate gene-frequency differences due to natural selection? Simple theory of populations connected by immigration suggests that when the selection coefficient s is much greater than the per capita rate of migration among populations m, then local adaptation can result in significant gene-frequency shifts (Wright 1969). As a result, species with potentially high gene flow, such as many marine species, are thought to provide fewer examples of local adaptation, and these are likely to be limited to examples where selection on single loci is particularly strong (Place & Powers 1979; Schmidt & Rand 1999). How often do open populations of marine species with high potential gene flow show adaptive genetic differentiation?

Case Study

The acorn barnacle (*Balanus glandula*) is a sessile marine species in the high intertidal zone of the west coast of North America. It releases swimming larvae that spend three to four weeks in the plankton, drifting with prevailing currents until settlement occurs on intertidal rocks. The species lives from Baja, California, to Alaska across a strong thermal gradient. How likely is it that adaptation to local conditions can occur? The width of an ideal, stable genetic cline is proportional to $(V/s)^{1/2}$, where V is the variance of migration distances from parent to offspring, and s is the selection differential (Sotka & Palumbi 2006, see also Slatkin 1973). For dispersal distances of 50, 100, and 200 km, this relationship suggests that selection differentials of only 1–2% can generate a stable cline in gene frequencies across 500 to 2,700 km (Box Fig. 5–1). Because the thermal gradient across which *B. glandula* lives occurs over this spatial scale, it is possible that relatively weak selection can generate stable clines and produce local genetic adaptation.

Estimates of gene-frequency shifts at nuclear and mitochondrial loci show evidence of such a cline along the California coast (Sotka et al. 2004; Sotka & Palumbi 2006), with cline widths of approximately 500 km. Furthermore, an analysis of six more nuclear loci show significant genetic structure at four of the six loci and significant clines at three loci (Jacobs-Palmer & Palumbi, unpublished data, Box Fig. 5–2). Differentiation at many loci could be due to selection acting individually locus by locus (see Endler 1977 for a discussion of cline evolution and maintenance). Alternatively, southern and northern populations may have diverged in allopatry, producing locally adapted gene pools that are coming together in the present day through migration. The spatially variable alleles diverged long ago (Wares & Cunningham 2005), suggesting that the current cline may be ancient

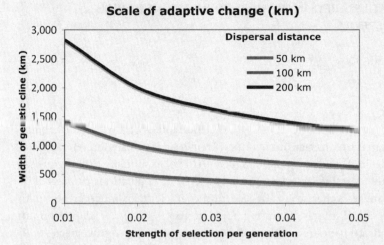

Box Figure 5–1: Width of a stable genetic cline resulting from the balance of dispersal distance and natural selection. Dispersal distances between 50 and 200 km, probably a reasonable range for planktonically dispersing barnacles (Kinlan & Gaines 2003; Sotka et al. 2004) and selection of 1–2% per generation can produce genetic clines 500–2,700 km in width (see the values within the red box). The observed genetic cline in the barnacle *Balanus glandula* is approximately 500 km wide (Sotka et al. 2004).

in origin. In either case, selection is implicated in the origin or maintenance of the cline.

Evidence of selection on gene frequencies in *B. glandula* over time is weak. There are only subtle shifts in gene frequency from larvae to settled juvenile

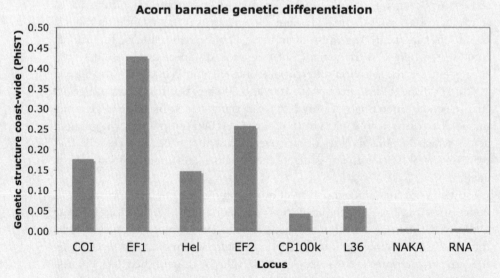

Box Figure 5–2: Genetic differentiation along the central California coast in eight protein-coding loci from the barnacle *Balanus glandula*. Six of eight loci show highly significant shifts along the coast, and five of these have strong clinal signatures with cline widths of approximately 450–550 km. Loci are protein-coding regions from cytochrome oxidase I, elongation factor 1, ribonucleic acid (RNA) helicase, elongation factor 2, a cement gland protein, the ribosomal protein L36, a sodium potassium ATPase, and RNA polymerasse II (Jacobs-Palmer, Galindo, & Palumbi, in preparation).

to adult, and no clear change in northern versus southern alleles in adults that have settled in low versus high intertidal locations (Wares & Cunningham 2005). Yet the theory just outlined suggests that clines of the width we observe in this species can be sustained by mild selection that alters fitness by only 1–2%. Such low rates of selection are unlikely to be visible to any but the most stringent ecological analysis, yet can generate strong genetic patterns in the face of active dispersal.

Evidence that a continuously distributed, high-dispersal marine species has evolved distinct genetic clines across a thermal gradient has implications for the likely impact of coastal climate change. West Coast species distributions are already shifting in response to warming seawater temperatures (Sagarin et al. 1999), and southern species of intertidal barnacles are immigrating into central California locations such as Monterey Bay (Barry et al. 1995). It may be that southern alleles of species also are moving north if selection coefficients are changing. In such cases, biogeographic shifts of species, a commonly tallied index of global change effects in ecosystems (Root et al. 2005; Parmesan 2006), may only capture some of the biological changes associated with global warming.

REFERENCES

Barry JP, Baxter CH, Sagarin RD, Gilman SE (1995) Climate-related, long-term faunal changes in a California rocky intertidal community. *Science*, **267**, 672–675.

Endler JA (1977) *Geographic Variation, Speciation, and Clines*. Princeton University Press, Princeton, NJ.

Kinlan B, Gaines S (2003) Propagule dispersal in marine and terrestrial environments: a community perspective. *Ecology*, **84**, 2007–2020.

Parmesan C (2006) Ecological and evolutionary responses to recent climate change. *Annual Review of Ecology Evolution and Systematics*, **37**, 637 669.

Place AR, Powers DA (1979) Genetic-variation and relative catalytic efficiencies – lactate dehydrogenase-B allozymes of Fundulus heteroclitus. *Proceedings of the National Academy of Sciences USA*, **76**, 2354–2358.

Root TL, MacMynowski DP, Mastrandrea MD, Schneider SH (2005) Human-modified temperatures induce species changes: joint attribution. *Proceedings of the National Academy of Sciences USA*, **102**, 7465–7469.

Sagarin RD, Barry JP, Gilman SE, Baxter CH (1999) Climate-related change in an intertidal community over short and long time scales. *Ecological Monographs*, **69**, 465–490.

Schmidt PS, Rand DM (1999) Intertidal microhabitat and selection at MPI: interlocus contrasts in the northern acorn barnacle, *Semibalanus balanoides*. *Evolution*, **53**, 135–146.

Slatkin M (1973) Gene flow and selection in a cline. *Genetics*, **75**, 733–756.

Sotka EE, Palumbi SR (2006) The use of genetic clines to estimate dispersal distances of marine larvae. *Ecology*, **87**, 1094–1103.

Sotka EE, Wares JP, Barth JB, Grosberg RK, Palumbi SR (2004) Strong genetic clines and geographic variation in gene flow in the rocky intertidal barnacle *Balanus glandula*. *Molecular Ecology*, **13**, 2143–2156.

Wares JP, Cunningham CW (2005) Isolation before glaciation in *Balanus glandula*. *Biological Bulletin*, **208**, 60–68.

Wright S (1969) *Evolution and the Genetics of Populations* (*Vol. 2*): *The Theory of Gene Frequencies*. University of Chicago, Chicago.

to new suited environments. How these mechanisms are acting during ongoing climatic change remains to be investigated in detail, but these processes are critical to predicting how species will adapt to changing environments. Because of the changes in levels of gene flow, we anticipate strong differences between species having continuous distribution and species with scattered distribution. The rate of adaptive differentiation may also be quite different between the leading edge and the rear end of distribution. As suggested by predictive models of bioclimatic envelopes (Thuiller et al. 2005), populations at the northern and eastern limit will be at the leading edge of range shifts and may benefit from immigrating genes via pollen flow from southern latitudes. In contrast, fitness of populations at the southern limit (rear end) may suffer from immigrating genes with lower fitness. The dynamics of differentiation need to be considered under different spatial settings of populations, and genetic associations with different allelic effects should be integrated into conservation and management strategies.

REFERENCES

Aitken SN, Yeaman S, Holliday JA, Wang T, Curtis-McLane S (2008) Adaptation, migration, or extirpation: climate change outcomes for tree populations. *Evolutionary Applications*, 1, 95–111.

Bialozyt R, Ziegenhagen B, Petit RJ (2006) Contrasting effects of long-distance seed dispersal on genetic diversity during range expansion. *Journal of Evolutionary Biology*, 19, 12–20.

Brewer S, Cheddadi R, De Beaulieu JL, Reille M (2002) The spread of deciduous *Quercus* throughout Europe since the last glacial period. *Forest Ecology and Management*, 156, 27–48.

Brewer S, Hely-Alleaume C, Cheddadi R et al. (2005) Postglacial history of Atlantic oakwoods: context, dynamics and controlling factors. *Botanical Journal of Scotland*, 57, 41–57.

Bulmer MG (1980) *The Mathematical Theory of Quantitative Genetics*. Oxford University Press, Oxford.

Bürger R, Krall C (2004) Quantitative genetic models and changing environments. In: *Evolutionary Conservation Biology* (eds. Ferrière R, Dieckmann U, Couvet D), pp. 171–187. Cambridge University Press, Cambridge, UK.

Bürger R, Lynch M (1995) Evolution and extinction in a changing environment: a quantitative genetic analysis. *Evolution*, 49, 151–163.

Carroll SP, Hendry AP, Reznick DN, Fox CW (2007) Evolution on ecological time scales. *Functional Ecology*, 21, 387–393.

Cheddadi R, De Beaulieu JL, Jouzel J et al. (2005) Similarity of vegetation dynamics during interglacial periods. *Proceedings of the National Academy of Sciences USA*, 39, 13939–13943.

Cheddadi R, Vendramin GG, Litt T et al. (2006) Imprints of glacial refugia in the modern diversity of *Pinus Sylvestris*. *Global Ecology and Biogeography*, 15, 271–282.

Ducousso A, Guyon JP, Kremer A (1996) Latitudinal and altitudinal variation of bud burst in western populations of sessile oak (*Quercus petraea* (Matt.) Liebl.). *Annales des Sciences Forestières*, 53, 775–782.

Ducousso A, Louvet JM, Jarret P, Kremer A (2005) Geographic variations of sessile oaks in French provenance tests. In: *Proceedings of the Joint Meeting of IUFRO Working Groups Genetic of Oaks and Improvement and Silviculture of Oaks* (eds. Rogers R, Ducousso A, Kanazashi A), pp. 128–138. Tsukuba, Japan. FFPRI (Forestry and Forest Products Research Institute) Scientific Meeting Report 3.

Eckerle M, Chakari A, Meyruis P, Zonneweld KAF (1996) Paleoclimatic reconstruction of the last deglaciation (18–8ka BP) in the Adriatic Sea region; a land–sea correlation based on palynological evidence. *Palaeogeography, Palaeoclimatology, Palaeoecology*, 122, 89–106.

Gingerich P (2001) Rates of evolution on the time scale of the evolutionary process. *Genetica*, **112–113**(1), 127–144.

Hall D, Luquez V, Garcia VM et al. (2007) Adaptive population differentiation in phenology across a latitudinal gradient in European Aspen (*Populus tremula*, L.): a comparison of neutral markers, candidate genes, and phenotypic traits. *Evolution*, **61**, 2849–2860.

Hamrick JL (2004) Response of forest trees to global environmental change. *Forest Ecology and Management*, **197**, 323–335.

Hendry AP, Kinnison MT (1999) Perspective: the pace of modern life, measuring rates of contemporary microevolution. *Evolution*, **53**, 1637–1651.

Hengeveld R (1989) *Dynamics of Biological Invasions*. Chapman & Hall, London.

Heuertz M, De Paoli E, Källmann T et al. (2006) Multilocus patterns of nucleotide diversity, linkage disequilibrium, and demographic history of Norway Spruce. *Genetics*, **174**, 2095–2105.

Houle D, Morikawa B, Lynch M (1996) Comparing mutational variabilities. *Genetics*, **145**, 1467–1483.

Huntley B, Birks HJB (1983) *An Atlas of Past and Present Pollen Maps for Europe: 0–13000 Years Ago*. Cambridge University Press, London.

Jensen JS (2000) Provenance variation in phenotypic traits in *Quercus robur* and *Quercus petraea* in Danish provenance trials. *Scandinavian Journal of Forest Research*, **15**, 297–308.

Jensen JS, Hansen JK (2008) Geographical variation in phenology of *Quercus petraea* (Matt.) Liebl. and *Quercus robur* L. oak grown in a greenhouse. *Scandinavian Journal of Forest Research*, **23**, 179–188.

Kleinschmit J (1993) Intraspecific variation of growth and adaptive traits in European oak species. *Annales des Sciences Forestières*, **50**, 166s–186s.

König AO (2005) Provenance research: evaluating the spatial pattern of genetic variation. In: *Conservation and Management of Forest Genetic Resources in Europe* (eds. Geburek Th, Turok J), pp. 275–238. Zvolen, Slovakia.

Kremer A, editor (2002) Range-wide distribution of chloroplast DNA diversity and pollen deposits in European white oaks: inferences about colonisation routes and management of oak genetic resources. *Forest Ecology and Management*, **156**, 1–223.

Kremer A (2007) How well can existing forests withstand climate change? In: *Climate Change and Forest Genetic Diversity: Implications for Sustainable Forest Management in Europe* (eds. Koskela J, Buck A, Teissier du Cros E), pp. 3–17. Bioversity International, Rome.

Kremer A, Kleinschmit J, Cottrell J et al. (2002) Is there a correlation between chloroplastic and nuclear divergence, or what are the roles of history and selection on genetic diversity in European oaks? *Forest Ecology and Management*, **156**, 75–87.

Kremer A, Zanetto A, Ducousso A (1997) Multilocus and multitrait measures of differentiation for gene markers and phenotypic traits. *Genetics*, **145**, 1229–1241.

Latta RG (1998) Differentiation of allelic frequencies at quantitative trait loci affecting locally adaptive traits. *The American Naturalist*, **151**, 283–292.

Latta RG (2003) Gene flow, adaptive population divergence and comparative population structure across loci. *New Phytologist*, **161**, 51–58.

Le Corre V, Dumolin-Lapègue S, Kremer A (1997a) Genetic variation at allozyme and RAPD loci in sessile oak *Quercus petraea* (Matt.) Liebl.: the role of history and geography. *Molecular Ecology*, **6**, 519–529.

Le Corre V, Machon N, Petit RJ, Kremer A (1997b) Colonization with long-distance seed dispersal and genetic structure of maternally inherited genes in forest trees: a simulation study. *Genetics Research*, **69**, 117–125.

Le Corre V, Kremer A (1998) Cumulative effects of founding events during colonisation on genetic diversity and differentiation in an island and stepping-stone model. *Journal of Evolutionary Biology*, **11**, 495–512.

Le Corre V, Kremer A (2003) Genetic variability at neutral markers, quantitative trait loci and trait in a subdivided population under selection. *Genetics*, **164**, 2005–2019.

Liepe K (1993) Growth chamber trial on frost hardiness and field trial on bud burst of sessile oak (*Quercus petraea* Liebl.) and *Q. robur* L. *Annales des Sciences Forestières*, **50**, 208–214.

Lynch M (1990) The rate of morphological evolution in mammals from the standpoint of the neutral expectation. *The American Naturalist*, **136**, 727–741.

Magri D, Vendramin GG, Comps B et al. (2006) A new scenario for the quaternary history of European beech populations: palaeobotanical evidence and genetic consequences. *New Phytologist*, **171**, 199–221.

Mariette S, Cottrell J, Csaikl UM et al. (2002) Comparison of levels of genetic diversity detected with AFLP and microsatellite markers within and among mixed Q. *petraea* (Matt.) Liebl. and Q. *robur* L. stands. *Silvae Genetica*, **51**, 72–79.

McKay JK, Latta RG (2002) Adaptive population divergence: markers, QTLs and traits. *Trends in Ecology and Evolution*, **17**, 285–291.

Nathan R (2006) Long-distance dispersal in plants. *Science*, **313**, 786–788.

Petit RJ, Brewer S, Bordacs S et al. (2002a) Identification of refugia and post-glacial colonisation routes of European white oaks based on chloroplast DNA and fossil pollen evidence. *Forest Ecology and Management*, **156**, 49–74.

Petit RJ, Csaikl UM, Bordacs S et al. (2002b) Chloroplast DNA variation in European white oaks: phylogeography and patterns of diversity based on data from over 2,600 populations. *Forest Ecology and Management*, **156**, 5–26.

Petit RJ, Hampe A (2006) Some evolutionary consequences of being a tree. *Annual Review of Ecology Evolution and Systematics*, **37**, 187–214.

Petit RJ, Hu FS, Dick CK (2008) Forests of the past: a window to future changes. *Science*, **320**, 1450–1452.

Petit RJ, Pineau E, Demesure B et al. (1997) Chloroplast DNA footprints of postglacial recolonization by oaks. *Proceedings of the National Academy of Sciences USA*, **94**, 9996–10001.

Rabineau M, Berne S, Olivet JL et al. (2006) Paleo sea levels reconsidered from direct observation of paleo shoreline position during Glacial Maxima (for the last 500,000 yr). *Earth and Planetary Science Letters*, **252**, 119–137.

Reich PB, Oleksyn J (2008) Climate warming will reduce growth and survival of Scots pine except in the far north. *Ecology Letters*, **11**, 588–597.

Sato M, Waxman D (2008) Adaptation to slow environmental change, with apparent anticipation to selection. *Journal of Theoretical Biology*, **252**, 166–172.

Savolainen O, Pyhajarvi T, Knurr T (2007) Gene flow and local adaptation of trees. *Annual Review of Ecology, Evolution and Systematics*, **38**, 595–619.

Schueler S, Schlünzen KH (2006) Modeling of oak pollen dispersal on the landscape level with a mesoscale atmospheric model. *Environmental Modeling and Assessment*, **11**, 179–194.

Schueler S, Schlünzen KH, Scholz F (2005) Viability and sensitivity of oak pollen and its implications for pollen-mediated gene flow. *Trends in Ecology and Evolution*, **19**, 154–161.

Smith SA, Donoghue MJ (2008) Rates of molecular evolution are linked to life history in flowering plants. *Science*, **322**, 86–89.

Stockwell CA, Hendry AP, Kinnison MT (2003) Contemporary evolution meets conservation biology. *Trends in Ecology and Evolution*, **18**, 94–101.

Streiff R, Ducousso A, Lexer C et al. (1999) Pollen dispersal inferred from paternity analysis in a mixed oak stand of *Quercus robur* L. and Q. *petraea* (Matt.) Liebl. *Molecular Ecology*, **8**, 831–841.

Thiede J (1978) A glacial Mediterranean. *Nature*, **276**, 680–683.

Thuiller W, Lavorel S, Araujo MB, Sykes MT, Prentice JC (2005) Climate change threats to plant diversity in Europe. *Proceedings of the National Academy of Sciences USA*, **102**, 8245–8250.

Turelli M (1984) Heritable genetic variation via mutation–selection balance: Lerch's zeta meets the abdominal bristle. *Theoretical Population Biology*, **25**, 138–193.

Valbuena-Carabana M, Gonzalez-Martinez SC, Sork VL et al. (2005) Gene flow and hybridisation in a mixed oak forest (*Quercus pyrenaica* Willd. and *Quercus petraea* (Matts.) Liebl.) in central Spain. *Heredity*, **95**, 457–485.

Zanetto A, Kremer A (1995) Geographical structure of gene diversity in *Quercus petraea* (Matt.) Liebl. I. Monolocus patterns of variation. *Heredity*, **75**, 506–517.

Zanetto A, Roussel G, Kremer A (1994) Geographical variation of inter-specific differentiation between *Quercus robur* L. and *Quercus petraea* (Matt.) Liebl. *Forest Genetics*, **1**, 111–123.

6 Association genetics, population genomics, and conservation: Revealing the genes underlying adaptation in natural populations of plants and animals

Krista M. Nichols and David B. Neale

INTRODUCTION

Understanding the genetic basis of complex adaptive traits is key to understanding how natural and anthropomorphic factors have influenced and will influence the shape of genetic diversity and trajectory of evolution in natural populations. Complex adaptive traits are quantitative traits – those that vary on a continuous scale, and even more generally, are sometimes defined as traits that are expressed as a function of products from multiple genes (Falconer & MacKay 1996; Roff 1997; Lynch & Walsh 1998). Although classical quantitative genetics has revealed the genetic basis to numerous morphological, physiological, and life history traits in plants and animals, the actual genes (loci) and allelic variation with loci underlying key functional differences among organisms remain unknown. Understanding the genes involved in species- and population-level diversity can provide important tools (i.e., genetic markers) for resource managers that are charged with conservation, management, and restoration of natural populations. In this chapter, our examples and review are focused on non-model, non-domesticated organisms as it is the diversity in natural populations, shaped by the natural processes of evolution, with which natural resource managers are most concerned.

Population genetics has undoubtedly been one of the most important fields in the conservation, management, and restoration of native plant and animal species. Together with ecological and life history information, "neutral" genetic markers, or those mirroring the neutral demographic processes of natural populations, are important tools for the delineation of management units or evolutionary significant units for conservation and management. Loci that have been shaped by natural selection, in the process of adaptive population divergence, can however exhibit levels of differentiation markedly different than neutral loci (Leinonen et al. 2008; Vali et al. 2008; Nosil et al. 2009). Until recently, understanding the genetic basis of complex traits has been limited to model

We acknowledge the input and comments from a variety of people in the writing of this chapter. KN thanks Ben Hecht, Sunnie McCalla, Garrett McKinney, Ashley Chin-Baarstad, and John Colletti for useful comments. We both thank Yue Chen and Alex Wong for assistance with literature review and the reference database (Table 6–1).

genetic markers
genes

—— causal mutation

Figure 6-1: Schematic representation of LD among genetic markers, genes, and a causal mutation. Gray and white contiguous blocks represent haplotype blocks with significant LD, and within
those blocks are genes. Detecting significant genotype–phenotype associations will depend on
LD between genetic markers and the causal mutation. In this case, the causal mutation (gray star)
is in LD with several markers (black stars). Note that the causal mutation is not itself surveyed in
the study, but by linkage with markers, a significant genotype–phenotype association would be
observed with the markers typed in the study.

organisms, which are easily reared in a laboratory or common garden environment, and for which genetic and genomic resources were available. With the
expansion of genomics and statistical genetics in the last decade and numerous
genetic tools capable of surveying large numbers of genetic markers or genes in
the genome, the identification of candidate gene and genome regions underlying ecologically relevant traits is now possible. The identification of genes and
allelic variation at genes functionally linked to ecologically variable phenotypes
is tractable for even non-model species, and results from association genetics
studies have great promise in providing added resolution in defining units for
conservation and management. These studies, together with classical quantitative genetic approaches, reveal that genes or markers linked to genes underlying
adaptive population divergence can give very different signatures of population
divergence, and that, even in the face of low to moderate gene flow observed
at neutral genetic markers, natural selection can maintain adaptive population
divergence at genes underlying ecologically relevant traits (Leinonen et al. 2008).

In all cases, the ability to identify genes or genome regions associated with adaptive population differentiation or within population diversity relies on linkage
disequilibrium (LD), also called gametic phase disequilibrium, between markers
surveyed and the causal mutation(s) (Fig. 6–1). LD is the nonrandom association
of alleles at different genes or loci (Slatkin 2008). LD can arise between loci that
are physically unlinked as a consequence of population genetic processes such
as genetic drift, population subdivision, population bottlenecks, inbreeding, and
epistasis; the magnitude of LD among physically linked loci is a function of
the amount of recombination among linked loci (see Slatkin 2008 for a review).
The concept of LD is an extremely important one in association genetics as the
extent of LD between observed markers and the causal genetic variant partially
responsible for the observed phenotypic variation will largely dictate the power
of different methods in revealing genes or genome regions associated with the
trait of interest. If LD is high over long stretches of the genome, fewer markers
are needed for association genetics, but the likelihood of identifying the genetic
variant responsible for phenotypic variation is much lower. If LD is low across
the genome or is found in only short haplotype blocks, many more markers
will be needed to detect associations across the whole genome; however, with
shorter blocks of LD across the genome, the task of identifying the causal genetic
variant becomes much easier. In rare cases is the causal mutation responsible for
the phenotypic variability surveyed or observed in initial analyses in non-model
organisms.

Table 6–1. A sampling of major reviews of the methods and utility of association genetics

Topic	Phenotype? quantified?	Pedigree/ crosses?	References
Population genomics & neutrality tests Tests for signatures of natural selection	No	No	Nielsen 2001, 2005; Luikart et al. 2003; Schlotterer 2003; Storz 2005; Vasemagi & Primmer 2005; Thornton et al. 2007; Li et al. 2008; Stinchcombe & Hoekstra 2008
LD mapping Genotype–phenotype association studies in populations of unknown pedigree	Yes	No	Gupta et al. 2005; Stinchcombe & Hoekstra 2008; Weir 2008
QTL analysis Detection of genes or genome regions associated with phenotypes of interest in pedigreed populations	Yes	Yes	Wu et al. 2002; Erickson et al. 2004; Slate 2005; Stinchcombe & Hoekstra 2008

The literature is replete with reviews of the theoretical and statistical approaches and methods used for association genetics and the identification of genes or genome regions that have been shaped by natural selection (Table 6–1). We do not intend to exhaustively recapitulate prior reviews but rather provide a general overview with relevance to natural or free-living populations, paying special attention to the advantages and challenges of these methods in non-model, non-domesticated natural or free-living plant and animal populations. We take the definition of natural population in terms of the genetic dissection of ecologically relevant traits as defined by Slate (2005), who provides an excellent overview and review of quantitative trait loci (QTL) mapping methodologies and empirical results in natural populations of animals. Slate (2005) defines a *natural population* as one that is "descended from recently sampled individuals of a non-domesticated origin" excluding "model organisms that have been reared in the laboratory for many generations." This definition is particularly important as we review the primary literature for the identification of QTL in natural populations as few association genetic studies have been conducted in un-manipulated, non-domesticated populations of organisms. Although we recognize that natural populations of model organisms have been explored, we limit our review to those species that are not long-standing model organisms (i.e., *Drosophila*, *Arabidopsis*, crops, and domesticated animals) and to those traits that have ecological significance in natural populations.

METHODS FOR DETECTING GENES FOR ECOLOGICALLY RELEVANT PHENOTYPES

Here, we detail methods that take two major approaches in the identification of genes or genome regions significantly associated with adaptive divergence among individuals and populations. The first group of methods (quantitative genetic approaches), including LD or association mapping and QTL analyses, rely on surveys of molecular markers and measures of known phenotypes of interest. The second group of methods (population genetic approaches), called hitchhiking mapping, does not necessarily require measurement of phenotypes on all individuals but rather aims to identify molecular markers showing unusual patterns of population genetic differentiation (i.e., outliers) between populations of interest.

Quantitative genetic approaches

LD or association mapping

LD or association mapping reveals genes or genome regions that are significantly associated with specific phenotypes in natural populations of organisms of unknown relationship (see Neale & Ingvarsson 2008; Stinchcombe & Hoekstra 2008; Weir 2008 for review). LD is among the intuitively simplest of tests for genotype–phenotype associations; individuals sampled from natural populations are evaluated for their phenotype in some type of replicated genetic test, genotyped for polymorphisms in a subset of candidate genes or throughout the genome, and genotype–phenotype associations are tested in the absence of linkage mapping or known family relationships. Because family relationships and population subdivision alone can lead to false-positive associations between genotype and phenotype, ad hoc, multivariate methods are used to account for population subdivision and relatedness using a subset of "neutral" genetic markers (Pritchard et al. 2000; Yu et al. 2006; Zhao et al. 2007). These methods eliminate false-positive associations that arise simply because of population subdivision but can also eliminate true phenotype–genotype associations that covary with population subdivision. Because association mapping uses relatively simple linear mixed models, additional fixed and random effects can be incorporated into models testing for genotype–phenotype associations to account for phenotypic variation among environments, sexes, year classes, and so forth, in addition to variation occurring as a result of population subdivision (Yu et al. 2006; Stich et al. 2008; Yu et al. 2008). There are two main approaches for LD or association mapping, and these are categorized into tests for association with specific candidate genes (or candidate regions) of interest, or genome-wide tests for association.

Candidate gene approaches. For non-model organisms lacking a genome sequence or significant genomic resources, the candidate gene approach for association mapping offers more immediate and simple tests for association with phenotypes of interest. Originally devised for tests of association in complex human diseases, numerous statistical tests or approaches have been devised for tests for association between candidate gene polymorphisms and phenotypes (see Long & Langley 1999; Balding 2006 for reviews). Genes with known roles in particular suites of life history, physiological, behavioral, or morphological traits in model organisms or better-studied taxonomic groups may provide the best candidates for similar traits in non-model organisms (Fitzpatrick et al. 2005). Even in the absence of significant genomic sequence resources, motifs in candidate genes that are conserved across taxa can be used to identify primers to isolate the homologous gene sequences in non-model organisms of interest (Krutovsky et al. 2007). Moreover, with massively parallel sequencing, candidate genes, whole transcriptomes, and whole genomes can be used for single nucleotide polymorphism (SNP) detection even in non-model organisms (Ellegren 2008). The candidate-gene approach has been particularly successful across taxonomic groups and offers the greatest promise for initial association mapping studies in non-model organisms. In some cases, genome regions identified from QTL mapping studies in controlled crosses of the same or related species would provide information on candidate regions for association studies. The disadvantage of the candidate-gene approach is that for some traits, a reasonably viable set of candidate genes is not available without pursuing genome-wide approaches such as whole genome expression or transcriptome studies.

Genome-wide association approaches. Genome-wide tests for genotype–phenotype associations are so far limited to model organisms for which significant genomic resources are available. For genome-wide tests of association, suites of markers distributed across the genome are tested for genotype–phenotype associations. In most cases, the position or order of these markers across the genome is known from linkage mapping or genome-sequencing efforts. A genome-wide scan, then, gives an overview of the patterns of genotype–phenotype associations along the chromosomes. Although deemed "genome-wide" approaches, a true whole genome approach would sample all polymorphisms at the genomic level, and this is a monumental task even in fully sequenced genomes. With LD among closely linked loci, it is not necessary to sample every polymorphism in the genome. In fully sequenced organisms, the extent of LD across the genome can be evaluated to determine, on average, the size of haplotype blocks in the genome, as depicted in Fig. 6–1. From this information, representative markers from those regions (sometimes called "tag SNPs") can be used for whole-genome approaches in LD mapping (Carlson et al. 2004). In non-model organisms, obtaining information on the size and genomic distribution of haplotype blocks across the genome is a huge undertaking in itself. In non-model organisms, the most promising approach for genome-wide association studies may come in surveying associations in large numbers of candidate genes or expressed sequences identified from transcriptome sequencing (gene-space scan).

There are a number of advantages of LD mapping in natural populations when compared to QTL mapping and population genomics approaches. First, although LD mapping can account for relatedness among individuals using neutral markers, complete and known family relationships among individuals in the sampled population(s) are not necessary as they are for QTL studies in natural populations. In many organisms, pedigrees cannot be determined directly by observation and thus rely on time-consuming and expensive efforts to reconstruct pedigree information using molecular markers (Blouin 2003; Pemberton 2008). Compared to QTL mapping in the more traditional sense of one or a few crosses, LD mapping has the advantage of surveying many more recombinants in the population, offering finer resolution for the possible detection of the causal mutation(s) responsible for variation in phenotype. For non-model organisms with few genomic resources available, candidate-gene association mapping offers the greatest promise for tests of phenotype–genotype associations. The major disadvantage of LD mapping in non-model organisms is the time and expense required to survey sequence polymorphisms for the development of SNP markers either in few candidate genes or on a genome-wide level.

Quantitative trait loci (QTL) analysis

QTL analyses seek to identify genes or genome regions significantly associated with phenotypes of interest in known crosses or pedigreed populations of plants and animals. The application, tools, and use of QTL analysis for natural populations are reviewed by Erickson and colleagues (2004) and Slate (2005); for a comprehensive overview of design and analysis of QTL, readers are referred to Doerge and coworkers (1997) and Lynch and Walsh (1998). Briefly, the tools required for this type of analysis include individuals produced in a known breeding scheme or of known relationships in a pedigreed population, molecular marker genotypes of these progeny for markers distributed across the genome, and phenotypes of interest measured in individuals from the breeding scheme or pedigree. The observed amount of recombination between markers used for genotyping is used to order markers into linkage groups, which are used as a framework for statistical tests of genotype–phenotype associations. By observing the cosegregation or inheritance of molecular marker genotypes with phenotypes of interest within the context of this linkage map, genome regions that are significantly associated with variation in the phenotypes are identified.

The type of breeding scheme used for QTL analysis in natural populations is largely related to the question(s) of interest. In model organisms, QTL analyses are commonly conducted in progeny produced from inbred line crosses, maximizing the amount of LD between marker genotype and phenotypes of interest. For outbred populations of interest, QTL analyses are conducted in crosses between individuals with divergent phenotypes or can be conducted in pedigreed populations. For questions regarding genes involved in speciation or reproductive isolation between divergent populations, crosses made between species or populations are necessary to dissect the architecture of quantitative traits, unless a natural hybrid zone can be identified. For questions regarding genes underlying phenotypic variation within populations, although crosses between individuals with divergent phenotypes can be made, analysis in the full or partial pedigree of the population would sample more of the genetic and phenotypic diversity within the population, taking into account the genetic relationships among all pairs of individuals (Slate et al. 1999; George et al. 2000; Pemberton 2008).

There are both advantages and disadvantages in using QTL analysis compared to other methods for association genetics. The advantage of the QTL approach in known, single-generation crosses is that LD between polymorphic markers and phenotypic traits are maximized as a result of observing many fewer recombination events in a single cross when compared to multiple generations and multiple crosses in a pedigree. Because LD is maximized (i.e., gray and white contiguous blocks of LD are longer in Fig. 6–1), many fewer markers are needed to perform QTL analysis; however, because LD is maximized, the likelihood that a QTL analysis will identify the causal mutation responsible for a proportion of the phenotypic trait variation is low. Moreover, because few individuals are selected for crossing, mutations at some loci associated with phenotypic variation may go undetected if markers linked to or the actual causal mutation are not polymorphic in the few individuals that were drawn for crosses from the larger population. As a result, crosses may not capture some of the significant causal variants for phenotypic variation that may be found if the entire population or populations are sampled. Progeny from crosses manipulated by the experimenter are generally reared and phenotyped in a laboratory or common garden, where the effects of environment can be controlled. Controlling the environment is an advantage for the detection of QTL in that trait variance due to environmental effects is minimized, but for some traits, the environment-dependent expression of the trait is an important context for studies interesting to ecologists, evolutionary biologists, and conservation biologists. In contrast, QTL analyses in pedigreed populations take advantage of the larger amount of recombination that has occurred among generations and families of individuals with different phenotypic trait values. Because of this reduced level of LD among phenotypic traits and causal mutations (i.e., gray and white contiguous blocks are shorter in Fig. 6–1) as a function of sampling more meioses in the population, QTL analyses in natural populations will require much larger sample sizes and many more markers to detect the same QTL that may have been observed in line crosses. In QTL analysis, pedigree information is directly included in tests for genotype–phenotype associations and is more accurate in defining shared coancestry among individuals than are methods used in LD mapping to account for kinship (Pemberton 2008). In both approaches, the power and precision to detect and localize loci underlying quantitative traits will depend on the number of markers chosen, the amount of recombination events or LD observed between markers, and the effect size of individual loci (Doerge et al. 1997; Lynch & Walsh 1998; Doerge 2002). Because recombination rates and LD are unique not only to species but also to specific regions of chromosomes, providing a magic number for the number of individuals and markers to choose for genome-wide approaches is not possible. Lynch and Walsh (1998) detail calculations for the numbers of individuals and markers to use in genome-wide QTL analysis with specific QTL effect sizes and desired accuracy of mapping QTL. QTL analyses in crosses made in non-model organisms in the laboratory or common garden environments are numerous, but the use of this approach in natural or free-living populations is limited to systems where the pedigree is known or can be estimated from parentage analysis using molecular markers.

Population genetic approaches

Hitchhiking mapping and outlier analysis
Population genomics is the assessment of population genetic parameters at large numbers of loci distributed across the genome, with the aim of identifying loci that have been shaped by natural selection (Schlotterer 2002, 2003; Luikart et al.

2003; Storz 2005; Thornton et al. 2007). Whereas much of the genome will reflect patterns of neutral genetic variation attributed to mutation and demographic processes, "outlier analysis" identifies loci in the genome showing unusually high or low patterns of variation among populations due to the effects of natural selection. This approach is also called hitchhiking mapping and rests on the idea that strong directional or divergent selection for an advantageous allele creates strong LD with closely linked loci, and that as an advantageous allele approaches fixation, a decrease in heterozygosity at closely linked loci will also be observed (Maynard Smith & Haigh 1974). LD around the beneficial mutation is strongest and more extensive when said mutation is a new mutation immediately acted upon by positive selection; when natural selection shapes standing genetic variation, the extent of LD of the beneficial mutation with unlinked loci will depend on the amount of neutral variation that has accumulated in the region (a function of the effective population size) and recombination (Przeworski et al. 2005). Several tests have been devised for tests of these signatures of natural selection in the genome. In all cases, population genomic tests are most powerful for detection of directional or divergent selection on new mutations, which instantly creates LD at linked neutral sites. Detecting signatures of selection on standing genetic variation is more difficult as diversity about the causal mutation is greater due to neutral evolution prior to the onset of directional selection.

Among all of the approaches reviewed herein, the tools required for population genomics are the simplest: Individuals from populations of interest are genotyped at polymorphic markers (amplified fragment length polymorphism [AFLP], microsatellite, or SNP) across the genome, population genetic parameters are calculated based on allele frequencies within and across sampled populations, and signatures of unusually high or low patterns of genetic diversity within and between populations are revealed with statistical tests. Population genetic parameters used for detection of outlier loci include: 1) F_{st} showing unusually large or small levels of population subdivision compared to most loci sampled (Lewontin & Krakauer 1973; Vitalis et al. 2001; Beaumont & Balding 2004; Beaumont 2005); 2) lnRV, which captures the natural log of the ratio of variance in microsatellite repeat number or allele sizes between populations (Schlotterer 2002); 3) lnRH, which captures the natural log of the ratio of expected heterozygosity between populations (Kauer et al. 2003); and 4) the Ewens–Watterson neutrality test, which tests for excess or deficits in expected homozygosity (Watterson 1977). Tests for outliers are made either empirically by the identification of outliers in the distributions of the population genetics parameters mentioned earlier in text, or by comparing the distribution of these test statistics to distributions of the same statistics generated by coalescent simulations under a model of neutral evolution and particular demographic scenarios (Teshima et al. 2006).

In non-model organisms, particularly organisms for which little or no genomic sequence information exists, the outlier analysis approach is among the easiest to perform as anonymous genetic markers such as AFLPs and microsatellites can readily be used. Moreover, as sequencing costs decline with the rapid development of new sequencing technologies, generation of genomic sequences in non-model organisms will become quite tractable (Ellegren 2008; see also Chapter 4 by DeWoody and colleagues). The outlier analysis approach does not require the collection of phenotype data, a time-consuming and difficult task for particularly complex phenotypes. As with the LD mapping and QTL approaches, the ability to sample all loci in the genome for signatures of natural selection will depend on the extent of LD among closely linked loci. Unfortunately, because outliers can occur as false positives or false negatives owing to the large numbers of tests performed and possible violation of simple assumptions of demography (Simonsen et al. 1995; Teshima et al. 2006), it is necessary to follow up with candidate outliers with additional validation to determine if the region linked to the markers indeed shows patterns of sequence variation consistent with directional selection and is functionally linked to phenotypes or life history traits.

Neutrality tests with sequence or SNP data

For single or few loci, tests of neutrality are based on sequence or SNP data. In some cases, outliers identified in population genomic studies are followed up with tests for signatures of natural selection using sequence information in candidate regions, and, in others, candidate genes are used. These tests can be roughly broken down into three categories: 1) tests within and among populations of the same species ("polymorphism tests"); 2) tests among species ("divergence tests"); and 3) joint tests of population and species level variation ("joint polymorphism and divergence tests") (Nielsen 2001, 2005; Walsh 2008). In all cases, the neutral model of evolution serves as the null hypothesis. For within-species analyses, site frequency spectrum of polymorphisms or haplotype diversity is compared against neutral expectations; examples of these types of tests include Tajima's D and Fu and Li's D and F tests (Nielsen 2001, 2005; Walsh 2008). Population genetic polymorphism tests are subject to strong assumptions about population demography and often have low power compared to divergence and joint tests (Simonsen et al. 1995; Nielsen 2001, 2005; Zhai et al. 2009). Many divergence tests and joint tests evaluate whether nonsynonymous-to-synonymous substitution rates (d_N/d_S) in genes deviate from those expected under neutrality and are not subject to false positives due to demographic processes (Nielsen 2001, 2005; Zhai et al. 2009). One popular joint test is called the McDonald–Kreitman test, which evaluates the d_N/d_S ratio within and between species. Another common joint test is the Hudson–Kreitman–Aguade test, which compares sequence variation within versus between species, with the idea that within- and between-species divergence under neutral expectations will depend only on mutation rate. In most cases, tests for signatures of natural selection using sequence data use multiple tests and approaches to verify whether the null hypothesis of neutrality can be rejected. Nielsen (2005) offers an excellent review of the effects of different scenarios of natural selection (directional and balancing selection, selective sweeps) on within- and between-species variability.

Tests for neutrality on one or few loci have an obvious advantage for non-model organisms and the same advantages as candidate-gene association tests. Sequence data are readily obtainable from non-model species for few candidate loci. One limitation of this approach, as mentioned earlier, includes false rejection of the null, neutrality hypothesis due to violation of assumptions of equilibrium demography when polymorphism-based tests are used. Tests based on divergence are inherently testing hypotheses about strong or repeated selection among species, whereas polymorphism tests within species detect recent selection. Readers are directed to Zhai and colleagues (2009) and Teshima and coworkers (2006) for excellent reviews and simulations of the power of outlier and neutrality tests for testing for signatures of natural selection.

IDENTIFICATION OF GENES UNDERLYING ADAPTIVE TRAITS: EXAMPLES IN PLANTS AND ANIMALS

Although most association genetic studies have been conducted in model organisms, the transfer of these tools to related non-model, non-domesticated, natural or free-living populations has allowed the genetic dissection of ecologically relevant traits, with potentially important implications for conservation and management application. In many cases, the real power in the identification of genes associated with ecologically relevant traits comes from combining these approaches (Vasemagi & Primmer 2005; Neale & Ingvarsson 2008; Stinchcombe & Hoekstra 2008). In the next sections, we give some examples of how these different approaches have been successful in the identification of genes and, in some cases, the causal mutations, responsible for a large proportion of phenotypic or ecotypic variability in non-model or natural or free-living populations of animals and plants.

Genome-wide association and QTL studies in animals

No genome-wide association studies have been conducted in natural or free-living populations of animals, but a few QTL studies have been published for natural or free-living populations of animal species (Table 6–2). Published QTL studies in natural or free-living populations are limited to long-term data sets derived from carefully tracked pedigrees in populations of large mammal species, namely red deer (*Cervus elaphus*; Slate et al. 2002) and Soay sheep (*Ovis aries*; Beraldi et al. 2007a,b). Linkage maps have been developed for several other free-living or natural populations and will serve as an important resource for QTL and LD mapping in those species.

Most genome-wide studies of genotype–phenotype associations for ecologically relevant traits have occurred in crosses manipulated by researchers in the laboratory, using QTL analyses (Table 6–2). These studies have identified QTL for morphological variation, behavior phenotypes (including host preference–mediating ecological speciation and mate preference), disease resistance, and other physiological or life history traits. QTL mapping, as mentioned earlier in text, is not a means to an end and rarely identifies the causal mutation(s) underlying phenotypic variation, but it provides important information on genome regions to further test for associations with phenotypes of interest using LD mapping or in tests for signatures of natural selection. In fact, QTL studies provide an important top-down approach in the identification of candidate regions and gene sets for candidate gene association and tests for natural selection (Tables 6–3 and 6–4, respectively). For example, in lake whitefish, QTL identified for growth and morphological characters in manipulated crosses between pelagic and benthic forms

Table 6–2. Examples of QTL analyses in non-model, non-domesticated animal species of ecological significance

Species	Common name	Morphology	Physiology	Behavioral	Life history & fitness
Amphibians & reptiles					
Ambystoma mexicanum	Mexican axolotl		Voss & Shaffer 2000		
Fish					
Astyanax mexicanus	Cavefish	Protas et al. 2006, 2008; Gross et al. 2009	Protas et al. 2008		
Coregonus clupeaformis	Lake whitefish	Rogers & Bernatchez 2005			
Gasterosteus aculeatus	Three-spined stickleback	Peichel et al. 2001; Colosimo et al. 2004; Shapiro et al. 2004; Albert et al. 2008; Miller et al. 2007			
Labeotropheus fuelleborni × *Metriaclima zebra*	Cichlid spp.	Streelman et al. 2003			
Oncorhynchus mykiss	Rainbow trout	Nichols et al. 2004, 2008; Zimmerman et al. 2005	Jackson et al. 1998; Ozaki et al. 2001; Robison et al. 2001; Martyniuk et al. 2003; Nichols et al. 2003, 2007, 2008; O'Malley et al. 2003; Zimmerman et al. 2004; Perry et al. 2005; Sundin et al. 2005; Drew et al. 2007		Martyniuk et al. 2003; Haidle et al. 2008
Oreochromis mossambicus × *O. aureus*	Tilapia		Cnaani et al. 2004; Moen et al. 2004		
Salmo salar	Atlantic salmon		Houston et al. 2008; Moen et al. 2005; Ozaki et al. 2005; Reid et al. 2005; Moghadam et al. 2007		Moghadam et al. 2007

(continued)

Table 6–2 (continued)

Species	Common name	Morphology	Physiology	Behavioral	Life-history & fitness
Insects					
Acyrthosiphon pisum	Pea aphid			Hawthorne & Via 2001; Via & Hawthorne 2002	Hawthorne & Via 2001; Via & Hawthorne 2002
Aedes aegypti	Mosquito		Zhong et al. 2006		
Anopheles gambiae	Mosquito		Menge et al. 2006		
Anopheles gambiae × *A. arabiensis*	Mosquito				Slotman et al. 2004
Apis mellifera	Honeybee			Rueppell et al. 2004, 2006; Hunt et al. 2007	
Bombus terrestris	Bumblebee		Wilfert et al. 2007a,b		Wilfert et al. 2007b
Culex pipiens × *C. quinquefasciatus*	Mosquito				Mori et al. 2007
Heliconius cydno × *H. pachinus*	Butterfly			Kronforst et al. 2006	
Heloconius melpomene	Butterfly	Baxter et al. 2009			
Laupala paranigra × *L. kohalensis*	Hawaiian cricket			Shaw et al. 2007	
Tribolium castaneum	Red flour beetle		Zhong et al. 2003, 2005		Zhong et al. 2005
Mammals					
Cervus elaphus	Red deer	Slate et al. 2002			Slate et al. 2002
Ovis aries	Soay sheep	Beraldi et al. 2007b	Beraldi et al., 2007a,b		Beraldi et al. 2007b

Table 6–3. Summary of selected candidate-gene association studies in non-model, non-domesticated animal species

Species	Common name	Trait	Candidate genes	Reference
Amphibians & reptiles				
Ambystoma mexicanum	Axolotl	Metamorphic timing	$THR\alpha$, $THR\beta$	Voss et al. 2003
Aspidoscelis inornata	Little striped whiptail	Body color	MC1R	Rosenblum et al. 2004
Thamnophis sirtalis	Garter snakes	Tetrodotoxin resistance	tsNa(V)1.4	Geffeney et al. 2005
Birds				
Acrocephalus arundinaceus	Great reed warbler	Parasite load	MHCI	Westerdahl et al. 2005
Anser c. caerulescens	Snow goose	Plumage color	MC1R	Mundy et al. 2004
Coereba flaveola	Bananaquit	Plumage color	MC1R	Theron et al. 2001
Parus major	Great tit	Personality	DRD4a	Fidler et al. 2007
Passer domesticus	House sparrow	Disease resistance	MHCIIb	Bonneaud et al. 2006
Stercorarius parasiticus	Arctic skuas	Plumage color	MC1R	Mundy et al. 2004
Insects				
Bicyclus anynana	Butterfly	Eyespot size	Distal-less (Dll)	Beldade et al. 2002
Solenopsis invicta	Fire ant	Social & mating system	Gp-9	Ross & Keller 1998; Krieger & Ross 2002
Fishes				
Astyanax mexicanus	Cavefish	Body coloration	MC1R	Gross et al. 2009
Gadus morhua	Atlantic cod	Muscle fiber number; growth and condition; migration behavior	PanI	Johnston & Andersen 2008; Jonsdottir et al. 2008; Pampoulie et al. 2008
Gasterosteus aculetus	Three-spined stickleback	Body armor plates	Eda	Cano et al. 2006; Kitano et al. 2008
Metriaclima zebra	Zebra mbuna cichlid	Body coloration	c-ski[a]	Streelman et al. 2003
Mammals				
Canis lupus	Gray wolf	Coat color	K locus	Anderson et al. 2009
Chaetodipus intermedius	Pocket mice	Coat color	MC1R, Agouti	Nachman et al. 2003
Ovis aries	Sheep	Coat color	MC1R; TYRP1	Deng et al. 2009; Gratten et al. 2007[b]
Peromyscus polionotus	Oldfield mouse	Coat color	MC1R, Agouti	Mullen & Hoekstra 2008
Other invertebrates				
Mya arenaria	Soft shell clam	Paralytic shellfish poisoning resistance	rNav 1.2a	Bricelj et al. 2005

[a] Gene linked to causal mutation.
[b] See also case study in this chapter.

Table 6–4. Examples of population genomics and tests for neutrality in natural populations of animals

Species	Common name	No. of loci or genes studied	Marker type	Reference
Amphibians & reptiles				
Rana spp.	Frogs	1	Candidate gene	Tennessen & Blouin 2008
Rana temporaria	Common frog	Many	AFLP	Bonin et al. 2006
Birds				
Falco naumanni	Lesser kestrel	1	Candidate gene (MHC)	Alcaide et al. 2008
Rupicapra rupicapra	Alpine chamois	1	Candidate gene (MHC)	Mona et al. 2008
Fishes				
Cichlid spp.	Cichlids	Many	SNPs	Loh et al. 2008
Clupea harengus	Atlantic herring	12	Microsatellites	Watts et al. 2008
Coregonus clupeaformis	Lake whitefish	Many	AFLP	Campbell & Bernatchez 2004
Fundulus heteroclitus	Mummichog	1	Candidate gene	Powers & Schulte 1998
Gadus morhua	Atlantic cod	Many, 11	SNPs; microsatellites	Nielsen et al. 2006; Moen et al. 2008
Gasterosteus aculeatus	Three-spined stickleback	15	Microsatellites	Raeymaekers et al. 2007; Makinen et al. 2008a,b; Barrett et al. 2008
		Many	Microsatellites, candidate genes	
		109	Microsatellites, candidate gene	
		1	Microsatellites, candidate gene	
Haplochromine/Tilapiine spp.	Cichlid spp.	82	Genes	Gerrard & Meyer 2007
Haplochromis spp., *Oreochromis niloticus*, *Astatotilapia burtoni*	African cichlids	Many	Candidate genes	Gerrard & Meyer 2007
Oncorhynchus tshawytscha	Chinook salmon	11	Microsatellite, candidate gene	O'Malley et al. 2007
Salmo salar	Atlantic salmon	14	Microsatellites, candidate genes	Vasemagi et al. 2005a,b
		95	Microsatellites	
Telmatherina prognatha	Sailfin silversides	573	AFLP	Herder et al. 2008
Telmatherina antoniae				
Theragra chalcogramma	Walleye pollock	38	Microsatellites, allozymes, candidate gene	Canino et al. 2005

Insects				
Acyrthosiphon pisum	Pea aphid	45	AFLP, EST	Via & West 2008
Apis mellifera	Honeybee	Many	SNPs	Zayed & Whitfield 2008
Colias eurytheme	Butterfly	1	Candidate gene	Wheat et al. 2006
		1	Candidate gene	Watt 1977
Melitaea cinxia	Glanville fritillary butterfly	1	Candidate gene	Orsini et al. 2009
Neochlamisus bebbianae	Leaf beetle	Many	AFLP	Egan et al. 2008
Timema cristinae	Walking sticks	Many	AFLP	Nosil et al. 2008
Zeiraphera diniana	Larch budmoth	Many	AFLP	Emelianov et al. 2004
Mammals				
Arvicola terrestris	Water vole	2	Candidate genes (MHC)	Bryja et al. 2007
Gracilinanus microtarsus and *Marmosops incanus*	American mouse opossums	1	Candidate gene (MHC)	Meyer-Lucht et al. 2008
Oryctolagus cuniculus	European rabbit	25	Allozymes	Campos et al. 2008
Ovis dalli	Wild sheep	3	Candidate genes	Worley et al. 2006
Peromyscus maniculatus	Deer mice	18	Allozymes	Storz & Dubach 2004
Peromyscus maniculatus	Deer mice	2	Candidate genes	Storz & Kelly 2008
Peromyscus polionotus	Oldfield mouse	2	Candidate genes	Mullen & Hoekstra 2008
Peromyscus spp.	Mice	10–37	Allozymes	Storz & Nachman 2003
Peromyscus spp.	Mice	1	Candidate gene	Gering et al. 2009
Other invertebrates				
Mytilus edulis	Mussel	11	Microsatellite	Faure et al. 2008
Crassostrea virginica	Oyster	Many	AFLP	Murray & Hare 2006
Littorina saxatilis	Marine snail	Many	AFLP	Wilding et al. 2001
		Many	AFLP	Galindo et al. 2009
		14	Candidate regions	Wood et al. 2008

colocalize with markers showing signatures of natural selection (i.e., "outlier" behavior) between sympatric pairs of these morphotypes found in several lakes in Quebec (Campbell & Bernatchez 2004; Rogers & Bernatchez 2005). In three-spined sticklebacks, the colocalization of QTL to the ectodysplasin gene provided the foundation for tests of *Eda* polymorphism and signatures of natural selection in natural populations exhibiting variation in lateral plate numbers (Colosimo et al. 2005).

To our knowledge, no genome-wide LD or association mapping studies have yet been conducted in free-living or natural populations of animals. Important first steps for genome-wide association studies, however, include the generation of genome-wide sequence and SNP variation information, as well as examination of the extent of LD within species. In some non-model species, these resources are beginning to emerge; these resources include information on the extent of LD in wild mice (Laurie et al. 2007), red deer (Slate & Pemberton 2007), and collared flycatchers (Backstrom et al. 2006b), as well as linkage maps in wild bird populations including the great reed warbler (Akesson et al. 2007), collared flycatcher (Backstrom et al. 2006a), and zebra finch (Stapley et al. 2008).

Candidate-gene association studies in animals

The candidate-gene approach is by far the approach most used, and holds the most promise for use in non-model, natural populations of organisms. In most cases, candidate genes are chosen based on their known role for particularly morphological, behavioral, physiological, or life history traits in other taxa. Comparative genomics and identification of genes by whole-genome expression studies are also avenues for the identification of candidate genes. In non-model animal species, many candidate-gene studies have been focused on body-color polymorphism (see Protas & Patel 2008 for review) and disease resistance and mate choice as it relates to major histocompatibility complex (MHC) loci polymorphisms (Piertney & Oliver 2006) (Table 6–3). Examples include genetic polymorphism in coat color genes in mice species that are adapted to different environments (Nachman et al. 2003; Steiner et al. 2007), in hemoglobin genes in mice adapted to different altitudes (Storz et al. 2007), in color genes associated with albinism in cavefish (Protas et al. 2006), and in plumage coloration involved in mate choice (Mundy et al. 2004). Although the adaptive significance is not apparent, a gene associated with coat-color polymorphism in Soay sheep (Gratten et al. 2007) has also been identified and appears to segregate in the population with linked fitness-related traits (see Box 6 case study). In fewer cases, candidate genes are identified for further study based on whole-genome approaches such as QTL or association mapping. For example, polymorphism in the *Eda* gene found in QTL for plate morph in three-spined stickleback is associated with plate morphs in wild marine and freshwater populations (Colosimo et al. 2005; Barrett et al. 2008; Kitano et al. 2008). The fact that most candidate genes have been identified outside of whole-genome approaches is more likely due to the extensive resources and time needed to conduct a genome-wide study and to the extensive genomic resources needed to follow up with QTL studies to identify genes within QTL regions.

BOX 6: UNRAVELING COUNTERINTUITIVE EVOLUTIONARY TRENDS: COAT COLOR IN SOAY SHEEP

Jake Gratten, Alastair J. Wilson, Allan F. McRae, Dario Beraldi, Peter M. Visscher, Josephine M. Pemberton, and Jon Slate

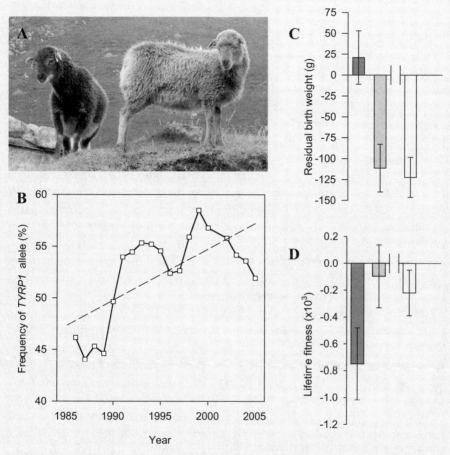

Box Figure 6–1: (a) The two coat-color morphs in Soay sheep. (b) Increase in frequency of the *Tyrp1* T allele over a twenty-year period (linear regression, slope of $+0.49\%$/year, $r^2 = 0.390$, $p = 0.004$). (c) Mean birth-weight differential of GG sheep (dark gray bar) and TT sheep (light gray bar), in each case relative to GT sheep, and of light sheep (white bar) relative to dark sheep. (d) Mean lifetime fitness differential of *Tyrp1* genotypes and coat-color phenotypes.

Background

Evolutionary biologists sometimes report that heritable traits under directional selection fail to evolve as predicted (Merila et al. 2001). A polymorphism for coat color in a wild population of Soay sheep (Box Fig. 6–1, a) is determined by a single nonsynonymous G → T substitution in the tyrosinase-related protein 1 gene (*Tyrp1*); GG homozygotes and GT heterozygotes have dark coats, whereas TT homozygotes are light (Gratten et al. 2007). Dark sheep are heavier than light sheep, and body size is positively correlated with fitness (Wilson et al. 2006). Therefore, it is surprising that the recessive light allele (T), which is not ancestral, has reached a frequency of approximately 0.50 and has not declined in frequency during twenty years of intensive monitoring (Box Fig. 6–1, b). Why does coat color, which has a simple genetic basis, show a counterintuitive evolutionary trajectory?

Case Study

More than 2,500 Soay sheep living between 1985 and 2005 were typed at the *TYRP1* causative mutation, and genotype data were integrated with pedigree, life history, and body size data (Gratten et al. 2008). First, associations between *Tyrp1* and body size were analyzed using an "animal model" approach, whereby polygenic effects on body size were modeled as a random effect, independently of *Tyrp1* genotype (fixed effect). Similar models were constructed with coat-color phenotype instead of *Tyrp1* genotype. Both color ($F_{1,2201.5} = 26.03$, $p < .0001$, $n = 2,370$) and *Tyrp1* ($F_{2,1623.1} = 8.96$, $p = .0001$, $n = 1,757$) explained significant variation in birth weight (Box Fig. 6–1, c) and body size later in life. Dark sheep were heavier than light sheep, and the G allele was partially dominant for body weight (such that GG $\geq$ GT > TT). These findings were supported by a transmission disequilibrium test (TDT: $F_{1,421} = 4.60$, $p = .034$), a form of combined association and linkage mapping that eliminates possible causes of spurious association such as undetected population structure or admixture. Thus, the relationship between body size and *Tyrp1* is due to genetic linkage, and dark sheep really are expected to be fitter than light sheep, all else being equal.

Next, associations between *Tyrp1*/coat color and fitness were analyzed using animal models and TDTs. Color was not associated with lifetime fitness, despite the fact that dark sheep were heavier than light sheep. *Tyrp1* genotype was associated, however, with lifetime fitness (Box Fig. 6–1, d; animal model: $F_{2,1336} = 4.03$, $p = .020$, $n = 1,355$). It is intriguing that fitness differences between *Tyrp1* genotypes were not those predicted by effects on body size. There was a cryptic difference between phenotypically indistinguishable homozygous (GG) and heterozygous (GT) dark sheep, with GG sheep being less fit than either GT dark sheep or TT light sheep. This association was also confirmed to be due to linkage (TDT; $F_{1,427} = 6.87$, $p = .010$, $n = 492$).

What do these results mean? The *Tyrp1* gene is associated with both body size and fitness, either directly or because it is in LD with tightly linked genes that affect the focal traits. These genes appear to act antagonistically because GG sheep are large but less fit, TT sheep carry alleles that confer small body size (but greater fitness), and GT sheep are relatively large and fit. Although body size is under directional selection, a localized negative genetic correlation in the vicinity of *Tyrp1* means that large body size alleles in this part of the genome are associated with decreased fitness. Overall, sheep carrying the T allele, irrespective of their size, are favored, which may explain why T has increased in frequency. These associations were able to be detected only when coat color genotype rather than phenotype was analyzed, due to the cryptic fitness difference between the two categories of dark sheep. The lessons from this study are that 1) an evolutionary response to selection can be modulated by local genetic correlations between linked genes, and 2) studying the underlying genotype of a trait may be necessary to understand its evolutionary dynamics. Although the conservation genetic implications of this research are less immediate, the work does illustrate the fact that making management decisions on the basis of one trait may have unpredictable consequences on other genetically correlated traits.

REFERENCES

Gratten J, Beraldi D, Lowder BV et al. (2007) Compelling evidence that a single nucleotide substitution in *TYRP1* is responsible for coat-colour polymorphism in a free-living population of Soay sheep. *Proceedings of the Royal Society of London-Series B: Biological Sciences*, **274**, 619–626.

Gratten J, Wilson AJ, McRae R et al. (2008) A localized negative genetic correlation constrains microevolution of coat color in wild sheep. *Science*, **319**, 318–320.

Merila J, Sheldon BC, Kruuk LEB (2001) Explaining stasis: microevolutionary studies in natural populations. *Genetica*, **112**, 199–222.

Wilson AJ, Pemberton JM, Pilkington JG et al. (2006) Environmental coupling of selection and heritability limits evolution. *PLoS Biology*, **4**, 1270–1275.

Tests for signatures of natural selection in animals

In natural animal populations, tests for signatures of natural selection have identified outlier loci associated with ecological specialization, speciation, and adaptation in a wide range of species (Table 6–4). Tests for signatures of natural selection at loci throughout the genome ("population genomics") have been one of the most rapidly developing areas for the identification of genes underlying adaptive population divergence. Because population genomics requires multiple testing and has the potential for the identification of false positives and negatives under certain selection and demographic scenarios (see *Hitchhiking mapping and outlier analysis*), the best supported evidence for true positives in outlier tests are those loci that can also be validated by linkage to QTL or by follow-up studies evaluating the identity of genes and nature of sequence variation functionally associated with traits of interest. For example, outliers identified by population genomics in sympatric ecotypes of lake whitefish are often colocalized to QTL regions for morphological and physiological differences among ecotypes (Campbell & Bernatchez 2004; Rogers & Bernatchez 2005). Outliers identifying footprints of selection in three-spined stickleback colocalize to QTL for morphotypic variation in that species (Makinen et al. 2008a). Anonymous AFLP marker loci identified as outliers among parapatric populations of *Littorina* gastropods (Wilding et al. 2001) have been used to isolate genomic sequence for finer-level sequence analysis of genes and genome regions under natural selection (Wood et al. 2008).

Genome-wide association and QTL studies in plants

Large-scale candidate gene or genome-wide association studies in plants have, until recently, been restricted to the model plant *Arabidopsis thaliana* (Zhao et al. 2007) or the domesticated plant *Zea mays* ssp. *mays* (Yu & Buckler 2006). There is an extensive literature on QTL mapping in forest trees, however (Table 6–5). We classify trees as plants from natural population versus domesticated plants because in just about every case, forest-tree QTL mapping studies begin with

Table 6–5. Summary of QTL analyses conducted in non-model, non-domesticated plant species

Scientific name	Common name	Forest Trees					
		Growth	Phenology	Disease resistance	Cold hardiness	Drought tolerance	Wood property
Castanea sativa	Sweet chestnut	Casasoli et al. 2004, 2006	Casasoli et al. 2004				
Cryptomeria japonica	Japanese cryptomeria	Yoshimaru et al. 1998					Kuramoto et al. 2000
Eucalyptus globulus	Tasmanian blue gum	Marques et al. 2002; Kirst et al. 2004; Thamarus et al. 2004; Bundock et al. 2008	Bundock et al. 2008	Bundock et al. 2008			Bundock et al. 2008
Eucalyptus grandis × *Eucalyptus urophylla*	Grand eucalyptus × Timor mountain gum	Grattapaglia et al. 1995, 1996; Verhaegen et al. 1997; Marques et al. 2002; Missiaggia et al. 2005					Grattapaglia et al. 1996; Byrne et al. 1997a,b; Verhaegen et al. 1997; Kirst et al. 2005
Eucalyptus nitens	Shining gum	Byrne et al. 1997a			Byrne et al. 1997b		
Eucalyptus tereticornis	Forest red gum	Marques et al. 2002					
Fagus sylvatica	European beech	Scalfi et al. 2004					
Larix decidua × *Larix kaempferi*	European larch × Japanese larch						Arcade et al. 2002
Pinus caribaea × *Pinus elliottii*	Caribbean pine × Slash pine	Shepherd et al. 2006					
Pinus elliottii × *Pinus palustris*	Slash pine × Longleaf pine	Shepherd et al. 2006					Shepherd et al. 2003

Species	Common name						
Pinus pinaster	Maritime pine	Plomion et al. 1996; Brendel et al. 2002; Chagne et al. 2003					Markussen et al. 2003; Pot et al. 2006
Pinus radiata	Monterey pine	Emebiri et al. 1997, 1998a,b					Kumar et al. 2000; Devey et al. 2004
Pinus sylvestris	Scots pine	Lerceteau et al. 2000	Hurme et al. 1997, 2000	Lind et al. 2007	Hurme et al. 1997		
Pinus taeda	Loblolly pine	Kaya et al. 1999; Chagne et al. 2003; Gwaze et al. 2003; Williams et al. 2007					Weng et al. 2002
Populus davidiana	Shan Yang	Kim et al. 2004					
Populus tementosa	Chinese white poplar	Zhang et al. 2006					Zhang et al. 2006
Populus trichocarpa × Populus deltoides	Black cottonwood × Eastern cottonwood	Wu & Stettler 1994; Bradshaw & Stettler 1995; Wu et al. 1997, 1998; Wu 1998; Li et al. 1999; Ferris et al. 2002; Wullschleger et al. 2005; Rae et al. 2006, 2007, 2008	Bradshaw & Stettler 1995; Li et al. 1999; Chen et al. 2002; Frewen et al. 2000	Newcombe & Bradshaw 1996; Newcombe et al. 1996; Jorge et al. 2005; Tagu et al. 2005		Tschaplinski et al. 2006	
Pseudotsuga menziesii	Douglas-fir	Jermstad et al. 2003	Jermstad et al. 2001a		Jermstad et al. 2001a; Wheeler et al. 2005		
Quercus petraea	Sessile oak	Gailing et al. 2005					Saintagne et al. 2004

(continued)

Table 6-5 (continued)

Forest Trees

Scientific name	Common name	Growth	Phenology	Disease resistance	Cold hardiness	Drought tolerance	Wood property
Quercus robur	English oak	Scotti-Saintagne et al. 2004a, 2005; Casasoli et al. 2006	Scotti-Saintagne et al. 2004a; Gailing et al. 2005			Parelle et al. 2007; Brendel et al. 2008	Saintagne et al. 2004
Salix dasyclados × *Salix viminalis*	Mao Zhi Liu × Basket willow	Ronnberg-Wastljung et al. 2005; Weih et al. 2006	Ronnberg-Wastljung et al. 2005; Weih et al. 2006			Ronnberg-Wästljung et al. 2005	
Salix viminalis × *Salix schwerinii*	Basket willow × Common Osier	Tsarouhas et al. 2002, 2003, 2004	Tsarouhas et al. 2002, 2003, 2004			Tsarouhas et al. 2004	

Herbaceous Plants

Scientific name	Common name	Fitness	Floral morphology	Herbivory	Heavy metal tolerance
Aquilegia formosa × *Aquilegia pubescens*	Western columbine × Sierra columbine		Hodges et al. 2002		
Arabidopsis halleri	N/A				Courbot et al. 2007; Willems et al. 2007
Arabidopsis lyrata	Lyre-leaved rock-cress			Heidel et al. 2006	
Iris fulva × *Iris brevicaulis*	Copper iris × Zigzag iris	Martin et al. 2005, 2006	Bouck et al. 2007		
Mimulus	Monkeyflower	Lin 2000; Hall et al. 2006	Lin & Ritland 1997; Schemske & Bradshaw 1999; Lin 2000; Bleiweiss 2001; Fishman et al. 2002; Hall et al. 2006		

mapping population parent trees that have not resulted from any more than one generation of phenotypic selection from natural populations. The number of studies from herbaceous, natural plant populations is much less extensive (Table 6–5). The complex traits of study in forest-tree QTL mapping studies fall into six broad categories (growth, phenology, disease resistance, cold hardiness, drought, and wood property). All of these traits can be considered "adaptive," although growth and wood property are generally considered "agronomic" and not specifically "adaptive." Early-generation QTL mapping studies in trees often used rather small population sizes (~100), in which the number of QTLs detected was likely underestimated and the size of effects overestimated. Later studies using population sizes of 500 or more probably provide better estimates of QTL number and effect.

QTL mapping studies in forest trees share many of the same approaches and results. Mapping population parent trees are nearly always highly heterozygous and not inbred. Using a highly heterozygous, non-inbred population results in not all QTL loci segregating, and thus being detectable in individual crosses. Nevertheless, a large proportion of the total phenotypic variance for a trait can be accounted for from individual crosses, although the sizes of individual QTL effects are generally small (1–3%) (Wheeler et al. 2005). The QTL approach is rather powerful for identifying the number of QTLs, their chromosomal regions, and the sizes of their effects; however, the resolution of map position is generally quite crude (10–20 cM), so for large genomes lacking reference sequences, the path to positional cloning of QTLs is long, expensive, and not easily justified. Therefore, the genes underlying adaptive-trait QTLs in forest trees remain unknown.

The situation in herbaceous, natural plant QTL mapping is somewhat different (Table 6 5). Here, the traits of interest are often those leading to speciation events such as floral morphology, and thus hybrid crosses are used to maximize QTL segregation. The number of QTLs for such traits is generally quite few, and the sizes of their individual effects are high, justifying positional cloning of such QTLs that will now be greatly facilitated by the genome sequencing *Arabidopsis lyrata*, *Aquilegia*, and *Mimulus*. Species such as *A. lyrata* and *Boechera stricta* will be good systems for discovering individual genes underlying complex adaptive traits using combined population genetic, QTL mapping, and association approaches.

Candidate-gene association studies in plants

Association studies in natural plant systems are candidate-gene–based due to the lack of reference genome sequences. Studies have been published for four forest-tree species and two herbaceous species (Table 6–6). The studies in *Populus*, *Eucalyptus*, and *A. lyrata* included only one candidate gene each, whereas the *Pinus*, *Pseudotsuga*, and *Zea mays* ssp. *mays* included many candidate genes each. All species are characterized by a rapid decay of LD, particularly the conifer species (Neale & Savolainen 2004), so the search for associations is challenging, but when an association is found, it is quite likely that the polymorphism is within the gene determining the complex trait (or at least closely linked). Full gene-space candidate-gene association studies are experimentally and economically tractable

Table 6–6. Examples of candidate-gene association analyses in plant species

Species	Common name	Trait	Candidate genes	Reference
Trees				
Eucalyptus nitens	Shining gum	Wood properties	*CCR*	Thumma et al 2005
Pinus taeda	Loblolly pine	Drought tolerance	*dhn-1, dhn-2, lp3-1, wrky-like, sod-chl*	Gonzalez-Martinez et al. 2008
		Wood properties	*cad, sams-2, comt-2, dhn-2, lp3–3, 4cl, ccr-1, α-tubulin, ccoaomt-1, agp-6, agp-like, c3h-1, c4h-1, c4h-2, cesA3β*	Gonzalez-Martinez et al. 2007
Pseudotsuga menziesii	Douglas-fir	Growth phenology cold-tolerance	*60s RPL31a*, CN639236.1 (guanine nucleotide-binding protein), ES421311.1 (hypothetical protein), Pm_CL783Contig1 (SOUL heme-binding family protein), *4CL1, LEA-EMB11*, CN637339.1 (hypothetical protein), CN638489.1 (α-expansin), sSPcDFD040B03103 (MADS-box transcription factor), CN637306.1 (MYB-like transcription factor), *f3h2*, Pm_CL234Contig1 (rab GTPase)	Eckert et al., 2009b
Herbaceous plants				
Arabidopsis lyrata	Lyre-leaved rock-cress	Herbivory	*GL1*	Kivimaki et al. 2007
Zea mays ssp. *parviglumis*	Balsas teosinte	Domestication	*d8, id1, tb1, te1, ts2, zap1, zen1, zfl2, ba1, elm1, ids1, ra1, ra2, su1, tb1, te1, td1, zagl1, zfl1, zfl2, ZmCIR1, ZmGI*	Weber et al. 2007, 2008

to perform with current generation sequencing and SNP genotyping technologies, and a large number of studies in a variety of forest-tree and herbaceous-plant species are now underway.

Tests for signatures of natural selection in plants

Population genetic approaches (i.e., tests of neutrality and outlier analysis) have been applied to large numbers of genes in the model plant *Arabidopsis thaliana* and in the domesticated crop (*Zea mays* ssp. *mays*). In a review by Wright and Gaut (2005), it is reported that as many as 20% of the genes may be under some form of selection, although that number is likely an overestimate. In natural plant populations, there are fewer studies and few genes have been evaluated (Table 6–7). Early resequencing studies in which tests of neutrality were performed included

Table 6–7. Examples of tests of neutrality or outlier analysis in natural plant populations

Species	Common name	No. of genes studied	Marker type	Reference
Trees				
Abies kawakamii	Kawakami fir	1	SNP, microsatellites, cDNA	Shih et al. 2007
Betula pendula	European white birch	2	Microsatellites	Jarvinen et al. 2003
Cathaya argyrophylla	Yin Shan	8	SNP, mitochondrial DNA	Wang & Ge 2006
Cryptomeria japonica	Sugi	7	SNP	Kado et al. 2003
Cunninghamia konishii	China fir	1	SNP	Hwang et al. 2003
Cunninghamia lanceolata	China fir	1	SNP	Hwang et al. 2003
Picea abies	Norway spruce	1, 22	SNP	Guillet-Claude et al. 2004; Heuertz et al. 2006
Picea glauca	White spruce	47	SNP	Namroud et al. 2008
Picea mariana	Black spruce	2	SNP	Guillet-Claude et al. 2004
Pinus lambertiana	Sugar pine	1	SNP	Jermstad et al. 2006
Pinus pinaster	Maritime pine	8	SNP	Pot et al. 2005
Pinus radiata	Monterey pine	8	SNP	Pot et al. 2005
Pinus sylvestris	Scots pine	1, 2, 14	SNP	Dvornyk et al. 2002; Garcia-Gil et al. 2003; Wachowiak et al. 2009
Pinus taeda	Loblolly pine	19, 18	SNP	Brown et al. 2004; Gonzalez-Martinez et al. 2006a
Populus tremula	European aspen	1, 5, 1	SNP	Ingvarsson 2005; Ingvarsson et al. 2006; Garcia & Ingvarsson 2007
Pseudotsuga menziesii	Douglas-fir	18, 121	SNP	Krutovsky & Neale 2005; Eckert et al. 2009a
Quercus petraea	Durmast oak	2	Microsatellites, SCARS, AFLP	Scotti-Saintagne et al. 2004b
Quercus robur	English oak	2	Microsatellites, SCARS, AFLP	Scotti-Saintagne et al. 2004b
Taxodium distichum	Bald cypress	4	SNP	Kado et al. 2006
Herbaceous plants				
Helianthus annuus	Sunflower	9	SNP	Liu & Burke 2006
Hordeum vulgare ssp. *spontaneum*	Barley	1, 9, 18, 877	SNP	Morrell et al. 2003, 2005; Rostoks et al. 2005; Jones et al. 2008
Oryza rufipogon and *Oryza nivara*	Rice	1, 10	SNP	Wang et al. 2007; Zhu et al. 2007
Persea americana	Avocado	4	SNP	Chen et al. 2008
Solanum ssp.	Tomato	8, 14	SNP	Roselius et al. 2005; Arunyawat et al. 2007

SCARS = sequence characterized amplified regions

one to no more than twenty genes. Based on the small sample of genes, it was not possible to gain an estimate of what proportion of these genomes might be under selection. Recent studies (Eckert et al. 2009a; Song et al. 2009) have reported neutrality tests for nearly 100 or more genes. These studies, combined with the earlier studies, suggest that approximately 10% of the genes may be under selection (Neale 2007). Thus, candidate-gene resequencing and tests of neutrality are efficient approaches toward identifying candidate genes for association studies that might underlie complex adaptive traits in natural plant populations. Furthermore, there is often a functional basis for candidate genes underlying a complex trait (Gonzalez-Martinez et al. 2006b; Eckert et al. 2009b). With the exception of *Populus trichocarpa* and *A. lyrata*, plants from natural populations lack a reference genome sequence to facilitate gene resequencing, although many have fairly rich expressed sequence tag (EST) databases. The newest generation of sequencing technologies makes it experimentally and economically possible to resequence large numbers of genes from natural plant systems.

The outlier approach has been applied to only a couple of natural plant populations to identify candidate genes (Scotti-Saintagne et al. 2004b; Namroud et al. 2008). The oak (Scotti-Saintagne et al. 2004b) and spruce (Namroud et al. 2008) studies identified 12% and 14% outlier loci, respectively. These percentages are consistent with estimates from neutrality testing of the proportion of genes under selection.

CASE STUDY: QTL, ASSOCIATION GENETICS, AND TESTS FOR NATURAL SELECTION IN A NATURAL FOREST-TREE POPULATION

As an animal example is provided in the boxed case study within this chapter, here we provide another example using the forest tree, Douglas-fir, as a case study for how combined population and quantitative genetic approaches can be used to discover the genes underlying a complex adaptive trait in a non-model and non-domesticated plant. Douglas-fir is a long-lived, woody perennial with limited genetic resources; it is not an organism that generally would be thought of as having attributes for easy identification of the genes underlying a complex adaptive trait. We show, however, that the combined population and quantitative genetic approaches we have outlined in this chapter can be applied to an organism such as Douglas-fir and how the knowledge derived can be applied in resource management strategies to help mitigate the impacts of climate change.

The adaptive complex traits of interest were bud phenology and cold-hardiness. Douglas-fir has a broad and ecologically diverse habitat in western North America. There is an extensive literature on the genetics of phenology and cold-hardiness in Douglas-fir based on a common garden approach (Campbell & Sorensen 1979; Aitken & Adams 1996, 1997; Rehfeldt 1997; Anekonda et al. 2000; St. Clair et al. 2005; St. Clair 2006). These studies clearly demonstrate the genetic control (high heritability) and adaptive patterns of variation across complex ecological landscapes. We surmised that phenology and cold-hardiness in Douglas-fir might then be good target complex adaptive traits to apply population and quantitative genetics approaches to finding the underlying genes.

The first step was to apply QTL mapping. A three-generation outbred pedigree was constructed, and the clonally propagated F$_2$ offspring were planted at two different test-site locations (Jermstad et al. 2001a,b). A restriction fragment length polymorphism (RFLP) linkage map was constructed (Jermstad et al. 1998), and the progeny were evaluated for bud phenology and cold-hardiness. Several QTLs for each of these traits were detected and mapped. Because the size of the segregating population was relatively small, however, it was likely that some QTLs were undetected and the sizes of individual QTL effects were overestimated. The parent trees were then re-mated to develop a much larger (~500) clonally replicated F$_2$ segregating population. In this experiment, however, the progeny were grown under experimental treatment conditions so that specific environmental cues (winter chill, spring heat sum, photoperiod, and moisture stress treatments) by QTL interactions could be estimated (Jermstad et al. 2003; Fig. 6–2). The goal of this aspect of the experiment was to identify QTLs interacting with specific cues from the environment and thus potentially giving clues as to the specific gene underlying the QTL. These QTL mapping experiments provided the first indications of the number of QTLs affecting bud phenology and cold-hardiness in Douglas-fir and their approximate locations in the genome, but the low-level resolution of their map position provided little indication of the specific genes underlying the QTL. A small number of candidate genes were mapped to the QTL maps, but again the resolution was rather crude (Wheeler et al. 2005).

In the next phase, the population-genomics approach was used to help identify candidate genes for cold-hardiness. In two studies, lists of 18 candidate genes (Krutovsky & Neale 2005) and 121 candidate genes (Eckert et al. 2009a) were developed based primarily on their function in *A. thaliana*. Amplicons from these candidate genes were resequenced in a small ($n = 24$) diversity panel to discover SNPs. The sequence polymorphism database developed from resequencing could then be used to estimate measures of nucleotide diversity and divergence and perform tests of neutrality. From these tests, six genes departed from neutrality and revealed signatures of selective sweeps (Table 6–8; Eckert et al. 2009a). In the next phase, these genes and others were tested for association with bud phenology and cold-hardiness to provide the quantitative genetic line of evidence that the genes underlying adaptive trait QTLs are now known.

An association mapping study was designed to test for association between SNPs in 117 candidate genes, including the 6 genes identified from the population-genomics approach (Table 6–8), and 21 adaptive-trait phenotypes, including bud phenology and cold-hardiness (Eckert et al. 2009b). An association population of 700 open-pollinated families from Douglas-fir trees sampled throughout the states of Washington and Oregon was assembled. Progeny from these families were grown in a randomized common garden, and all 21 phenotypes were evaluated. A maternal breeding value was estimated for each trait and each family. Next, an Illumina GoldenGate genotyping chip was designed that contained 384 SNPs from the 117 candidate genes. All 700 mother trees were genotyped for all 384 SNPs. The phenotype–genotype data set used included 21 traits and 228 high-quality SNP genotypes. A general linear model was used to test for associations between SNPs and the traits measured, and 30 significant associations were found (Eckert et al. 2009b). There were not, however, any

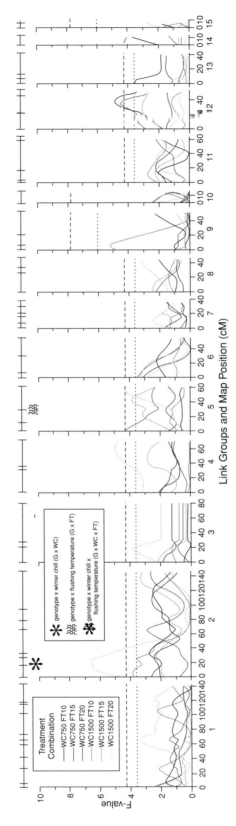

Figure 6–2: Terminal bud flush (TBF$_{GI}$) was used as a surrogate trait to measure growth initiation in the spring. The overwintering bud is released from dormancy and growth is initiated. Seven QTLs for TBF were detected in the growth initiation experiment (TBF$_{GI}$). QTLs were found on six linkage groups (LGs 2, 3, 4, 5, 12, and 4) and were detected in five of the six treatment (T) combinations. Only two QTL × T interactions were found, one for winter chill on LG2 and one for flushing temperature on LG5. The interaction detected on LG5 is located at a marker that is intermediate between two QTLs detected by interval mapping (from Jermstad et al. 2003). *See Color Plate 7ii.*

Table 6–8. A list of candidate genes putatively affected by directional natural selection (from Eckert et al. [2009a])

Locus	Gene product	Result[a]
Compound DHEW test		
Pm_CL908Contig1	GRAM-containing/ABA-responsive protein	$p_D = .001, p_H < .001, p_{EW} = .080$
ES420171.1	Cold-regulated plasma membrane protein	$p_D = .009, p_H = .050, p_{EW} = .035$
ES420250.1	Dehydrin-like protein	$p_D = .072, p_H = .083, p_{EW} = .042$
CN634517.1	Lumenal-binding protein	$p_D = .034, p_H = .148, p_{EW} = .076$
Polymorphism-to-divergence		
Pm_CL61Contig1	Cyclosporin A-binding protein	$k = 0.32$
Pm_CL908Contig1	GRAM-containing/ABA-responsive protein	$k = 0.58$
CN638556.1	Transcription regulation protein	$k = 0.41$
Synonymous-to-nonsynonymous divergence		
Pm_CL922Contig1	Thaumatin-like protein	Ka/Ks = 14.48, $\theta_\pi / D_{xy} = 0.087$
CN634677.1	LRR receptor-like protein kinase	Ka/Ks = 10.78, $\theta_\pi / D_{xy} = 0.066$

[a] Results for the *DHEW* test are given as *p* values for each of the component tests (D = Tajima's D; H = Fay and Wu's H; EW = Ewen–Watterson test) comprising the joint test. Values for the EW test are one minus the left-tailed probabilities (cf.). Listed are *p* values for drift within a constant size population. Loci were significant when demographic models included in the simulations are bolded. For polymorphism-to-divergence tests, parameter estimates for a maximum likelihood implementation of the Hudson-Kreitman-Aguadé test are listed. Estimates of k are from a nested model where all three putative targets of selection are allowed to have free parameters. The parameter k specifies the level of elevation ($k > 1$) or reduction ($k < 1$) in diversity relative to divergence. Ka/Ks values were considered extreme when greater than 5.

genes in common between the population genomic approach and the association approach. This study was based on just 121 genes, so that when it was repeated with a large number of genes, one would expect to find many genes in common between approaches. These genes would be those most likely to be underlying complex adaptive traits and be under natural selection in populations of Douglas-fir.

THE GENES OF ADAPTIVE DIFFERENTIATION: UTILITY FOR CONSERVATION AND MANAGEMENT

Although genetics has historically been used to infer relationships among populations and species from "neutral" genetic information, adding information regarding the genetic architecture and genes involved in adaptive phenotypic diversification has great promise for conservation and management of natural, free-living populations. Bonin and colleagues (2007) describe a new index of population adaptation using results from population-genomic approaches and have found that diversity estimates from neutral and adaptive sets of loci are uncorrelated and tell different stories about the standing genetic diversity within and between populations. The idea that neutral and adaptive indices of diversity show different patterns is not new, however, and is an important consideration for the future of conservation genetics (Crandall et al. 2000; Merila & Crnokrak 2001; Reed & Frankham 2003; Kohn et al. 2006; Leinonen et al. 2008). The goal of most conservation programs for wild populations of organisms has been to maintain

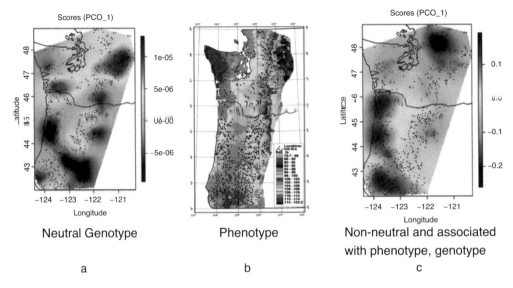

Figure 6–3: Patterns of neutral genetic diversity (a), phenotypic diversity (b), and adaptive genetic diversity (c). *See Color Plate IX.*

genetic diversity, while preserving local adaptations (Moritz 2002; Bonin et al. 2007). In other words, most conservation programs recognize both the phenotypic differences among populations and the demographic processes revealed by population genetic analysis in evaluating long-term population viability and delineating units for conservation (see Moritz 2002 for a review).

One of the key components required for the maintenance of genetic diversity is in obtaining baseline information about the genetic diversity in species and populations of interest so that in the face of anthropogenic impacts and environmental change, the influences of this change on genetic diversity may be monitored (Schwartz et al. 2007; Hoffmann & Willi 2008). Many natural plant and animal populations are threatened by the effects of environmental change. Populations that are currently adapted to a geographic region may no longer be adapted to that location due to changes in temperature, moisture availability, and other environmental factors. It is therefore important to develop detailed and precise descriptions of standing adaptive genetic variation in plant and animal populations so that monitoring activities can be implemented to detect genetic changes in populations. Incorporating genetic information from candidate regions associated with adaptive traits has been historically difficult, as the information has simply not been available for non-model species; however, this trend is changing as studies begin to reveal the genes and shifts in allele frequencies at those genes in response to environmental changes (Hoffmann & Willi 2008). Monitoring the changes in allele frequencies of genes underlying adaptive phenotypes, after being identified, is relatively straightforward. In an example from the Douglas-fir case study earlier in text, the patterns of diversity within non-neutral, phenotype-associated candidate genes show significant similarity to the patterns of phenotypic variation (Fig. 6–3). In contrast, there is little similarity in patterns of variation between neutral genetic variation and phenotypic variation in this system. It can be imagined how land managers might use geographic information system (GIS)-type applications to lay standing patterns of adaptive genetic variation over predicted environmental patterns (*sensu* Joost et al. 2007, temperature, moisture, etc.) and to develop strategies for assisted migration of genotypes to ensure adaptation in the face of climate change. It is clear that the population and quantitative genomic approaches to understanding adaptive genetic variation in natural plant and animal populations will be of great value in genomically assisted gene-resource conservation and management strategies to mitigate the negative effects of environmental change.

The inclusion of genes underlying adaptive phenotypes will become imperative in conservation genetics, but much work remains on the details of which and how many "adaptive" loci to include in conservation genetic analyses. Hoffman and Willi (2008) review recent theoretical advances in this area and suggest that using loci that explain more than 5% of the phenotypic variation within and among populations will be useful in identifying shifts in allele frequencies in response to environmental change. Just as research on the number of loci and alleles per loci has been important in population genetics using "neutral" loci, the selection of genetic loci that contribute to a significant portion of the phenotypes describing differences among individuals within and across populations and species boundaries will be an active area of ongoing and future research.

REFERENCES

Aitken SN, Adams WT (1996) Genetics of fall and winter cold hardiness of coastal Douglas-fir in Oregon. *Canadian Journal of Forest Research-Revue Canadienne De Recherche Forestiere*, **26**, 1828–1837.

Aitken SN, Adams WT (1997) Spring cold hardiness under strong genetic control in Oregon populations of *Pseudotsuga menziesii var. menziesii. Canadian Journal of Forest Research-Revue Canadienne De Recherche Forestiere*, **27**, 1773–1780.

Akesson M, Hansson B, Hasselquist D, Bensch S (2007) Linkage mapping of AFLP markers in a wild population of great reed warblers: importance of heterozygosity and number of genotyped individuals. *Molecular Ecology*, **16**, 2189–2202.

Albert AYK, Sawaya S, Vines TH et al. (2008) The genetics of adaptive shape shift in stickleback: pleiotropy and effect size. *Evolution*, **62**, 76–85.

Alcaide M, Edwards SV, Negro JJ, Serrano D, Tella JL (2008) Extensive polymorphism and geographical variation at a positively selected MHC class IIB gene of the lesser kestrel (*Falco naumanni*). *Molecular Ecology*, **17**, 2652–2665.

Anderson TM, vonHoldt BM, Candille SI et al. (2009) Molecular and evolutionary history of melanism in North American gray wolves. *Science*, **323**, 1339–1343.

Anekonda TS, Adams WT, Aitken SN et al. (2000) Genetics of cold hardiness in a cloned full-sib family of coastal Douglas-fir. *Canadian Journal of Forest Research-Revue Canadienne De Recherche Forestiere*, **30**, 837–840.

Arcade A, Faivre-Rampant P, Paques LE, Prat D (2002) Localisation of genomic regions controlling microdensitometric parameters of wood characteristics in hybrid larches. *Annals of Forest Science*, **59**, 607–615.

Arunyawat U, Stephan W, Stadler T (2007) Using multilocus sequence data to assess population structure, natural selection, and linkage disequilibrium in wild tomatoes. *Molecular Biology and Evolution*, **24**, 2310–2322.

Backstrom N, Brandstrom M, Gustafsson L et al. (2006a) Genetic mapping in a natural population of collared flycatchers (*Ficedula albicollis*): conserved synteny but gene order rearrangements on the avian Z chromosome. *Genetics*, **174**, 377–386.

Backstrom N, Ovarnstrom A, Gustafsson L, Ellegren H (2006b) Levels of linkage disequilibrium in a wild bird population. *Biology Letters*, **2**, 435–438.

Balding DJ (2006) A tutorial on statistical methods for population association studies. *Nature Reviews Genetics*, **7**, 781–791.

Barrett RDH, Rogers SM, Schluter D (2008) Natural selection on a major armor gene in threespine stickleback. *Science*, **322**, 255–257.

Baxter SW, Johnston SE, Jiggins CD (2009) Butterfly speciation and the distribution of gene effect sizes fixed during adaptation. *Heredity*, **102**, 57–65.

Beaumont MA (2005) Adaptation and speciation: what can F st tell us? *Trends in Ecology & Evolution*, **20**, 435–440.

Beaumont MA, Balding DJ (2004) Identifying adaptive genetic divergence among populations from genome scans. *Molecular Ecology*, **13**, 969–980.

Beldade P, Brakefield PM, Long AD (2002) Contribution of distal-less to quantitative variation in butterfly eyespots. *Nature*, **415**, 315–318.

Beraldi D, McRae AF, Gratten J et al. (2007a) Quantitative trait loci (QTL) mapping of resistance to strongyles and coccidia in the free-living Soay sheep (*Ovis aries*). *International Journal for Parasitology*, **37**, 121–129.

Beraldi D, McRae AF, Gratten J et al. (2007b) Mapping quantitative trait loci underlying fitness-related traits in a free-living sheep population. *Evolution*, **61**, 1403–1416.

Bleiweiss R (2001) Mimicry on the QT(L): genetics of speciation in *Mimulus*. *Evolution*, **55**, 1706–1709.

Blouin MS (2003) DNA-based methods for pedigree reconstruction and kinship analysis in natural populations. *Trends in Ecology & Evolution*, **18**, 503–511.

Bonin A, Nicole F, Pompanon F, Miaud C, Taberlet P (2007) Population adaptive index: a new method to help measure intraspecific genetic diversity and prioritize populations for conservation. *Conservation Biology*, **21**, 697–708.

Bonin A, Taberlet P, Miaud C, Pompanon F (2006) Explorative genome scan to detect candidate loci for adaptation along a gradient of altitude in the common frog (*Rana temporaria*). *Molecular Biology and Evolution*, **23**, 773–783.

Bonneaud C, Perez-Tris J, Federici P, Chastel O, Sorci G (2006) Major histocompatibility alleles associated with local resistance to malaria in a passerine. *Evolution*, **60**, 383–389.

Bouck A, Wessler SR, Arnold ML (2007) Qtl analysis of floral traits in Louisiana Iris hybrids. *Evolution*, **61**, 2308–2319.

Bradshaw HD Jr, Stettler RF (1995) Molecular genetics of growth and development in populus. IV. Mapping QTLs with large effects on growth, and phenology traits in a forest tree. *Genetics*, **139**, 963–973.

Brendel O, Le Thiec D, Scotti-Saintagne C et al. (2008) Quantitative trait loci controlling water use efficiency and related traits in *Quercus robur* L. *Tree Genetics & Genomes*, **4**, 263–278.

Brendel O, Pot D, Plomion C, Rozenberg P, Guehl JM (2002) Genetic parameters and QTL analysis of delta13C and ring width in maritime pine. *Plant Cell and Environment*, **25**, 945–953.

Bricelj VM, Connell L, Konoki K et al. (2005) Sodium channel mutation leading to saxitoxin resistance in clams increases risk of PSP. *Nature*, **434**, 763–767.

Brown GR, Gill GP, Kuntz RJ, Langley CH, Neale DB (2004) Nucleotide diversity and linkage disequilibrium in loblolly pine. *Proceedings of the National Academy of Sciences USA*, **101**, 15255–15260.

Bryja J, Charbonnel N, Berthier K, Galan M, Cosson JF (2007) Density-related changes in selection pattern for major histocompatibility complex genes in fluctuating populations of voles. *Molecular Ecology*, **16**, 5084–5097.

Bundock PC, Potts BM, Vaillancourt RE (2008) Detection and stability of quantitative trait loci (QTL) in *Eucalyptus globulus*. *Tree Genetics & Genomes*, **4**, 85–95.

Byrne M, Murrell JC, Owen JV et al. (1997a) Identification and mode of action of quantitative trait loci affecting seedling height and leaf area in Eucalyptus nitens. *Theoretical and Applied Genetics*, **94**, 674–681.

Byrne M, Murrell JC, Owen JV, Williams ER, Moran GF (1997b) Mapping of quantitative trait loci influencing frost tolerance in Eucalyptus nitens. *Theoretical and Applied Genetics*, **95**, 975–979.

Campbell D, Bernatchez L (2004) Generic scan using AFLP markers as a means to assess the role of directional selection in the divergence of sympatric whitefish ecotypes. *Molecular Biology and Evolution*, **21**, 945–956.

Campbell RK, Sorensen FC (1979) New basis for characterizing germination. *Journal of Seed Technology*, **4**, 24–34.

Campos R, Storz JF, Ferrand N (2008) Evidence for contrasting modes of selection at inter-acting globin genes in the European rabbit (*Oryctolagus cuniculus*). *Heredity*, **100**, 602–609.

Canino MF, O'Reilly PT, Hauser L, Bentzen P (2005) Genetic differentiation in walleye pollock (*Theragra chalcogramma*) in response to selection at the pantophysin (PanI) locus. *Canadian Journal of Fisheries and Aquatic Sciences*, **62**, 2519–2529.

Cano JM, Matsuba C, Makinen H, Merila J (2006) The utility of QTL-linked markers to detect selective sweeps in natural populations – a case study of the EDA gene and a linked marker in threespine stickleback. *Molecular Ecology*, **15**, 4613–4621.

Carlson CS, Eberle MA, Rieder MJ et al. (2004) Selecting a maximally informative set of single-nucleotide polymorphisms for association analyses using linkage disequilibrium. *American Journal of Human Genetics*, **74**, 106–120.

Casasoli M, Derory J, Morera-Dutrey C et al. (2006) Comparison of quantitative trait loci for adaptive traits between oak and chestnut based on an expressed sequence tag consensus map. *Genetics*, **172**, 533–546.

Casasoli M, Pot D, Plomion C et al. (2004) Identification of QTLs affecting adaptive traits in *Castanea sativa* Mill. *Plant Cell and Environment*, **27**, 1088–1101.

Chagne D, Brown G, Lalanne C et al. (2003) Comparative genome and QTL mapping between maritime and loblolly pines. *Molecular Breeding*, **12**, 185–195.

Chen H, Morrell PL, de la Cruz M, Clegg MT (2008) Nucleotide diversity and linkage disequilibrium in wild avocado (*Persea americana mill.*). *Journal of Heredity*, **99**, 382–389.

Chen THH, Howe GT, Bradshaw HD (2002) Molecular genetic analysis of dormancy-related traits in poplars. *Weed Science*, **50**, 232–240.

Cnaani A, Zilberman N, Tinman S, Hulata G, Ron M (2004) Genome-scan analysis for quantitative trait loci in an F-2 tilapia hybrid. *Molecular Genetics and Genomics*, **272**, 162–172.

Colosimo PF, Hosemann KE, Balabhadra S et al. (2005) Widespread parallel evolution in sticklebacks by repeated fixation of ectodysplasin alleles. *Science*, **307**, 1928–1933.

Colosimo PF, Peichel CL, Nereng K et al. (2004) The genetic architecture of parallel armor plate reduction in threespine sticklebacks. *PLoS Biology*, **2**, 635–641.

Courbot M, Willems G, Motte P et al. (2007) A major quantitative trait locus for cadmium tolerance in *Arabidopsis halleri* colocalizes with HMA4, a gene encoding a heavy metal ATPase. *Plant Physiology*, **144**, 1052–1065.

Crandall KA, Bininda-Emonds OR, Mace GM, Wayne RK (2000) Considering evolutionary processes in conservation biology. *Trends in Ecology and Evolution*, **15**, 290–295.

Deng WD, Shu W, Yang SL, Shi XW, Mao HM (2009) Pigmentation in black-boned sheep (*Ovis aries*): association with polymorphism of the MC1R gene. *Molecular Biology Reports*, **36**, 431–436.

Devey ME, Carson SD, Nolan MF et al. (2004) QTL associations for density and diameter in *Pinus radiata* and the potential for marker-aided selection. *Theoretical and Applied Genetics*, **108**, 516–524.

Doerge RW (2002) Mapping and analysis of quantitative trait loci in experimental popula-tions. *Nature Reviews Genetics*, **3**, 43–52.

Doerge RW, Zeng ZB, Weir BS (1997) Statistical issues in the search for genes affecting quantitative traits in experimental populations. *Statistical Science*, **12**, 195–219.

Drew RE, Schwabl H, Wheeler PA, Thorgaard GH (2007) Detection of QTL influencing cortisol levels in rainbow trout (*Oncorhynchus mykiss*). *Aquaculture*, **272**, S183–S194.

Dvornyk V, Sirvio A, Mikkonen M, Savolainen O (2002) Low nucleotide diversity at the pal1 locus in the widely distributed *Pinus sylvestris*. *Molecular Biology and Evolution*, **19**, 179–188.

Eckert A, Wegrzyn J, Pande B et al. (2009a) Multilocus patterns of nucleotide diversity and divergence reveal positive selection at candidate genes related to cold-hardiness in coastal Douglas-fir (*Pseudotsuga menziesii var. menziesii*). *Genetics*, **183**, 289–298.

Eckert AJ, Bower AD, Wegrzyn JL et al. (2009b) Association genetics of coastal Douglas-fir (*Pseudotsuga menziesii var. menziesii, Pinaceae*). I. Cold-hardiness related traits. *Genetics*, **182**, 1289–1302.

Egan SP, Nosil P, Funk DJ (2008) Selection and genomic differentiation during ecological speciation: isolating the contributions of host association via a comparative genome scan of *Neochlamisus bebbianae* leaf beetles. *Evolution*, **62**, 1162–1181.

Ellegren H (2008) Sequencing goes 454 and takes large-scale genomics into the wild. *Molecular Ecology*, 17, 1679–1681.

Emebiri LC, Devey ME, Matheson AC, Slee MU (1997) Linkage of RAPD markers to NESTUR, a stem growth index in *radiata* pine seedlings. *Theoretical and Applied Genetics*, **95**, 119–124.

Emebiri LC, Devey ME, Matheson AC, Slee MU (1998a) Age-related changes in the expression of QTLs for growth in *radiata* pine seedlings. *Theoretical and Applied Genetics*, **97**, 1053–1061.

Emebiri LC, Devey ME, Matheson AC, Slee MU (1998b) Interval mapping of quantitative trait loci affecting NESTUR, a stem growth efficiency index of *radiata* pine seedlings. *Theoretical and Applied Genetics*, **97**, 1062–1068.

Emelianov I, Marec F, Mallet J (2004) Genomic evidence for divergence with gene flow in host races of the larch budmoth. *Proceedings of the Royal Society of London Series B-Biological Sciences*, **271**, 97–105.

Erickson DL, Fenster CB, Stenoien HK, Price D (2004) Quantitative trait locus analyses and the study of evolutionary process. *Molecular Ecology*, **13**, 2505–2522.

Falconer D, MacKay T (1996) *Introduction to Quantitative Genetics*, 4th edn. Longman Group Ltd., Essex, England.

Faure MF, David P, Bonhomme F, Bierne N (2008) Genetic hitchhiking in a subdivided population of *Mytilus edulis*. *BMC Evolutionary Biology*, **8**, 164.

Ferris R, Long L, Bunn SM et al. (2002) Leaf stomatal and epidermal cell development: identification of putative quantitative trait loci in relation to elevated carbon dioxide concentration in poplar. *Tree Physiology*, **22**, 633–640.

Fidler AE, van Oers K, Drent PJ et al. (2007) Drd4 gene polymorphisms are associated with personality variation in a passerine bird. *Proceedings of the Royal Society of London Series B-Biological Sciences*, **274**, 1685–1691.

Fishman L, Kelly AJ, Willis JH (2002) Minor quantitative trait loci underlie floral traits associated with mating system divergence in *Mimulus*. *Evolution*, **56**, 2138–2155.

Fitzpatrick MJ, Ben-Shahar Y, Smid HM et al. (2005) Candidate genes for behavioural ecology. *Trends in Ecology & Evolution*, **20**, 96–104.

Frewen BE, Chen THH, Howe GT et al. (2000) Quantitative trait loci and candidate gene mapping of bud set and bud flush in Populus. *Genetics*, **154**, 837–845.

Gailing O, Kremer A, Steiner W, Hattemer HH, Finkeldey R (2005) Results on quantitative trait loci for flushing date in oaks can be transferred to different segregating progenies. *Plant Biology (Stuttgart)*, **7**, 516–525.

Galindo J, Moran P, Rolan-Alvarez E (2009) Comparing geographical genetic differentiation between candidate and noncandidate loci for adaptation strengthens support for parallel ecological divergence in the marine snail *Littorina saxatilis*. *Molecular Ecology*, **18**, 919–930.

Garcia MV, Ingvarsson PK (2007) An excess of nonsynonymous polymorphism and extensive haplotype structure at the PtABI1B locus in European aspen (*Populus tremula*): a case of balancing selection in an obligately outcrossing plant? *Heredity*, **99**, 381–388.

Garcia-Gil MR, Mikkonen M, Savolainen O (2003) Nucleotide diversity at two phytochrome loci along a latitudinal cline in *Pinus sylvestris*. *Molecular Ecology*, **12**, 1195–1206.

Geffeney SL, Fujimoto E, Brodie ED, Brodie ED, Ruben PC (2005) Evolutionary diversification of TTX-resistant sodium channels in a predator–prey interaction. *Nature*, **434**, 759–763.

George AW, Visscher PM, Haley CS (2000) Mapping quantitative trait loci in complex pedigrees: a two-step variance component approach. *Genetics*, **156**, 2081–2092.

Gering EJ, Opazo JC, Storz JF (2009) Molecular evolution of cytochrome b in high- and low-altitude deer mice (genus *Peromyscus*). *Heredity*, **102**, 226–235.

Gerrard DT, Meyer A (2007) Positive selection and gene conversion in SPP120, a fertilization-related gene, during the east African Cichlid fish radiation. *Molecular Biology and Evolution*, **24**, 2286–2297.

Gonzalez-Martinez SC, Ersoz E, Brown GR, Wheeler NC, Neale DB (2006a) DNA sequence variation and selection of tag single-nucleotide polymorphisms at candidate genes for drought-stress response in *Pinus taeda* L. *Genetics*, **172**, 1915–1926.

Gonzalez-Martinez SC, Huber D, Ersoz E, Davis JM, Neale DB (2008) Association genetics in *Pinus taeda* L. II. Carbon isotope discrimination. *Heredity*, **101**, 19–26.

Gonzalez-Martinez SC, Krutovsky KV, Neale DB (2006b) Forest-tree population genomics and adaptive evolution. *New Phytologist*, **170**, 227–238.

Gonzalez-Martinez SC, Wheeler NC, Ersoz E, Nelson CD, Neale DB (2007) Association genetics in *Pinus taeda* L. I. Wood property traits. *Genetics*, **175**, 399–409.

Grattapaglia D, Bertolucci FLG, Penchel R, Sederoff RR (1996) Genetic mapping of quantitative trait loci controlling growth and wood quality traits in *Eucalyptus grandis* using a maternal half-sib family and RAPD markers. *Genetics*, **144**, 1205–1214.

Grattapaglia D, Bertolucci FL, Sederoff RR (1995) Genetic mapping of QTLs controlling vegetative propagation in *Eucalyptus grandis* and *E. urophylla* using a pseudo-testcross strategy and RAPD markers. *Theoretical and Applied Genetics*, **90**, 933–947.

Gratten J, Beraldi D, Lowder BV et al. (2007) Compelling evidence that a single nucleotide substitution in TYRP1 is responsible for coat-colour polymorphism in a free-living population of Soay sheep. *Proceedings of the Royal Society of London Series B-Biological Sciences*, **274**, 619–626.

Gross JB, Borowsky R, Tabin CJ (2009) A novel role for Mc1r in the parallel evolution of depigmentation in independent populations of the cavefish *Astyanax mexicanus*. *PLoS Genetics*, **5**(1), e1000326.

Guillet-Claude C, Isabel N, Pelgas B, Bousquet J (2004) The evolutionary implications of knox-I gene duplications in conifers: correlated evidence from phylogeny, gene mapping, and analysis of functional divergence. *Molecular Biology and Evolution*, **21**, 2232–2245.

Gupta PK, Rustgi S, Kulwal PL (2005) Linkage disequilibrium and association studies in higher plants: present status and future prospects. *Plant Molecular Biology*, **57**, 461–485.

Gwaze DP, Zhou Y, Reyes-Valdes MH, Al-Rababah MA, Williams CG (2003) Haplotypic QTL mapping in an outbred pedigree. *Genetical Research*, **81**, 43–50.

Haidle L, Janssen JE, Gharbi K et al. (2008) Determination of quantitative trait loci (QTL) for early maturation in rainbow trout (*Oncorhynchus mykiss*). *Marine Biotechnology*, **10**, 579–592.

Hall MC, Basten CJ, Willis JH (2006) Pleiotropic quantitative trait loci contribute to population divergence in traits associated with life-history variation in *Mimulus guttatus*. *Genetics*, **172**, 1829–1844.

Hawthorne DJ, Via S (2001) Genetic linkage of ecological specialization and reproductive isolation in pea aphids. *Nature*, **412**, 904–907.

Heidel AJ, Clauss MJ, Kroymann J, Savolainen O, Mitchell-Olds T (2006) Natural variation in MAM within and between populations of *Arabidopsis lyrata* determines glucosinolate phenotype. *Genetics*, **173**, 1629–1636.

Herder F, Pfaender J, Schliewen UK (2008) Adaptive sympatric speciation of polychromatic "roundfin" sailfin silverside fish in Lake Matano (Sulawesi). *Evolution*, **62**, 2178–2195.

Heuertz M, De Paoli E, Kallman T et al. (2006) Multilocus patterns of nucleotide diversity, linkage disequilibrium and demographic history of Norway spruce [*Picea abies* (L.) Karst]. *Genetics*, **174**, 2095–2105.

Hodges SA, Whittall JB, Fulton M, Yang JY (2002) Genetics of floral traits influencing reproductive isolation between *Aquilegia formosa* and *Aquilegia pubescens*. *American Naturalist*, **159**, S51–S60.

Hoffmann AA, Willi Y (2008) Detecting genetic responses to environmental change. *Nature Reviews Genetics*, **9**, 421–432.

Houston RD, Haley CS, Hamilton A et al. (2008) Major quantitative trait loci affect resistance to infectious pancreatic necrosis in Atlantic salmon (*Salmo salar*). *Genetics*, **178**, 1109–1115.

Hunt GJ, Amdam GV, Schlipalius D et al. (2007) Behavioral genomics of honeybee foraging and nest defense. *Naturwissenschaften*, **94**, 247–267.

Hurme P, Repo T, Savolainen O, Paakkonen T (1997) Climatic adaptation of bud set and frost hardiness in Scots pine (*Pinus sylvestris*). *Canadian Journal of Forest Research – Revue Canadienne de Recherche Forestiere*, **27**, 716–723.

Hurme P, Sillanpaa MJ, Arjas E, Repo T, Savolainen O (2000) Genetic basis of climatic adaptation in Scots pine by Bayesian quantitative trait locus analysis. *Genetics*, **156**, 1309–1322.

Hwang SY, Lin TP, Ma CS ct al. (2003) Postglacial population growth of *Cunninghamia konishii (Cunn*. *intitd* from phylogeographical and mismatch analysis of chloroplast DNA variation. *Molecular Ecology*, **12**, 2689–2695.

Ingvarsson PK (2005) Nucleotide polymorphism and linkage disequilibrium within and among natural populations of European Aspen (*Populus tremula* L., *Salicaceae*). *Genetics*, **169**, 945–953.

Ingvarsson PK, Garcia MV, Hall D, Luquez V, Jansson S (2006) Clinal variation in phyB2, a candidate gene for day-length-induced growth cessation and bud set, across a latitudinal gradient in European aspen (*Populus tremula*). *Genetics*, **172**, 1845–1853.

Jackson TR, Ferguson MM, Danzmann RG et al. (1998) Identification of two QTL influencing upper temperature tolerance in three rainbow trout (*Oncorhynchus mykiss*) half-sib families. *Heredity*, **80**, 143–151.

Jarvinen P, Lemmetyinen J, Savolainen O, Sopanen T (2003) DNA sequence variation in BpMADS2 gene in two populations of *Betula pendula*. *Molecular Ecology*, **12**, 369–384.

Jermstad KD, Bassoni DL, Jech KS et al. (2003) Mapping of quantitative trait loci controlling adaptive traits in coastal Douglas fir. III. Quantitative trait loci-by-environment interactions. *Genetics*, **165**, 1489–1506.

Jermstad KD, Bassoni DL, Jech KS, Wheeler NC, Neale DB (2001a) Mapping of quantitative trait loci controlling adaptive traits in coastal Douglas-fir. I. Timing of vegetative bud flush. *Theoretical and Applied Genetics*, **102**, 1142–1151.

Jermstad KD, Bassoni DL, Wheeler NC et al. (2001b) Mapping of quantitative trait loci controlling adaptive traits in coastal Douglas-fir. II. Spring and fall cold-hardiness. *Theoretical and Applied Genetics*, **102**, 1152–1158.

Jermstad KD, Bassoni DL, Wheeler NC, Neale DB (1998) A sex-averaged genetic linkage map in coastal Douglas-fir (*Pseudotsuga menziesii [Mirb.] Franco var 'menziesii'*) based on RFLP and RAPD markers. *Theoretical and Applied Genetics*, **97**, 762–770.

Jermstad KD, Sheppard LA, Kinloch BB et al. (2006) Isolation of a full-length CC-NBS-LRR resistance gene analog candidate from sugar pine showing low nucleotide diversity. *Tree Genetics & Genomes*, **2**, 76–85.

Johnston IA, Andersen O (2008) Number of muscle fibres in adult Atlantic cod varies with temperature during embryonic development and pantophysin (PanI) genotype. *Aquatic Biology*, **4**, 167–173.

Jones H, Leigh FJ, Mackay I et al. (2008) Population-based resequencing reveals that the flowering time adaptation of cultivated barley originated east of the fertile crescent. *Molecular Biology and Evolution*, **25**, 2211–2219.

Jonsdottir IG, Marteinsdottir G, Pampoulie C (2008) Relation of growth and condition with the Pan I locus in Atlantic cod (*Gadus morhua* L.) around Iceland. *Marine Biology*, **154**, 867–874.

Joost S, Bonin A, Bruford MW et al. (2007) A spatial analysis method (SAM) to detect candidate loci for selection: towards a landscape genomics approach to adaptation. *Molecular Ecology*, **16**, 3955–3969.

Jorge V, Dowkiw A, Faivre-Rampant P, Bastien C (2005) Genetic architecture of qualitative and quantitative *Melampsora larici-populina* leaf rust resistance in hybrid poplar: genetic mapping and QTL detection. *New Phytologist*, **167**, 113–127.

Kado T, Ushio Y, Yoshimaru H, Tsumura Y, Tachida H (2006) Contrasting patterns of DNA variation in natural populations of two related conifers, *Cryptomeria japonica* and *Taxodium distichum* (*Cupressaceae sensu lato*). *Genes & Genetic Systems*, **81**, 103–113.

Kado T, Yoshimaru H, Tsumura Y, Tachida H (2003) DNA variation in a conifer, *Cryptomeria japonica* (*Cupressaceae sensu lato*). *Genetics*, **164**, 1547–1559.

Kauer MO, Dieringer D, Schlotterer C (2003) A microsatellite variability screen for positive selection associated with the "Out of Africa" habitat expansion of *Drosophila melanogaster*. *Genetics*, **165**, 1137–1148.

Kaya Z, Sewell MM, Neale DB (1999) Identification of quantitative trait loci influencing annual height- and diameter-increment growth in loblolly pine (*Pinus taeda L.*). *Theoretical and Applied Genetics*, **98**, 586–592.

Kim YY, Kang BY, Choi HS et al. (2004) Identification of QTL (quantitative trait loci) associated with 2-year growth traits of full-sib progenies in *Populus davidiana* dode based on composite interval mapping. *Journal of Korean Forestry Society*, **93**, 251–264.

Kirst M, Basten CJ, Myburg AA, Zeng Z-B, Sederoff RR (2005) Genetic architecture of transcript-level variation in differentiating xylem of a eucalyptus hybrid. *Genetics*, **169**, 2295–2303.

Kirst M, Myburg AA, De Leon JPG et al. (2004) Coordinated genetic regulation of growth and lignin revealed by quantitative trait locus analysis of cDNA microarray data in an interspecific backcross of eucalyptus. *Plant Physiology (Rockville)*, **135**, 2368–2378.

Kitano J, Bolnick DI, Beauchamp DA et al. (2008) Reverse evolution of armor plates in the threespine stickleback. *Current Biology*, **18**, 769–774.

Kivimaki M, Karkkainen K, Gaudeul M, Loe G, Agren J (2007) Gene, phenotype and function: GLABROUS1 and resistance to herbivory in natural populations of *Arabidopsis lyrata*. *Molecular Ecology*, **16**, 453–462.

Kohn MH, Murphy WJ, Ostrander EA, Wayne RK (2006) Genomics and conservation genetics. *Trends in Ecology & Evolution*, **21**, 629–637.

Krieger MJB, Ross KG (2002) Identification of a major gene regulating complex social behavior. *Science*, **295**, 328–332.

Kronforst MR, Young LG, Kapan DD et al. (2006) Linkage of butterfly mate preference and wing color preference cue at the genomic location of wingless. *Proceedings of the National Academy of Sciences USA*, **103**, 6575–6580.

Krutovsky KV, Elsik CG, Matvienko M, Kozik A, Neale DB (2007) Conserved ortholog sets in forest trees. *Tree Genetics & Genomes*, **3**, 61–70.

Krutovsky KV, Neale DB (2005) Nucleotide diversity and linkage disequilibrium in cold-hardiness- and wood quality-related candidate genes in Douglas fir. *Genetics*, **171**, 2029–2041.

Kumar S, Spelman RJ, Garrick DJ et al. (2000) Multiple-marker mapping of wood density loci in an outbred pedigree of *radiata* pine. *Theoretical and Applied Genetics*, **100**, 926–933.

Kuramoto N, Kondo T, Fujisawa Y et al. (2000) Detection of quantitative trait loci for wood strength in *Cryptomeria japonica*. *Canadian Journal of Forest Research*, **30**, 1525–1533.

Laurie CC, Nickerson DA, Anderson AD et al. (2007) Linkage disequilibrium in wild mice. *PLoS Genetics*, **3**, 1487–1495.

Leinonen T, O'Hara RB, Cano JM, Merila J (2008) Comparative studies of quantitative trait and neutral marker divergence: a meta-analysis. *Journal of Evolutionary Biology*, **21**, 1–17.

Lerceteau E, Plomion C, Andersson B (2000) AFLP mapping and detection of quantitative trait loci (QTLs) for economically important traits in *Pinus sylvestris*: a preliminary study. *Molecular Breeding*, **6**, 451–458.

Lewontin RC, Krakauer J (1973) Distribution of gene frequency as a test of theory of selective neutrality of polymorphisms. *Genetics*, **74**, 175–195.

Li J, Su X, Zhang Q, Louis Z (1999) Detection of QTLs for growth and phenology traits of poplar using RAPD markers. *Forest Research*, **12**, 111–117.

Li YF, Costello JC, Holloway AK, Hahn MW (2008) "Reverse ecology" and the power of population genomics. *Evolution*, **62**, 2984–2994.

Lin JZ (2000) The relationship between loci for mating system and fitness-related traits in *Mimulus* (*Scrophulariaceae*): a test for deleterious pleiotropy of QTLs with large effects. *Genome*, **43**, 628–633.

Lin JZ, Ritland K (1997) Quantitative trait loci differentiating the outbreeding *Mimulus guttatus* from the inbreeding *M. platycalyx*. *Genetics*, **146**, 1115–1121.

Lind M, Dalman K, Stenlid J, Karlsson B, Olson A (2007) Identification of quantitative trait loci affecting virulence in the basidiomycete *Heterobasidion annosum* s.l. *Current Genetics*, **52**, 35–44.

Liu AZ, Burke JM (2006) Patterns of nucleotide diversity in wild and cultivated sunflower. *Genetics*, **173**, 321–330.

Loh YHE, Katz LS, Mims MC et al. (2008) Comparative analysis reveals signatures of differentiation amid genomic polymorphism in Lake Malawi ci lilily *Genome Biology*, 9(7), R113

Long AD, Langley CH (1999) The power of association studies to detect the contribution of candidate genetic loci to variation in complex traits. *Genome Research*, **9**, 720–731.

Luikart G, England PR, Tallmon D, Jordan S, Taberlet P (2003) The power and promise of population genomics: from genotyping to genome typing. *Nature Reviews Genetics*, **4**, 981–994.

Lynch M, Walsh B (1998) *Genetics and Analysis of Quantitative Traits*. Sinauer Associates, Sunderland, MA.

Makinen HS, Cano M, Merila J (2008a) Identifying footprints of directional and balancing selection in marine and freshwater three-spined stickleback (*Gasterosteus aculeatus*) populations. *Molecular Ecology*, **17**, 3565–3582.

Makinen HS, Shikano T, Cano JM, Merila J (2008b) Hitchhiking mapping reveals a candidate genomic region for natural selection in three-spined stickleback chromosome VIII. *Genetics*, **178**, 453–465.

Markussen T, Fladung M, Achere V et al. (2003) Identification of QTLs controlling growth, chemical and physical wood property traits in *Pinus pinaster* (Ait.). *Silvae Genetica*, **52**, 8–15.

Marques CM, Brondani RPV, Grattapaglia D, Sederoff R (2002) Conservation and synteny of SSR loci and QTLs for vegetative propagation in four Eucalyptus species. *Theoretical and Applied Genetics*, **105**, 474–478.

Martin NH, Bouck AC, Arnold ML (2005) Loci affecting long-term hybrid survivorship in Louisiana irises: implications for reproductive isolation and introgression. *Evolution*, **59**, 2116–2124.

Martin NH, Bouck AC, Arnold ML (2006) Detecting adaptive trait introgression between *Iris fulva* and *I. brevicaulis* in highly selective field conditions. *Genetics*, **172**, 2481–2489.

Martyniuk CJ, Perry GML, Mogahadam HK, Ferguson MM, Danzmann RG (2003) The genetic architecture of correlations among growth-related traits and male age at maturation in rainbow trout. *Journal of Fish Biology*, **63**(3), 746–764.

Maynard Smith J, Haigh J (1974) The hitch-hiking effect of a favourable gene. *Genetical Research*, **23**, 23–35.

Menge DM, Zhong DB, Guda T et al. (2006) Quantitative trait loci controlling refractoriness to *Plasmodium falciparum* in natural *Anopheles gambiae* mosquitoes from a malaria-endemic region in western Kenya. *Genetics*, **173**, 235–241.

Merila J, Crnokrak P (2001) Comparison of genetic differentiation at marker loci and quantitative traits. *Journal of Evolutionary Biology*, **14**, 892–903.

Meyer-Lucht Y, Otten C, Puttker T, Sommer S (2008) Selection, diversity and evolutionary patterns of the MHC class II DAB in free-ranging neotropical marsupials. *BMC Genetics*, **9**, 39.

Miller CT, Beleza S, Pollen AA et al. (2007) Cis-regulatory changes in kit ligand expression and parallel evolution of pigmentation in sticklebacks and humans. *Cell*, **131**, 1179–1189.

Missiaggia AA, Piacezzi AL, Grattapaglia D (2005) Genetic mapping of Eef1, a major effect QTL for early flowering in *Eucalyptus grandis*. *Tree Genetics & Genomes*, **1**, 79–84.

Moen T, Agresti JJ, Cnaani A et al. (2004) A genome scan of a four-way tilapia cross supports the existence of a quantitative trait locus for cold tolerance on linkage group 23. *Aquaculture Research*, **35**, 893–904.

Moen T, Hayes B, Nilsen F et al. (2008) Identification and characterisation of novel SNP markers in Atlantic cod: evidence for directional selection. *BMC Genetics*, **9**, 18.

Moen T, Munck H, Raya LG (2005) A genome scan reveals a QTL for resistance to infectious salmon anaemia in Atlantic salmon (*Salmo salar*). *Aquaculture*, **247**, 25–26.

Moghadam HK, Poissant J, Fotherby H et al. (2007) Quantitative trait loci for body weight, condition factor and age at sexual maturation in Arctic charr (*Salvelinus alpinus*): comparative analysis with rainbow trout (*Oncorhynchus mykiss*) and Atlantic salmon (*Salmo salar*). *Molecular Genetics and Genomics*, **277**, 647–661.

Mona S, Crestanello B, Bankhead-Dronnet S et al. (2008) Disentangling the effects of recombination, selection, and demography on the genetic variation at a major histocompatibility complex class II gene in the alpine chamois. *Molecular Ecology*, **17**, 4053–4067.

Mori A, Romero-Severson J, Severson DW (2007) Genetic basis for reproductive diapause is correlated with life history traits within the Culex pipiens complex. *Insect Molecular Biology*, **16**, 515–524.

Moritz C (2002) Strategies to protect biological diversity and the evolutionary processes that sustain it. *Systematic Biology*, **51**, 238–254.

Morrell PL, Lundy KE, Clegg MT (2003) Distinct geographic patterns of genetic diversity are maintained in wild barley (*Hordeum vulgare ssp spontaneum*) despite migration. *Proceedings of the National Academy of Sciences USA* **100**, 10812–10817.

Morrell PL, Toleno DM, Lundy KE, Clegg MT (2005) Low levels of linkage disequilibrium in wild barley (*Hordeum vulgare ssp spontaneum*) despite high rates of self-fertilization. *Proceedings of the National Academy of Sciences USA*, **102**, 2442–2447.

Mullen LM, Hoekstra HE (2008) Natural selection along an environmental gradient: a classic cline in mouse pigmentation. *Evolution*, **62**, 1555–1569.

Mundy NI, Badcock NS, Hart T et al. (2004) Conserved genetic basis of a quantitative plumage trait involved in mate choice. *Science*, **303**, 1870–1873.

Murray MC, Hare MP (2006) A genomic scan for divergent selection in a secondary contact zone between Atlantic and Gulf of Mexico oysters, *Crassostrea virginica*. *Molecular Ecology*, **15**, 4229–4242.

Nachman MW, Hoekstra HE, D'Agostino SL (2003) The genetic basis of adaptive melanism in pocket mice. *Proceedings of the National Academy of Sciences USA*, **100**, 5268–5273.

Namroud MC, Beaulieu J, Juge N, Laroche J, Bousquet J (2008) Scanning the genome for gene single nucleotide polymorphisms involved in adaptive population differentiation in white spruce. *Molecular Ecology*, **17**, 3599–3613.

Neale DB (2007) Genomics to tree breeding and forest health. *Current Opinion in Genetics & Development*, **17**, 539–544.

Neale DB, Ingvarsson PK (2008) Population, quantitative and comparative genomics of adaptation in forest trees. *Current Opinion in Plant Biology*, **11**, 149–155.

Neale DB, Savolainen O (2004) Association genetics of complex traits in conifers. *Trends in Plant Science*, **9**, 325–330.

Newcombe G, Bradshaw HD Jr (1996) Quantitative trait loci conferring resistance in hybrid poplar to *Septoria populicola*, the cause of leaf spot. *Canadian Journal of Forest Research*, **26**, 1943–1950.

Newcombe G, Bradshaw HD Jr, Chastagner GA, Stettler RF (1996) A major gene for resistance to *Melampsora medusae f. sp. deltoidae* in a hybrid poplar pedigree. *Phytopathology*, **86**, 87–94.

Nichols KM, Bartholomew J, Thorgaard GH (2003) Mapping multiple genetic loci associated with Ceratomyxa shasta resistance in *Oncorhynchus mykiss*. *Diseases of Aquatic Organisms*, **56**, 145–154.

Nichols KM, Broman KW, Sundin K et al. (2007) Quantitative trait loci x maternal cytoplasmic environment interaction for development rate in *Oncorhynchus mykiss*. *Genetics*, **175**, 335–347.

Nichols KM, Felip A, Wheeler PA, Thorgaard GH (2008) The genetic basis of smoltification-related traits in *Onchorhynchus mykiss*. *Genetics*, **179**, 1559–1575.

Nichols KM, Wheeler PA, Thorgaard GH (2004) Quantitative trait loci analyses for meristic traits in *Oncorhynchus mykiss*. *Environmental Biology of Fishes*, **69**, 317–331.

Nielsen R (2001) Statistical tests of selective neutrality in the age of genomics. *Heredity*, **86**, 641–647.

Nielsen R (2005) Molecular signatures of natural selection. *Annual Review of Genetics*, **39**, 197–218.

Nielsen EE, Hansen MM, Meldrup D (2006) Evidence of microsatellite hitch-hiking selection in Atlantic cod (*Gadus morhua L.*): implications for inferring population structure in nonmodel organisms. *Molecular Ecology*, **15**, 3219–3229.

Nosil P, Egan SP, Funk DJ (2008) Heterogeneous genomic differentiation between walking stick ecotypes: "isolation by adaptation" and multiple roles for divergent selection. *Evolution*, **62**, 316–336.

Nosil P, Funk DJ, Ortiz-Barrientos D (2009) Divergent selection and heterogeneous genomic divergence. *Molecular Ecology*, **18**, 375–402.

O'Malley KG, Camara MD, Banks MA (2007) Candidate loci reveal genetic differentiation between temporally divergent migratory runs of Chinook salmon (*Oncorhynchus tshawytscha*). *Molecular Ecology*, **16**, 4930–4941.

O'Malley KG, Sakamoto T, Danzmann RG, Ferguson MM (2003) Quantitative trait loci for spawning date and body weight in rainbow trout: testing for conserved effects across ancestrally duplicated chromosomes. *Journal of Heredity*, **94**, 273–284.

Orsini L, Wheat CW, Haag CR et al. (2009) Fitness differences associated with Pgi SNP genotypes in the Glanville fritillary butterfly (*Melitaea cinxia*). *Journal of Evolutionary Biology*, **22**, 367–375.

Ozaki A, de Leon FG, Glebe B et al. (2005) Identification of QTL for resistance to *Cryptobia salmositica* infection in Atlantic salmon (*Salmo salar*): a model for pathogen disease resistance QTL? *Aquaculture*, **247**, 27–27.

Ozaki A, Sakamoto T, Khoo S et al. (2001) Quantitative trait loci (QTLs) associated with resistance/susceptibility to infectious pancreatic necrosis virus (IPNV) in rainbow trout (*Oncorhynchus mykiss*). *Molecular Genetics and Genomics*, **265**, 23–31.

Pampoulie C, Jakobsdottir KB, Marteinsdottir G, Thorsteinsson V (2008) Are vertical behaviour patterns related to the pantophysin locus in the Atlantic cod (*Gadus morhua L.*)? *Behavior Genetics*, **38**, 76–81.

Parelle J, Zapater M, Scotti-Saintagne C et al. (2007) Quantitative trait loci of tolerance to waterlogging in a European oak (*Quercus robur L.*): physiological relevance and temporal effect patterns. *Plant Cell and Environment*, **30**, 422–434.

Peichel CL, Nereng KS, Ohgi KA et al. (2001) The genetic architecture of divergence between threespine stickleback species. *Nature*, **414**, 901–905.

Pemberton JM (2008) Wild pedigrees: the way forward. *Proceedings of the Royal Society of London-Series B: Biological Sciences*, **275**, 613–621.

Perry GML, Ferguson MM, Sakamoto T, Danzmann RG (2005) Sex-linked quantitative trait loci for thermotolerance and length in the rainbow trout. *Journal of Heredity*, **96**, 97–107.

Piertney SB, Oliver MK (2006) The evolutionary ecology of the major histocompatibility complex. *Heredity*, **96**, 7–21.

Plomion C, Durel CE, O'Malley DM (1996) Genetic dissection of height in maritime pine seedlings raised under accelerated growth conditions. *Theoretical and Applied Genetics*, **93**, 849–858.

Pot D, McMillan L, Echt C et al. (2005) Nucleotide variation in genes involved in wood formation in two pine species. *New Phytologist*, **167**, 101–112.

Pot D, Rodrigues J-C, Rozenberg P et al. (2006) QTLs and candidate genes for wood properties in maritime pine (*Pinus pinaster Ait.*). *Tree Genetics & Genomes*, **2**, 10–24.

Powers DA, Schulte PM (1998) Evolutionary adaptations of gene structure and expression in natural populations in relation to a changing environment: a multidisciplinary approach to address the million-year saga of a small fish. *Journal of Experimental Zoology*, **282**, 71–94.

Pritchard JK, Stephens M, Rosenberg NA, Donnelly P (2000) Association mapping in structured populations. *American Journal of Human Genetics*, **67**, 170–181.

Protas M, Tabansky I, Conrad M et al. (2008) Multi-trait evolution in a cave fish, *Astyanax mexicanus*. *Evolution & Development*, **10**, 196–209.

Protas ME, Hersey C, Kochanek D et al. (2006) Genetic analysis of cavefish reveals molecular convergence in the evolution of albinism. *Nature Genetics*, **38**, 107–111.

Protas ME, Patel NH (2008) Evolution of coloration patterns. *Annual Review of Cell and Developmental Biology*, **24**, 425–446.

Przeworski M, Coop G, Wall JD (2005) The signature of positive selection on standing genetic variation. *Evolution,* **59,** 2312–2323.

Rae AM, Ferris R, Tallis MJ, Taylor G (2006) Elucidating genomic regions determining enhanced leaf growth and delayed senescence in elevated CO2. *Plant Cell and Environment,* **29,** 1730–1741.

Rae AM, Pinel MPC, Bastien C et al. (2008) QTL for yield in bioenergy *Populus*: identifying GxE interactions from growth at three contrasting sites. *Tree Genetics & Genomes,* **4,** 97–112.

Rae AM, Tricker PJ, Bunn SM, Taylor G (2007) Adaptation of tree growth to elevated CO2: quantitative trait loci for biomass in *Populus. New Phytologist,* **175,** 59–69.

Raeymaekers JAM, Van Houdt JKJ, Larmuseau MHD, Geldof S, Volckaert FAM (2007) Divergent selection as revealed by P-ST and QTL-based F-ST in three-spined stickleback (*Gasterosteus aculeatus*) populations along a coastal-inland gradient. *Molecular Ecology,* **16,** 891–905.

Reed DH, Frankham R (2003) Correlation between fitness and genetic diversity. *Conservation Biology,* **17,** 230–237.

Rehfeldt GE (1997) Quantitative analyses of the genetic structure of closely related conifers with disparate distributions and demographics: the *Cupressus arizonica* (*Cupressaceae*) complex. *American Journal of Botany,* **84,** 190–200.

Reid DP, Szanto A, Glebe B, Danzmann RG, Ferguson MM (2005) QTL for body weight and condition factor in Atlantic salmon (*Salmo salar*): comparative analysis with rainbow trout (*Oncorhynchus mykiss*) and Arctic charr (*Salvelinus alpinus*). *Heredity,* **94,** 166–172.

Robison BD, Wheeler PA, Sundin K, Sikka P, Thorgaard GH (2001) Composite interval mapping reveals a major locus influencing embryonic development rate in rainbow trout (*Oncorhynchus mykiss*). *Journal of Heredity,* **92,** 16–22.

Roff DA (1997) *Evolutionary Quantitative Genetics.* Chapman & Hall, New York.

Rogers SM, Bernatchez L (2005) Integrating QTL mapping and genome scans towards the characterization of candidate loci under parallel selection in the lake whitefish (*Coregonus clupeaformis*). *Molecular Ecology,* **14,** 351–361.

Ronnberg-Wastljung AC, Glynn C, Weih M (2005) QTL analyses of drought tolerance and growth for a *Salix dasyclados* x *Salix viminalis* hybrid in contrasting water regimes. *Theoretical and Applied Genetics,* **110,** 537–549.

Roselius K, Stephan W, Stadler T (2005) The relationship of nucleotide polymorphism, recombination rate and selection in wild tomato species. *Genetics,* **171,** 753–763.

Rosenblum EB, Hoekstra HE, Nachman MW (2004) Adaptive reptile color variation and the evolution of the Mc1r gene. *Evolution,* **58,** 1794–1808.

Ross KG, Keller L (1998) Genetic control of social organization in an ant. *Proceedings of the National Academy of Sciences USA,* **95,** 14232–14237.

Rostoks N, Mudie S, Cardle L et al. (2005) Genome-wide SNP discovery and linkage analysis in barley based on genes responsive to abiotic stress. *Molecular Genetics and Genomics,* **274,** 515–527.

Rueppell O, Chandra SBC, Pankiw T et al. (2006) The genetic architecture of sucrose responsiveness in the honeybee (*Apis mellifera* L.). *Genetics,* **172,** 243–251.

Rueppell O, Pankiw T, Nielsen DI et al. (2004) The genetic architecture of the behavioral ontogeny of foraging in honeybee workers. *Genetics,* **167,** 1767–1779.

Saintagne C, Bodenes C, Barreneche T et al. (2004) Distribution of genomic regions differentiating oak species assessed by QTL detection. *Heredity,* **92,** 20–30.

Scalfi M, Troggio M, Piovani P et al. (2004) A RAPD, AFLP and SSR linkage map, and QTL analysis in European beech (*Fagus sylvatica* L.). *Theoretical and Applied Genetics,* **108,** 433–441.

Schemske DW, Bradshaw HD Jr (1999) Pollinator preference and the evolution of floral traits in monkeyflowers (*Mimulus*). *Proceedings of the National Academy of Sciences USA,* **96,** 11910–11915.

Schlotterer C (2002) A microsatellite-based multilocus screen for the identification of local selective sweeps. *Genetics,* **160,** 753–763.

Schlotterer C (2003) Hitchhiking mapping – functional genomics from the population genetics perspective. *Trends in Genetics,* **19,** 32–38.

Schwartz MK, Luikart G, Waples RS (2007) Genetic monitoring as a promising tool for conservation and management. *Trends in Ecology & Evolution*, **22**, 25–33.

Scotti-Saintagne C, Bertocchi E, Barreneche T (2005) Quantitative trait loci mapping for vegetative propagation in pedunculate oak. *Annals of Forest Science*, **62**, 369–374.

Scotti-Saintagne C, Bodenes C, Barreneche T et al. (2004a) Detection of quantitative trait loci controlling bud burst and height growth in *Quercus robur L. Theoretical and Applied Genetics*, **109**, 1648–1659.

Scotti-Saintagne C, Mariette S, Porth I et al. (2004b) Genome scanning for interspecific differentiation between two closely related oak species (*Quercus robur* L. and *Q. petraea* (*Matt.*) *Liebl.*). *Genetics*, **168**, 1615–1626.

Shapiro MD, Marks ME, Peichel CL et al. (2004) Genetic and developmental basis of evolutionary pelvic reduction in threespine sticklebacks. *Nature*, **428**, 717–723.

Shaw KL, Parsons YM, Lesnick SC (2007) QTL analysis of a rapidly evolving speciation phenotype in the Hawaiian cricket *Laupala*. *Molecular Ecology*, **16**, 2879–2892.

Shepherd M, Cross M, Dieters MJ et al. (2003) Genetics of physical wood properties and early growth in a tropical pine hybrid. *Canadian Journal of Forest Research*, **33**, 1923–1932.

Shepherd M, Huang S, Eggler P et al. (2006) Congruence in QTL for adventitious rooting in *Pinus elliottii* x *Pinus caribaea* hybrids resolves between and within-species effects. *Molecular Breeding*, **18**, 11–28.

Shih FL, Hwang SY, Cheng YP, Lee PF, Lin TP (2007) Uniform genetic diversity, low differentiation, and neutral evolution characterize contemporary refuge populations of Taiwan fir (*Abies kawakamii, Pinaceae*). *American Journal of Botany*, **94**, 194–202.

Simonsen KL, Churchill GA, Aquadro CF (1995) Properties of statistical tests of neutrality for DNA polymorphism data. *Genetics*, **141**, 413–429.

Slate J (2005) Quantitative trait locus mapping in natural populations: progress, caveats and future directions. *Molecular Ecology*, **14**, 363–379.

Slate J, Pemberton JM (2007) Admixture and patterns of linkage disequilibrium in a free-living vertebrate population. *Journal of Evolutionary Biology*, **20**, 1415–1427.

Slate J, Pemberton JM, Visscher PM (1999) Power to detect QTL in a free-living polygynous population. *Heredity*, **83**, 327–336.

Slate J, Visscher PM, MacGregor S et al. (2002) A genome scan for quantitative trait loci in a wild population of red deer (*Cervus elaphus*). *Genetics*, **162**, 1863–1873.

Slatkin M (2008) Linkage disequilibrium – understanding the evolutionary past and mapping the medical future. *Nature Reviews Genetics*, **9**, 477–485.

Slotman M, della Torre A, Powell JR (2004) The genetics of inviability and male sterility in hybrids between *Anopheles gambiae* and *An. arabiensis*. *Genetics*, **167**, 275–287.

Song BH, Winsor AJ, Schmid KJ et al. (2009) Multilocus patterns of nucleotide diversity, population structure and linkage disequilibrium in *Boecchera stricta*, a wild relative of *Arabidopsis*. *Genetics*, **181**, 1021–1033.

St. Clair JB (2006) Genetic variation in fall cold hardiness in coastal Douglas-fir in western Oregon and Washington. *Canadian Journal of Botany-Revue Canadienne De Botanique*, **84**, 1110–1121.

St. Clair JB, Mandel NL, Vance-Boland KW (2005) Genecology of Douglas-fir in western Oregon and Washington. *Annals of Botany*, **96**, 1199–1214.

Stapley J, Birkhead TR, Burke T, Slate J (2008) A linkage map of the zebra finch *Taeniopygia guttata* provides new insights into avian genome evolution. *Genetics*, **179**, 651–667.

Steiner CC, Weber JN, Hoekstra HE (2007) Adaptive variation in beach mice produced by two interacting pigmentation genes. *PLoS Biology*, **5**, 1880–1889.

Stich B, Mohring J, Piepho HP et al. (2008) Comparison of mixed-model approaches for association mapping. *Genetics*, **178**, 1745–1754.

Stinchcombe JR, Hoekstra HE (2008) Combining population genomics and quantitative genetics: finding the genes underlying ecologically important traits. *Heredity*, **100**, 158–170.

Storz JF (2005) Using genome scans of DNA polymorphism to infer adaptive population divergence. *Molecular Ecology*, **14**, 671–688.

Storz JF, Dubach JM (2004) Natural selection drives altitudinal divergence at the albumin locus in deer mice, *Peromyscus maniculatus*. *Evolution*, **58**, 1342–1352.

Storz JF, Kelly JK (2008) Effects of spatially varying selection on nucleotide diversity and linkage disequilibrium: insights from deer mouse globin genes. *Genetics*, **180**, 367–379.

Storz JF, Nachman MW (2003) Natural selection on protein polymorphism in the rodent genus *Peromyscus*: evidence from interlocus contrasts. *Evolution*, **57**, 2628–2635.

Storz JF, Sabatino SJ, Hoffmann FG et al. (2007) The molecular basis of high-altitude adaptation in deer mice. *PLoS Genetics*, **3**, 448–459.

Streelman JT, Albertson RC, Kocher TD (2003) Genome mapping of the orange blotch colour pattern in cichlid fishes. *Molecular Ecology*, **12**, 2465–2471.

Sundin K, Brown KH, Drew RE et al. (2005) Genetic analysis of a development rate QTL in backcrosses of clonal rainbow trout, *Oncorhynchus mykiss*. *Aquaculture*, **247**, 75–83.

Tagu D, Bastien C, Faivre-Rampant P et al. (2005) Genetic analysis of phenotypic variation for ectomycorrhiza formation in an interspecific F1 poplar full-sib family. *Mycorrhiza*, **15**, 87–91.

Tennessen JA, Blouin MS (2008) Balancing selection at a frog antimicrobial peptide locus: fluctuating immune effector alleles? *Molecular Biology and Evolution*, **25**, 2669–2680.

Teshima KM, Coop G, Przeworski M (2006) How reliable are empirical genomic scans for selective sweeps? *Genome Research*, **16**, 702–712.

Thamarus K, Groom K, Bradley A et al. (2004) Identification of quantitative trait loci for wood and fibre properties in two full-sib pedigrees of *Eucalyptus globulus*. *Theoretical and Applied Genetics*, **109**, 856–864.

Theron E, Hawkins K, Bermingham E, Ricklefs RE, Mundy NI (2001) The molecular basis of an avian plumage polymorphism in the wild: a melanocortin-1-receptor point mutation is perfectly associated with the melanic plumage morph of the bananaquit, *Coereba flaveola*. *Current Biology*, **11**, 550–557.

Thornton KR, Jensen JD, Becquet C, Andolfatto P (2007) Progress and prospects in mapping recent selection in the genome. *Heredity*, **98**, 340–348.

Thumma BR, Nolan MF, Evans R, Moran GF (2005) Polymorphisms in cinnamoyl CoA reductase (CCR) are associated with variation in microfibril angle in *Eucalyptus spp*. *Genetics*, **171**, 1257–1265.

Tsarouhas V, Gullberg U, Lagercrantz U (2002) An AFLP and RFLP linkage map and quantitative trait locus (QTL) analysis of growth traits in *Salix*. *Theoretical and Applied Genetics*, **105**, 277–288.

Tsarouhas V, Gullberg U, Lagercrantz U (2003) Mapping of quantitative trait loci controlling timing of bud flush in *Salix*. *Hereditas (Lund)*, **138**, 172–178.

Tsarouhas V, Gullberg U, Lagererantz U (2004) Mapping of quantitative trait loci (QTLs) affecting autumn freezing resistance and phenology in *Salix*. *Theoretical and Applied Genetics*, **108**, 1335–1342.

Tschaplinski TJ, Tuskan GA, Sewell MM et al. (2006) Phenotypic variation and quantitative trait locus identification for osmotic potential in an interspecific hybrid inbred F-2 poplar pedigree grown in contrasting environments. *Tree Physiology*, **26**, 595–604.

Vali U, Einarsson A, Waits L, Ellegren H (2008) To what extent do microsatellite markers reflect genome-wide genetic diversity in natural populations? *Molecular Ecology*, **17**, 3808–3817.

Vasemagi A, Gross R, Paaver T et al. (2005a) Analysis of gene associated tandem repeat markers in Atlantic salmon (*Salmo salar L.*) populations: implications for restoration and conservation in the Baltic Sea. *Conservation Genetics*, **6**, 385–397.

Vasemagi A, Nilsson J, Primmer CR (2005b) Expressed sequence tag-linked microsatellites as a source of gene-associated polymorphisms for detecting signatures of divergent selection in Atlantic salmon (*Salmo salar L.*). *Molecular Biology and Evolution*, **22**, 1067–1076.

Vasemagi A, Primmer CR (2005) Challenges for identifying functionally important genetic variation: the promise of combining complementary research strategies. *Molecular Ecology*, **14**, 3623–3642.

Verhaegen D, Plomion C, Gion JM et al. (1997) Quantitative trait dissection analysis in Eucalyptus using RAPD markers: 1. Detection of QTL in interspecific hybrid progeny, stability of QTL expression across different ages. *Theoretical and Applied Genetics,* **95**, 597–608.

Via S, Hawthorne DJ (2002) The genetic architecture of ecological specialization: correlated gene effects on host use and habitat choice in pea aphids, *American Naturalist* **160**, S76–S88.

Via S, West J (2008) The genetic mosaic suggests a new role for hitchhiking in ecological speciation. *Molecular Ecology,* **17**, 4334–4345.

Vitalis R, Dawson K, Boursot P (2001) Interpretation of variation across marker loci as evidence of selection. *Genetics,* **158**, 1811–1823.

Voss SR, Prudic KL, Oliver JC, Shaffer HB (2003) Candidate gene analysis of metamorphic timing in ambystomatid salamanders. *Molecular Ecology,* **12**, 1217–1223.

Voss SR, Shaffer HB (2000) Evolutionary genetics of metamorphic failure using wild-caught vs. laboratory axolotls (*Ambystoma mexicanum*). *Molecular Ecology,* **9**, 1401–1407.

Wachowiak W, Balk PA, Savolainen O (2009) Search for nucleotide diversity patterns of local adaptation in dehydrins and other cold-related candidate genes in Scots pine (*Pinus sylvestris L.*). *Tree Genetics & Genomes,* **5**, 117–132.

Walsh B (2008) Using molecular markers for detecting domestication, improvement, and adaptation genes. *Euphytica,* **161**, 1–17.

Wang HW, Ge S (2006) Phylogeography of the endangered *Cathaya argyrophylla* (Pinaceae) inferred from sequence variation of mitochondrial and nuclear DNA. *Molecular Ecology,* **15**, 4109–4122.

Wang S, Zhu QH, Guo XY et al. (2007) Molecular evolution and selection of a gene encoding two tandem microRNAs in rice. *FEBS Letters,* **581**, 4789–4793.

Watt WB (1977) Adaptation at specific loci. 1. Natural selection on phosphoglucose isomerase of Colias butterflies – biochemical and population aspects. *Genetics,* **87**, 177–194.

Watterson GA (1977) Heterosis or neutrality? *Genetics,* **85**, 789–814.

Watts PC, O'Leary D, Cross MC et al. (2008) Contrasting levels of genetic differentiation among putative neutral microsatellite loci in Atlantic herring *Clupea harengus* populations and the implications for assessing stock structure. *Hydrobiologia,* **606**, 27–33.

Weber A, Clark RM, Vaughn L et al. (2007) Major regulatory genes in maize contribute to standing variation in teosinte (*Zea mays ssp parviglumis*). *Genetics,* **177**, 2349–2359.

Weber AL, Briggs WH, Rucker J et al. (2008) The genetic architecture of complex traits in teosinte (*Zea mays ssp parviglumis*): new evidence from association mapping. *Genetics,* **180**, 1221–1232.

Weih M, Ronnberg-Wastljung AC, Glynn C (2006) Genetic basis of phenotypic correlations among growth traits in hybrid willow (*Salix dasyclados x S-viminalis*) grown under two water regimes. *New Phytologist,* **170**, 467–477.

Weir BS (2008) Linkage disequilibrium and association mapping. *Annual Review of Genomics and Human Genetics,* **9**, 129–142.

Weng C, Kubisiak TL, Nelson CD, Stine M (2002) Mapping quantitative trait loci controlling early growth in a (longleaf pine × slash pine) × slash pine BC1 family. *Theoretical and Applied Genetics,* **104**, 852–859.

Westerdahl H, Waldenstrom J, Hansson B et al. (2005) Associations between malaria and MHC genes in a migratory songbird. *Proceedings of the Royal Society of London – Series B: Biological Sciences,* **272**, 1511–1518.

Wheat CW, Watt WB, Pollock DD, Schulte PM (2006) From DNA to fitness differences: sequences and structures of adaptive variants of Colias phosphoglucose isomerase (PGI). *Molecular Biology and Evolution,* **23**, 499–512.

Wheeler NC, Jermstad KD, Krutovsky K et al. (2005) Mapping of quantitative trait loci controlling adaptive traits in coastal Douglas-fir. IV. Cold-hardiness QTL verification and candidate gene mapping. *Molecular Breeding,* **15**, 145–156.

Wilding CS, Butlin RK, Grahame J (2001) Differential gene exchange between parapatric morphs of *Littorina saxatilis* detected using AFLP markers. *Journal of Evolutionary Biology,* **14**, 611–619.

Wilfert L, Gadau J, Baer B, Schmid-Hempel P (2007a) Natural variation in the genetic architecture of a host-parasite interaction in the bumblebee *Bombus terrestris*. *Molecular Ecology*, **16**, 1327–1339.

Wilfert L, Gadau J, Schmid-Hempel P (2007b) The genetic architecture of immune defense and reproduction in male *Bombus terrestris* bumblebees. *Evolution*, **61**, 804–815.

Willems G, Drager DB, Courbot M et al. (2007) The genetic basis of zinc tolerance in the metallophyte *Arabidopsis halleri ssp. halleri* (*Brassicaceae*): an analysis of quantitative trait loci. *Genetics*, **176**, 659–674.

Williams CG, Reyes-Valdes MH, Huber DA (2007) Validating a QTL region characterized by multiple haplotypes. *Theoretical and Applied Genetics*, **116**, 87–94.

Wood HM, Grahame JW, Humphray S, Rogers J, Butlin RK (2008) Sequence differentiation in regions identified by a genome scan for local adaptation. *Molecular Ecology*, **17**, 3123–3135.

Worley K, Carey J, Veitch A, Coltman DW (2006) Detecting the signature of selection on immune genes in highly structured populations of wild sheep (*Ovis dalli*). *Molecular Ecology*, **15**, 623–637.

Wright SI, Gaut BS (2005) Molecular population genetics and the search for adaptive evolution in plants. *Molecular Biology and Evolution*, **22**, 506–519.

Wu R, Bradshaw HD, Jr., Stettler RF (1997) Molecular genetics of growth and development in *Populus* (*Salicaceae*). V. Mapping quantitative trait loci affecting leaf variation. *American Journal of Botany*, **84**, 143–153.

Wu R, Bradshaw HD, Jr., Stettler RF (1998) Developmental quantitative genetics of growth in *Populus*. *Theoretical and Applied Genetics*, **97**, 1110–1119.

Wu R, Stettler RF (1994) Quantitative genetics of growth and development in *Populus*. I. A three-generation comparison of tree architecture during the first 2 years of growth. *Theoretical and Applied Genetics*, **89**, 1046–1054.

Wu RL (1998) Genetic mapping of QTLs affecting tree growth and architecture in *Populus*: implication for ideotype breeding. *Theoretical and Applied Genetics*, **96**, 447–457.

Wu RL, Ma CX, Casella G (2002) Joint linkage and linkage disequilibrium mapping of quantitative trait loci in natural populations. *Genetics*, **160**, 779–792.

Wullschleger SD, Yin TM, DiFazio SP et al. (2005) Phenotypic variation in growth and biomass distribution for two advanced-generation pedigrees of hybrid poplar. *Canadian Journal of Forest Research*, **35**, 1779–1789.

Yoshimaru H, Ohba K, Tsurumi K et al. (1998) Detection of quantitative trait loci for juvenile growth, flower bearing and rooting ability based on a linkage map of sugi (*Cryptomeria japonica D. Don*). *Theoretical and Applied Genetics*, **97**, 45–50.

Yu JM, Buckler ES (2006) Genetic association mapping and genome organization of maize. *Current Opinion in Biotechnology*, **17**, 155–160.

Yu JM, Holland JB, McMullen MD, Buckler ES (2008) Genetic design and statistical power of nested association mapping in maize. *Genetics*, **178**, 539–551.

Yu JM, Pressoir G, Briggs WH et al. (2006) A unified mixed-model method for association mapping that accounts for multiple levels of relatedness. *Nature Genetics*, **38**, 203–208.

Zayed A, Whitfield CW (2008) A genome-wide signature of positive selection in ancient and recent invasive expansions of the honey bee *Apis mellifera*. *Proceedings of the National Academy of Sciences USA*, **105**, 3421–3426.

Zhai WW, Nielsen R, Slatkin M (2009) An investigation of the statistical power of neutrality tests based on comparative and population genetic data. *Molecular Biology and Evolution*, **26**, 273–283.

Zhang D, Zhang Z, Yang K (2006) QTL analysis of growth and wood chemical content traits in an interspecific backcross family of white poplar (*Populus tomentosa* × *P-bolleana*) × *P-tomentosa*. *Canadian Journal of Forest Research*, **36**, 2015–2023.

Zhao KY, Aranzana MJ, Kim S et al. (2007) An Arabidopsis example of association mapping in structured samples. *PLoS Genetics*, **3**, 71–82.

Zhong DB, Menge DM, Temu EA, Chen H, Yan GY (2006) Amplified fragment length polymorphism mapping of quantitative trait loci for malaria parasite susceptibility in the yellow fever mosquito *Aedes aegypti*. *Genetics*, **173**, 1337–1345.

Zhong DB, Pai A, Yan GY (2003) Quantitative trait loci for susceptibility to tapeworm infection in the red flour beetle. *Genetics*, **165**, 1307–1315.

Zhong DB, Pai A, Yan GY (2005) Costly resistance to parasitism: evidence from simultaneous quantitative trait loci mapping for resistance and fitness in *Tribolium castaneum*. *Genetics,* **169**, 2127–2135.

Zhu QH, Zheng XM, Luo JC, Gaut BS, Ge S (2007) Multilocus analysis of nucleotide variation of *Oryza sativa* and its wild relatives: severe bottleneck during domestication of rice. *Molecular Biology and Evolution,* **24**, 875–888.

Zimmerman AM, Evenhuis JP, Thorgaard GH, Ristow SS (2004) A single major chromosomal region controls natural killer cell-like activity in rainbow trout. *Immunogenetics,* **55**, 825–835.

Zimmerman AM, Wheeler PA, Ristow SS, Thorgaard GH (2005) Composite interval mapping reveals three QTL associated with pyloric caeca number in rainbow trout, *Oncorhynchus mykiss*. *Aquaculture,* **247**, 85–95.

7 Hybridization in threatened and endangered animal taxa: Implications for conservation and management of biodiversity

Kelly R. Zamudio and Richard G. Harrison

Hybridization between species often results in offspring that are less fit than pure parental forms and may result in selection for traits that enhance prezygotic barriers to gene flow (a process known as *reinforcement*). Evolutionary biologists also recognize a role for interspecific hybridization in promoting the evolution of novel forms (Stebbins 1950; Anderson & Stebbins 1954; Harrison 1993a,b; Rhymer & Simberloff 1996; Arnold 1997). Natural hybridization occurs relatively frequently among divergent populations or species of plants and animals (Stace 1975; Barton & Bengtsson 1986; Grant & Grant 1992; Mallet 2005), and, despite the relative rarity of hybrids in any single population, only a few are needed to allow the exchange of advantageous alleles between species (i.e., *introgression*). The historical admixture of genomes has also contributed to speciation, especially in plants but also in some animal taxa (Arnold 1997; Dowling & Secor 1997; Mallet 2005; Gompert et al. 2006; Mavárez et al. 2006; Grant & Grant 2008).

In conservation biology, the predominant view is that introgressive hybridization is a cause of extinction (Rhymer & Simberloff 1996; Wolf et al. 2001), especially when hybridization events are anthropogenically mediated. Previously allopatric species are now commonly brought into contact due to anthropogenic introductions (Pysek et al. 1995). Likewise, previously parapatric or sympatric species now commonly hybridize due to human-induced changes in the environment that reduce elements of habitat heterogeneity that once contributed to diversification and maintained species independence (e.g., through ecological isolation; see Candolin et al. 2006; Hendry et al. 2006; Seehausen et al. 2008). Some have argued that this "smoothing" of landscapes is equivalent to a "speciation reversal," a process that will become increasingly common in changing environments (Seehausen et al. 2008).

Our view of whether hybridization benefits or erodes biodiversity will necessarily be influenced by the context in which hybridization occurs and the degree

We thank Andrew DeWoody and the other editors for the opportunity to contribute to this volume and for their efforts in organizing a network of researchers to discuss these critical issues in conservation biology. Two anonymous reviewers commented on and improved earlier versions of the manuscript. Richard Harrison thanks the University of St. Andrews for hosting him during his sabbatical leave. Conservation genetic and hybridization studies in our labs have been funded by the National Science Foundation, the President's Council of Cornell Women, and the College of Arts and Sciences, Cornell University.

to which hybridization has the potential to reduce species diversity (Allendorf et al. 2001; Price & Muir 2008). Hybridization events are initiated and facilitated by many different anthropogenic activities, including introductions of previously allopatric species, alteration of species' ranges, or more subtle changes to habitats that break down reproductive barriers between sympatric species (see Box 7) These different contexts make it especially challenging for conservation biologists to manage endangered species that are threatened by hybridization and to prevent extinction of the parental forms. Allendorf et al. (2001) categorized cases of hybridization, with a focus on the cause of the hybridization event (natural or anthropogenic) and, within each of those categories, the extent of introgression between hybridizing taxa. Natural hybrid taxa fall into hybridization Types I–III, with increasing degrees of introgression. The three categories of anthropogenic hybrids (Types IV–VI) also reflect increasing degree of introgression, with Type IV hybridization events producing only F1 offspring and no or limited backcrossing, and Type V and VI cases forming hybrid swarms with partial or complete admixture, respectively (Allendorf et al. 2001). This system for ranking the "severity" of hybridization served as an alarm for conservation biologists and a framework for determining the loss of "pure" parental species in conservation efforts (Allendorf et al. 2001).

Categorizing the degree of introgressive hybridization is a useful starting point when analyzing a posteriori the outcome of hybridization events that have already occurred. The challenge in conservation biology, however, is to develop predictions based on ecological or species-typical correlates that might allow us to detect the cases of hybridization that are most likely to result in negative outcomes for endangered or threatened species. Hybridization can result in the decline or loss of species when hybrids exhibit reduced fitness relative to parental forms and the endangered species, often far less abundant than the species with which it hybridizes, becomes "demographically swamped" due to low recruitment. Alternatively, hybrids may show no reduction in fitness and may replace pure parental forms through "genetic swamping" (Ellstrand & Elam 1993; Levin et al. 1996; Rhymer & Simberloff 1996). The ideal situation for conservation biology would be if potentially hybridizing species could be identified a priori, allowing time for research and mitigation efforts that specifically targeted environmental or species-specific barriers to introgression after the species came into contact. In most cases of hybridizing endangered taxa, we do not have a thorough understanding of the dynamics of hybridization, and this limits our opportunities to identify mechanisms that might be used in conservation efforts (Wolf et al. 2001). As we explore the increasing number of cases of anthropogenically mediated hybridization in species that are already considered to be of conservation concern, we should focus on particular characteristics of species, or the contexts of hybridizing species pairs, that might promote or restrict introgression (Wolf et al. 2001; Randler 2006). Those parameters might lend themselves to manipulation in the future, as our knowledge of how to assess and manage hybridization increases.

In this chapter, we review the recent literature on hybridization between an endangered or threatened taxon and a related nonendangered animal species. Specifically, we examine how many documented cases of hybridization threaten

BOX 7: MATING OPPORTUNITIES IN ANIMAL HYBRID ZONES

Marjorie Matocq

Problem

When potentially hybridizing taxa overlap spatially, asymmetry in densities between taxa or sex-ratio biases within taxa can result in a form of reproductive interference (Gröning and Hochkirch 2008) that promotes hybridization. Hubbs (1955) first recognized that when one species is scarce while another is abundant, individuals of the rare species may have difficulty locating fertile conspecific mates and, by chance alone, will mate with heterospecifics. Even in systems where females, for example, show strong mate preference, rare females or those that remain unpaired after all conspecific males have paired may eventually pair with heterospecific males rather than go unmated (Wirtz 1999; Veen et al. 2001). Asymmetry in densities or the availability of mates is likely to be particularly common at the leading edge of expansion or invasion fronts (e.g., Malmos et al. 2001) and may be an important factor contributing to hybridization in taxa of conservation concern.

Case Study

"Hybrid zone dynamics and species replacement between *Orconectes* crayfishes in a northern Wisconsin lake" (Perry et al. 2001a).

The rusty crayfish, *O. rusticus*, is native to parts of Ohio, Kentucky, and Indiana but has been distributed broadly across North America by humans. This distribution has resulted in the widespread extirpation of resident crayfish and changes in community structure. *O. rusticus* was introduced to Trout Lake, Wisconsin, around 1979 at a time when *O. propinquus* comprised 90% of the native crayfish fauna of the lake. Although *O. rusticus* is displacing *O. propinquus*, the two species readily hybridize at the expansion front with 95% of F_1 hybrids resulting from crosses between *O. rusticus* females and *O. propinquus* males. Although fecundity and/or viability differences between crosses and mate preference differences may contribute to the observed cytonuclear pattern, estimates of the mating asymmetry coefficient predict that 80% of hybrids should have an *O. rusticus* mitochondrion. The mating asymmetry arises because *O. rusticus* females are 3.5 times more common than *O. rusticus* males at the expansion front (Box Fig. 7–1). This sex bias likely leads to difficulty in obtaining conspecific matings by *O. rusticus* females and results in an opportunity for heterospecific matings.

On the one hand, ecological replacement of *O. propinquus* is being hastened by hybridization because individuals of this species are at a competitive disadvantage not only to pure *O. rusticus* but also to *O. rusticus*/*O. propinquus* hybrids. On the other hand, hybridization is allowing the conservation of some *O. propinquus* alleles, as these are found in otherwise largely *O. rusticus* individuals far from the area of ongoing hybridization (Perry et al. 2001b). Paradoxically, then, hybridization is both hastening the extirpation of *O. propinquus* and allowing conservation of at least some of its genome.

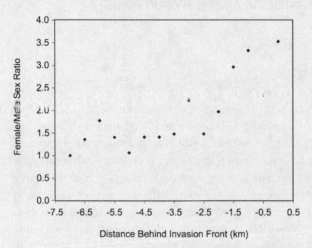

Box Figure 7–1: Female-to-male sex ratio (corrected for trapping bias) of *O. rusticus* behind their invasion front into the range of *O. propinquus* in Trout Lake, Wisconsin. Lack of conspecific mating opportunities for *O. rusticus* females at the invasion front (zero of the *x*-axis) contributes to hybridization with *O. propinquus*. Modified from figure 6 of Perry et al. (2001a).

The case of *Orconectes* crayfish demonstrates that the ecological and demographic context in which mating decisions are made can profoundly impact the degree and direction of hybridization at the expansion front of invasions.

REFERENCES

Gröning J, Hochkirch A (2008) Reproductive interference between animal species. *Quarterly Review of Biology*, **83**(3), 257–282.

Hubbs CL (1955) Hybridization between fish species in nature. *Systematic Zoology*, **4**, 1–20.

Malmos KB, Sullivan BK, Lamb T (2001) Calling behavior and directional hybridization between two toads (*Bufo microscaphus* × *B. woodhousii*) in Arizona. *Evolution*, **55**(3), 626–630.

Perry WL, Feder JL, Dwyer G, Lodge DM (2001a) Hybrid zone dynamics and species replacement between Orconectes crayfishes in a northern Wisconsin lake. *Evolution*, **55**, 1153–1166.

Perry WL, Feder JL, Lodge DM (2001b) Implications of hybridization between introduced and resident Orconectes crayfishes. *Conservation Biology*, **15**(6), 1656–1666.

Veen T, Borge T, Griffith SC et al. (2001) Hybridization and adaptive mate choice in flycatchers. *Nature*, **411**, 45–50.

Wirtz P (1999) Mother species–father species: unidirectional hybridization in animals with female choice. *Animal Behaviour*, **58**, 1–12.

to lead to the loss of diversity and whether specific factors – such as the ecological context for hybridization, demographic factors, or the evolution of isolating mechanisms – might predict the degree of threat or provide guidance for how we should best manage potential admixture. Our survey underscores the important role of anthropogenic change as a promoter of genetic admixture in endangered taxa (Wolf et al. 2001; Seehausen et al. 2008; Brede et al. 2009). We found that hybridizing endangered species typically have smaller population sizes than the hybridizing nonendangered relatives, and that the breakdown of specialized habitat use contributes to most cases of hybridization. Although these demographic and ecological correlates of hybridization are commonly recognized, we found a surprising lack of information about the relative fitness of hybrids, the degree of backcrossing, and the presence or absence of pre- or post-zygotic isolating barriers. These characteristics determine the dynamics of hybridization and are critical data for efforts to predict and, if necessary, control the extent of hybridization. Finally, we found cases where hybridized endangered taxa showed higher fitness in anthropogenically modified environments, suggesting that some degree of hybridization might benefit endangered taxa by providing greater adaptive potential with which to respond to environmental change.

As conservation biologists, we need to establish clear benchmarks for the management of hybridizing species that include proactive measures to establish standardized methods of detection (Miller et al. 2003; Randi 2008), estimates of the consequences of historical and recent hybridization (Allendorf et al. 2001), as well as the ecological and evolutionary contexts for the hybridization events. We close with recommendations for areas of study that might increase our ability to predict the severity of threats due to hybridization in animal taxa and with possible evolutionary mechanisms that might be exploited in the management of these systems.

CASE STUDY

We surveyed primary literature databases for well-documented instances of hybridization involving threatened or endangered taxa. We developed specific criteria for inclusion of studies in our review and included only cases reported in the literature that 1) demonstrated conclusively the presence of hybridization in a focal taxon that was considered threatened or endangered, with appropriate analyses of reference populations as well as suspected hybrid populations; 2) involved cases of hybridization between natural and/or domesticated animal species; 3) made use of appropriate markers (molecular or morphological) with sufficient sample sizes for detection and characterization of hybridization; and/or 4) reported data on the context for hybridization in the focal taxon, including frequency of hybridization, whether hybridization was facilitated by anthropogenic habitat disturbance or by other anthropogenic or natural mechanisms. These criteria reduce the number of cases included in our study but allow us to examine specific ecological and evolutionary aspects that might contribute specifically to hybridization in threatened wildlife.

We surveyed the literature from 1998 through 2008 using Web of Science literature keyword searches. For each published study that met our inclusion criteria, we collected available data on potential parameters that might be predictors or correlates of hybridization. First, we explored the demographic or ecological contexts for hybridization in each case. Specifically, we asked 1) whether the hybridization event was naturally occurring or anthropogenically facilitated; 2) which hybridization category (I–VI) applied to each case, according to the degree of introgression of the two taxa; 3) whether any information existed on the population sizes of the two hybridizing taxa; and 4) whether the hybridizing taxa were habitat specialists either in the zone of sympatry or in the original range of the parental forms. Second, we asked whether published studies on those cases of hybridization reported key parameters that might promote an understanding of the most likely outcome of hybridization for each endangered species. Specifically, we gathered information on the mating dynamics between the parental forms and the relative fitness of hybrid offspring. When the data were available, we scored each hybridization case for 5) the fitness (survival and reproductive success) of hybrid offspring (more fit, less fit, or equal in fitness to parental types); 6) whether the hybrid zone showed evidence of backcrossing with parental types; 7) whether the hybrid zone showed evidence of asymmetry in introgression; and 8) whether mating or reproductive success in the hybrid zone indicated the evolution of pre- or post-zygotic barriers to reproduction between the hybridizing pairs. If quantitative data on a specific aspect of hybridizing taxa were not present in the articles reviewed, we listed that category as "unknown."

We used our survey results to identify hybridization dynamics involving threatened and endangered species that could contribute to our understanding of the causes, correlates, and possible outcomes of these events. First, we assessed the proportion of natural and anthropogenically mediated cases of hybridization and compared the degree of introgression among our samples of endangered or threatened taxa. We then compared cases of natural and anthropogenic hybridization to address whether they differed in degree of habitat specialization and in population size. Third, we examined cases where hybrid taxa are more fit than parental forms and asked whether hybrid fitness was highest in anthropogenically modified habitats. Finally, we quantified the cases with evidence for potential isolation mechanisms and asked whether pre- or post-zygotic isolating mechanisms were more common. Combined, our survey data summarize the most common evolutionary aspects and ecological correlates of hybridization in endangered and threatened animal taxa. Our survey also directs attention to specific strategies that could be employed to reduce introgression when it is a threat to specific taxa of conservation concern and challenges us to consider cases in which hybridization might be a useful management tool in the preservation of biodiversity.

RESULTS

Anthropogenic and natural hybridization in endangered animal taxa

We found sixty-two articles that discussed hybridization involving at least one endangered or threatened animal species. Of these, twenty-five reported on species that met our specific criteria for this study (listed by taxonomic group in Table 7–1). The remaining thirty-seven articles were excluded from further analyses because they either reported on hybridization in endangered plant species

Plate I: Book contributors at an October 2008 meeting, held at the John S. Wright Forestry Center (Purdue University). (See Preface.)

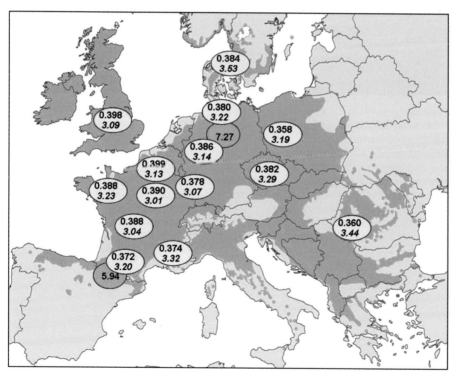

Plate II: Geographic range of *Quercus petraea* and distribution of genetic diversity based on nuclear data (according to Zanetto & Kremer 1995, figs. 2 and 3, and unpublished data). Data in yellow circles correspond to thirteen isoenzymatic loci and represent observed heterozygosity (in bold characters) and allelic richness (in italic). Each circle is the mean value obtained from two to eleven populations. Data in pink circles correspond to nucleotide diversity (π values) obtained from nine candidate genes of bud burst, from fifty sequences in North Germany, and fifty sequences in the Pyrénées. Distribution map was obtained from Euforgen (http://www.bioversityinternational.org/networks/euforgen/Euf_Distribution_Maps.asp). (See Figure 5–3.)

Community & Ecosystem Heritability in
Populus angustifolia

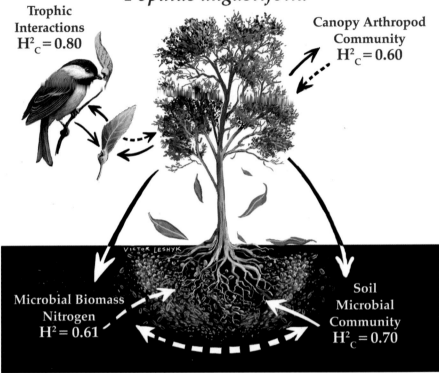

Trophic Interactions $H^2_C = 0.80$

Canopy Arthropod Community $H^2_C = 0.60$

Microbial Biomass Nitrogen $H^2 = 0.61$

Soil Microbial Community $H^2_C = 0.70$

VICTOR LESHYK

Plate III: Community and ecosystem phenotypes of individual tree genotypes of *P. angustifolia* show broad-sense heritability (H^2). Significant heritability of the canopy arthropod community, soil microbial community, trophic interactions between birds and insects, and soil nutrient pools demonstrate how trait variation in a foundation tree can structure communities and ecosystem processes (H^2_C for community traits; H^2 for single ecosystem trait). Because soil microorganisms mediate many ecosystem processes, including litter decomposition and rates of nutrient mineralization, the formation of these communities may feed back to affect plant fitness. Solid lines indicate known interactions; dashed lines indicate possible interactions. Quantitative genetic patterns such as these argue that genomic approaches will enhance our understanding of how interacting community members influence ecosystem processes. Figure from Whitham and coworkers (2008). (See Figure 3–2.)

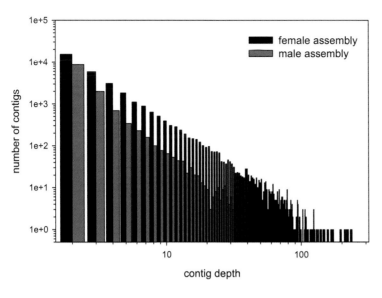

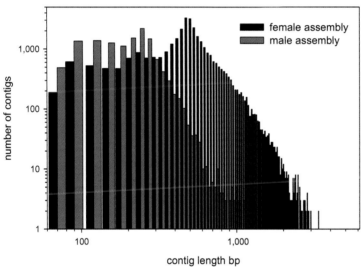

Plate IV: The data from 454 runs can be used to generate thousands of long contigs that each consists of many individual sequencing reads. From Hale et al. (2010). (See Figure 4–2.)

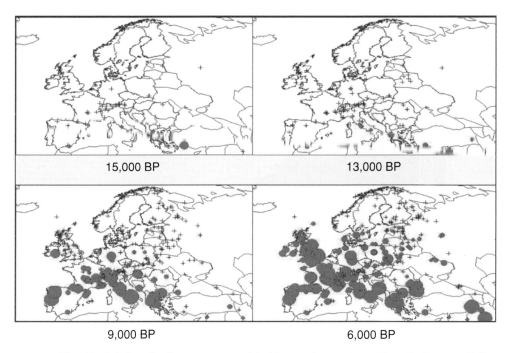

Plate V: Variation of pollen percentages of deciduous oaks as extracted from the European Pollen Database (according to Brewer et al. 2002, Fig. 2). Crosses indicate locations of sites; sizes of the red circles indicate percentages varying from 1% to 50% (1%, 10%, 25%, and 50%). (See Figure 5–1.)

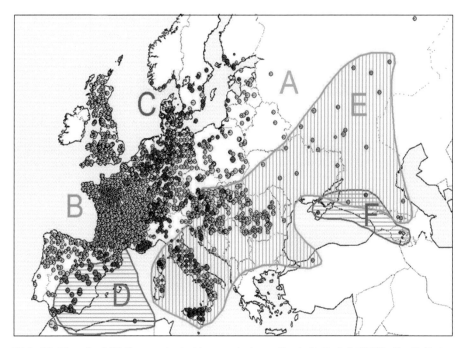

Plate VI: Map of cpDNA lineages in deciduous oaks (according to Petit et al. 2002b, Fig. 3). Forty-eight different haplotypes were identified that cluster in six major lineages indicated by colors and letters. (See Figure 5–2.)

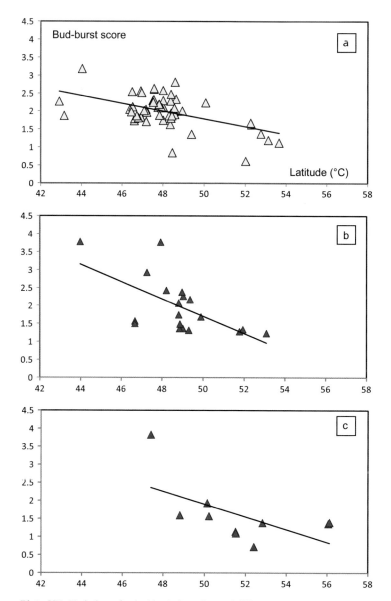

Plate VII: Variation of apical bud phenology of different populations of *Quercus petraea* in provenance tests. The analysis was done separately for each different maternal lineage to which modern populations belong. (a) yellow (B lineage) according to Fig. 5–2; (b) blue (A lineage) according to Fig. 5–2; (c) red (C lineage) according to Fig. 5–2. Bud development was recorded according to bud burst scores varying from 0 (dormant) to 5 (elongating leaves). (See Figure 5–4.)

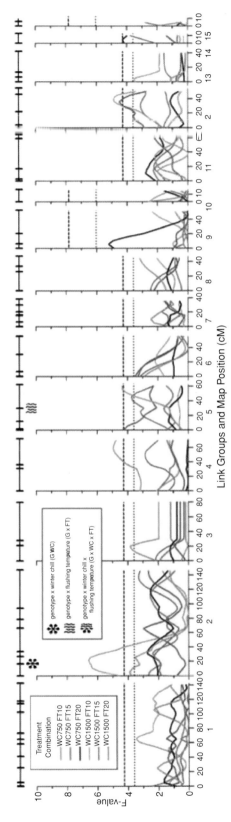

Plate VIII: Terminal bud flush (TBF_GI) was used as a surrogate trait to measure growth initiation in the spring. The overwintering bud is released from dormancy, and growth is initiated. Seven QTL for TEF were detected in the growth initiation experiment (TBF_GI). QTL were found on six linkage groups (LGs 2, 3, 4, 5, 12, and 14) and were detected in five of the six treatment (T) combinations. Only two QTL × T interactions were found, one for winter chill on LG2 and one for flushing temperature on LG5. The interaction detected on LG5 is located at a marker that is intermediate between two QTL detected by interval mapping (from Jermstad et al. 2003). (See Figure 6–2.)

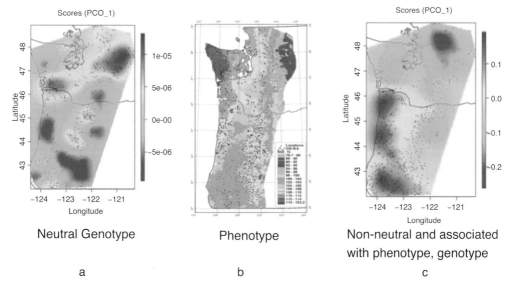

Neutral Genotype	Phenotype	Non-neutral and associated with phenotype, genotype
a	b	c

Plate IX: Patterns of neutral genetic diversity (a), phenotypic diversity (b), and adaptive genetic diversity (c). (See Figure 6–3.)

a	b

Plate X: The golden-cheeked warbler (a) requires mature oak-juniper forests for nesting habitat. The black-capped vireo (b) nests in early-successional shrubs and edge habitat (photos by Kelly Barr). (See Figure 9–3.)

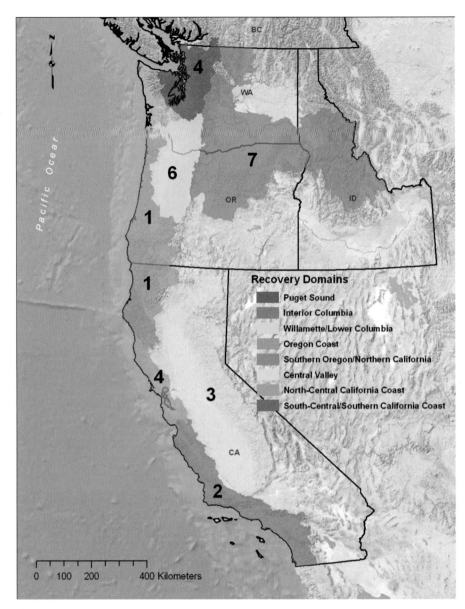

Plate XI: ESA recovery domains for Pacific salmon; the number of listed salmon ESUs in each domain is shown in large black numerals. (See Figure 10–1.)

Table 7–1. Cases of hybridization between endangered or threatened animal taxa (E/T taxon) and a congener or subspecies (other taxon), and the references for data on factors preventing or promoting hybridization in each case

Case	Taxon	E/T taxon	Other taxon	Source(s)
1	Invertebrate	*Acropora pulchra*	*A. millepora*	1
2	Invertebrate	*Orconectes propinquus*	*O. rusticus*	2
3	Fish	*Cyprinodon pecosensis*	*C. variegatus*	3,4
4	Fish	*Oncorhynchus clarki lewisi*	*O. mykiss*	5
5	Fish	*Oncorhynchus clarki henshawi*	*O. mykiss*	6
6	Fish	*Chasmistes brevirostris*	*Catostomus snyderi*	7
7	Fish	*Pseudorasbora pumila*	*P. parva*	8
8	Amphibian	*Ambystoma californiense*	*A. tigrinum*	9,10
9	Reptile	*Aspidoscelis dixoni*	*A. tigris*	11
10	Bird	*Vermivora chrysoptera*	*V. pinus*	12
11	Bird	*Strix occidentalis caurina*	*S. varia*	13
12	Bird	*Puffinus mauretanicus*	*P. yelkouan*	14
13	Bird	*Cyanoramphus forbesi*	*C. novazelandiae*	15
14	Bird	*Oxyura leucocephala*	*O. jamaicensis*	16
15	Mammal	*Canis rufus*	*C. latrans*	17,18,19
16	Mammal	*Canis lycaon*	*C. lupus*	20
17	Mammal	*Canis lupus*	*C. familiaris*	21
18	Mammal	*Bubalus arnee*	*B. bubalis*	22
19	Mammal	*Felis silvestris silvestris*	*F. catus*	23
20	Mammal	*Lynx canadensis*	*L. rufus*	24
21	Mammal	*Arctocephalus tropicalis*	*A. gazella*	25

1: Willis et al. 2006; 2: Perry et al. 2001; 3: Rosenfield et al. 2004; 4: Rosenfield & Kodric-Brown 2003; 5: Allendorf et al. 2004; 6: Peacock & Kirchoff 2004; 7: Tranah & May 2006; 8: Konishi & Takata 2004; 9: Riley et al. 2003; 10: Fitzpatrick & Shaffer 2007a; 11: Cole et al. 2007; 12: Vallender et al. 2007; 13: Haig et al. 2004; 14: Genovart et al. 2007; 15: Tompkins et al. 2006; 16: Muñroz-Fuentes et al. 2007; 17: Miller et al. 2003; 18: Fredrickson & Hedrick 2006; 19: Phillips et al. 2003; 20: Kyle et al. 2006; 21: Verardi et al. 2006; 22: Flamand et al. 2003; 23: Oliveira et al. 2008; 24: Schwartz et al. 2004; 25: Kingston & Gwilliam 2007.

or, in cases of animal hybridization, lacked sufficiently detailed data on the identity of parental species or the degree of introgression in the hybrid zone. For three species pairs, more than one article was identified from the literature survey (Table 7–1), so we focused on twenty-one distinct examples of hybridization in this chapter.

Of the twenty-one cases included in our study, only four (19%) were cases of natural hybridization among parapatric or sympatric taxa (Types I–III). Those four cases show variable levels of introgression between the nonendangered and endangered members of the species pair. Not surprisingly, our survey shows that animal taxa that are listed as threatened or endangered are more frequently involved in anthropogenically mediated hybridization; sixteen (76%) of the cases reviewed involve cases of hybridization attributable to anthropogenic habitat change, due to either introductions or shifts in ranges or habitats of the species involved. In one case, the cause of hybridization was unknown (Table 7–2). Fifteen of the sixteen hybridization cases thought to be anthropogenically mediated show extensive introgression between species, with production of fertile F1 and evidence for backcrossing with parental forms (hybridization Types V and VI).

Table 7–2. Survey of ecological correlates of introgression in twenty-one documented cases of hybridization involving endangered and/or threatened taxa. In each case, we identified the cause of hybridization (natural or anthropogenic), hybridization category (I–VI), differences in population size, habitat use, hybrid fitness, evidence for backcrossing, asymmetric introgression, or the presence of pre- and/or post-zygotic barriers to reproduction between the two species

	Taxon	Species	Cause	Cat.	Pop. Size	Habitat	Hybrid fitness	Backcrosses	Asymmetry	Prezygotic	Postzygotic
1	Invertebrate	A. pulchra	Natural	III	Unknown	Specialized	More	Unknown	Unknown	Yes	No
2	Invertebrate	O. propinquus	Anthro	V	Smaller	Same	Unknown	Yes	Yes	Unknown	Unknown
3	Fish	C. pecosensis	Anthro	V/VI	Smaller	Specialized	More	Yes	No	Yes	No
4	Fish	O. c. lewisi	Anthro	V/VI	Smaller	Same	Less	Yes	No	Yes	Yes
5	Fish	O. c. hanshawi	Anthro	V	Larger	Specialized	Unknown	Unknown	No	Unknown	Unknown
6	Fish	C. brevirostris	Natural	II	Unknown	Same	Unknown	Yes	No	Yes	No
7	Fish	P. pumila	Anthro	VI	Smaller	Same	Less	No	Yes	Yes	Yes
8	Amphibian	A. californiense	Anthro	V	Larger	Specialized	More	Yes	Yes	Yes	No
9	Reptile	A. dixoni	Anthro	IV	Smaller	Specialized	Less	No	Yes	Yes	Yes
10	Bird	V. chrysoptera	Anthro	V/VI	Smaller	Specialized	More	Yes	Yes	Yes	No
11	Bird	S. o. caurina	Anthro	V	Smaller	Specialized	Less	Yes	Yes	Yes	No
12	Bird	P. mauretanicus	Natural	II	Smaller	Unknown	Unknown	Yes	Unknown	Yes	No
13	Bird	C. forbesi	Anthro	V	Smaller	Unknown	More	Unknown	Unknown	Unknown	Unknown
14	Bird	O. leucocephala	Anthro	V	Smaller	Unknown	Less Fit	Yes	Yes	Yes	Yes
15	Mammal	C. rufus	Anthro	V/VI	Smaller	Specialized	Less Fit	Yes	Unknown	Yes	Yes
16	Mammal	C. lycaon	Natural	III	Smaller	Specialized	Unknown	Yes	Unknown	Unknown	Unknown
17	Mammal	C. lupus	Anthro	V	Smaller	Specialized	Unknown	Yes	No	Unknown	Unknown
18	Mammal	B. arnee	Anthro	V	Smaller	Specialized	Unknown	Unknown	No	Unknown	Unknown
19	Mammal	F. s. silvestris	Anthro	V	Smaller	Specialized	Unknown	Yes	Yes	Unknown	Unknown
20	Mammal	L. canadensis	Unknown	IV (?)	Smaller	Specialized	Unknown	No (?)	Yes	Unknown	Unknown
21	Mammal	A. tropicalis	Anthro	V	Unknown	Same	Unknown	Yes	Yes	No	Yes

Restricting our literature survey to the last decade captures much of the most recent conservation genetics literature but excludes some of the "classic" cases of hybridization discussed in the earlier literature (see review by Rhymer & Simberloff 1996). Our survey included seven of the more than twenty cases summarized by Rhymer and Simberloff (1996). Of the cases that we "missed," virtually all involve hybridization of an endangered species with an introduced, feral, or domestic relative. In nearly all of these cases, hybridization and backcrossing have led to extensive admixture. Therefore, our results are consistent with those obtained previously.

Ecological correlates of hybridization

Of the four cases of natural hybridization in our study, two involve hybridization of endangered taxa with population ranges smaller than that of the nonendangered species; two of the endangered species are also habitat specialists (Table 7–2). These features are more common among the cases of anthropogenically mediated hybridization; of the fifteen cases for which data are available, thirteen (87%) involve an endangered or threatened species with a smaller range size than the nonendangered hybridizing congener. In ten (67%) of those cases, the endangered taxon shows specialization to or preferential use of a particular habitat type that is more restricted in distribution than that of a more generalist congener.

Hybrid zone dynamics in endangered taxa

Data on the breeding dynamics within hybrid zones are more difficult to obtain and, therefore, are not as frequently available as the data on ecological or demographic correlates of hybridization (Table 7–2). Among our four cases of naturally hybridizing threatened species, data on the fitness of hybrids were known in only one case in which hybrid fitness was higher than that of the parental species. In three of the cases of natural hybridization, hybrids were known to backcross with parental species. Finally, data on the presence or absence of pre- and postzygotic barriers to gene flow were available for three of those cases, but only one of the endangered species showed evidence of partial prezygotic isolation (Table 7–2).

Among the sixteen anthropogenically mediated hybridization cases, we found data on the fitness of hybrids relative to parental forms in ten cases. In six of those cases, hybrids were less fit than parental lineages; in four, they showed higher fitness due to higher survival, reproductive success, or both. Likewise, data on the mechanisms determining the extent of introgression were also sparse. For eleven of the cases, we found evidence for backcrossing between F1 hybrids and parental forms; for nine of the cases, the literature showed evidence of asymmetric introgression, with disproportionate loss of "pure" genomes in the endangered or threatened species. Finally, seven of the cases included in our study showed some evidence of partial barriers to gene flow among the hybridizing taxa. Six of the species pairs showed evidence of postzygotic barriers, whereas only one revealed the presence of prezygotic barriers to gene flow.

DISCUSSION

What data are missing in studies of hybridization in endangered taxa?

One of the most obvious results of our survey is the absence of data on the local dynamics and evolutionary processes occurring within hybrid zones of endangered or threatened species (Table 7–2). In most of these cases, the direction and extent of hybridization has been characterized. Likewise, the ecological context for hybridization (e.g., whether the hybridization occurred because of changes in habitat distribution or an accidental introduction of the nonendangered hybridizing species) is usually understood (Table 7–2). What are lacking are data on the evolutionary consequences of such changes and the processes operating in the recently established hybrid zones. Most notably, the fitness of hybrids relative to parental types and presence of pre- and post-zygotic barriers to gene flow in the hybridizing pair are important predictors of evolutionary outcome that are often not known. The reason for paucity of data in these evolutionary categories is likely the difficulty in obtaining these kinds of data on rare and endangered species. For example, quantifying the relative fitness of hybrids would ideally include identification of hybrid offspring and a comparative study of their survival and reproductive success within the hybrid population. Likewise, barriers to interspecific reproduction can result from many different mechanisms, including differences in prezygotic behavioral mate choice or phenology and postzygotic mechanisms such as hybrid inviability (Harrison 1990; Coyne & Orr 1997). Thus, identifying the presence of reproductive barriers and characterizing the underlying mechanism(s) can be a complex task. These data are critical, however, if we are to understand the factors promoting hybridization and exploit them to manage hybridization in endangered species. The empirical and analytical methods to characterize the occurrence, degree, and direction of introgression due to hybridization are now readily available and increasing in sophistication (Manel et al. 2005; Randi 2008). We now need to focus on studies and methods to characterize the mechanisms that have led to those patterns.

Predictors and correlates of hybridization in endangered taxa

A recent review of hybridization (Mallet 2005) suggests that we should be "suspicious" of arguments that hybridization is most often a result of environmental disturbance. The fact that the majority of hybridization cases between a threatened species and a nonendangered relative are anthropogenically facilitated, however, underscores the important effects that habitat modification can have on species recognition and assortative mating (Seehausen et al. 2008). Environmental change that alters patterns of habitat heterogeneity may profoundly influence the extent to which habitat isolation persists as a barrier to gene exchange and may compromise the independence of lineages in early stages of divergence.

The cases included in our survey also corroborate previous surveys of hybridization in nonendangered animal taxa and highlight the importance of demography as a predictor of hybridization (Randler 2006; Brede et al. 2009). A consistent correlate of hybridization is small population size of one species (in our case, the endangered one; Table 7–2). Threatened and endangered species often are listed because of small local populations and/or low overall abundance; thus, the observed correlation is not surprising. This demographic pattern, however, has important implications for the predominant direction of hybridization: Individuals of rare species are less likely to encounter conspecifics and thus more likely to hybridize with members of sympatric, closely related species. Where endangered species have become fully sympatric with nonendangered (and more common) relatives, "genetic swamping" of the less common parental lineage may result (Wolf et al. 2001) (i.e., pure parentals will gradually disappear entirely, to be replaced with individuals of mixed ancestry). Likewise, F1 hybrids will tend to backcross with the more common parental type, further reducing the numbers of individuals with substantial proportions of the endangered species' genome. In such cases, allele frequencies of the rare subpopulation will gradually tend to the allele frequencies of the large subpopulation, leading, in extreme cases, to the complete disappearance of the sink subpopulation. In cases of parapatry, where there is a distinct hybrid zone between the endangered species and its relative, the hybrid zone may move, pushed forward by "weight of numbers" (Barton & Hewitt 1981). As a consequence, the "pure" endangered species will gradually disappear. Similarly, in cases of local introductions within the range of an endangered species, hybridization and backcrossing may lead to an expanding area of admixed individuals and a corresponding decrease in the range of the "pure" endangered species.

With few exceptions, the endangered species we surveyed show smaller population sizes in contact zones and also relatively high degrees of gene exchange; most fall within hybrid categories V (widespread introgression) and VI (complete introgression/hybrid swarm) (Allendorf et al. 2001). Therefore, the scenarios outlined in the previous paragraph may not be uncommon. The propensity for endangered species to have small population sizes increases their risk of genetic swamping and makes it especially challenging to devise practical and efficient management plans for populations threatened by hybridization. Selective removal of the nonendangered species and of hybrid offspring is one method that has been applied in attempts to preserve the genetic integrity of endangered populations (Miller et al. 2003; Adams & Waits 2007). Unfortunately, in practice, this management approach may reduce the rate of gene exchange (Stoskopf et al. 2005) but is unlikely to eliminate the occurrence of hybridization in most cases. First, selective removal of the nonendangered species or their hybrid offspring will need to be continuous because the source population for individuals of the nonendangered species is usually large compared with the population of the endangered species. Second, removal of hybrid offspring may, in fact, simply further reduce the population size of alleles from the endangered species, if matings between pure parentals of this species are rare. Thus, reduction of what is often already a relatively small population may result in higher probabilities of stochastic extinctions of alleles from the endangered species. In terms of demographics and the correlated hybrid zone dynamics, the decks are stacked against endangered species; selective removal of hybrids and hybridizing nonendangered species will not likely have a permanent effect on limiting genetic introgression, although it may serve as a temporary measure.

A second important ecological correlate of hybridization in endangered and nonendangered animal taxa is habitat specialization (Seehausen et al. 2008). Most of the surveyed species, for which we have information on habitat preferences and use, show some degree of differential habitat use when compared to close relatives (Table 7–2). Although the environmental differences are sometimes subtle (MacCallum et al. 1998; Haig et al. 2004), they appear, in retrospect, to have been sufficient barriers to maintain independent evolutionary trajectories for incipient species. If anthropogenic habitat changes specifically erode those barriers, the likelihood of independent coexistence of species will decrease. The nature of anthropogenic change and the historical relationship between species will also have consequences for predicting and preventing hybridization. Commonly, two potentially hybridizing species exist parapatrically or sympatrically and maintain divergent populations through use of different microhabitats (i.e., ecological isolation). These closely related and ecologically divergent species will be particularly susceptible to changes in the environment because their evolutionary origins were dependent on recent selection for divergent adaptation to localized environments (Seehausen et al. 2008). The genetic integrity of many endangered species continues to depend on specialization in these heterogeneous habitats (see Chapter 9 by Leberg and colleagues). Thus, changes that alter the distribution of microhabitats or decrease the specificity of habitat use by one of the hybridizing species (typically the species that is less specialized) will increase the incidence of hybridization. This is one arena in which mitigation and intervention could play a significant role. If a major factor promoting hybridization is the "smoothing" or "equalization" of environments, then it should be possible to identify particular environmental parameters that are crucial for maintaining reproductive isolation and manage those to favor the persistence of endangered species. In some generalized cases, such as global climate change, which alter the distribution of appropriate habitats (Parmesan & Yohe 2003; Thomas et al. 2004) or alter phenologies of species in detrimental ways (Both et al. 2006), restoring those critical environmental elements may be difficult, if not impossible. However, in other cases, anthropogenic changes are specifically linked to aspects that alter the likelihood of species discrimination or encounter (Seehausen et al. 1997; Fitzpatrick & Shaffer 2007b). In these cases, specific environmental characteristics may promote the successful encounter of conspecific individuals within mosaic hybrid zones.

A well-characterized example of habitat-dependent hybridization is found in the California tiger salamander (*Ambystoma californiense*), an endangered species that hybridizes with the more widespread barred tiger salamander (*Ambystoma tigrinum mavortium*) following anthropogenic introductions of the barred tiger salamander within the breeding range of the formerly allopatric California tiger salamander (Riley et al. 2003; Fitzpatrick & Shaffer 2007b). The introduced species has spread from various introduction points, and the hybrid zone currently occurs throughout the Salinas Valley of California (Fitzpatrick & Shaffer 2007b). This landscape has been modified for cattle farming and agriculture, resulting in the proliferation of perennial or ephemeral cattle ponds, both of which have longer hydroperiods than the highly ephemeral vernal pools that were historically the natural breeding habitat of *A. californiense* (Fitzpatrick & Shaffer 2004).

Detailed genetic characterization of populations in different ponds show that populations in natural vernal pools are the least admixed (Fitzpatrick & Shaffer 2004), suggesting that characteristics of those breeding ponds promote some form of environment-dependent reproductive isolation through either individual mate choice within the breeding pond or reduced population size of the introduced species (Fitzpatrick & Shaffer 2004, 2007b). If specific environmental parameters controlling hybridization can be identified, restoration of habitats can attempt to mimic the environments that best match conservation goals for each species. Experimental control of hybridization through "environmental engineering" would be an excellent opportunity to evaluate the success of intervention and may be one mode of intervention that lends itself well to experimental adaptive management. Engineering the environment to reduce hybridization may not require complete restoration of historical habitats because in some endangered species, outbreeding occurs only when conspecifics are not available for mating or are present at low population densities (Randler 2006; Willis et al. 2006). Landscape restoration with focus on specific habitat characteristics required by endangered taxa is already widely applied in conservation efforts (Bruggeman & Jones 2008) and can be successfully encouraged through incentive programs that benefit land owners and users who voluntarily manage their properties to enhance biodiversity (Drechsler et al. 2007).

Hybrid zone dynamics in endangered taxa

Studies of hybrid zone dynamics in nonendangered taxa have provided important insights into the evolution of barriers to interspecific reproduction (Dowling & Secor 1997; Mallet 2005). Reproductive isolation arises from pre- or post-zygotic mechanisms that prevent the exchange of genes among two genetically independent populations or species (Coyne & Orr 2004). Theory suggests that selection against less fit hybrids can reinforce the evolution of prezygotic barriers to gene flow (Dobzhansky 1951); although there is some evidence for this process, the evolution of prezygotic isolation will differ depending on whether the two lineages are sympatric or allopatric and whether other selective regimens might also favor isolation barriers (Coyne & Orr 1989, 1997; Hurt et al. 2005). A critical question in conservation biology is whether such prezygotic barriers to gene flow will evolve sufficiently rapidly in cases of hybridization of endangered species to effectively reduce potential threats of hybridization (Fredrickson & Hedrick 2006). Unfortunately, the data to address this question are still sparse (Table 7–2). Given the limited data for reinforcement in presumably long-standing hybrid zones (Marshall et al. 2002), conservation biologists should not, however, rely on this evolutionary mechanism to provide protection against the erosion of the genetic integrity of endangered species. If new data provide evidence that reinforcement can occur on an ecological timescale, we will need to revisit this issue and the role it might play in conservation of hybridizing species.

In contrast, conspecific sperm precedence (CSP) may reduce the rate at which rare species disappear in the face of hybridization with a more common relative. CSP describes a phenomenon in which females that mate with both conspecific and heterospecific males produce offspring sired predominantly or exclusively by the conspecific male (Howard 1999). CSP is known to be an effective barrier to gene exchange in natural hybrid zones (e.g., Gregory & Howard 1994), and laboratory studies have shown that this barrier persists even when females mate multiply with heterospecific males and only once with a conspecific male (Gregory & Howard 1993). If females of an endangered species mate multiple times and if CSP exists, then even occasional encounters with a conspecific male may allow females to produce substantial numbers of "pure" offspring (i.e., not hybrids).

Can hybridization in threatened or endangered species promote conservation?

Conservation and management of biodiversity will become especially difficult with continued changes in global climate and habitats and in the face of other novel challenges, such as emergent diseases and invasive species (Harvell et al. 2002; Blehert et al. 2009). The landscape for endangered species is shifting rapidly, requiring more frequent conservation assessments and more intensive and evidence-based conservation approaches (Pullin et al. 2004; Sutherland et al. 2004). One active area of research now concerns the evolutionary potential of species challenged by climate or habitat change (Addo-Bediako et al. 2000; Bernardo & Spotila 2006) and the extent to which species have the ability to adapt to environmental changes (Ghalambor et al. 2007). Skeptics emphasize the short time frames of anthropogenic habitat change and the relatively slow evolutionary response of most phenotypes (Parmesan 2006); however, some recent examples of adaptations to changing climatic conditions indicate that individual genetic variation for resilient phenotypes may not be as limited as previously thought (Willi et al. 2006; Rako et al. 2007). Although the jury is still out on the potential for evolutionary adaptations to keep pace with global climate change to any appreciable degree, it is clear that understanding the evolutionary processes that have given rise to diversity in currently endangered plant and animal lineages may help us to understand the factors that have contributed to their resilience (Willi et al. 2006; Willis et al. 2006; Ellstrand 2009; Stelkens & Seehausen 2009). Hybridization of closely related species, although rare at ecological time scales, has apparently played an important role in the diversification of animal lineages (Seehausen et al. 2003; Willis et al. 2006). If hybridization and introgression allow adaptation to new environments, then in theory this process could also contribute to persistence of lineages in the face of environmental change (Seehausen 2004).

We identified cases in our survey in which hybridization specifically contributed to better performance of hybrid offspring in modified or changing environments. One of the best documented cases of hybridization in endangered animal taxa is in reef corals of the genus *Acropora* (Willis et al. 2006). Species of this genus hybridize readily in no-choice experimental crosses (Willis et al. 1997; van Oppen et al. 2002), but hybrids are produced at lower frequencies in sperm-choice trials among conspecifics (Willis et al. 2006), suggesting partial prezygotic isolating barriers with differential gamete use based on compatibility and sperm availability. Thus, the mating system of *Acropora* allows for a type of bet-hedging; when conspecific sperm is available, it is used preferentially (as mentioned earlier in our discussion of CSP), but in cases where matings are scarce (e.g., as in marginal or range edge habitats), hybridization permits the production of some fertile offspring (Willis et al. 2006). Reef corals in the genus *Acropora* are also spatially segregated, with different species occupying reef crests, slopes, and inner reef flats (Willis et al. 2006). The hybrids of *Acropora pulchra* and *A. millepora* grow faster and survive better in the environmentally variable reef crest and reef flat habitats not occupied by the parental species (Willis et al. 2006), suggesting that hybridization may be an important process for adaptation to new environments.

An example of apparent adaptive hybridization has also been documented in spadefoot toads (Pfennig 2007), in which interspecific differences in development rate between these inhabitants of ephemeral ponds lead to environment-dependent mate choice and facultative hybridization. Neither of the two species involved is endangered, but this example clearly focuses attention on what should be the "unit" that is the target of conservation efforts. If conservation efforts are solely devoted to maintaining the presumably integrated genomes of existing species, then hybridization clearly poses a threat. However, if maintaining allelic diversity at individual loci (or chromosomal regions) is worthy of our attention, then in some cases hybridization can be a "good" strategy and might, in fact, prevent local extinction.

Another well-documented example of increased adaptive potential in hybridized species includes increased resilience of small, endangered populations to novel disease challenges (Tompkins et al. 2006). With changes in global climate and habitat integrity come increases in the global threat of emerging infectious diseases (Daszak et al. 2000; Harvell et al. 2002; Mydlarz et al. 2006), which will likely have the highest impact on small, isolated populations (Altizer et al. 2003; Reid et al. 2003). In the context of disease, some degree of hybridization may favor species survival through increased diversity of the endangered species' gene pool and higher resilience to novel disease challenges. One such case is found in the endangered Forbes' parakeet (*Cyanoramphus forbesi*). This species hybridizes with the more widespread red-crowned parakeet (*C. novaezelandiae*), resulting in viable hybrid offspring with increased innate and cell-mediated immune responses (Chan et al. 2006; Tompkins et al. 2006). This increase in immune function was evident even in individuals showing few red-crowned parakeet phenotypes and thus was likely the result of backcrosses with Forbes' parental types. This immune advantage of individuals of mixed ancestry could, if properly managed, increase individual survivorship and viability of small remaining populations of Forbes' parakeets (Tompkins et al. 2006). This increase in resilience to emergent infectious diseases, however, would indeed come at a cost to the genetic integrity of the endangered parental species.

These examples of documented benefits of hybridization for endangered taxa underscore the complexity of managing endangered taxa while taking into account both ecological and evolutionary mechanisms for species and allele persistence. Hybridization could potentially benefit endangered species because they frequently exist in small, isolated populations that typically have lower genetic diversity (Ellstrand & Elam 1993; Spielman et al. 2004). Hybridization can also benefit evolving lineages more generally because it serves as a mechanism for admixing genomes and increasing the adaptive potential and persistence of species in stochastic environments. The challenge for conservation biologists is to measure the trade-offs inherent in managing species for "genetic purity" in a rapidly changing environment and to decide whether managing populations for hybridization could be an option for some endangered species. Conserving diversity does not necessarily mean preserving the status quo, especially during periods of rapid environmental change.

Bridging basic and applied studies of hybridization for conservation of endangered taxa

The challenges for conservation biologists in the current era of rapid habitat and global change are daunting (Parmesan & Yohe 2003). In terms of hybridization and the genetic admixture of genomes of endangered species, empirical evidence and predictions for future climate change suggest that we should prepare for an increasingly admixed world, where anthropogenically mediated opportunities for introductions and reduced habitat heterogeneity will become more common (Pysek et al. 1995; Williamson 1996). Proactive management of endangered populations and species will be the key to conserving current diversity as well as the adaptive potential required for future lineage diversification. The mechanisms promoting lineage diversification and maintenance of incipient species in natural settings might seem difficult to re-create, but understanding how those mechanisms have worked in natural diversifying lineages can help us determine whether completely curbing hybridization will be a productive strategy, and what evolutionary mechanisms might be important for managing these cases based on each specific case. Fortunately, cases involving hybridization can rest on decades of empirical and theoretical studies of hybridization of nonendangered species (Harrison 1993a; Arnold 1997) that provide an important framework for our predictions about the outcome of hybridization in endangered taxa.

Two important points arise from our survey of species threatened by hybridization. The first is that the challenges of admixture and genetic swamping of endangered species offer an excellent opportunity to bridge the gap between our knowledge of evolutionary processes and the outcome of conservation actions applied at ecological time scales; however, the data required to do so are still, for the most part, lacking. To best exploit possible incipient mate recognition or mate-selection mechanisms in the prevention or enhancement of hybridization, we need to focus data collection in ways that document both the genetics and demography of hybridization. We can best manage endangered species if we know the genetic mechanisms that promote or restrict hybridization, such as the extent of hybrid unfitness or hybrid advantage, the pattern of sperm use, and possible fitness consequences of recombined genomes.

Our survey also underscores the fact that hybridization can be a threat to the continued persistence of endangered species but, not uncommonly, we see potential benefits of this process, especially when the threatened species must persist in a rapidly changing landscape or environment. The genetic rescue effects from isolated populations suffering detrimental effects of genetic erosion have exploited admixture for decades, primarily in highly managed systems (Madsen et al. 1999; Allendorf et al. 2001). The examples we highlighted indicate that conservation biologists need to consider hybridization in all its dimensions and ask whether some level of hybridization and introgression is acceptable, given what we know of the costs and benefits in each case. Clearly, the costs will not be considered too high by most conservation biologists if the alternative to hybridization is extinction, yet the thresholds for what are "acceptable" degrees of hybridization have not yet been fully explored (Daniels & Corbett 2003; Haig et al. 2004; Fitzpatrick & Shaffer 2007b). We will clearly benefit from a more holistic view of the evolutionary processes that contribute to the origin and maintenance of genetic diversity. The conservation status of global biodiversity is constantly changing, in large part due to the widespread effects of human-induced environmental alteration. Thus, this view may allow us to better understand and even exploit the evolutionary processes evident in endangered species, to stabilize the decline of species worldwide, and to better manage the processes that have produced this diversity.

REFERENCES

Adams JR, Waits LP (2007) An efficient method for screening faecal DNA genotypes and detecting new individuals and hybrids in the red wolf (*Canis rufus*) experimental population area. *Conservation Genetics*, **8**, 123–131.

Addo-Bediako A, Chown SL, Gaston KJ (2000) Thermal tolerance, climatic variability and latitude. *Proceedings of the Royal Society of London-Series B: Biological Sciences*, **267**, 739–745.

Allendorf FW, Leary RF, Hitt NP et al. (2004) Intercrosses and the U.S. Endangered Species Act: should hybridized populations be included as Westslope cutthroat trout? *Conservation Biology*, **18**, 1203–1213.

Allendorf FW, Leary RF, Spruell P, Wenburg JK (2001) The problems with hybrids: setting conservation guidelines. *Trends in Ecology and Evolution*, **16**, 613–622.

Altizer S, Harvell D, Friedle E (2003) Rapid evolutionary dynamics and disease threats to biodiversity. *Trends in Ecology and Evolution*, **18**, 589–596.

Anderson E, Stebbins GL Jr (1954) Hybridization as an evolutionary stimulus. *Evolution*, **8**, 378–388.

Arnold ML (1997) *Natural Hybridization and Evolution.* Oxford University Press, Oxford.

Barton N, Bengtsson BO (1986) The barrier to genetic exchange between hybridising populations. *Heredity*, **57**, 357–376.

Barton N, Hewitt GM (1981) Hybrid zones and speciation. In: *Evolution and Speciation: Essays in Honor of M. J. D. White* (eds. Atchley WR, Woodruff DS), pp. 109–145. Cambridge University Press, Cambridge.

Bernardo J, Spotila JR (2006) Physiological constraints on organismal response to global warming: mechanistic insights from clinically varying populations and implications for assessing endangerment. *Biology Letters*, **2**, 135–139.

Blehert DS, Hicks AC, Behr M et al. (2009) Bat white-nose syndrome: an emerging fungal pathogen? *Science,* **323**, 227.

Both C, Bouwhuis S, Lessells CM, Visser ME (2006) Climate change and population declines in a long-distance migratory bird. *Nature*, **441**, 81–83.

Brede N, Sandrock C, Straile D et al. (2009) The impact of human-made ecological changes on the genetic architecture of Daphnia species. *Proceedings of the National Academy of Sciences USA*, **106**, 4758–4763.

Bruggeman DJ, Jones ML (2008) Should habitat trading be based on mitigation ratios derived from landscape indices? A model-based analysis of compensatory restoration options for the red-cockaded woodpecker. *Environmental Management*, **42**, 591–602.

Candolin U, Salesto T, Evers M (2006) Changed environmental conditions weaken sexual selection in sticklebacks. *Journal of Evolutionary Biology*, **20**, 233–239.

Chan CH, Ballantyne K, Aikman H et al. (2006) Genetic analysis of interspecific hybridisation in the world's only Forbes' parakeet (*Cyanoramphus forbesi*) natural population. *Conservation Genetics*, **7**, 493–506.

Cole CJ, Painter CW, Dessauer HC, Taylor HL (2007) Hybridization between the endangered unisexual gray-checkered whiptail lizard (*Aspidoscelis dixoni*) and the bisexual western whiptail lizard (*Aspidoscelis tigris*) in southwestern New Mexico. *American Museum Novitates*, 1–31, no. 3555.

Coyne J, Orr HA (1989) Patterns of speciation in *Drosophila*. *Evolution*, **43**, 362–381.

Coyne J, Orr HA (1997) "Patterns of speciation in *Drosophila*" revisited. *Evolution*, **51**, 295–303.

Coyne J, Orr HA (2004) *Speciation*. Sinauer, Sunderland, MA.

Daniels MJ, Corbett L (2003) Redefining introgressed protected mammals: when is a wildcat a wild cat and a dingo a wild dog? *Wildlife Research*, **30**, 213–218.

Daszak P, Cunningham AA, Hyatt AD (2000) Emerging infectious diseases of wildlife – threats to biodiversity and human health. *Science*, **287**, 443–449.

Dobzhansky T (1951) *Genetics and the Origin of Species*. Columbia University Press, New York.

Dowling TE, Secor CL (1997) The role of hybridization and introgression in the diversification of animals. *Annual Review of Ecology and Systematics*, **28**, 593–619.

Drechsler M, Johst K, Ohl C, Wätzold F (2007) Designing cost-effective payments for conservation measures to generate spatiotemporal habitat heterogeneity. *Conservation Biology*, **21**, 1471–1486.

Ellstrand N (2009) Evolution of invasiveness in plants following hybridization. *Biological Invasions*, **11**, 1089–1091.

Ellstrand NC, Elam DR (1993) Population genetic consequences of small population size: implications for plant conservation. *Annual Review of Ecology and Systematics*, **30**, 539–563.

Fitzpatrick BM, Shaffer HB (2004) Environment-dependent admixture dynamics in a tiger salamander hybrid zone. *Evolution*, **58**, 1282–1293.

Fitzpatrick BM, Shaffer HB (2007a) Hybrid vigor between native and introduced salamanders raises new challenges for conservation. *Proceedings of the National Academy of Sciences USA*, **104**, 15793–15798.

Fitzpatrick BM, Shaffer HB (2007b) Introduction history and habitat variation explain the landscape genetics of hybrid tiger salamanders. *Ecological Applications*, **17**, 598–608.

Flamand JRB, Vankan D, Gairhe KP, Duong H, Barker JSF (2003) Genetic identification of wild Asian water buffalo in Nepal. *Animal Conservation*, **6**, 265–270.

Fredrickson RJ, Hedrick PW (2006) Dynamics of hybridization and introgression in red wolves and coyotes. *Conservation Biology*, **20**, 1272–1283.

Genovart M, Oro D, Juste J, Bertorelle G (2007) What genetics tell us about the conservation of the critically endangered Balearic shearwater? *Biological Conservation*, **137**, 283–293.

Ghalambor CK, McKay JK, Carroll SP, Reznick DN (2007) Adaptive versus non-adaptive phenotypic plasticity and the potential for contemporary adaptation in new environments. *Functional Ecology*, **21**, 394–407.

Gompert Z, Fordyce JA, Forister ML, Shapiro AM, Nice CC (2006) Homoploid hybrid speciation in an extreme habitat. *Science*, **314**, 1923–1925.

Grant BR, Grant PR (2008) Fission and fusion of Darwin's finches populations. *Philosophical Transactions of the Royal Society of London-Series B: Biological Sciences*, **363**, 2821–2829.

Grant PR, Grant BR (1992) Hybridization of bird species. *Science*, **256**, 193–197.

Gregory PG, Howard DJ (1993) Laboratory hybridization studies of *Allonemobius fasciatus* and *A. socius* (Orthoptera: Gryllidae). *Annals of the Entomological Society of America*, **86**, 694–701.

Gregory PG, Howard DJ (1994) A post-insemination barrier to fertilization isolates two closely related ground crickets. *Evolution*, **48**, 705–710.

Haig SM, Mullins TD, Forsman ED, Trail PW, Wennerberg L (2004) Genetic identification of Spotted Owls, Barred Owls, and their hybrids: legal implications of hybrid identity. *Conservation Biology*, **18**, 1347–1357.

Harrison RG (1990) Hybrid zones: windows on the evolutionary process. *Oxford Surveys in Evolutionary Biology*, **7**, 69–128.

Harrison RG (1993a) *Hybrid Zones and the Evolutionary Process*. Oxford University Press, Oxford.

Harrison RG (1993b) Hybrids and hybrid zones: historical perspective. In: *Hybrid Zones and the Evolutionary Process*, pp. 3–12. Oxford University Press, Oxford.

Harvell CD, Mitchell CE, Ward JA et al. (2002) Climate warming and disease risks for terrestrial and marine biota. *Science*, **296**, 2158–2162.

Hendry AP, Grant PR, Grant BR et al. (2006) Possible human impacts on adaptive radiation: beak size bimodality in Darwin's finches. *Proceedings of the Royal Society of London-Series B: Biological Sciences*, **273**, 1887–1894.

Howard DJ (1999) Conspecific sperm and pollen precedence and speciation. *Annual Review of Ecology and Systematics*, **30**, 109–132.

Hurt CR, Farzin M, Hedrick PW (2005) Premating, not postmating, barriers drive genetic dynamics in experimental hybrid populations of the endangered Sonoran topminnow. *Genetics*, **171**, 655–662.

Kingston JJ, Gwilliam J (2007) Hybridization between two sympatrically breeding species of fur seal at Iles Crozet revealed by genetic analysis. *Conservation Genetics*, **8**, 1133–1145.

Konishi M, Takata K (2004) Size-dependent male-male competition for a spawning substrate between *Pseudorasbora parva* and *Pseudorasbora pumila*. *Ichthyological Research*, **51**, 184–187.

Kyle CJ, Johnson AR, Patterson BR et al. (2006) Genetic nature of eastern wolves: past, present and future. *Conservation Genetics*, **7**, 273–287.

Levin DA, Francisco-Ortega JK, Jansen RK (1996) Hybridization and the extinction of rare plant species. *Conservation Biology*, **10**, 10–16.

MacCallum CJ, Nurnberger B, Barton NH, Szymura JM (1998) Habitat preference in the Bombina hybrid zone in Croatia. *Evolution*, **52**, 227–239.

Madsen T, Shine R, Olsson M, Wittzell H (1999) Conservation biology: restoration of an inbred adder population. *Nature*, **402**, 34–35.

Mallet J (2005) Hybridization as an invasion of the genome. *Trends in Ecology and Evolution*, **20**, 229–237.

Manel S, Gaggiotti OE, Waples RS (2005) Assignment methods: matching biological questions with appropriate techniques. *Trends in Ecology and Evolution*, **20**, 136–142.

Marshall JL, Arnold ML, Howard DJ (2002) Reinforcement: the road not taken. *Trends in Ecology and Evolution*, **17**, 558–563.

Mavárez J, Salazar CA, Bermingham E et al. (2006) Speciation by hybridization in Heliconius butterflies. *Nature*, **441**, 868–871.

Miller CR, Adams JR, Waits LP (2003) Pedigree-based assignment tests for reversing coyote (*Canis latrans*) introgression into the wild red wolf (*Canis rufus*) population. *Molecular Ecology*, **12**, 3287–3301.

Muñoz Fuentes V, Vilà C, Green AJ, Negro JJ, Sorenson MD (2007) Hybridization between white-headed ducks and introduced ruddy ducks in Spain. *Molecular Ecology*, **16**, 629–638.

Mydlarz LD, Jones LE, Harvell CD (2006) Innate immunity, environmental drivers, and disease ecology of marine and freshwater invertebrates. *Annual Review of Ecology, Evolution, and Systematics*, **37**, 251–288.

Oliveira R, Godinho R, Randi E, Ferrand N, Alves PC (2008) Molecular analysis of hybridisation between wild and domestic cats (*Felis silvestris*) in Portugal: implications for conservation. *Conservation Genetics*, **9**, 1–11.

Parmesan C (2006) Ecological and evolutionary responses to recent climate change. *Annual Review of Ecology, Evolution, and Systematics*, **37**, 637–669.

Parmesan C, Yohe G (2003) A globally coherent fingerprint of climate change impacts across natural systems. *Nature*, **421**, 37–42.

Peacock MM, Kirchoff V (2004) Assessing the conservation value of hybridized cutthroat trout populations in the Quinn River drainage, Nevada. *Transactions of the American Fisheries Society*, **133**, 309–325.

Perry WL, Feder JL, Lodge DM (2001) Implications of hybridization between introduced and resident Orconectes crayfishes. *Conservation Biology*, **15**, 1656–1666.

Pfennig KS (2007) Facultative mate choice drives adaptive hybridization. *Science*, **318**, 965–967.

Phillips MK, Henry VG, Kelly BT (2003) Restoration of the red wolf. In: *Wolves: Behavior, Ecology and Conservation* (eds. Mech LD, Boitani L), pp. 272–288. University of Chicago Press, Chicago.

Price DK, Muir C (2008) Conservation implications of hybridization in Hawaiian picture-winged *Drosophila*. *Molecular Phylogenetics and Evolution*, **47**, 1217–1226.

Pullin AS, Knight TM, Stone DA, Charman K (2004) Do conservation managers use scientific evidence to support their decision-making? *Biological Conservation*, **119**, 245–252.

Pysek PK, Prach K, Rejmanek M, Wade M (1995) *Plant Invasions: General Aspects and Special Problems*. SPB Academic Publishing, Amsterdam.

Rako L, Blacket MJ, McKechnie SW, Hoffmann AA (2007) Candidate genes and thermal phenotypes: identifying ecologically important genetic variation for thermotolerance in the Australian *Drosophila* melanogaster cline. *Molecular Ecology*, **16**, 2948–2957.

Randi E (2008) Detecting hybridization between wild species and their domesticated relatives. *Molecular Ecology*, **17**, 285–293.

Randler C (2006) Behavioural and ecological correlates of natural hybridization in birds. *Ibis*, **148**, 459–467.

Reid JM, Arcese P, Keller LF (2003) Inbreeding depresses immune responses in song sparrows (*Melospiza melodia*): direct and inter-generational effects. *Proceedings of the Royal Society of London-Series B: Biological Sciences*, **270**, 2151–2157.

Rhymer JM, Simberloff D (1996) Extinction by hybridization and introgression. *Annual Review of Ecology and Systematics*, **27**, 83–109.

Riley SPD, Shaffer HB, Voss SR, Fitzpatrick BM (2003) Hybridization between a rare, native tiger salamander (*Ambystoma californiense*) and its introduced congener. *Ecological Applications*, **13**, 1263–1275.

Rosenfield JA, Kodric-Brown A (2003) Sexual selection promotes hybridization between Pecos pupfish, *Cyprinodon pecosensis*, and sheepshead minnow, *C. variegatus*. *Journal of Evolutionary Biology*, **16**, 595–606.

Rosenfield JA, Nolasco S, Lindauer S, Sandoval C, Kodric-Brown A (2004) The role of hybrid vigor in the replacement of Pecos pupfish by its hybrids with sheepshead minnow. *Conservation Biology*, **18**, 1589–1598.

Schwartz MK, Pilgrim KL, McKelvey KS et al. (2004) Hybridization between Canada lynx and bobcats: genetic results and management implications. *Conservation Genetics*, **5**, 349–355.

Seehausen O (2004) Hybridization and adaptive radiation. *Trends in Ecology and Evolution*, **19**, 198–207.

Seehausen O, Koetsier E, Schneider MV et al. (2003) Nuclear markers reveal unexpected genetic variation and a Congolese-Nilotic origin of the Lake Victoria cichlid species flock. *Proceedings of the Royal Society of London-Series B: Biological Sciences*, **270**, 129–137.

Seehausen O, Takimoto G, Roy D, Jokela J (2008) Speciation reversal and biodiversity dynamics with hybridization in changing environments. *Molecular Ecology*, **17**, 30–44.

Seehausen O, van Alphen JJM, Witte F (1997) Cichlid fish diversity threatened by eutrophication that curbs sexual selection. *Science*, **277**, 1808–1811.

Spielman D, Brook BW, Frankham R (2004) Most species are not driven to extinction before genetic factors impact them. *Proceedings of the National Academy of Sciences USA*, **101**, 15261–15264.

Stace CA (1975) *Hybridization and the Flora of the British Isles*. Academic Press, New York.

Stebbins GL (1950) *Variation and Evolution in Plants*. Columbia University Press, New York.

Stelkens R, Seehausen O (2009) Genetic distance between species predicts novel trait expression in their hybrids. *Evolution*, **63**, 884–897.

Stoskopf MK, Beck K, Fazio BB et al. (2005) Implementing recovery of the red wolf – integrating research scientists and managers. *Wildlife Society Bulletin*, **33**, 1145–1152.

Sutherland WJ, Pullin AS, Dolman PM, Knight TM (2004) The need for evidence-based conservation. *Trends in Ecology and Evolution*, **19**, 305–308.

Thomas CD, Cameron A, Green RE et al. (2004) Extinction risk from climate change. *Nature*, **427**, 145–148.

Tompkins DM, Mitchell RA, Bryant DM (2006) Hybridization increases measures of innate and cell-mediated immunity in an endangered bird species. *Journal of Animal Ecology*, **75**, 559–564.

Tranah GJ, May B (2006) Patterns of intra- and interspecies genetic diversity in Klamath River basin suckers. *Transactions of the American Fisheries Society*, **135**, 306–316.

Vallender R, Robertson RJ, Friesen VL, Lovette IJ (2007) Complex hybridization dynamics between golden-winged and blue-winged warblers (*Vermivora chrysoptera* and *Vermivora pinus*) revealed by AFLP, microsatellite, intron and mtDNA markers. *Molecular Ecology*, **16**, 2017–2029.

van Oppen MJH, Willis BL, van Rheede T, Mikller DJ (2002) Spawning times, reproductive compatibilities and genetic structuring in the *Acropora aspera* group: evidence for natural hybridization and semipermeable species boundaries in corals. *Molecular Ecology*, **11**, 1363–1376.

Verardi A, Lucchini V, Randi E (2006) Detecting introgressive hybridization between free-ranging domestic dogs and wild wolves (*Canis lupus*) by admixture linkage disequilibrium analysis. *Molecular Ecology*, **15**, 2845–2855.

Willi Y, Van Buskirk J, Hoffmann AA (2006) Limits to the adaptive potential of small populations. *Annual Review of Ecology, Evolution, and Systematics*, **37**, 433–458.

Williamson MH (1996) *Biological Invasions*. Chapman & Hall, London; New York.

Willis BL, van Oppen MJH, Miller DJ, Vollmer SV, Ayre DJ (2006) The role of hybridization in the evolution of reef corals. *Annual Review of Ecology, Evolution, and Systematics*, **37**, 489–517.

Willis BL, Babcock R, Harrison P, Wallace C (1997) Experimental hybridization and breeding incompatibilities within the mating systems of mass spawning corals. *Coral Reefs*, **16**, S53–S65.

Wolf DE, Takebayashi N, Rieseberg LH (2001) Predicting the risk of extinction through hybridization. *Conservation Biology*, **15**, 1039–1053.

8 Pollen and seed movement in disturbed tropical landscapes

J. L. Hamrick

INTRODUCTION

A characteristic of modern landscapes worldwide is that continuous habitats have become, largely due to human activities, mosaics of remnant habitat fragments embedded in an urban or agricultural matrix (see Chapter 9 by Leberg and colleagues). Landscape fragmentation can have distinct ecological (e.g., species extinction), demographic (e.g., lowered reproduction and elevated mortality), and genetic (e.g., less genetic diversity and increased inbreeding) consequences. In northern temperate regions, human impacts on natural landscapes date back several thousand years but, in most tropical landscapes, widespread human disturbance is more recent, with the heaviest impacts occurring during the last fifty years. Disturbance and fragmentation of once continuous habitats can have immediate, short-term, and long-term consequences for the management and conservation of genetic diversity within tropical plant species.

Immediate consequences

Landscape fragmentation has immediate consequences for the levels and distribution of genetic diversity that are not dependent on population genetic processes acting across subsequent generations. Three factors can have immediate effects on the genetic composition of fragmented populations: 1) the proportion of the original population that is removed, 2) the number of individuals that survive in each fragment, and 3) patterns of genetic variation present within natural populations prior to fragmentation (e.g., clusters of related individuals).

If genetic diversity within the original, continuous population was distributed at random prior to fragmentation, genetic diversity within each fragment will be a function of the number of individuals within the fragment; the smaller the size of the remnant population, the lower the genetic diversity maintained within the population. One way to consider this effect is to calculate the probability that a locus polymorphic in the original population is polymorphic in a remnant population. If R is the probability of maintaining a polymorphic locus, p_i is the frequency of the ith allele, and N is the remnant population size, then $R = 1 - \Sigma p_i^{2N}$ (Hedrick 2000). From this relationship, the more uneven allele frequencies are in the original population, the larger the number of founders needed to ensure the retention of polymorphism at the locus. When there is inbreeding in the original population (as is the case for many plant species), and genotype frequencies are not in Hardy–Weinberg proportions, the probability of maintaining polymorphism at a locus in the remnant population is given by $R = 1 - \Sigma P_i^N$, where P_i is the frequency of the various homozygous classes (Hedrick 2000). Thus, the probability of maintaining polymorphism in a remnant population for inbreeding species is less than that for outcrossing species. Finally, Nei and colleagues (1975) demonstrated that the size of the founder population has more impact on the maintenance of rare alleles, whereas the temporal duration (i.e., the number of generations) of small population sizes has a greater influence on overall levels of genetic diversity.

If spatial genetic structure exists prior to fragmentation, it becomes necessary to consider not only the number of individuals in a fragment but also the spatial scale of fragmentation and pre-existing genetic structure (Nason et al. 1997). The distribution of genetic variation may be associated with landscape features (e.g., slope aspect, soil type, or moisture) that produce nonrandom distributions of adaptive genetic variation (Nason et al. 1997). If remnant fragments are not randomly distributed relative to landscape structure, genetic variation that adapts the species to habitats that are not represented in the fragments could be lost at a higher rate than that expected for neutral genetic variation. For example, fragments are more likely to occur on steep slopes, riparian areas, or ridge tops. Such spatial heterogeneity could have important fitness consequences and could hinder population-restoration efforts in habitats that are not well represented within remnant fragments.

Even in relatively uniform landscapes, genetic variation may not be randomly distributed. There is increasing evidence for many plant species that limited seed dispersal and patterns of recruitment (e.g., gap colonization) produce nonrandom patterns of genetic diversity (Vekemans & Hardy 2004; Dick et al. 2008). Thus, the distribution of genetic variation within and among remnant populations will depend on the spatial scale of fragmentation relative to the spatial scale of pre-existing neighborhoods (in the sense of Wright 1951). If a fragment contains several neighborhoods, then genetic diversity will be largely partitioned within rather than among fragments, whereas if fragments are smaller than a neighborhood, there would be relatively less genetic diversity within fragments and relatively more variation among fragments (Nason et al. 1997). In general, then, fragmentation at a given spatial scale will have different genetic implications for species whose genetic diversity is structured at varying spatial scales.

Short-term effects

There are two interrelated population genetic factors that primarily affect the genetic composition of small, isolated populations: genetic drift and inbreeding. Genetic drift is the loss of genetic diversity via the random loss of alleles, and it is a function of effective population size (N_e, see Box 8–1) and the degree of genetic isolation (i.e., the rate of gene flow, m). Wright (1951) demonstrated that if $N_e m < 1$, genetic drift would reduce genetic diversity within the population, where $N_e m$ is the number of migrants per generation. In contrast, if $N_e m > 4$, allele frequencies in remnant populations will change little due to genetic drift.

BOX 8–1: EFFECTIVE POPULATION SIZE

J. L. Hamrick

Effective population size (N_e) is envisioned as an ideal population of N diploid individuals produced each generation from a random sample of $2N$ gametes that has the same rate of increase in homozygosity or gene frequency change or the actual population under consideration has. There are several different effective numbers that are recognized, but the two most often described are the inbreeding effective number (N_{ei}) and the variance effective number (N_{ev}). The inbreeding effective number affects the probability of homozygosity due to common ancestry, whereas the variance effective number influences the amount of allele frequency change per generation. Deviations of either effective number (N_e) from the census number of adults (N) can result from temporal variation in population numbers and/or from greater than binomial variability in progeny number per parent.

Inbreeding within a remnant population is a function of the pre-existing level of relatedness among individuals within the population, as well as the effective population size and the magnitude of gene flow. Random mating in a small population of effective size N_e individuals will eventually lead to a loss of heterozygosity at a rate of approximately $1/2N_e$ per generation. If there is gene flow into a population at rate m, the equilibrium inbreeding coefficient (F_e) is given by:

$$F_e = \frac{(1-m)^2}{2N_e - (2N_e - 1)(1-m)^2} \text{(Li 1976)}$$

In other words, gene flow increases N_e by a factor of $1/(1-m)$. Thus, if population sizes are small and gene flow is limited among remnant populations, we should see changes in allele frequencies within populations, the loss of rare alleles from individual populations, and an increase in inbreeding within the populations. We should also see an increase in genetic differentiation among remnant populations within a disturbed landscape. At the overall landscape level, however, we should see little loss of genetic diversity.

Long-term effects of fragmentation

Long term effects of habitat fragmentation will depend on the degree of genetic isolation experienced by fragmented populations. If gene flow has been consistently limited, each fragment should drift to fixation for one allele or another at a locus. Overall genetic diversity (with the exception of a few low-frequency alleles) within the landscape will not be lost, however. In contrast, the distribution of genetic diversity within the landscape may have changed (i.e., genetic differentiation among fragments may have increased greatly). If, as is likely, fragmented populations are subsequently destroyed or the population within the fragment goes extinct, then overall genetic diversity within the landscape will decrease.

This could result In the loss of locally adapted genes or coadapted gene complexes (Allard 1988). If, in contrast, genetic connectivity is maintained among fragments or with larger, more continuous populations, genetic diversity within remnant populations will be maintained ($N_em > 4$) or decrease only slightly ($1 < N_em < 4$). As a result, the subsequent loss of fragment populations should produce little overall loss of genetic diversity within the landscape (see Box 8–2).

PREDICTIONS

The question arises as to whether one can predict the genetic consequences for species that have survived a fragmentation event. Estimates of three parameters should allow predictions of how the genetic composition of a fragmented landscape would change. First, it is essential to have accurate estimates of gene-flow rates (pollen and seed) into fragmented populations. As discussed earlier in text, the immigration of genes into a population will increase effective population sizes, thereby slowing genetic drift and reducing inbreeding. Second, estimates are needed of male and female reproductive success of each adult. The greater the variance of reproductive success among adults, the lower effective population size, resulting in more rapid changes in allele frequencies due to genetic drift and increases in homozygosity due to inbreeding. Estimates of the rate of gene flow and effective population sizes can be combined (i.e., N_em) to estimate the number of migrants per generation from which the expected effects of genetic drift and inbreeding can be calculated (Wright 1951). Finally, estimates of relatedness among adults within remnant populations are required to estimate the likelihood of mating between related individuals (i.e., biparental inbreeding). Fortunately, a single molecular marker–based parentage analysis of several remnant populations can provide accurate estimates of these parameters. In this chapter, I will primarily focus on estimates of the rates of pollen flow for tropical plants (mostly trees) within fragmented tropical landscapes. For technical and analytical reasons, estimates of seed movement are only now beginning to appear in the tropical biology literature (e.g., Jones et al. 2005; Hardesty et al. 2006). As a result, I will focus less on gene flow via seed. In addition, much of the available indirect data (Petit et al. 2005) indicates that gene flow via seeds is often lower than that by pollen (but see the section on seed dispersal).

ESTIMATES OF POLLEN FLOW

There are several approaches that have been employed to measure pollen-flow rates between populations (e.g., observing pollinator movements, tracking pollen marked with dyes, identifying individuals with unique alleles), but these traditional approaches all have problems regarding the accurate quantification of gene flow and/or the relative reproductive success of individuals. More recently, molecular genetic markers (especially allozymes and microsatellites) have been used to identify the parents of seeds or juvenile individuals. After an individual's pedigree has been established, estimates of a number of population genetic

BOX 8–2: ALLELIC RECHARGE IN POPULATIONS RECOVERING FROM BOTTLENECK EVENTS

Joseph D. Busch, Jennifer McCreight, and Peter M. Waser

Problem

Surprising processes can sometimes mitigate the loss of genetic diversity. For example, some animals demonstrate a compensatory increase in the distance traveled by successful dispersers when population size decreases. If populations are not truly isolated, increased immigration at low population density has the potential to recharge allelic diversity. Conservation decisions based on demographic data alone may not accurately assess the potential for such recharge.

Case Study

The banner-tailed kangaroo rat, *Dipodomys spectabilis*, is a highly philopatric mammal. Natal dispersal is generally limited to less than 100 meters. Vacant dens are a limiting resource for dispersing juveniles, whose survival is only half that of philopatric individuals (Jones et al. 1988). In years of low population density, however, females become less philopatric, dispersing about 3 times farther than males (Waser et al. 2006).

The demography of two southeastern Arizona populations, Rucker and Portal, is well known from a mark–recapture study (Waser & Ayers 2003). Censuses capture approximately 98% of the individuals in each population, allowing nearly complete documentation of their genetic composition. Both populations have experienced demographic bottlenecks during the last twenty-five years (Box Fig. 8–2–1). At Rucker, the 2002–2003 bottleneck was short, lasting just one to two generations. In contrast, the Portal population experienced a two-decade decline, culminating in extreme bottlenecks in 1994 and 2004.

Contemporary bottleneck-elucidation methods that rely on the loss of rare microsatellite alleles fail to detect these known demographic crashes (Busch et al. 2007). In these populations, bottleneck signatures are extremely short-lived because absent and rare alleles are rapidly replaced by cryptic immigrants. For instance, the number of alleles at Rucker remained fairly stable from 1994 through 2007. The bottleneck of 2002–2003 resulted in some lost variation, but new alleles soon entered the population. Portal shows more year-to-year variation in the number of alleles, reflecting the fact that its population size has twice fallen below ten animals. During the years following these demographic troughs, the number of alleles rose sharply, demonstrating a high level of gene flow from neighboring, unsampled populations. In this way, Portal maintained allelic diversity at levels similar to those at Rucker, despite its much smaller numbers.

An increase in female dispersal during low-density years may enhance mitochondrial gene flow as well. In the 1994 bottleneck, Portal had just four females, each with a unique control-region haplotype. In the succeeding four years of strong population growth, the number of haplotypes more than tripled (from four to fourteen). These new mitochondria must have arrived due to female

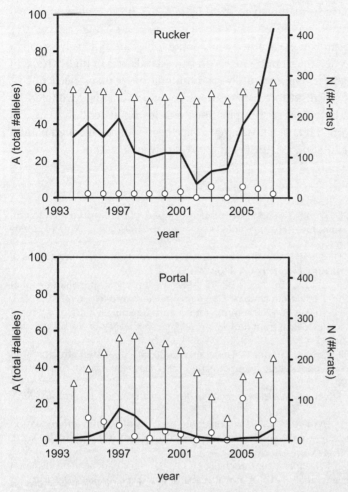

Box Figure 8–2–1: Replenishment of microsatellite alleles occurs during years of bottleneck recovery in two populations of banner-tailed kangaroo rats. Open triangles with vertical lines show the total number of alleles from eight microsatellite loci (left vertical axis); circles denote the number of new alleles entering the population that year; solid lines symbolize the total number of adults and juveniles known to be alive in August of each year (right vertical axis).

immigration (mutation in such a small breeding population is an unlikely source of new mitochondrial diversity). None of the Portal males in 1998 had unique haplotypes, reinforcing the significant role that female movements played in the population's genetic recovery.

The process leading to increased gene flow at low population density is primarily the increased survival of dispersers at low density (Jones et al. 1988). Young animals may also delay dispersal when habitats are "saturated"; both tendencies cause estimated neighborhood size to increase at low density (Busch et al. 2009).

In many animals, emigration rates increase with density. Models by McCarthy (1999), however, have shown that as density increases, so does the proportion of emigrants that die, and those that survive are increasingly the individuals lucky enough to find an opening nearby.

Other animals have been shown to exhibit increases in gene flow into low-density populations (Rabenold et al. 1991; Keller et al. 2001). Of course, this stabilizing mechanism can exist only when populations are not truly isolated. Genetic approaches document bottlenecks on true islands and in highly isolated continental populations. Opportunities for immigration perhaps need not be common, but they do need to exist. Without genetic exchange facilitated by immigrants, critically endangered populations risk losing genetic diversity, as seen in other kangaroo rat taxa whose ranges have been reduced through habitat conversion and degradation (Metcalf et al. 2001, Loew et al. 2005).

REFERENCES

Busch JD, Waser PM, DeWoody JA (2007) Recent demographic bottlenecks are not accompanied by a genetic signature in banner-tailed kangaroo rats. *Molecular Ecology*, **16**, 2450–2462.

Busch JD, Waser PM, DeWoody JA (2009) The influence of density and sex on patterns of fine-scale genetic structure. *Evolution*, **63**, 2302–2314.

Jones WT, Waser PM, Elliott LF, Link NE, Bush BB (1988) Philopatry, dispersal and habitat saturation in the banner-tailed kangaroo rat *Dipodomys spectabilis*. *Ecology*, **69**, 1466–1473.

Keller LF, Jeffery KJ, Arcese P et al. (2001) Immigration and the ephemerality of a natural population bottleneck: evidence from molecular markers. *Proceedings of the Royal Society of London-Series B: Biological Sciences*, **268**, 1387–1394.

Loew SS, Williams DF, Ralls K, Pilgrim K, Fleischer RC (2005) Population structure and genetic variation in the endangered giant kangaroo rat (*Dipodomys ingens*). *Conservation Genetics*, **6**, 495–510.

McCarthy, MA (1999) Effects of competition on natal dispersal distance. *Ecological Modelling*, **114**, 305–310.

Metcalf AE, Nunney L, Hyman BC (2001) Geographic patterns of genetic differentiation within the restricted range of the endangered Stephens' kangaroo rat *Dipodomys stephensi*. *Evolution*, **55**, 1233–1244.

Rabenold PP, Rabenold KN, Piper WH, Minchella DJ (1991). Density-dependent dispersal in social wrens: genetic analysis using novel matriline markers. *Animal Behaviour*, **42**, 144–146.

Waser PM, Ayers JM (2003) Microhabitat use and population decline in banner-tailed kangaroo rats. *Journal of Mammalogy*, **84**, 1031–1043.

Waser PM, Busch JD, McCormick CR, DeWoody JA (2006) Parentage analysis detects cryptic precapture dispersal in a philopatric rodent. *Molecular Ecology*, **15**, 1929–1937.

parameters can be quantified for natural plant populations. Examples include mating patterns within populations, male and female reproductive success, gene flow into populations, and effective population sizes.

There are at least four factors that influence our ability to accurately identify an individual's parents. The number of polymorphic loci, the number of alleles per locus, and the evenness of allele frequencies all significantly affect one's ability to exclude individual adults as possible parents of a seed or seedling. Finally, the number of candidate parents will also influence the magnitude of parental ambiguity. Fortunately, for most tropical tree species, densities of reproductive adults are low. Thus, the small, discrete, remnant populations that characterize many highly disturbed tropical landscapes provide an ideal setting for the estimation of parameters needed to predict the genetic consequences of landscape fragmentation.

Table 8–1. Pollen flow rates and maximum documented distances of pollen movement across disturbed landscapes for several tropical plant species

Species	Pollination system	Rate of pollen flow (%)	Maximum distance (m)	Reference
Bursera simaruba (Burseraceae)	Insects	73	1,000	1
Cecropia obtusifolia (Moraceae)	Wind	27; 9	6,000; 14,000	2
Ceiba pentandra (Malvaceae)	Bats	–	18,600	3
Cordia alliodora (Boraginaceae)	Insects	3	280	4
Dinizia excelsa (Fabaceae)	Bees	68	3,200	5
Dipteryx panamensis (Fabaceae)	Bees	12	2,300	6
Enterolobium cyclocarpum (Fabaceae)	Moths, bees	75	2,800	7
Ficus spp. (Moraceae)	Wasps	90	14,200	8
Gliricidia sepium (Fabaceae)	Bees	6	275	9
Guaiacum sanctum (Zygophyllaceae)	Bees	60	4,000	10
Hymenaea courbaril (Fabaceae)	Bats	55	1,900	11
		52	800	12
Laelia rubescens (Orchidaceae)	Hummingbirds	34	1,580+	13
Pithecellobium elegans (Fabaceae)	Bees, etc.	15	350+	14
Spondias mombin (Anacardiaceae)	Bees	60	1,000	15
Swietenia humilis (Meliaceae)	Small butterflies, bees	68	4,500	16
Symphonia globulifera (Clusiaceae)	Hummingbirds	14	350+	17

All species were trees except *L. rubescens*, which is an epiphytic orchid. 1: Dunphy & Hamrick (2007); 2: Kaufman et al. (1998); 3: Gribel, cited in Ward et al. (2005); 4: Boshier et al. (1995); 5: Dick (2001b); 6: Hanson et al. (2008); 7: Hamrick (2001), Hamrick unpublished; 8: Nason et al. (1998); 9: Dawson et al. (1997); 10: Fuchs (2007); 11: de Lacerda et al. (2008); 12: Dunphy et al. (2004); 13: Trapnell & Hamrick (2005); 14: Chase et al. (1996); 15: Nason & Hamrick (1997); 16: White et al. (2002); 17: Aldrich & Hamrick (1998).

I focus here on studies that have used molecular genetic markers to estimate pollen flow rates into remnant populations of tropical trees located in highly disturbed landscapes. Results of parentage analyses of pollen flow within disturbed tropical landscapes are available for fifteen species of tropical trees and one epiphytic orchid (Table 8–1). Although there is considerable variation among species in terms of pollen flow rates into recipient populations, and the maximum estimated distance of pollen movement varies greatly, in general, these studies agree that there are high rates of effective pollen movement across disturbed landscapes (Table 8–1). In most cases, estimates of pollen flow should be reasonably accurate. In contrast, estimates of maximum pollen-movement distances are almost certainly constrained by the physical dimensions of the study areas. Another

Table 8–2. Pollen flow rates and distances for *S. humilis* within a fragmented Honduras landscape (after White et al. 2002)

Population	Number of adults	Isolation distance (km)	% Pollen flow	Mean min. pollen flow distance (km)
Los Tablas[a]	97	–	36.0	–
Brutus/Jicarito	44	1.1	47.0	3.1
Jinte	22	1.1	38.3	1.7
Tablas Plains	7	1.2	68.4	1.0
Tree 501[b]	1	1.4	100.0	>4.5

[a] A 68-ha plot within a 500-ha forest
[b] An isolated tree

limitation of several studies is that they were conducted in one site (i.e., population) and during a single reproductive event. As a result, we know relatively little about the effects of specific landscape features on the rates and distances of pollen flow (Sork & Smouse 2006). Analyses of a single site also preclude comparisons between pollen flow rates and distances in disturbed and undisturbed landscapes, limiting our ability to determine the effects of landscape fragmentation on pollen movement. It is perhaps as important that we have even less understanding of how these parameters vary among reproductive events. (For example, do pollen flow rates and distances vary with flowering-tree densities? Does variation in flowering phenology among years affect pollen flow rates? Does temporal variation in pollinator availability or composition affect pollen flow rates?)

Studies that include multiple sites (e.g., *Spondias mombin*, Nason & Hamrick 1997; *Swietenia humilis*, White et al. 2002; *Laelia rubescens*, Trapnell & Hamrick 2005; *Bursera simaruba*, Dunphy & Hamrick 2007) indicate that pollen flow rates vary among remnant populations within the same landscape. Nason and Hamrick (1997) studied pollen flow into two continuous forest populations of *S. mombin* located on Barro Colorado Island (BCI), Panama, and several island populations located in the surrounding Lake Gatun. Rates of pollen flow into the continuous forest populations did not vary (0.44 versus 0.46), whereas estimates of pollen flow into four island populations varied from 0.60 to 1.00, with the largest population (twenty-two adults) having the lowest rate of pollen flow and the three smaller populations (four to eight adults) having nearly 100% pollen flow. A similar pattern was observed by White and colleagues (2002) for four fragmented populations of *S. humilis* from Honduras (Table 8–2). Pollen flow varied from 38 to 68%, with the smallest fragments having the highest pollen flow rates. Trapnell and Hamrick (2005) found that pollen flow varied from 20 to 83% across eleven populations of the epiphytic orchid, *L. rubescens*. Populations with high numbers of flowering inflorescences experienced less pollen flow than did populations with less flowering. A similar pattern (pollen flow varied from 0.47 to 1.00) was seen across five populations of *B. simaruba* studied in a fragmented Puerto Rican landscape (Dunphy & Hamrick 2007). Thus, in every study in which multiple fragments have been analyzed, fragments with fewer adult individuals experienced higher rates of pollen flow. Pollen limitation could occur in the smaller populations due to low numbers of local pollen donors. The existence of self-incompatibility systems in many tropical trees (Bawa 1974;

Opler et al. 1980) could further limit levels of legitimate pollen. As a result, if immigrant pollen deposition was equivalent in large and small fragments, lower levels of local pollen in the smaller populations would result in higher rates of effective pollen flow. Unfortunately, most of the studies cited earlier in the text do not provide the detail needed to test this possibility. An exception is the study on *L. rubescens* by Trapnell and Hamrick (2005), who demonstrated that overall reproductive success was lower in populations with fewer inflorescences. In populations with larger floral displays (and lower pollen flow rates), pollinators apparently stay within the population long enough to facilitate multiple within-population pollinations.

There are relatively few studies comparing pollen flow into clusters of trees in continuous forests (or larger fragments) with pollen flow into small fragments. Thus, it is difficult to actually document the effects of habitat disturbance on the genetic connectivity of individuals and populations. The study by Aldrich and Hamrick (1998) on *Symphonia globulifera* compared pollen and seed flow into three fragments within a disturbed premontane landscape with a similarly sized plot within a nearby 235-ha fragment. They found that pollen and seed movement into the continuous forest plot was lower (68%) than that seen in the fragmented landscape plots (89%). Nason and Hamrick (1997), in their study of *S. mombin*, demonstrated that pollen flow into two plots located in the continuous forest of BCI averaged approximately 45%, whereas pollen flow into five small island populations averaged 89%. The study by White and colleagues (2002) of *S. humilis* in Honduras indicated a similar pattern with the high-density, continuous forest plot experiencing the lowest pollen flow (36%, Table 8–2). These results may indicate that pollen vectors vary between disturbed and undisturbed populations (e.g., Dick, 2001b, see case study) or that pollinator behavior varies among disturbed and undisturbed sites (e.g., Aldrich and Hamrick, 1998; see case study).

There are only two studies in Table 8–1 that measured pollen flow over multiple years. Hamrick (2001, and unpublished data) estimated pollen flow into six *Enterolobium cyclocarpum* populations located in a disturbed tropical dry forest in the Guanacaste province of Costa Rica over a three-year period. Pollen flow estimates varied an average of 11.4% for the same sites, although pollen flow rates into specific sites varied by as much as 43%. Trapnell and Hamrick (2004) measured pollen flow into six populations of the epiphytic orchid, *L. rubescens*, over two reproductive events (1999 and 2000). Differences in pollen flow rates per population between the two years ranged from 1 to 43% and averaged 13.2%. The major difference seen between these two years was that the distance of pollen movement averaged 279 m in 1999 and 519 m in 2000. The greater pollen-movement distance seen in 2000 was associated with a 37% increase in the number of flowering inflorescences.

Only a few studies attempted to estimate the effective number of adults (N_e) within recipient fragments (see Box 8–1). This value is needed to calculate $N_e m$, which allows direct estimates of changes in genetic structure across remnant populations within a fragmented landscape [e.g., $F_{ST} = 1/(1 + 4N_e m)$]. To estimate N_e, an accurate estimate of male and female reproductive success is needed. Significant heterogeneity in reproductive success in either sex would substantially reduce N_e for remnant populations. Calculations of $N_e m$ were made by Apsit 1998 and Apsit and Hamrick (unpublished data) for three populations of *E. cyclocarpum*, by Aldrich and Hamrick (1998) for two populations of *S. globulifera*, and by Dunphy and coworkers (2004) and Dunphy and Hamrick (2007) for *Hymenaea courbaril* and *B. simaruba*, respectively. For each study, parentage analyses were used to calculate relative male reproductive success in each population. Direct estimates of relative fruit production were used to estimate female reproductive success for each individual. For *E. cyclocarpum*, effective population sizes were equal to approximately 60% of the actual number of adults. When combined with the gene flow rate (= $\frac{1}{2}$ pollen flow), $N_e m$ values ranged from 0.46 to 1.12. Equilibrium estimates of F_{ST} ranged from 0.182 to 0.352. The current F_{ST} value for the landscape studied was 0.051. Thus, whereas estimated pollen flow for these three *E. cyclocarpum* sites was approximately 50%, low adult densities and even lower N_e values led to a predicted increase in genetic differentiation among populations. Aldrich and Hamrick (1998) calculated variance-effective population sizes for fragmented populations ($N_{ev} = 7.3$) and for the continuous forest plot ($N_{ev} = 16.7$). Accounting for the proportion of migration, $N_e m$ was 2.8 for the fragmented forest and 9.4 for the continuous forest plot. Thus, genetic drift would be 3 times as strong in the remnant forest patches as in the continuous forest. Dunphy and Hamrick (2007) obtained estimates of $N_e m$ for five remnant populations of *B. simaruba* located in highly fragmented southwestern Puerto Rico. Fragment population sizes ranged from three to nine adults, whereas $N_e m$ ranged from 0.71 to 3.00. It is interesting that two populations of bat-pollinated *Hymenaea courbaril* with four and six flowering individuals had estimated $N_e m$ values of only 0.40 and 0.63 (Dunphy et al. 2004). These examples demonstrate that if we are to accurately predict the effects of landscape fragmentation on the levels and distribution of genetic diversity, estimates of male and female reproductive success must be obtained in addition to estimates of pollen flow rates.

SEED DISPERSAL IN DISTURBED LANDSCAPES

Because assignment of maternity to dispersed seeds or seedlings has been technically difficult, studies of seed movement in fragmented tropical landscapes have lagged behind studies that document pollen movement (Sork & Smouse 2006; Dick et al. 2008). Only two of the four studies in Table 8–3 provide direct estimates of seed flow rates into forest fragments. Hanson and colleagues (2007), in a study of bat-dispersed *Dipteryx panamensis*, estimated that 14% of the seeds captured in a forest fragment originated outside the fragment. Mean seed-movement distance into the fragment was 710 m, whereas mean seed movement within the recipient fragment was 113 m. Aldrich and Hamrick (1998) estimated seed-flow rates into three 1-ha fragments and a 1-ha continuous forest plot of *S. globulifera*. The mean rate of seed movement into the fragments was 85%, whereas the continuous forest plot experienced 59% seed immigration.

Seed dispersal can also result in the establishment of new populations (Hamrick & Nason 1996) or the expansion of fragmented populations into the surrounding matrix. Born and coworkers (2008) documented that the wind-dispersed African

Table 8–3. Estimated mean and maximum distances of seed movement in disturbed landscapes

Species	Dispersal agent	Mean distance (m)	Maximum distance (m)	Reference
Aucoumea klaineana (Burseraceae)	Wind	118	258	1
Dipteryx panamensis (Fabaceae)	Bats	198	853	2
Iriartea deltoidea (Arecaceae)	Mammals and birds	248	2,350	3
Symphonia globulifera (Clusiaceae)	Bats	–	350	4

1: Born et al. (2008); 2: Hanson et al. (2007); 3: Sezen et al. (2007); 4: Aldrich & Hamrick (1998).

tree, *Aucoumea klaineana*, dispersed 258 m into open savanna, whereas the large mammal- and bird-dispersed neotropical palm, *Iriartea deltoidea*, dispersed up to 2,350 m into abandoned agricultural land in northeastern Costa Rica (Sezen et al. 2007). It is apparent that like pollen dispersal, seed dispersal for at least some tropical tree species is sufficient to maintain genetic connectivity among remnant patches within highly disturbed landscapes. In addition, colonization of abandoned farmland has the potential to receive propagules from several remnant populations, greatly increasing genetic diversity.

GENETIC RELATEDNESS WITHIN FRAGMENTS

Many tropical trees have significant levels of genetic relatedness (i.e., family structure) within undisturbed forests (Hamrick & Nason 1996; Dick et al. 2008) indicating that when a forest landscape is broken into relatively small fragments, individuals remaining within the fragments would be related to one another. When matings occur within such fragments, biparental inbreeding will occur immediately after the fragmentation event. Three species listed in Table 8–1 (*E. cyclocarpum*, *L. rubescens*, and *S. globulifera*) have significant spatial genetic structure within remnant populations in highly disturbed landscapes. The degree to which inbreeding increases within remnant populations immediately after fragmentation would be a function of fragment size and/or the number of neighborhoods (i.e., groups of related individuals) within the remnant population. If the fragment contains a single neighborhood, inbreeding should be higher in the fragment than in the original population due to the absence of pollen movement over moderate distances (i.e., inter-neighborhood) within the undisturbed forest (e.g., Aldrich & Hamrick, 1998).

CASE STUDIES

The five case studies that follow were chosen to illustrate a specific point regarding pollen movement between remnant populations located within highly disturbed (>50% forest removal) tropical landscapes.

S. humilis (Meliaceae)

White and colleagues (2002) used four highly variable microsatellite loci to mea-sure pollen flow in a landscape fragmented by pastures in the Punta Raton region of Honduras. *S. humilis* is a medium-sized tree that ranges along the Pacific low-lands of Mexico and Central America. It is one of the three species of *Swietenia* that are commonly known as American mahogany (Styles 1981). Its unisexual flowers are pollinated by small butterflies, bees, and other generalist insects. The study site consisted of three remnant populations with forty-four, twenty-two, and seven trees, and a single isolated tree. A larger, approximately 500-ha forest was located within approximately 1,600 m of the nearest fragment. All adult trees in the fragments were genotyped for the four microsatellite loci (exclusion probability = 0.983), as were ninety-seven trees within a 68-ha area of the con-tinuous forest. White and colleagues (2002) analyzed thirty progeny from each of seventeen adult trees (twelve from fragments, five from the continuous forest plot). The authors used a fractional paternity procedure (Devlin et al. 1988) to assign paternity to progeny arrays of each adult.

Pollen flow rates were lowest (36%) for trees located within the continuous forest plot and highest (68.4%) into the smallest fragment (Table 8–2). The isolated tree received all of its pollen from the continuous forest that was 4.5 km away. Minimum distance of pollen movement into the three fragmented populations ranged from 1.6 to 3.1 km. Although White and colleagues (2002) state that there was a trend of near-neighbor mating within remnant popula-tions, smaller fragment populations had a higher proportion of long-distance pollen flow. White and coworkers (2002) conclude that landscape fragmentation had probably increased long-distance pollen movement. They also argue that although they noted the absence of low-frequency alleles (White et al. 1999), the high rates of gene flow in this fragmented landscape should "...restore, main-tain, or even augment the levels of genetic variation..." within this *S. humilis* metapopulation. Their results agree with those of Cespedes and coworkers (2003), who found from a study of *S. macrophylla* in Costa Rica that successional sites have more alleles than adjacent populations of older adults, indicating long-distance pollen flow.

Dinizia excelsa (Fabaceae)

Dick (2001a) used five microsatellite loci to examine pollen movement between *D. excelsa* trees located in open pastures, forest fragments, and a continuous forest in the Biological Dynamics of Forest Fragments Project (Lovejoy & Bierregaard 1990) located approximately 90 km north of Manaus, Brazil. *D. excelsa* is a large canopy tree endemic to Brazil. Its natural density in undisturbed forest is about one adult per 6 ha (Dick 2001a), and it is often left standing in open pastures when the continuous forest is clear cut. It has small hermaphroditic flowers that are typically visited by small insects. Dick (2001a) used rope-climbing techniques to examine pollinators of seven trees located in the three habitat types. From twenty-five to fifty-five seeds from each of eleven adult trees were analyzed, and the paternity of each seed was assigned by comparing maternal and progeny

Table 8–4. Mean rates of gene flow into three forest fragments of *S. globulifera* compared to rates of gene flow into a 1-hectare plot embedded in a continuous forest (after Aldrich 1997 and Aldrich & Hamrick 1998)

Source	Fragment mean	Forest plot
Within site	0.023	0.156
Haploid immigration	0.170	0.333
Diploid immigration	0.807	0.511
Total immigration	0.892	0.678

genotypes with genotypes of thirty-six potential pollen parents. Direct observations of insect visitation indicated that only native insects visited trees within the continuous forest, and fourteen species of stingless bees were the primary pollinators (Dick 2001a). In disturbed landscapes, Africanized honeybees greatly outnumbered all native insects.

Genetic analyses demonstrated that maternal trees within pastures experienced approximately 10% lower rates of outcrossing (84.5 versus 95%) than did trees in continuous forest. Also, mean pollinator-movement distances for trees located in pasture was 1,288 m and was much longer than the mean pollinator-movement distance in a 10-ha fragment (417 m). Both estimates greatly exceeded the mean distance between trees (235 and 50 m, respectively). Dick (2001b) concluded that the much greater presence of Africanized honeybees in disturbed habitats led to differences in mean pollen-movement distances between pasture and forest trees. This finding is consistent with the observations of Roubik (1989), who demonstrated that the foraging range of Africanized honeybees was 212 km², whereas that of native stingless bees was 12.5 km². Dick (2001b) concluded that pasture trees are fully interactive with trees in the remnant patches, and may actually act as " . . . stepping stones for gene flow between fragmented populations. . . . "

S. globulifera (Clusiaceae)

Aldrich and Hamrick (1998) used three, highly variable microsatellite loci to reconstruct parentage of *S. globulifera* seedlings established in three 1-ha fragments of premontane forest in southern Costa Rica (see Table 8–4). *S. globulifera* is a canopy tree distributed throughout Central and South America, although it probably originated in western Africa (Dick et al. 2003). Its red, hermaphroditic flowers are pollinated primarily by hummingbirds, and its single-seeded fruits are distributed by bats that drop seeds below their feeding roosts, which, in fragmented landscapes, are primarily located in remnant forest patches. The study site was located in a recently fragmented (i.e., last fifteen to forty years) landscape located at an approximately 1,000-m elevation (Aldrich 1997). A 38.5-ha circular plot was established that included three remnant-forest patches in a pasture matrix that contained occasional *S. globulifera* adults. A control plot (3.2 ha) was established in an undisturbed forest (235 ha) located nearby. All adults, saplings, and most seedlings of *S. globulifera* were mapped and genotyped, and parental origins of seedlings and saplings were determined.

Only 2.3% of the seedlings had both parents within the remnant populations; 17% had one parent and 81% had both parents outside of the population (Table 8–4). The total rate of gene flow was 89%. Within the continuous forest plot, approximately 16% of the seedlings had both parents within the 1-ha plot, and 33 and 51% of the seedlings had one or both parents, respectively, outside of the plot. Total gene flow into the forest plot was 68%. Selfing rates of seedlings within the continuous forest (0.098) and remnant forest plots (0.114) were consistently lower than values for seedlings produced by pasture trees ($s = 0.261$). Furthermore, two adults located within the pastures produced 52.5% of the new seedlings within the fragments. As a result, although there were forty-three adults in the pasture and remnant forests, the effective population size (N_e) was approximately seven individuals. The number of migrants per generation (N_em) for remnant populations was 2.8 compared to 9.4 for the continuous forest plot.

These results can be explained by the effect of fragmentation on the behavior of the hummingbird pollinators and fruit-eating bats. Release from competition of the pasture trees led to much more flowering, which caused the hummingbirds to change their foraging behavior from traplining in undisturbed forests to territorial behavior in the pastures (e.g., Linhart 1973; Linhart & Feinsinger 1980). This change in hummingbird behavior led to more geitonogamous selfing within the canopies of pasture trees. Because bats rarely eat in the fruiting trees where they forage, fruits are carried to feeding roosts located in the remnant forest patches where the seeds are dropped. As a result of the changes in pollinator and frugivore behavior, an apparently demographically healthy population of seedlings actually may experience serious genetic problems in subsequent generations.

E. cyclocarpum (Fabaceae)

Hamrick and colleagues (2001, unpublished data) used fifteen polymorphic allozyme loci to identify the pollen parents of seeds produced by *E. cyclocarpum* adults from four study sites located in Guanacaste Province, Costa Rica. *E. cyclocarpum* is a dominant tree of tropical dry forests. Its range extends from Mexico to northern South America. *E. cyclocarpum* is a mimosoid legume that produces inflorescences of flowers that are receptive for a single night. Its chief pollinators are moths, although some pollen may be dispersed by bees in the early morning (O. Rocha, personal communication). Pollen of many mimosoid legumes is dispersed as a unit (i.e., a polyad). As a result, seeds of individual fruits have a single pollen donor (i.e., seeds within a fruit are full-sibs). This phenomenon allows the inference of the multilocus genotype of each pollen donor. Analyses of a large number of fruits allow descriptions of the distribution of pollen movement within a study site and estimates of pollen flow rates. The four study sites, contained within a 40-by-40 km area, were chosen based on the degree of habitat disturbance. Within each site, several maternal trees, some in clusters and some isolated, were selected. Fifty to sixty pods per tree were collected, and all the seeds in each pod were analyzed at fifteen polymorphic allozyme loci. Data from two of the four study sites, Palo Verde National Park (PV) and Hacienda Solimar (SLM), are presented here.

PV was established in the late 1970s. Prior to that, it was a working ranch with large tracts of primary forest interspersed with cleared pastures. Today, PV

Table 8–5. Mean pollen-movement parameters for six fragmented populations and seven isolated trees of *E. cyclocarpum* at PV and for thirteen trees at SLM

| Parameter | PV | | SLM |
	Within sites	Isolated trees	Individual trees
Number of trees	4.2	7	13
Rates of pollen flow (%)	75.4	100.0	91[b]
Estimated number of unique pollen parents	147.2 (50.3)[a]	91.7	92.2
Effective number of pollen parents (# pods)	14.9 (27.2)	22.2 (33.0)	40.6 (60)
Breeding population radius (m)	1,523 (905)[a]	1,527	822

[a] Value per tree
[b] Proportion of pollen originating from beyond the eight to twelve nearest neighbors

consists of primary forests that are predominantly located on dry limestone hills with successional forests located in flatter areas where pastures existed previously. *E. cyclocarpum* are generally found on the lower slopes and as large trees in secondary forests. Densities of *E. cyclocarpum* are quite low, with mean distances between nearest neighbors averaging 247 m. Six clusters of adult trees containing three to six individuals and seven isolated trees (by >500 m) were designated for paternity analyses. SLM was cleared for pastureland in 1955 and is presently a working ranch with large, scattered *E. cyclocarpum* within pastures and smaller adults distributed along semipermanent stream courses. Density of *E. cyclocarpum* is much higher at SLM than at PV, with nearest neighbors averaging 52 m from each other. Thirteen maternal trees were designated for SLM.

Results of the full-sib paternity analyses (Table 8–5) indicate that adults in the more disturbed habitat (SLM) had a much larger effective number of pollen parents (40.6) than did those in the currently less disturbed PV site (14.9). Rates of immigrant pollen flow from beyond the immediate vicinity of the maternal trees were higher at SLM than at PV (91 versus 75.4%). The radius of the per-tree breeding area at SLM is somewhat smaller than at PV (822 versus 905 m). Isolated trees at PV receive pollen from longer distances (1,500 versus 900 m) than do trees located within clusters. The longest documented pollen-movement distance was 2,800 m.

Generally, trees in disturbed and undisturbed habitats receive pollen from a high number of pollen parents, with mean average pollen-flow distances of close to 1 kilometer. Contrary to predictions, isolated trees (by >500 m) receive pollen from more pollen donors and from longer distances than do trees within clusters, whether they are in primary forest or in highly disturbed landscapes. The differences seen between SLM and PV are consistent with the high conspecific densities and more open-canopy conditions at SLM.

E. cyclocarpum has also been shown (Gonzales et al. 2010) to maintain significant spatial genetic structure within fragmented landscapes. In these landscapes, clusters of related individuals are not found in the matrix (usually pasture) surrounding maternal trees. Rather, related individuals occur in recruitment "safe sites" (streamsides, broken topography) that are usually at some distance from the maternal plants. Evidently, its fruits are consumed by cattle and horses in the pastures (Janzen, 1982; Janzen & Martin, 1982; JLH, personal observation), and seeds are later defecated where newly germinated seedlings are protected from grazing. Seedlings that germinate near maternal individuals in the open pastures are readily consumed or trampled by livestock that seek the shade of adult trees (JLH, personal observation).

Guaiacum sanctum (Zygophyllaceae)

Fuchs (2007) and Fuchs and Hamrick (2010, in review) used thirteen polymorphic allozyme loci to determine the paternity of seeds from twenty-one adult *G. sanctum* near Cuajiniquil in northwest Costa Rica. *G. sanctum* is an evergreen-forest understory tree that is predominantly found on limestone outcrops in tropical dry forests (Gonzales-Rivas et al. 2006). It is distributed from Costa Rica to northern Mexico and southern Florida and throughout the Greater Antilles (Holdridge & Poveda 1975). In Costa Rica, the main pollinators currently are Africanized honeybees, but wasps and solitary bees also visit its purple flowers.

The study site is a highly disturbed landscape composed of small patches of forest embedded in an agricultural matrix (pastures and sugar-cane plantations). The twenty-one adult trees have a disjunct distribution with nine adults grouped within a northern cluster and eleven trees within the southern cluster approximately three kilometers away. There is a single isolated tree located approximately midway between the two clusters. Other *G. sanctum* may be located within inaccessible areas separated by six kilometers from the sampled individuals. Progeny from the twenty-one adult trees were analyzed for the thirteen allozyme loci, and the male reproductive success of the twenty-one adults was estimated via fractional paternity assignment (Gerber et al. 2003).

Pollen-movement distances estimated from singly sired seeds were bimodal (i.e., within versus between clusters) and averaged 2,085 m. Approximately 16% of the seeds were sired by unidentified adults outside of the two clusters leading to estimates of pollen-flow distances of more than four kilometers. The isolated tree was the most successful pollen donor (14%). This tree sired 7.8% of the seeds from the southern mothers and 5.4% of the seeds from northern mothers, leading to the conclusion that this isolated tree facilitated pollen movement between the two clusters by serving as a stepping stone between the two sites. Pollinators in one cluster may be blocked from perceiving the other cluster by landscape features but can perceive the isolated individual. After they move to the isolated tree, they can perceive the second cluster.

CONSERVATION IMPLICATIONS

Typically, most conservation and management efforts are focused on the preservation of large tracts of relatively pristine habitat where ecological and evolutionary processes can proceed without significant human perturbations. These objectives probably should be given highest priority by the conservation community but, too often, little attention is given to the conservation and management of disturbed habitats. Realistically, the earth has reached the point at which disturbed landscapes now occupy more area than pristine landscapes in most temperate and tropical regions. A case, therefore, can be made for actively managing disturbed landscapes to maximize both species diversity and the genetic diversity that occurs within species. As this chapter has shown, it is likely that for many long-lived tropical plant species, the majority of the genetic diversity within species should still remain within the landscape, although, given calculated levels of $N_e m$, genetic diversity may be more structured than in undisturbed sites. These empirical observations provide hope that with active, biologically informed management, high levels of genetic diversity can be maintained for many tropical plant species.

At the simplest level of management, it is likely that remnant forests will serve as a primary source of propagules for the restoration of native species in degraded landscapes. Genetic changes that have occurred due to fragmentation (e.g., increased inbreeding or reduced genetic diversity) would be introduced into the restored areas. As a result, restoration efforts may fail or may not perform up to expectations. In a worst-case scenario, such failures may call into question whether habitat restoration is feasible or desirable with native species, leading, perhaps, to the planting of introduced species. In addition, remnant populations could serve as a source of propagules for the recolonization of adjacent abandoned lands. As a result, the footprint of genetic changes that have occurred as a result of fragmentation may be visible for several generations in these successional areas. Genetic analyses could be used to identify remnant populations that have maintained the highest levels of genetic diversity and have experienced the least inbreeding. These populations could then be used preferentially as sources of propagules for landscape-restoration efforts.

An understanding of the genetic connectivity occurring within fragmented landscapes can be used to identify remnant populations and isolated individuals that are essential to the maintenance of genetic diversity within the landscape. Management strategies could be developed that would maximize genetic connectivity by ensuring the preservation of keystone populations and individuals. Furthermore, simulations based on empirical data could identify locations that, if restored, would increase genetic connectivity within the landscape. In essence, empirical data on the patterns and levels of pollen and seed movement could be used to develop networks of populations and individuals that maximize the maintenance of genetic diversity for a species or suite of species with high commercial or ecological value. With this management scenario, genetic markers would be used to monitor genetic connectivity and diversity within these managed landscapes.

CONCLUSIONS

Fragmentation of once continuous forests does not appear to limit pollen movement as has been predicted. This generalization may, however, depend on the species involved. To date, pollen-flow studies indicate that bats and many insect pollinators fly several hundred meters across open, disturbed habitats. The available data, although still limited, indicate that pollen movement across fragmented landscapes may be greater than that seen in undisturbed sites (e.g., *E. cyclocarpum*, *D. excelsa*, and *S. humilis*). Trees in isolated fragments and individual trees successfully set seeds. There is information, however, for *E. cyclocarpum* that indicates that seeds produced by isolated trees are inferior to seeds produced by forest trees with more near-neighbors (Rocha & Aguilar 2001). More studies with this level of detail are needed. Also, we do not know the long-term effects on the pollinators of having to fly greater distances between trees.

It should be apparent from the studies discussed that to make accurate predictions of the effects of landscape fragmentation on the levels and distribution of genetic diversity, one must understand the natural history of the species involved. In particular, we need a better understanding of the plant species' pollinators and seed-dispersal agents in natural and disturbed habitats. Dick's study (2001b) of *D. excelsa* demonstrated that pollinators observed in natural forests were not the predominant pollinators in disturbed habitats, greatly affecting reproductive success and pollen movement distances. For most species, however, it is not known whether pollinators change when the landscape becomes fragmented. The study of *S. globulifera* by Aldrich and Hamrick (1998) goes one step further by demonstrating that the behavior of its pollinators and the seed-dispersal agents are modified in fragmented landscapes. These behavioral changes have dramatically affected the genetic composition of the next generation of *S. globulifera* in the remnant forest patches.

Janzen (1986) suggested that isolated trees left standing in open pastures or agricultural fields represented the "living dead" because they were unlikely to produce viable offspring. Several studies reviewed here demonstrate that isolated trees located well away from remnant forest patches continue to influence mating patterns within highly disturbed landscapes. These trees produce, in some cases, more seeds than do trees located in undisturbed forests (e.g., *S. globulifera* and *D. excelsa*), or they can serve as the pollen parents of trees located in remnant forest patches. In certain situations (e.g., *G. sanctum*, *D. excelsa*), isolated trees play a critical role at the landscape level by serving as stepping stones for the movement of genes between forest fragments. There is also some evidence that isolated trees contribute seeds and seedlings to forest fragments (e.g., *S. globulifera* and *E. cyclocarpum*). Removal of isolated trees from fragmented landscapes (i.e., "giving up on the landscape") could have a disproportionately negative impact on genetic connectivity within the landscape and could, for trees, lead to the long-term loss of genetic diversity at the landscape level. The development of informed management plans for fragmented landscapes could, in contrast, maintain natural levels of genetic diversity within species.

REFERENCES

Aldrich PR (1997) *Dispersal and the Scale of Fragmentation in Tropical Tree Populations*. Ph.D. Dissertation, The University of Georgia.

Aldrich PR, Hamrick JL (1998) Reproductive dominance of pasture trees in a fragmented tropical forest mosaic. *Science*, **281**, 103–105.

Allard RW (1988) Genetic changes associated with the evolution of adaptedness in cultivated plants and their wild progenitors. *Journal of Heredity*, **79**, 225–238.

Apsit VJ (1998) *Fragmentation and Pollen Movement in a Costa Rican Dry Forest Tree Species.* Ph.D. Dissertation, The University of Georgia.

Bawa KS (1974). Breeding systems of tree species of a lowland tropical community. *Evolution,* **28**, 85–92.

Born C, Hardy OJ, Chavallier MH et al. (2008) Small-scale spatial genetic structure in the Central African rainforest tree species, *Aucoumea klaineana*: a stepwise approach to infer the impact of limited gene dispersal, population history, and habitat fragmentation. *Molecular Ecology,* **17**, 2041–2050.

Boshier DH, Chase M, Bawa KS (1995) Population genetics of *Cordia alliodora* (Boraginaceae), a neotropical tree. 3: Gene flow, neighborhood, and population substructure. *American Journal of Botany,* **82**, 484–490.

Cespedes M, Gutierrez MV, Halbrook NM, Roche OJ (2003) Restoration of genetic diversity in the dry forest tree, *Swietenia macrophylla* (Meliaceae), after pasture abandonment in Costa Rica. *Molecular Ecology,* **12**, 3201–3212.

Chase MR, Moller C, Kesseli R, Bawa KS (1996) Distant gene flow in tropical trees. *Nature,* **383**, 398–399.

Dawson IK, Waugh R, Simons AJ, Powell W (1997) Simple sequence repeats provide a direct estimate of pollen-mediated gene dispersal in the tropical tree, *Gliricidia sepium. Molecular Ecology,* **6**, 179–183.

de Lacerda AEB, Kanashiro M, Sebbenn M (2008) Long-pollen movement and deviation of random mating in a low-density continuous population of a tropical tree, *Hymenaea courbaril*, in the Brazilian Amazon. *Biotropica,* **40**, 462–470.

Devlin B, Roeder K, Ellstrand N (1988) Fractional paternity assignment: theoretical development and comparisons to other methods. *Theoretical and Applied Genetics,* **76**, 369–380.

Dick C (2001a) Habitat change, African honeybees and fecundity in the Amazonian tree, *Dinizia excelsa* (Fabaceaea). In: *Lessons from Amazonia: The Ecology and Conservation of a Fragmented Tropical Forest* (eds. Bierregaard RO, Gascon C, Lovejoy TE, Mesquita R), pp. 146–157. Yale University Press, New Haven, CT.

Dick CW (2001b) Genetic rescue of remnant tropical trees by an alien pollinator. *Proceedings of the Royal Society of London-Series B: Biological Sciences,* **28**, 2391–2396.

Dick CW, Abdul-Salim K, Bermingham E (2003) Molecular systematics reveals cryptic tertiary diversification of a widespread tropical rainforest tree. *American Naturalist,* **162**, 691–703.

Dick CW, Hardy OJ, Jones FA, Petit RJ (2008) Spatial scales of pollen and seed-mediated gene flow in tropical rainforest trees. *Tropical Plant Biology,* **1**, 20–33.

Dunphy BK, Hamrick JL (2007) Estimation of gene flow into fragmented populations of *Bursera simaruba* (Burseraceae) in the dry-forest life zone of Puerto Rico. *American Journal of Botany,* **94**, 1786–1794.

Dunphy BK, Hamrick JL, Scwagerl J (2004) A comparison of direct and indirect measures of gene flow in the bat-pollinated tree, *Hymenaea courbaril*, in the dry-forest life zone of southwestern Puerto Rico. *International Journal of Plant Science,* **165**, 427–436.

Fuchs EJ (2007) *Population Genetics of the Endangered Tropical Tree,* Guaiacum sanctum *(Zygophyllaceae).* Ph.D. Dissertation, The University of Georgia.

Fuchs, EJ, Hamrick JL (2010) Mating system and pollen flow between remnant populations of the endangered tropical tree, *Guaiacum sanctum* (Zygophyllaceae). The importance of isolated trees. *Conservation Genetics*, in review.

Gerber S, Chabrier P, Kremer A (2003) FAMOZ: A software for parentage analysis using dominant, codominant, and uniparentally inherited markers. *Molecular Ecology Notes,* **3**, 479–481.

Gonzales E, Hamrick JL, Smouse PE, Trapnell DW, Peakall R (2010) The impact of landscape disturbance on spatial genetic structure in the Guanacaste tree, *Enterolobium cyclocarpum* (Fabaceae). *Journal of Heredity,* **101**, 133–143.

Gonzales-Rivas B, Figabu M, Gerhardt K, Castro-Marin G, Oden PC (2006) Species composition, diversity, and local uses of tropical dry deciduous and gallery forests in Nicaragua. *Biodiversity and Conservation,* **15**, 1509–1527.

Hamrick JL (2001) Breeding patterns of a tropical dry-forest tree species, *Enterolobium cyclocarpum*, in disturbed and undisturbed habitats. In: *Tropical Ecosystems, Structure, Diversity, and Human Welfare* (eds. Ganeshaiah KN, Uma Shaanker R, Bawa KS), pp. 291–294. Oxford and IBH Publication Co., New Delhi.

Hamrick JL, Nason JD (1996) Consequences of dispersal in plants. In: *Population Dynamics and Ecological Space and Time* (eds. Rhodes OE, Chesser RK, Smith MH), pp. 203–236. University of Chicago Press, Chicago.

Hanson T, Brunsfeld S, Finegan B, Waits L (2007) Conventional and genetic measures of seed dispersal for *Dipteryx panamensis* (Fabaceae) in continuous and fragmented Costa Rican rainforest. *Journal of Tropical Ecology*, **23**, 635–642.

Hanson TR, Brunsfeld SI, Finegan B, Waits LP (2008) Pollen dispersal and genetic structure of the tropical tree, *Dipteryx panamensis*, in a fragmented Costa Rican landscape. *Molecular Ecology*, **17**, 2060–2073.

Hardesty BD, Hubbell SP, Bermingham E (2006) Genetic evidence of frequent long-distance recruitment in a vertebrate-dispersed tree. *Ecology Letters*, **9**, 516–525.

Hedrick DW (2000) *Genetics of Populations*. Jones and Bartlett, Sudbury, MA.

Holdridge LR, Poveda LJ (1975) *Arboles de Costa Rica* (Vol. **1**). Cantro Cientifico Tropical, San Jose, Costa Rica.

Janzen DH (1982) Differential seed survival and passage rates in cows and horses, surrogate Pleistocene dispersal agents. *Oikos*, **38**, 150–156.

Janzen DH (1986) The future of tropical ecology. *Annual Reviews of Ecology and Systematics*, **17**, 305–324.

Janzen DH, Martin P (1982) Neotropical anachronisms: the fruits the gomphotheres ate. *Science*, **215**, 19–27.

Jones FA, Chen J, Weng GJ, Hubbell SP (2005) A genetic evaluation of seed dispersal in the neotropical tree, *Jacaranda copaia* (Bignoniaceae). *American Naturalist*, **166**, 543–555.

Kaufman SR, Smouse PE, Alvarez-Buylla ER (1998) Pollen-mediated gene flow and differential male reproductive success in a tropical pioneer tree, *Cecropia obtusifolia Bertal* (Moraceae). *Heredity*, **81**, 164–173.

Linhart YB (1973) Ecological and behavioral determinants of pollen dispersal in hummingbird-pollinated *Heliconia*. *American Naturalist*, **107**, 511–523.

Linhart YB, Feinsinger P (1980) Plant-hummingbird interactions: effects of island size and of specialization on pollination. *Journal of Ecology*, **68**, 745–760.

Lovejoy T, Bierregaard R (1990) Central Amazonian forests and the minimum critical size of ecosystems project. In: *Four Neotropical Forests* (ed. Gentry A), pp. 60–75. Yale University Press, New Haven, CT.

Nason JD, Aldrich PR, Hamrick JL (1997) Dispersal and the dynamics of genetic structure in fragmented tropical tree populations. In: *Tropical Forest Remnants: Ecology, Management, and Conservation in Fragmented Communities* (eds. Laurance WF, Bierregaard RO), pp. 304–320. University of Chicago Press, Chicago.

Nason JD, Hamrick JL (1997) Reproductive and genetic consequences of forest fragmentation: two case studies of neotropical canopy trees. *Journal of Heredity*, **88**, 264–276.

Nason JD, Herre EA, Hamrick JL (1998) The breeding structure of a tropical keystone plant resource. *Nature*, **391**, 685–687.

Nei M, Maruyama T, Chakrabarty R (1975) The bottleneck effect and genetic variability in populations. *Evolution*, **29**, 1–10.

Opler PA, Baker HG, Frankie GW (1980). Plant reproductive characteristics during secondary succession in neotropical lowland forest ecosystems. *Biotropica*, **12**, 40–46.

Petit RJ, Duminil J, Fineschi S, Hampe A, Vendramin GG (2005) Comparative organization of chloroplasts, mitochondrial and nuclear diversity in plant populations. *Molecular Ecology*, **14**, 689–702.

Rocha OJ, Aguilar G (2001) Reproductive biology of the dry-forest tree, *Enterolobium cyclocarpum Jacq.* (Guanacaste) in Costa Rica. *American Journal of Botany*, **88**, 1607–1614.

Roubik DW (1989) *Ecology and Natural History of Tropical Bees*. Cambridge University Press, New York.

Sezen UU, Chazdon RL, Holsinger KE (2007) Multigenerational genetic analysis of tropical secondary regeneration in a canopy palm. *Ecology*, **88**, 3065–3075.

Sork VL, Smouse PE (2006) Genetic analysis of landscape connectivity in tree populations. *Landscape Ecology*, **21**, 821–836.

Styles BT (1981). *Swietenioideae*. In: *Flora Neotropica Monograph No. 28: Meliaceae* (eds. Pennington TD, Styles BT, Taylor DAH), pp. 359–418. New York Botanical Garden, New York.

Trapnell DW, Hamrick JL (2005) Mating patterns and gene flow in the neotropical epiphytic orchid, *Laelia rubescens*. *Molecular Ecology*, **14**, 75–84.

Vekemans X, Hardy OJ (2004) New insights from fine-scale spatial genetic structure analyses in plant populations. *Molecular Ecology*, **13**, 921–935.

Ward M, Dick CW, Gribel R, Lowe AJ (2005) To self, or not to self . . . A review of outcrossing and pollen-mediated gene flow in neotropical trees. *Heredity*, **95**, 246–254.

White GM, Boshier DH, Powell W (1999) Genetic variation within a fragmented population of *Swietenia humilis Zucc. Molecular Ecology*, **8**, 1899–1909.

White GM, Boshier DH, Powell W (2002) Increased pollen flow counteracts fragmentation in a tropical dry-forest: an example from *Swietenia humilis Zuccarini*. *Proceedings of the National Academy of Sciences USA*, **99**, 2038–2042.

Wright S (1951) The genetical structure of populations. *Annals of Eugenics*, **15**, 323–354.

9 Implications of landscape alteration for the conservation of genetic diversity of endangered species

Paul L. Leberg, Giridhar N. R. Athrey, Kelly R. Barr,
Denise L. Lindsay, and Richard F. Lance

Humans have dramatically altered biotic communities around the world. During the process of converting forests, grasslands, and wetlands for agriculture, urban development, and transportation, the remnants of natural habitat have become increasingly fragmented. This fragmentation of habitat has many biological consequences. Foremost is the reduction of available habitat resulting in a reduced size of populations of species dependent on natural land covers (Andren 1994; Fahrig 1997, 2003). Fragmentation, though, is more than just habitat loss; it is also the division of remaining habitat into patches that experience at least partial isolation from other such fragments (Fahrig 2003). Fragmentation increases habitat edges and, consequently, the exposure of populations to the resulting alterations in microclimate and biota associated with edge environments (Suarez et al. 1998; Fagan et al. 1999). As fragments become more isolated, the frequency of movements between fragments is reduced. Furthermore, when localized extinctions in isolated fragments occur, immigration from neighboring fragments can be insufficient to allow recolonization (Tilman et al. 1994; Fahrig 2003). Extinctions associated with demographic stochasticity are expected to be more common in small fragments than in more continuous habitat tracts (Griffen & Drake 2008).

The consequences of fragmentation have the potential to influence genetic diversity in species requiring continuous tracts of habitat (see Chapter 8 by J. Hamrick). Localized population declines cause allele frequencies to drift; moreover, widespread population declines due to fragmentation across the range of a species may reduce its effective population size (Pannell & Charlesworth 2000; Alo & Turner 2005). With declines of census and effective population size, inhabitants of isolated fragments begin to resemble independent populations, become

We thank the many people who assisted with field sampling and logistics, especially G. Ekricht, R. Fain, C. Farquhar, V. Fazio, J. Gryzbowski, T. Hayden, S. Hodge, J. Karges, J. Kimball, L. Lapham, K. Moore, R. Myers, J. Neal, C. Pekins, I. Pollet, C. Sexton, S. Tweddale, G. Wampler, and D. Wolf. Access to field sites was granted by the U.S. Army, U.S. Fish and Wildlife Service, Texas Parks and Wildlife Department, The Nature Conservancy, Klondike Ranch, Quail Ridge Ranch, Panther Cave Ranch, Dobbs Run Ranch, and Environmental Defense. Funding was provided by the U.S. Army Basic Research Program's Environmental Quality and Installations Focus Area, as administered by the U.S. Army Engineer Research & Development Center. We thank the conference organizers and attendees, Purdue students and post-docs, and anonymous reviewers for comments on our presentation of this material and on earlier drafts of this chapter.

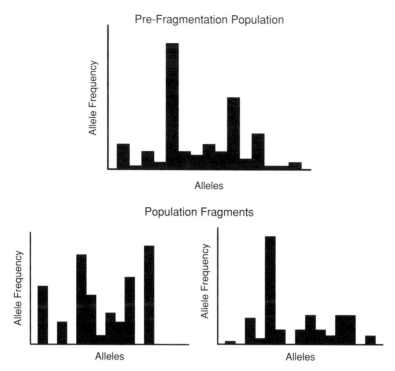

Figure 9–1: Effects of fragmentation on the distributions of allele frequencies. Prior to fragmentation, a population contains many alleles, most of which are at low frequencies. In each of the population fragments, drift results in a loss of some alleles and increased frequencies of others. Thus, allelic diversity and, to a lesser extent, heterozygosity will decrease in the population fragments, and genetic differences between fragments will increase.

more vulnerable to stochastic processes, lose genetic diversity, and increase in relatedness (Barrowclough 1980; Leberg 2005). Restricted gene flow can dramatically increase accumulation of genetic differences among populations (differentiation) through genetic drift (Fig. 9–1; but see also Box 8–2 by Busch, McCreight, & Waser). The loss of genetic diversity within fragments and the increase in differentiation among them are predicted by theory (Wright 1931, 1968) and have been the subject of many investigations (Leberg 1991; Streiff et al. 2005; Segelbacher et al. 2008). Similar processes associated with small population sizes can affect genetic variation in restored populations (see Box 9).

Population fragmentation should thus result in the reallocation of genetic diversity from within populations to differentiation among populations (Amos & Balmford 2001; Frankham 2005). Although there is considerable information on the influence of natural fragmentation on the distribution of genetic variation (Frankham 1996, 1997), less is known about how the relatively recent process of human-caused fragmentation has affected variation. It is possible that the time scale associated with recent fragmentation is often insufficient to either have resulted in redistribution of genetic variation or to observe the genetic signatures associated with long-term isolation of small patches of habitat (Olivieri et al. 2008; Shepard et al. 2008). Recent reviews of anthropogenic influences of genetic variation following fragmentation have focused on loss of diversity within fragments (DiBattista 2008; Kramer et al. 2008); the expectation of increased genetic

BOX 9: DUNE RESTORATION INTRODUCES GENETICALLY DISTINCT AMERICAN BEACHGRASS, *AMMOPHILA BREVILIGULATA*, INTO A THREATENED LOCAL POPULATION

Julie R. Etterson and Rebecca M. Holmstrom

Problem

Restoration of native vegetation has become a critical component of conservation biology in degraded or destroyed natural habitats. If our goal is to reconstitute self-sustaining populations, we should obtain restoration material from source populations that can survive and reproduce at the restoration site. Restored populations must also harbor sufficient genetic variability to permit ongoing evolutionary responses to a changing environment. Thus, an understanding of the genetic structure of natural populations as well as propagules used for restoration is essential. The reality on the ground, however, is that restoration projects often rely on nonlocal sources of seed or clonal propagules and have little or no information about the genetic structure of either native populations or restoration material.

Case Study

This study explored the potential impact of using nonlocal material for restoration within a threatened native population of the dune grass *Ammophila breviligulata* Fern. (Anderson 2006). For three decades, documented restorations along the shore of Lake Superior in Duluth, Minnesota, have relied on a single commercial source in Michigan to supply vegetative propagules. We used both molecular and common garden approaches to examine the extent to which reconstituted populations resemble the threatened remnant stands of *A. breviligulata*. Using highly variable inter-simple sequence repeat (ISSR) markers, we compared 1) molecular genetic diversity among well-established local populations; 2) restored populations, planted exclusively with commercial propagules five to fourteen years ago; and 3) a contemporary sample of nursery stock obtained from the commercial supplier. To compare ecotypic differences, we planted two common gardens that included vegetative propagules sampled from local stands and propagules obtained from the commercial supplier.

The molecular study showed that native populations are substantially more genetically diverse than are restored populations (Box Fig. 9–1). In fact, populations that had been restored for five to fourteen years were genetically identical based on these ISSR markers (Fant et al. 2008). It is surprising that this was also the only genotype detected in the commercially available beachgrass and was not found in the natural populations. Thus, restoration efforts of *A. breviligulata* have failed to replicate the genetic constitution of natural populations in terms of either genetic diversity or the composition of genotypes.

Does this extreme lack of diversity translate into low fitness? It is interesting that comparison of the vigor of local and nonlocal plants grown in Minnesota

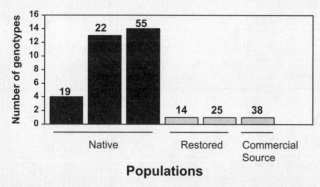

Box Figure 9–1: Number of multilocus genotypes based on three ISSR markers within American beachgrass populations, including: 1) three native Minnesota populations; 2) two populations restored five and fourteen years ago using propagules obtained from a commercial supplier in Michigan; and 3) contemporary accessions obtained from the same Michigan supplier. Sample sizes are shown above bars.

suggests that it does not (Holmstrom et al. in press). In almost every respect, Michigan plants were larger and more robust. Importantly, Michigan plants had higher survival rates, produced more biomass (including rhizomes), and flowered when native plants did not (Box Fig. 9–2). This finding may indicate that plants in the gardens of the commercial supplier have been subjected to unconscious or artificial selection through propagation of more robust and vigorous clones. The consequence in Minnesota may be that Michigan genotypes may displace or "swamp" the threatened Minnesota genotypes through vegetative spread, competition, and recruitment from seed. Ultimately, this displacement could lead to an overall loss of genetic diversity if introduced beachgrass stands become dominant. Surveys of the dune community indicate that this process is already underway with Michigan ecotypes appearing in places where no previous restorations have been documented.

This research highlights the value of both molecular and phenotypic data for informing restoration practices. Both approaches provide strong evidence that local Minnesota and introduced Michigan populations are genetically distinct. We found no concordance, however, between levels of molecular variation and ecotypic vigor. Specifically, molecular diversity was not a reliable predictor of plant fitness as measured by either asexual or sexual reproduction. Overall, our study underscores the importance of obtaining baseline genetic surveys of remnant native populations and restoration propagules before restoration efforts are initiated, especially when the populations are threatened or endangered. Careful documentation of restoration projects will allow long-term monitoring of restoration success and assessment of any impact that there might be on remnant natural populations.

REFERENCES

Anderson (Holmstrom) RM (2006) *The Use of Nonlocal Propagules for Habitat Restoration: Implications for a Threatened Population of* Ammophila breviligulata *in Minnesota*. MS Thesis, University of Minnesota, Duluth.

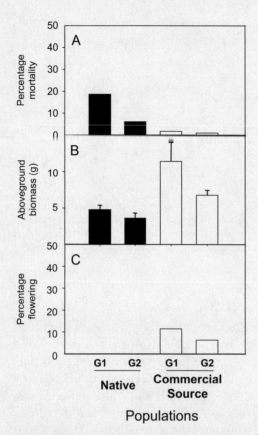

Box Figure 9–2: Comparison among American beachgrass plants obtained from a native Minnesota population and from a commercial supplier in Michigan. Common gardens were established for two years (G1) or one year (G2). G1 and G2 were composed of 140 and 117 propagules per population, respectively (N_{total} = 514).

Fant JB, Holmstrom RM, Sirkin E, Etterson JR, Masi S (2008) Genetic structure of threatened native populations and propagules used for restoration, in a clonal species, *Ammophila breviligulata* fern (American beachgrass). *Restoration Ecology*, **16**, 594–603.

Holmstrom RM, Etterson JR, Schimpf DJ (in press) Dune restoration introduces genetically distinct American beachgrass, *Ammophila breviligulata*, into a threatened local population. *Restoration Ecology*. DOI:111/j.1526-100X.2009.00593.x.

differentiation among population fragments was not examined. Genetic differentiation may occur more rapidly than loss of genetic diversity within populations (Avise 1992). We do not yet know if one of the expected genetic signatures of fragmentation is a more sensitive indicator of recent fragmentation than the other.

In addition to understanding how anthropogenic fragmentation affects the repartitioning of genetic diversity within and among populations, it would be helpful to know which attributes of species affect these responses. For example, would species rarity influence susceptibility to the genetic consequences of fragmentation? Rare or endangered species exist as small populations and may exhibit genetic signatures of fragmentation more rapidly than may common species if reduced population sizes, associated with rarity, make them more susceptible to genetic drift. Louy and colleagues (2007) report stronger genetic responses to fragmentation in an endangered butterfly than in more common species. If this result is observed across species, it raises the possibility that studies of abundant species might not greatly inform us about the consequences of fragmentation in species of conservation concern.

Vagility of a species is likely to be associated with gene flow; thus, species with large movement capabilities should be less susceptible to the effects of human-induced fragmentation (Packer et al. 1998; Grapputo et al. 2005; Keyghobadi et al. 2006; Mills et al. 2007). Large carnivores, birds, and some fishes have been documented traversing large distances, passing across urbanized areas or anthropogenic structures, and, in some cases, re-establishing connections between historically disconnected populations (Sacks et al. 2006). In a comparison of three butterfly species, reduced dispersal capability was associated with increased influence of fragmentation on genetic variation (Louy et al. 2007). It is not clear how consistent this pattern would be across species. For instance, some species capable of moving long distances exhibit restricted dispersal among patches of fragmented habitats (Doak 2000; Hansson et al. 2002).

We also do not yet know if there are habitat-specific effects of fragmentation on the responses of genetic variation in affected populations. The vast majority of studies of the ecological effects of anthropogenic fragmentation have focused on forested ecosystems. It is not surprising that the striking conversion of a vertically structured habitat such as forests into clear cuts or agricultural fields has been of intense interest to ecologists. Forests are also disproportionately represented in these studies because species depending on secondary growth or late-successional habitat types are often highly sensitive to changes in the habitat. Furthermore, it might be expected that species dependent on less structurally complex habitats, such as grasslands, deserts, or habitat edges, might be more capable of movement across human-influenced landscapes than those dependent on forests might be. We are unaware of evaluations of this issue with regard to the genetic consequences of fragmentation.

We have investigated the effects of fragmentation on the genetic structure of two species of endangered passerines: the golden-cheeked warbler, *Dendroica chrysoparia* (Lindsay et al. 2008), and black-capped vireo, *Vireo atricapilla* (Barr et al. 2008). These studies suggested that vagility and habitat requirements might not predict the influence of habitat fragmentation on the distribution of genetic variation within and among populations. Our work also raised the possibility that rare or endangered species might be more susceptible to fragmentation effects than more common species might be. Our first objective is to review our research on these two migratory songbirds relative to the genetic consequences of fragmentation. We also present results of a literature survey we conducted to assess the influence of species rarity, vagility, and habitat requirements on the genetic implications of anthropogenic fragmentation.

FRAGMENTATION EFFECTS IN TWO ENDANGERED PASSERINES

Few organisms have the potential for movement among isolated habitats as do birds. Most are capable of long-distance flight, and many species traverse long distances during annual migration. Despite these capabilities, some species exhibit

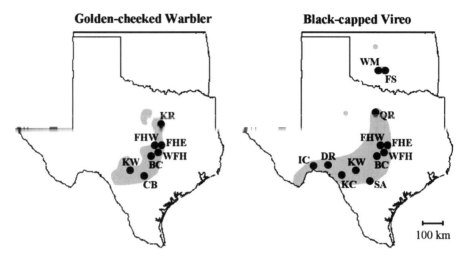

Figure 9–2: Map of sample site locations (dots) and the current breeding ranges (shaded) of the golden-cheeked warbler (modified from Lindsay et al. 2008) and black-capped vireo (modified from Barr et al. 2008) in Texas and Oklahoma. See Lindsay et al. (2008) and Barr et al. (2008) for detailed descriptions of the study sites and explanations of abbreviations.

a tendency not to cross short distances of unfavorable habitat. These reports deal primarily with forest-dwelling species (Desrochers & Hannon 1997; Harris & Reed 2002). We know little, however, about the ability of species dependent on other habitats to move across landscapes. For most avian species, we also know little about dispersal of individuals from their birthplaces to the sites where they will reproduce. The few studies that have addressed the genetic consequences of fragmentation in birds have come to a range of conclusions on the extent of fragmentation effects (e.g., Uimaniemi et al. 2000, 2003; Segelbacher et al. 2003; Pitra et al. 2004; Veit et al. 2005).

We are conducting ongoing investigations into the effects of landscape alteration on the conservation genetics of two endangered passerines, the golden-cheeked warbler and the black-capped vireo. These species experience considerable overlap in their distributions in central Texas (Fig. 9–2) and provide insights into how landscape processes occurring in the region are affecting two species with similar vagility but contrasting habitat requirements.

The golden-cheeked warbler

This migratory songbird (Fig. 9–3) is a federally listed endangered species with a breeding range confined to twenty-seven counties in central Texas (Fig. 9–2) and an estimated total population size of only 13,800 pairs (U.S. Fish and Wildlife Service [USFWS] 2005). Although Rappole and coworkers (2003) have suggested that low capacity of the wintering grounds in Central America might be the limiting factor in the viability of the golden-cheeked warbler, the species was listed as endangered after a large reduction in breeding habitat (Table 9–1) and population size (25%) was detected within a thirty-year period (USFWS 1992). The loss and fragmentation of breeding habitat may potentially affect the genetic variability of the species, an important factor in its prospects for long-term survival.

Table 9–1. Comparison of the biological and genetic aspects of the two species considered in our case study: namely, the golden-cheeked warbler and the black-capped vireo. Overall measures of genetic diversity (heterozygosity and allelic richness) and interpopulation differentiation among sampled populations (F_{ST}) are provided for comparison, in addition to the maximum differentiation recorded between any two sampled populations

	Golden-cheeked warbler	Black-capped vireo
Status	Endangered	Endangered
Year listed	1990	1987
Habitat type	Late successional	Early successional
Estimated habitat reduction	35%	30%
Population size at listing	Unknown	191 pairs
Present population size	13,800 pairs	10,047 pairs
Overall heterozygosity	0.77	0.74
Allelic richness	7.74	7.73
Maximum F_{ST} between two sample sites[a]	0.028 (KR-BC)[a]	0.053 (DR-FHE)
Overall F_{ST}	0.0043	0.021

[a] Sample site designations are KR (Klondike Ranch), BC (Balcones Canyonlands NWR), DR (Devils River), and FHE (Fort Hood East).

Golden-cheeked warblers are habitat specialists and exhibit relatively high site fidelity (Morse 1989). They require a specific breeding habitat of dense, mature stands of Ashe juniper (*Juniperus ashei*) and deciduous trees (mostly *Quercus* spp.), a climax community that normally develops along streambeds or limestone outcrops (Kroll 1980). Only mature Ashe junipers (aged twenty-five to fifty years) possess the shredding bark that is an essential component of nest construction. The increased reproductive success of golden-cheeked warblers with interior territories suggests a preference for the forest interior (Ladd & Gass 1999). In general, greater densities of golden-cheeked warblers are associated with larger continuous patches with less edge, taller average tree height with more height variability, and a higher density of deciduous trees (Wahl et al. 1990).

Optimal breeding habitat is considered to be in woodland patches of at least 100 ha (USFWS 1996). Although golden-cheeked warblers will use smaller habitat patches, the isolation of these patches reduces their availability, which reduces

a   b

Figure 9–3: The golden-cheeked warbler (a) requires mature oak-juniper forests for nesting habitat. The black-capped vireo (b) nests in early-successional shrubs and edge habitat (photos by Kelly Barr). *See Color Plate X.*

the mean reproductive success of the population (Rappole et al. 2003). Recent habitat losses have occurred due to urban expansion and the construction of large reservoirs (Texas Parks and Wildlife Department [TPWD] 1995). Historically, loss of nesting habitat was caused by the clearing of juniper-oak woodlands for livestock production, timber sale, and agriculture (TPWD 1995; Ware & Greis 2002). Over-browsing by wild ungulates, over-grazing by livestock, oak wilt fungus, fire suppression, and the resulting increase in juniper monocultures have been detrimental to golden-cheeked warblers (USFWS 1992; TPWD 1995). The warbler also suffers from brood parasitism by brown-headed cowbirds (*Molothrus ater*) (Ladd & Gass 1999), which may have increased with fragmentation.

Current recovery efforts include land preservation, habitat management, and cowbird-removal programs. Although recovery efforts for the golden-cheeked warbler are underway, the potential for loss of genetic diversity due to previous habitat fragmentation is high. It is estimated that continued viability of the golden-cheeked warbler requires 1,000 to 3,000 breeding pairs per population (USFWS 1996), and currently only two sites within the entire breeding range are thought to support such numbers (USFWS 2004). Most populations occur in much smaller numbers on fragments with no direct connection to the large breeding populations.

We sampled seven sites throughout its breeding range (Fig. 9–2) and obtained the genotypes of 107 individuals at nine microsatellite loci (Lindsay et al. 2008). Overall, levels of polymorphism were high (Table 9–1). Although one of the smallest populations we sampled (census size: thirty-five males at Klondike Ranch [KR]; Fig. 9–2) had the lowest levels of allelic variation, there was no statistical relationship between measures of within-population genetic variation and census population size. With the exception of those at KR, most populations had similar levels of allelic diversity and heterozygosity, regardless of their degree of isolation. Given that two of the populations were located in habitat fragments that were both small and relatively isolated from larger populations, it is possible that either the duration or extent of isolation may not be sufficient for genetic drift to result in loss of within-population variation.

There was evidence of genetic differentiation among sampled populations (Lindsay et al. 2008). Although the level of differentiation was small (Table 9–1), it was significantly greater than what would be expected for a panmictic population and higher than has been observed in other warblers over the same spatial scale (Gibbs et al. 2000; Clegg et al. 2003; Veit et al. 2005). Avian species might be expected to exhibit low genetic differentiation due to their high vagility, especially if habitat is continuous (Crochet 2000); however, the breeding habitat of golden-cheeked warblers is now heavily fragmented. Furthermore, their high breeding-site fidelity may also be of consequence for interpopulation differentiation. Males and females average 73 and 55% breeding site fidelity, respectively, with adult breeders dispersing, on average, less than 1 km as a result of spring migration from their wintering grounds (Ladd & Gass 1999). Initial dispersal of juveniles prior to fall migration is higher than movements by adults, averaging 9 km for males and 3 km for females (Ladd & Gass 1999). These movements do not appear to be sufficient, however, to overcome the isolating effects of fragmentation.

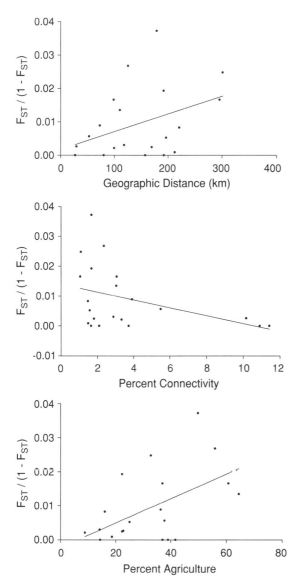

Figure 9–4: Relationship of genetic differentiation ($F_{ST}/(1 - F_{ST})$) with geographic distance, connectivity, and percent agricultural land among golden-cheeked warbler populations (modified from Lindsay et al. 2008).

Using Mantel tests, we found relationships between genetic differentiation and both habitat connectivity and geographic distance among populations (Fig. 9–4). Because connectivity and geographic distance were correlated, it is not possible to differentiate between the influence of isolation-by-distance and fragmentation on differentiation. Pair-wise comparisons between KR and the other sites contributed the most to the observed associations. It is interesting that the amount of agricultural land between sites has a strong association with genetic differentiation (Fig. 9–4). A decreased likelihood of birds crossing agricultural landscapes might explain the observed differentiation between KR and other sites because KR is isolated from other sites by extensive areas of agriculture. Other populations with similarly low levels of connectivity, but which were not in agricultural areas, did not experience comparably high levels of genetic differentiation. There was also limited evidence that sites separated by high degrees of urban development were more likely to be genetically differentiated. These observations suggest that the intervening habitat between fragments affects levels of gene flow in this highly vagile species. Previous studies have provided evidence that the type of land use between habitats can effect behavioral barriers to dispersal by forest-dwelling passerines, especially for those with specialized habitat requirements (Desrochers & Hannon 1997; Harris & Reid 2002).

We found no evidence that the golden-cheeked warbler has experienced either a substantial loss of genetic variation within populations or severe bottlenecks. If current trends in development and agriculture continue in central Texas, however, isolated populations of golden-cheeked warblers may suffer the effects of inbreeding and genetic drift. Any management effort should consider the significance of habitat connectivity for the maintenance of viable population units. Survival of isolated populations could be enhanced by creating better connections between remaining habitat fragments (Morse 1989).

The black-capped vireo

Over much of its range, this migratory songbird (Fig. 9–3) nests in early-successional habitat composed of low, dense shrubs (Graber 1961; Grzybowski 1995). The principal threats to the species' long-term persistence are increased abundance of the brown-headed cowbird, disruption of disturbance regimens by fire suppression, and conversion of habitat to other land uses such as croplands, pastures, and housing developments (Grzybowski 1995; Hunter et al. 2002). Thus, threats faced by black-capped vireos are very similar to those faced by golden-cheeked warblers, despite their contrasting habitat requirements (Table 9–1).

When the species was listed as endangered in 1987, the total U.S. population was thought to be approximately 190 breeding pairs. With protection and management, population estimates have increased to at least 6,000 singing males (Wilkins et al. 2006). Little is known about the size and history of the breeding population in Mexico, but it might be comparable to that found in the United States (Wilkins et al. 2006). The birds migrate a considerable distance between their wintering grounds along the Pacific coast of southwestern Mexico and their breeding grounds in northern Mexico, central Texas, and southern Oklahoma (Fig. 9–2). Fazio and colleagues (2004) found higher differentiation than is usually seen in songbirds among the four major extant concentrations of the species at the time of their study (1992). Since that time, vireo populations have dramatically increased and expanded across their range (Wilkins et al. 2006). It is possible that this rapid expansion in numbers is associated with a decrease in population structure.

Because of their ability to move great distances, black-capped vireos would not be expected to have interpopulation genetic differentiation. Despite this, a variety of population structures has been suggested for the species that could result in intersite genetic differentiation, including that of a metapopulation

(Grzybowski 1991; Wilkins et al. 2006) and a source–sink model (Fazio et al. 2004). Also contributing to predictions about population structure is the strong site fidelity of adult males, with approximately 96% returning annually to previously established territories (Graber 1961; Grzybowski 1995). Unfortunately, little is known about dispersal between the time of fledging and establishment of breeding territories.

We employed microsatellites developed by Barr and coworkers (2007) to assay black-capped vireos from twelve sites (Fig. 9–2), representing most of the known major breeding concentrations in the United States (Wilkins et al. 2006). There were no detectable differences between sites in population genetic diversity, and smaller population fragments did not have reduced variation (Barr et al. 2008). There was some evidence that one site (Balcones Canyon [BC]) had experienced a recent bottleneck, based on the approach of Cornuet and Luikart (1996). Although this population is presently composed of a large number of birds compared to many other sampled sites (100 singing males; Barr et al. 2008), there is anecdotal evidence that the area historically supported only a few pairs (C. Sexton, 2006, personal communication).

As was the case with golden-cheeked warblers, there was more evidence of population differentiation potentially due to the fragmentation of vireo habitat than there was of reductions of within-population genetic diversity (Barr et al. 2008, Table 9–1). Except for pairs of populations that were spatially proximate and connected by continuous, protected habitat, there was significant genetic differentiation among samples from most sites. No other microsatellite-based population studies have been conducted on other vireos, so we cannot compare this level of differentiation to closely related species. Most previous studies of migratory songbirds did not detect significant genetic differentiation between pairs of sites separated by 50–100 km (Arguedas & Parker 2000; Gibbs et al. 2000; Clegg et al. 2003; Veit et al. 2005). Therefore, results based on microsatellites parallel those based on allozymes (Fazio et al. 2004) in suggesting that genetic structure in this species is high compared to that of other songbirds.

Differentiation among black-capped vireo sample sites is substantially higher than that observed among sample sites of the golden-cheeked warbler (Table 9–1). If only the sites within the golden-cheeked warbler's range are considered (see Fig. 9–2), the level of genetic differentiation among black-capped vireo populations is double that of the populations of the golden-cheeked warbler. There are reasons to expect black-capped vireos to exhibit higher, rather than lower, gene flow than the golden-cheeked warblers among our sampled populations. Notwithstanding observations of high site fidelity, the rapid turnover of early-successional habitats necessitates that individuals have the ability to locate and colonize new habitats. Also, whereas golden-cheeked warblers require large patches of habitat, black-capped vireos are sometimes found along edges and in small stands capable of supporting a single territory (Wilkins et al. 2006). Therefore, it was surprising that even with the ability to inhabit edge and smaller patches (which are much more available than large, continuous blocks of habitat), black-capped vireos exhibit far greater genetic differentiation than do golden-cheeked warblers. In this

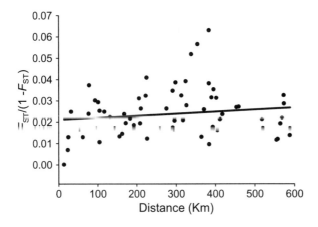

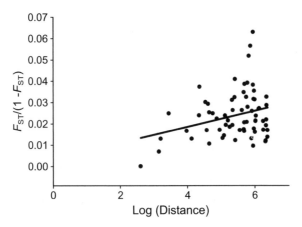

Figure 9–5: Relationship of genetic differentiation ($F_{ST}/(1 - F_{ST})$) with geographic distance and the logarithm of geographic distance among black-capped vireo populations (modified from Barr et al. 2008).

comparison, we find little support for the idea that a late-successional specialist should be more susceptible to the effects of fragmentation than a species that uses early-successional habitats.

Unlike those in the golden-cheeked warbler, Mantel tests in black-capped vireos did not detect a strong relationship between connectivity and genetic differentiation; however, available spatial data did not have the resolution to specifically identify suitable vireo habitat (Barr et al. 2008). We did observe that differentiation was more associated with the log of geographic distance than with untransformed distance – a pattern expected when gene flow is occurring in two dimensions (Rousset 1997). Considering the relatively linear distribution of black-capped vireos across Texas and Oklahoma (Fig. 9–5), the stronger relationship of genetic distance with the log of distance than with distance was surprising. Another explanation for the relationship between log geographic distance and genetic distance is that the log transformation emphasizes small-distance comparisons and de-emphasizes longer distances. This pattern would be expected where gene flow and genetic drift are differentially influential, with gene flow being of more consequence at small distances and genetic drift at large distances (Hutchison & Templeton 1999; Koizumi et al. 2006). When distances separating pairs of populations exceeded 100 km, there was little evidence that genetic differentiation increased with distance (Fig. 9–5).

It could be argued that the strong differentiation we detected is a remnant of an extended period of time during which population sizes were highly reduced. If re-established gene flow between populations has not recovered sufficiently or for a long enough period of time, rebounding populations may be retaining the signal of past founder events (Hutchison & Templeton 1999; Koizumi et al. 2006). Although we detected little evidence of severe genetic bottlenecks, there is still substantial differentiation among most sites separated by more than a few kilometers. It is possible that drift occurring after founder events and prolonged small size in some cases are sufficient to overcome the reduction of genetic differentiation that might otherwise occur when gene flow is high. This drift would disrupt patterns expected under an isolation-by-distance model but would not necessarily produce a genetic signal of strong bottlenecks.

Gene flow does not appear to be high enough to view the sampled locations as the kind of nonstructured populations seen in many migratory passerines. This finding was supported by both the detection of clusters of individuals via several assignment approaches and a method suggested by Waples and Gaggiotti (2006) for designating populations (Barr et al. 2008). Vireos do not appear to be dispersing across great distances in large numbers. Even where habitat is abundant, such as in the core of the distribution in central Texas, genetic differentiation exists among many sites. Furthermore, the rapid expansion of the species across much of its breeding range suggests that gene flow should be high. The existence of substantial genetic differentiation despite high potential vagility and population expansion argues for greater attention to dispersal behavior when considering the species' response to fragmentation.

Understanding the relationships among gene flow, conservation strategies, and fragmentation is critical for the management of this endangered species. Some sites that previously had none or small numbers of birds now have large contemporary populations that were potentially established by immigrants from far smaller populations. Therefore, small concentrations of breeding birds (acting as genetic stepping stones) might still be important for connectedness of larger sites.

Prior to European settlement, a spatial–temporal mosaic of rangelands, like those encompassing most of the distribution of the black-capped vireo, were maintained by fire, natural grazing, and other such disturbances (Fuhlendorf & Smeins 1997; Fuhlendorf & Engle 2001). It is likely that suitable habitat patches were available more or less continuously across the range of the black-capped vireo prior to the current disruption of disturbance regimens by fire suppression, the fragmentation of natural land cover by agriculture and urbanization, and the conservation practice of maintaining large areas of homogeneous vegetative cover (Fuhlendorf & Engle 2001, 2004). Areas large enough to accommodate small-scale disturbances are needed to allow such a dynamic shifting system to persist and thus maintain a stable amount of habitat in each successional state (DeAngelis & Waterhouse 1987). When suitable habitat falls below 30%, which is likely the case for black-capped vireos, colonization of habitat patches in shifting mosaics decreases dramatically (Wimberly 2006), even for highly mobile species. Given the extent of converted land use, suitable early-succession habitat is mostly limited to a small number of sites where there is active management. The structure of black-capped vireo populations may be in a state of flux, with the combination of low gene flow and historically small populations preempting migration–drift equilibrium. Future sampling is needed to determine whether the observed levels of population structure represent a signature of this transition or an equilibrium established through intensive management of this species.

Lessons from case studies

For both study species, we observed little evidence of reductions in genetic diversity, regardless of the size of the population fragment. There was some evidence that black-capped vireos might have experienced founder events associated with colonization of new habitat patches; however, the impact of these events on genetic diversity appeared to be low.

Despite high vagility, both endangered songbirds exhibit levels of differentiation higher than those observed in many other migratory birds with more continuous ranges. It is possible that factors associated with the large population declines experienced by these species have made them more susceptible to the influence of fragmentation on genetic differentiation than birds with larger, less-threatened populations.

Occupying similar distributions, these songbirds differ primarily in their dependence on late- and early-successional habitats. Yet, the black-capped vireo exhibits considerably higher levels of differentiation than does the golden-cheeked warbler. The genetic structure in the golden-cheeked warbler appears to be directly related to landscape alteration, with populations separated by agricultural or urban areas exhibiting greater differentiation than those separated by larger proportions of natural habitats. The causal factor is less clear in the black-capped vireo because edge and early-successional habitat are harder to quantify than is the abundance of forest; one might expect, however, that philopatry and founder effects associated with patch colonization might play a role in the observed levels of differentiation.

LITERATURE SURVEY OF FRAGMENTATION EFFECTS ON GENETIC VARIATION

To better understand how fragmentation affects genetic diversity, and the influences of habitat, dispersal ability, and conservation status of a species, we conducted a literature survey. We performed searches in Google Scholar and Web of Science using combinations of terms, including "genetic structure," "genetic diversity," "fragmentation," "anthropogenic," "human," and "recent." The search was limited to articles published in the last ten years to avoid potential temporal biases by combining earlier work, often based on few, less polymorphic loci, with later studies that generally examined larger numbers of individuals and loci. We did not include search terms related to conservation status, habitat, or dispersal ability because we wanted to minimize the possibility that our choice of search terms would bias outcomes of comparisons regarding these variables.

From the articles identified by this process, we retained those that estimated both genetic diversity within populations and genetic differentiation among populations, and we also drew conclusions about how anthropogenic fragmentation affected those components of genetic variation. When two or more species were examined in a study, the species were treated as separate observations. Species could appear in the database multiple times if they were the subject of several publications, but multiple studies on the same populations from the same group of investigators were treated as one observation. If the conclusions of these studies contrasted, we only considered the results of the study examining the largest number of loci and individuals.

Based on 116 articles, our survey yielded 120 observations on the response of genetic variation to fragmentation. For each of these observations, we noted if the authors had found decreased genetic diversity within populations or increased genetic differentiation among populations as a result of anthropogenic fragmentation. Authors made this determination of genetic effects of fragmentation in a variety of ways, including comparison of results to theoretical predictions, to the results from other studies, from direct comparisons of samples drawn from unfragmented and fragmented habitats or prefragmentation and postfragmentation periods. Although the strength of these inferences varied considerably, we did not take these differences into account. We felt that it was most reasonable to accept the conclusions of the authors, vetted by peer-review, at face value rather than to make our own assessment as to whether fragmentation had affected study populations.

The study organisms in each article were scored for having low or high dispersal ability relative to the spatial distances among samples. We also scored the study species with regard to its most commonly used habitat and identified three general classes for this comparison: forest, nonforest (including forest edges), and aquatic. In general, forests represent late-successional habitats that have been the focus of most of the ecological assessments of fragmentation. Nonforested habitats included marsh, brushlands, deserts, and grasslands. We also included species in this category if they tended to be associated with forest edge. We assigned species dependent primarily on aquatic habitats to their own common category. Study organisms were also classified by conservation status. When possible, determinations of vagility, habitat, and conservation status were made from information provided in the publication. In cases in which information was not provided, we based categorizations on general reference sources. Finally, we included whether the study organism was a plant, invertebrate, or vertebrate to determine if taxon influenced the effects of fragmentation on differentiation or within population diversity.

We used logistic regression to assess the influence of taxon, dispersal ability, habitat, and conservation status on genetic diversity and differentiation. Models were evaluated with Akaike's information criterion following Burnham and Anderson (2002). These analyses indicated no support for models containing

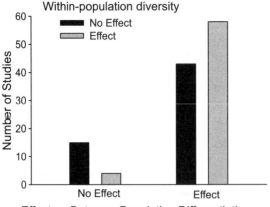

Figure 9–6: Numbers of studies where authors concluded that fragmentation did or did not have an effect on genetic diversity within populations or between populations of their study species. A higher proportion of studies in our review found effects of genetic differentiation (horizontal axis) than found reductions in within-population diversity (gray shaded bars).

interaction effects, so our discussion is limited to the main effects of conservation status, vagility, and habitat. Furthermore, because taxon did not interact with any of the main effects of interest, it is not discussed further.

Genetic responses to fragmentation

Genetic differentiation appeared to be a much more common response to fragmentation than did loss of genetic diversity. Of the studies examined, 84% concluded that fragmentation resulted in increased population differentiation, whereas only 51% detected the loss of diversity in population fragments (Fig. 9–6). In a meta-analysis, DiBattista (2008) found that within-population genetic diversity tended to be lost as a result of anthropogenic fragmentation. Because DiBattista (2008) did not evaluate genetic differentiation among populations but did examine effect sizes, it is hard to compare our conclusions.

We attribute little relevance to the high proportion of studies reporting a genetic fingerprint of fragmentation. There may be a reporting bias against studies of fragmented landscapes that do not detect loss of diversity or increased population differentiation. More interesting is the association between the proportion of studies reporting effects on within-population reductions in variation and those reporting increases in among-population differentiation. Almost all of the cases where loss of genetic diversity was detected, increased differentiation was also detected (94%; Fig. 9–6). In many cases where increased genetic differentiation among fragments was observed, however, there was no evidence of reduced genetic diversity within populations (43%; Fig. 9–6). It appears that differentiation among populations is more sensitive to fragmentation than is loss of genetic variation. This observation is in concordance with the conclusions of our investigations of golden-cheeked warblers and black-capped vireos that suggested that fragmentation had resulted in increased genetic differentiation among populations but had not resulted in losses of genetic diversity.

Why do different components of genetic diversity differ in their response to fragmentation? One possibility is the difference in the sensitivities between statistical approaches involved in assessing changes within and between population genetic diversity. The wide range of comparisons and statistical approaches used by authors to reach their conclusions makes an evaluation of this question difficult. Another explanation for the apparent contrast between the responses of genetic diversity and differentiation is how fragmentation affects levels of drift and inbreeding. Genetic differentiation among populations is most sensitive to drift, whereas some measures of loss of genetic diversity within populations, such as heterozygosity, are more directly affected by inbreeding (Leberg 2005). It has been well established that drift is more sensitive to recent decreases in population size, whereas inbreeding is most sensitive to prolonged bottlenecks (Nei et al. 1975; Crow & Denniston 1988; Leberg 2005).

Many of the fragmentation events included in the reviewed studies occurred in the relatively recent past, making it more likely to detect bottleneck effects on measures based on drift, such as interpopulation differentiation, than those based on changes in heterozygosity, which is often used to assess within-population diversity. Other measures of diversity, such as allelic richness, are more sensitive to drift than is heterozygosity (Nei et al. 1975; Leberg 1992; Spencer et al. 2000); however, such measures are often subject to sampling bias (Leberg 2002). It is worth noting that in our studies, where we controlled for differences in sampling effort, there was still little evidence for loss of genetic diversity within population fragments (Barr et al. 2008; Lindsay et al. 2008). Regardless of the cause of the contrast in sensitivity of within- and between-population genetic variation to fragmentation, we would encourage evaluations of both components when attempting to understand the genetic consequences of this important landscape process. Failure to detect losses of diversity within population fragments (Kramer et al. 2008) should not necessarily be interpreted as a lack of genetic consequences of fragmentation, as differentiation might be occurring prior to detectable loss of within-population diversity.

Conservation status

Initially, we wondered if fragmentation would have implications for the genetic structure of highly mobile birds, but our results raised the possibility that endangered species might be more susceptible to the genetic consequences of fragmentation than would more common species. In a comparison of closely related butterfly species, Louy and colleagues (2007) found the strongest fragmentation effects in the rarest species. Perhaps there is something about the biology of threatened taxa, such as small, effective population sizes or sensitivity to edge and area effects, that makes them more sensitive than common species to the genetic consequences of fragmentation. On the basis of our literature survey, we did not find strong evidence that authors reported more instances of reduced variation within fragments or increased differentiation between fragments for rare species than for common ones (Fig. 9–7).

It appears that conservation status does not have a large effect on the response of a species to genetic consequences of fragmentation. This finding does not argue

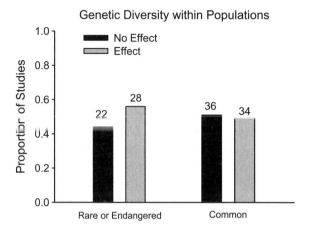

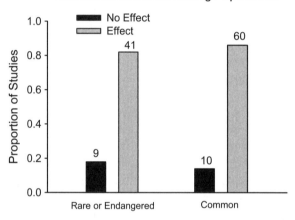

Figure 9–7: Proportions of studies of species of conservation concern ($N = 50$) and of common species ($N = 70$) reporting reduced genetic diversity within populations or increased genetic differentiation among populations in response to fragmentation. Sample sizes are located above bars.

against the observation of Louy and coworkers (2007) that fragmentation effects were more severe in a rare species than in a common species – we did not assess the magnitude of the genetic effects of fragmentation in our survey. Our results do suggest that common species should be useful models for understanding the genetic effects of fragmentation on rare or elusive species that might be more difficult to sample.

Vagility

We also evaluated how unique our observations of fragmentation effects in the black-capped vireo and golden-cheeked warbler were among species with high vagility. Species capable of long-distance dispersal should be less susceptible to the consequences of fragmentation than species capable of only short movements (Keitt 1997; Bohonak 1999; D'Eon et al. 2002). We found no evidence that species

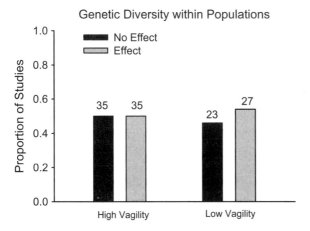

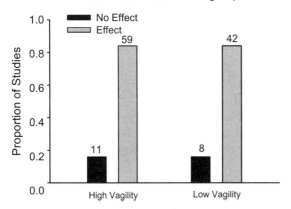

Figure 9–8: Proportions of studies of species classified as having high ($N = 70$) or low ($N = 50$) vagility reporting reduced genetic diversity within populations or increased genetic differentiation among populations in response to fragmentation. Sample sizes are located above bars.

categorized as having restricted movement capabilities exhibited more genetic effects of fragmentation than did species categorized as having high vagility (Fig. 9–8).

This result was quite unexpected, given the well-understood relationship between gene flow, which is related to vagility, and genetic differentiation (Wright 1943, 1968; Crow & Kimura 1970; Waples 1998). It is possible that variation in our ability to assess vagility was partially responsible for the failure to detect the expected decrease in fragmentation effects with the increased movement capability of study taxa. Of the three ecological and demographic criteria we evaluated, classifying organisms by their vagility presented the most difficulty. Authors' descriptions of the vagility of their study organisms relative to the spatial scale of their study were sometimes questionable, although we did not alter their classifications. In other cases, in which the authors provided little or no information about the vagility of their organisms, we may have made inaccurate assessments of movement capabilities.

It is likely that potential vagility is not necessarily correlated with realized dispersal and thus gene flow. Highly vagile species might experience reduced movement among fragments, even if the fragments are not separated by great distances. For example, many forest-dwelling birds are known to avoid crossing open habitats, even if they are easily capable of traversing the distances (Desrochers & Hannon 1997; Veit et al. 2005). For golden-cheeked warblers, we have speculated that this preference for natural land covers with some vertical structure is responsible for genetic differentiation among fragments (Lindsay et al 2008). It is also possible that sociality in some species might reduce movements between fragments (Cale 2003; Fraser et al. 2004; Rivers et al. 2005). Given the large number of apparently highly vagile species that exhibit at least some of the genetic consequences of fragmentation, we suggest the need for additional investigations into how factors such as interfragment land cover and social behavior might limit dispersal among fragments.

Habitat structure

The vast majority of studies of fragmentation have involved forested habitats isolated by areas where the forest cover was removed. Given the dramatic removal of vertical structure associated with forest fragmentation, it is not surprising that forests have been a focus of fragmentation research. With the preponderance of thought about fragmentation centered on forested habitats, however, it is reasonable to question whether the influence of fragmentation on genetic structure is similar in other species using other habitats. Approximately 62% of all the studies we examined dealt with forest-dwelling species, whereas species dependent on other terrestrial or aquatic habitats comprised 30 and 8% of the studies, respectively.

Models including our habitat classification provided information about whether authors reported reduced within-population genetic diversity in response to fragmentation (Fig. 9–9). Reports of fragmentation effects on within-population diversity were most common in aquatic habitats and least common in forested habitats. All of the studies involving aquatic species demonstrated increased genetic differentiation in response to fragmentation, compared to roughly 85% of the studies in terrestrial habitats (Fig. 9–9).

Species dependent on either aquatic habitats or nonforested terrestrial habitats appear to be more sensitive to the genetic effects of fragmentation than do forest-dwelling species. It is possible that this result is partially due to the large influence of dams on the aquatic systems studied for effects of fragmentation (Dallas et al. 2002; Alo & Turner 2005; Kawamura et al. 2007; Reid et al. 2008). The linear nature of streams and coastlines may also make species dependent on these habitats more susceptible to fragmentation than those in terrestrial environments, where dispersers may migrate to a fragment along several dimensions. It is unclear, however, why species dependent on less vertically structured terrestrial habitats would be more sensitive to fragmentation than forest-dwellers. This result is in concordance with our observation of greater evidence of genetic responses of black-capped vireos, an edge species, than that of golden-cheeked warblers, a forest-dependent species (Barr et al. 2008). Although fragmentation effects in forested systems have been heavily studied, managers and researchers

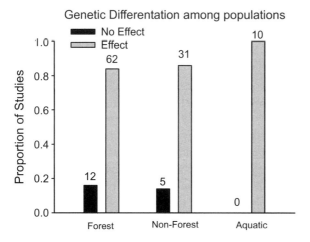

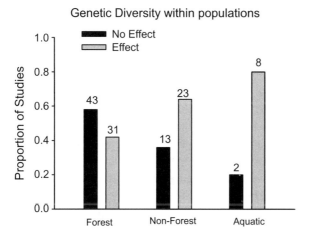

Figure 9–9: Proportions of studies of species classified as inhabiting forests ($N = 74$), nonforested terrestrial habitats ($N = 36$), or aquatic habitats ($N = 10$) reporting reduced genetic diversity within populations or increased genetic differentiation among populations in response to fragmentation. Sample sizes are located above bars.

should be cautious about applying those results to other systems as they might under-represent the consequences of fragmentation for species dependent on nonforested habitats.

CONCLUSIONS

The vast majority of studies on fragmented populations find evidence of increased levels of genetic differentiation and, to a lesser extent, loss of genetic diversity within population fragments. These observations are largely unaffected by the conservation status or movement capabilities of the study species. Aquatic species and species dependent on nonforested habitats appear to be more susceptible to the genetic consequences of fragmentation than are species dependent on forested habitats. These observations are generally concordant with the observations we have made in two highly vagile species of migratory, endangered

passerines. Clearly, our literature survey has limitations, including potential publication biases, a lack of control of the quality of the studies, and an inability to
quantify both the degree of fragmentation that has occurred and the magnitude
of genetic responses. Furthermore, few of the studies we examined included the
kind of experimental designs and controls that allow authors to make strong
inferences about the genetic effects of fragmentation.

The prevailing trend across a large number of studies, however, leaves little
doubt that fragmentation has altered the genetic structure of many species. It is
not clear if these changes will have long-term consequences for the survival of
many species as the nongenetic consequences of fragmentation, such as reduced
population sizes and increased exposure to habitat edges, may have more pressing
consequences for population viability (Templeton et al. 1990; Thomas 2000;
Watson 2003; Joshi et al. 2006; Kramer et al. 2008). This is not to say that the
effects of reduced genetic diversity within fragments, along with a reduction of
gene flow, do not have fitness consequences (Young et al. 1996; Gu et al. 2002;
Van Rossum et al. 2004; Johansson et al. 2007).

Given the apparent increases in differentiation in response to fragmentation,
assays of genetic variation might be most useful for assessing whether movement
is restricted among fragments. Such assessments are often prohibitively expensive
to accomplish with traditional ecological approaches. Thus, in addition to using
genetic markers to monitor changes in population size (Balloux & Lugon-Moulin
2002; Manel et al. 2003; Oyler & Leberg 2005), there should be greater emphasis
on the development of methodologies (Berry et al. 2004; Waser et al. 2006; Lee
et al. 2007; Zhan et al. 2007) for the monitoring of dispersal across landscapes.

Managers should be cognizant of the relatively high frequency of fragmentation effects on movement and subsequent genetic exchange, even for organisms
that appear to be capable of moving between isolated patches. Strong signatures of
differentiation found even among vagile organisms indicate that rates of dispersal may be much lower than might be expected based on natural history alone.
It is important to expand our perspective on managing species in fragmented
habitats to include species dependent on early-successional habitats as well as on
aquatic systems and to develop strategies that promote the connectivity among
such habitats.

REFERENCES

Alo D, Turner TF (2005) Effects of habitat fragmentation on effective population size in the
endangered Rio Grande silvery minnow. *Conservation Biology*, **19**, 1138–1148.

Amos W, Balmford A (2001) When does conservation genetics matter? *Heredity*, **87**, 257–
265.

Andren H (1994) Effects of habitat fragmentation on birds and mammals in landscapes
with different proportions of suitable habitat – a review. *Oikos*, **71**, 355–366.

Arguedas N, Parker PG (2000) Seasonal migration and genetic population structure in house
wrens. *The Condor*, **102**, 517–528.

Avise JC (1992) Molecular population-structure and the biogeographic history of a regional
fauna – A case-history with lessons for conservation biology. *Oikos*, **63**, 62–76.

Balloux F, Lugon-Moulin N (2002) The estimation of population differentiation with
microsatellite markers. *Molecular Ecology*, **11**, 155–165.

Barr KR, Dharmarajan G, Rhodes OE Jr, Lance R, Leberg PL (2007) Novel microsatellite
loci for the study of the black-capped vireo (*Vireo atricapilla*). *Molecular Ecology Notes*, **7**,
1067–1069.

Barr KR, Lindsay DL, Athrey G et al. (2008) Population structure in an endangered songbird: maintenance of genetic differentiation despite high vagility and significant population recovery. *Molecular Ecology*, **17**, 3628–3639.
Barrowclough GF (1980) Gene flow, effective population sizes, and genetic variance-components in birds. *Evolution*, **34**, 789–798.
Berry O, Tocher MD, Sarre SD (2004) Can assignment tests measure dispersal? *Molecular Ecology*, **13**, 551–561.
Bohonak AJ (1999) Dispersal, gene flow, and population structure. *Quarterly Review of Biology*, **74**, 21–45.
Burnham KP, Anderson DR (2002) *Model selection and multimodel inference: a practical information theoretic approach*, 2nd ed. Springer-Verlag, New York, NY.
Cale PG (2003) The influence of social behaviour, dispersal and landscape fragmentation on population structure in a sedentary bird. *Biological Conservation*, **109**, 237–248.
Clegg SM, Kelly JF, Kimura M, Smith TB (2003) Combining genetic markers and stable isotopes to reveal population connectivity and migration patterns in a neotropical migrant, Wilson's warbler (*Wilsonia pusilla*). *Molecular Ecology*, **12**, 819–830.
Cornuet JM, Luikart G (1996) Description and power analysis of two tests for detecting recent population bottlenecks from allele frequency data. *Genetics*, **144**, 2001–2014.
Crochet PA (2000) Genetic structure of avian populations – allozymes revisited. *Molecular Ecology*, **9**, 1463–1469.
Crow JF, Denniston C (1988) Inbreeding and variance effective population numbers. *Evolution*, **42**, 482–495.
Crow JF, Kimura M (1970) *An Introduction to Population Genetics Theory*. Harper & Row, New York.
D'Eon R, Glenn SM, Parfitt I, Fortin MJ (2002) Landscape connectivity as a function of scale and organism vagility in a real forested landscape. *Conservation Ecology*, 6. [online] URL: http://www.consecol.org/vol6/iss2/art10/
Dallas JF, Marshall F, Piertney SB, Bacon PJ, Racey PA (2002) Spatially restricted gene flow and reduced microsatellite polymorphism in the Eurasian otter *Lutra lutra* in Britain. *Conservation Genetics*, **3**, 15–29.
DeAngelis DL, Waterhouse JC (1987) Equilibrium and nonequilibrium concepts in ecological models. *Ecological Monographs*, **57**, 1–21.
Desrochers A, Hannon SJ (1997) Gap crossing decisions by forest songbirds during the post-fledging period. *Conservation Biology*, **11**, 1204–1210.
DiBattista JD (2008) Patterns of genetic variation in anthropogenically impacted populations. *Conservation Genetics*, **9**, 141–156.
Doak P (2000) Population consequences of restricted dispersal for an insect herbivore in a subdivided habitat. *Ecology*, **81**, 1828–1841.
Fagan WF, Cantrell RS, Cosner C (1999) How habitat edges change species interactions. *The American Naturalist*, **153**, 165–182.
Fahrig L (1997) Relative effects of habitat loss and fragmentation on population extinction. *Journal of Wildlife Management*, **61**, 603–610.
Fahrig L (2003) Effects of habitat fragmentation on biodiversity. *Annual Review of Ecology Evolution and Systematics*, **34**, 487–515.
Fazio VW, Miles DB, White MM (2004) Genetic differentiation in the endangered black-capped vireo. *Condor*, **106**, 377–385.
Frankham R (1996) Relationship of genetic variation to population size in wildlife. *Conservation Biology*, **10**, 1500–1508.
Frankham R (1997) Do island populations have less genetic variation than mainland populations? *Heredity*, **78**, 311–327.
Frankham R (2005) Genetics and extinction. *Biological Conservation*, **126**, 131–140.
Fraser DJ, Lippe C, Bernatchez L (2004) Consequences of unequal population size, asymmetric gene flow and sex-biased dispersal on population structure in brook charr (*Salvelinus fontinalis*). *Molecular Ecology*, **13**, 67–80.
Fuhlendorf SD, Engle DM (2001) Restoring heterogeneity on rangelands: ecosystem management based on evolutionary grazing patterns. *Bioscience*, **51**, 625–632.
Fuhlendorf SD, Engle DM (2004) Application of the fire-grazing interaction to restore a shifting mosaic on tallgrass prairie. *Journal of Applied Ecology*, **41**, 604–614.

Fuhlendorf SD, Smeins FE (1997) Long-term vegetation dynamics mediated by herbivores, weather and fire in a Juniperus-Quercus savanna. *Journal of Vegetation Science*, **8**, 819–828.

Gibbs HL, Dawson RJG, Hobson KA (2000) Limited differentiation in microsatellite DNA variation among northern populations of the yellow warbler: evidence for male-biased gene flow? *Molecular Ecology*, **9**, 2137–2147.

Graber JW (1961) Distribution, habitat requirements and life history of the black-capped vireo (*Vireo atricapilla*). *Ecological Monographs*, **31**, 25.

Grapputo A, Boman S, Lindstrom L, Lyytinen A, Mappes J (2005) The voyage of an invasive species across continents: genetic diversity of North American and European Colorado potato beetle populations. *Molecular Ecology*, **14**, 4207–4219.

Griffen BD, Drake JM (2008) Effects of habitat quality and size on extinction in experimental populations. *Proceedings of the Royal Society of London-Series B: Biological Sciences*, **275**, 2251–2256.

Grzybowski J (1991) *Black-capped Vireo* (Vireo atricapillus) *Recovery Plan*. U.S. Fish and Wildlife Service, Albuquerque, NM.

Grzybowski J (1995) Black-capped vireo (*Vireo atricapillus*). In: *Birds of North America* (eds. Poole A, Gill F). The American Ornithologist's Union, Washington, DC.

Gu WD, Heikkila R, Hanski I (2002) Estimating the consequences of habitat fragmentation on extinction risk in dynamic landscapes. *Landscape Ecology*, **17**, 699–710.

Hansson B, Bensch S, Hasselquist D, Nielsen B (2002) Restricted dispersal in a long-distance migrant bird with patchy distribution, the great reed warbler. *Oecologia*, **130**, 536–542.

Harris RJ, Reid JM (2002) Behavioral barriers to non-migratory movements of birds. *Annales Zoologici Fennici*, **39**, 275–290.

Hunter WC, Buehler DA, Canterbury RA, Confer JL, Hamel PB (2002) Conservation of disturbance-dependent birds in eastern North America. *Wildlife Society Bulletin*, **29**, 440–455.

Hutchison DW, Templeton AR (1999) Correlation of pairwise genetic and geographic distance measures: inferring the relative influences of gene flow and drift on the distribution of genetic variability. *Evolution*, **53**, 1898–1914.

Johansson M, Primmer CR, Merila J (2007) Does habitat fragmentation reduce fitness and adaptability? A case study of the common frog (*Rana temporaria*). *Molecular Ecology*, **16**, 2693–2700.

Joshi J, Stoll P, Rusterholz HP et al. (2006) Small-scale experimental habitat fragmentation reduces colonization rates in species-rich grasslands. *Oecologia*, **148**, 144–152.

Kawamura K, Kubota M, Furukawa M, Harada Y (2007) The genetic structure of endangered indigenous populations of the amago salmon, *Oncorhynchus masou ishikawae*, in Japan. *Conservation Genetics*, **8**, 1163–1176.

Keitt TH (1997) Stability and complexity on a lattice: coexistence of species in an individual-based food web model. *Ecological Modelling*, **102**, 243–258.

Keyghobadi N, Unger K, Weintraub JD, Fonseca DM (2006) Remnant populations of the regal fritillary (*Speyeria idalia*) in Pennsylvania: local genetic structure in a high gene flow species. *Conservation Genetics*, **7**, 309–313.

Koizumi I, Yamamoto S, Maekawa K (2006) Decomposed pairwise regression analysis of genetic and geographic distance reveals a metapopulation structure of stream-dwelling Dolly Varden charr. *Molecular Ecology*, **15**, 3175–3189.

Kramer AT, Ison JL, Ashley MV, Howe HF (2008) The paradox of forest fragmentation genetics. *Conservation Biology*, **22**, 878–885.

Kroll JC (1980) Habitat requirements of the golden-cheeked warbler: management implications. *Journal of Range Management*, **33**, 60–66.

Ladd C, Gass L (1999) Dendroica chrysoparia golden-cheeked warbler. *Birds of North America*, **420**, 1–24.

Leberg PL (1991) Influence of fragmentation and bottlenecks on genetic-divergence of wild turkey populations. *Conservation Biology*, **5**, 522–530.

Leberg PL (1992) Effects of population bottlenecks on genetic diversity as measured by allozyme electrophoresis. *Evolution*, **46**, 477–494.

Leberg PL (2002) Estimating allelic richness: effects of sample size and bottlenecks. *Molecular Ecology*, **11**, 2445–2451.

Leberg PL (2005) Genetic approaches for estimating the effective size of populations. *Journal of Wildlife Management*, **69**, 1385–1399.

Lee PLM, Luschi P, Hays GC (2007) Detecting female precise natal philopatry in green turtles using assignment methods. *Molecular Ecology*, **16**, 61–74.

Lindsay DL, Barr KR, Lance RF et al. (2008) Habitat fragmentation and genetic diversity of an endangered, migratory songbird, the golden-cheeked warbler (*Dendroica chrysoparia*). *Molecular Ecology*, **17**, 2122–2133.

Louy D, Habel JC, Schmitt T et al. (2007) Strongly diverging population genetic patterns of three skipper species: the role of habitat fragmentation and dispersal ability. *Conservation Genetics*, **8**, 671–681.

Manel S, Schwartz MK, Luikart G, Taberlet P (2003) Landscape genetics: combining landscape ecology and population genetics. *Trends in Ecology & Evolution*, **18**, 189–197.

Mills S, Hunt DH, Gomez A (2007) Global isolation by distance despite strong regional phylogeography in a small metazoan. *BMC Evolutionary Biology*, **7**, 1–10.

Morse DH (1989) *American Warblers: An Ecological and Behavioral Perspective*. Harvard University Press, Cambridge, MA.

Nei M, Maruyama T, Chakraborty R (1975) The bottleneck effect and genetic variability in populations. *Evolution*, **29**, 1–10.

Olivieri GL, Sousa V, Chikhi L, Radespiel U (2008) From genetic diversity and structure to conservation: genetic signature of recent population declines in three mouse lemur species (*Microcebus* spp.). *Biological Conservation*, **141**, 1257–1271.

Oyler S, Leberg PL (2005) Conservation genetics in wildlife biology. In: *Research and Management Techniques for Wildlife Investigations and Management* (ed. Braun C), pp. 632–657. Wildlife Society, Washington, DC.

Packer L, Taylor JS, Savignano DA et al. (1998) Population biology of an endangered butterfly, *Lycaeides melissa samuelis* (Lepidoptera; Lycaenidae): genetic variation, gene flow, and taxonomic status. *Canadian Journal of Zoology-Revue Canadienne De Zoologie*, **76**, 320–329.

Pannell JR, Charlesworth B (2000) Effects of metapopulation processes on measures of genetic diversity. *Philosophical Transactions of the Royal Society of London-Series B: Biological Sciences*, **355**, 1851–1864.

Pitra C, D'Aloia M-A, Lieckfeldt D, Combreau O (2004) Genetic variation across the current range of the Asian houbara bustard (*Chlamydotis undulate macqueenii*). *Conservation Genetics*, **5**, 205–215.

Rappole JH, King DI, Diez J (2003) Winter- vs. breeding-habitat limitation for an endangered avian migrant. *Ecological Applications*, **13**, 735–742.

Reid SM, Wilson CC, Mandrak NE, Carl LM (2008) Population structure and genetic diversity of black redhorse (*Moxostoma duquesnei*) in a highly fragmented watershed. *Conservation Genetics*, **9**, 531–546.

Rivers NM, Butlin RK, Altringham JD (2005) Genetic population structure of Natterer's bats explained by mating at swarming sites and philopatry. *Molecular Ecology*, **14**, 4299–4312.

Rousset F (1997) Genetic differentiation and estimation of gene flow from F-statistics under isolation by distance. *Genetics*, **145**, 1219–1228.

Sacks BN, Ernest HB, Boydston EE (2006) San Francisco's golden gate: a bridge between historically distinct coyote (*Canis latrans*) populations? *Western North American Naturalist*, **66**, 263–264.

Segelbacher G, Höglund J, Storch I (2003) From connectivity to isolation: genetic consequences of population fragmentation in capercaillie across Europe. *Molecular Ecology*, **12**, 1773–1780.

Segelbacher G, Manel S, Tomiuk J (2008) Temporal and spatial analyses disclose consequences of habitat fragmentation on the genetic diversity in capercaillie (*Tetrao urogallus*). *Molecular Ecology*, **17**, 2356–2367.

Shepard DB, Kuhns AR, Dreslik MJ, Phillips CA (2008) Roads as barriers to animal movement in fragmented landscapes. *Animal Conservation*, **11**, 288–296.

Spencer CC, Neigel JE, Leberg PL (2000) Experimental evaluation of the usefulness of microsatellite DNA for detecting demographic bottlenecks. *Molecular Ecology*, **9**, 1517–1528.

Streiff R, Audiot P, Foucart A, Lecoq M, Rasplus JY (2005) Genetic survey of two endangered grasshopper subspecies, *Prionotropis hystrix rhodanica* and *Prionotropis hystrix azami* (*Orthoptera, Pampagidae*): within- and between-population dynamics at the regional scale. *Conservation Genetics*, **7**, 331–344.

Suarez AV, Bolger DT, Case TJ (1998) Effects of fragmentation and invasion on native ant communities in coastal southern California. *Ecology*, **79**, 2041–2056.

Templeton A, Shaw K, Routman E, Davis SK (1990) The genetic consequences of habitat fragmentation. *Annals of the Missouri Botanical Garden*, **77**, 13–27.

Texas Parks and Wildlife Department (TPWD) (1995) Golden-cheeked warbler. PWD BK W7000–013.

Thomas CD (2000) Dispersal and extinction in fragmented landscapes. *Proceedings of the Royal Society of London-Series B: Biological Sciences*, **267**, 139–145.

Tilman D, May RM, Lehman CL, Nowak MA (1994) Habitat destruction and the extinction debt. *Nature*, **371**, 65–66.

Uimaniemi L, Orell M, Kvist L, Jokimäki J, Lumme J (2003) Genetic variation of the Siberian tit *Parus cinctus* populations at the regional level: a mitochondrial sequence analysis. *Ecography*, **26**, 98–106.

Uimaniemi L, Orell M, Mönkkönen M, Huhta E, Jokimäki J, Lumme J (2000) Genetic diversity in the Siberian jay *Perisoreus infaustus* in fragmented old-growth forests of Fennoscandia. *Ecography*, **23**, 669–677.

U.S. Fish and Wildlife Service (USFWS) (1992) *Golden-cheeked Warbler* (Dendroica chrysoparia) *Recovery Plan* (ed. Keddy-Hector DF), p. 97. USFWS, Albuquerque, NM.

USFWS (1996) Golden-cheeked warbler population and habitat viability assessment report.

USFWS (2004) Biological Opinion, Consultation Number 2-12-05-F-021.

USFWS (2005) Biological Opinion, Consultation Number 2-15-2004-F-0266.

Van Rossum F, De Sousa SC, Triest L (2004) Genetic consequences of habitat fragmentation in an agricultural landscape on the common *Primula veris*, and comparison with its rare congener, *P. vulgaris*. *Conservation Genetics*, **5**, 231–245.

Veit ML, Robertson RJ, Hamel PB, Friesen VL (2005) Population genetic structure and dispersal across a fragmented landscape in cerulean warblers (*Dendroica cerulea*). *Conservation Genetics*, **6**, 159–174.

Wahl R, Diamond D, Shaw D (1990) *The golden-cheeked warbler: a status review. Final report submitted to Office of Endangered Species*. USFWS, Albuquerque, NM.

Waples RS (1998) Separating the wheat from the chaff: patterns of genetic differentiation in high gene flow species. *Journal of Heredity*, **89**, 438–450.

Waples RS, Gaggiotti O (2006) What is a population? An empirical study of some genetic methods for identifying the number of gene pools and their degree of connectivity. *Molecular Ecology*, **15**, 1419–1439.

Ware D, Greis J (2002) Southern Forest Resource Assessment. USDA Forest Service.

Waser PM, Busch JD, McCormick CR, DeWoody JA (2006) Parentage analysis detects cryptic precapture dispersal in a philopatric rodent. *Molecular Ecology*, **15**, 1929–1937.

Watson DM (2003) Long-term consequences of habitat fragmentation – highland birds in Oaxaca, Mexico. *Biological Conservation*, **111**, 283–303.

Wilkins N, Powell RA, Conkey A, Snelgrove AG (2006) *Population status and threat analysis for the black-capped vireo*. US Fish and Wildlife Service, Region 2, Albuquerque, NM.

Wimberly MC (2006) Species dynamics in disturbed landscapes: when does a shifting habitat mosaic enhance connectivity? *Landscape Ecology*, **21**, 35–46.

Wright S (1931) Evolution in Mendelian populations. *Genetics*, **16**, 97–159.

Wright S (1943) Isolation by distance. *Genetics*, **28**, 114–138.

Wright S (1968) The theory of gene frequencies. In: *Evolution and the Genetics of Populations*, p. 511. The University of Chicago Press, Chicago.

Young A, Boyle T, Brown T (1996) The population genetic consequences of habitat fragmentation for plants. *Trends in Ecology & Evolution*, **11**, 413–418.

Zhan XJ, Zhang ZJ, Wu H et al. (2007) Molecular analysis of dispersal in giant pandas. *Molecular Ecology*, **16**, 3792–3800.

10 Integrating evolutionary considerations into recovery planning for Pacific salmon

Robin S. Waples, Michelle M. McClure, Thomas C. Wainwright, Paul McElhany, and Peter W. Lawson

Pacific salmon (see Table 10–1 for more information about terms in bold) enjoy iconic status in northwestern North America. As key components of both freshwater (Schindler et al. 2003) and marine (Beamish 2005) ecosystems, salmon play an important biological role in community structure and function. But salmon are no less crucial to the fabric of human societies. They have provided important food resources to Native Americans for at least 10,000 years (Butler & O'Connor 2004) and figure prominently in cultural, social, and economic traditions. Over the last ~200 years following European settlement, Pacific salmon have supported substantial commercial and sport fisheries, as well as continuing tribal harvest. Renowned for their long migrations and strong homing instinct, salmon have long been symbolic of Northwestern beauty and culture for human inhabitants of the region.

However, Pacific salmon also face a wide range of challenges to their persistence, due largely to major anthropogenic changes to their ecosystems (National Research Council 1996; Lackey et al. 2006). Urbanization, dams, road construction, harvesting, logging, mining, ranching, hatcheries, agriculture, invasive species, and other forms of habitat modification have all taken their toll on salmon populations. As a consequence, approximately 30% of historic salmon populations in the contiguous United States have been extirpated (Gustafson et al. 2007), and half of those that remain are formally protected under the U.S. Endangered Species Act (**ESA**) (Table 10–2).

The Pacific salmon listings are arguably the most complex ever undertaken by the ESA. Each species consists of hundreds, if not thousands, of local populations, many of which show evidence for local adaptations. Grouping this mass of diversity into units that can be effectively managed and administered is therefore a considerable challenge. Moreover, the intrinsic biological requirements of Pacific salmon – abundant and high-quality freshwater habitat; cool, clean water; productive estuarine and ocean ecosystems; unimpeded migratory routes between freshwater and marine environments – guarantee that decisions

Work described in this chapter is a result of the efforts of many people, both inside and outside of NOAA. A large number of documents related to salmon recovery planning in the Pacific Northwest can be found at http://www.nwfsc.noaa.gov/trt/index.cfm. A parallel process was developed in California, with its own flavors of technical challenges and practical solutions; information about those efforts can be found at http://swfsc.noaa.gov/textblock.aspx?Division=FED&id=2242.

Table 10–1. Glossary of terms and abbreviations associated with salmon recovery planning

---

DPS (Distinct Population Segment) – One of three categories of biodiversity (taxonomic species, named subspecies, and distinct population segments) that can be listed as a "species" under the ESA. The current ESA language recognizing the ability to list DPSs was introduced in a 1978 amendment and applies only to vertebrates.

ESA (U.S. Endangered Species Act) – Enacted in 1973, it is generally considered the most powerful environmental piece of legislation in the world.

ESU (Evolutionarily Significant Unit) – Generally, a level of biological organization below that of a full species but above that of local populations. The concept was originally intended to help identify zoo populations that merit special attention because of their evolutionary distinctiveness, but it has been applied more broadly to the problem of identifying conservation units in nature. A variety of approaches for defining ESUs has been proposed in the literature. The ESU framework for Pacific salmon was the first to be applied to natural populations and was developed to provide a biological framework for identifying salmon DPSs.

NOAA (National Oceanic and Atmospheric Administration) – The government agency charged with implementing the ESA for marine and anadromous species. ESA evaluations for terrestrial and freshwater species are conducted by the U.S. Fish and Wildlife Service.

Pacific salmon – Anadromous forms of the genus *Oncorhynchus*. North America has five species of Pacific salmon: pink, *O. gorbuscha*; chum, *O. keta*; coho, *O. kisutch*; sockeye, *O. nerka*; and Chinook, *O. tshawytscha*. We also use the term "Pacific salmon" to refer to steelhead, which is the anadromous form of rainbow trout, *O. mykiss*.

Recovery Domain – A geographic region (see Fig. 10–1) used to organize ESA recovery planning efforts for Pacific salmon. Each domain included a team of scientists (the TRT) and one or more policy groups with which the science teams interacted.

Strata – An intermediate level of biodiversity, between local populations and ESUs. In general, populations within a stratum share more physical and ecological features than do populations in different strata, so conserving multiple viable strata helps to ensure that the SS and D of VSP (see below) criteria are met at the ESU level. Fig. 10–2 shows the strata identified within the Oregon Coast ESU of coho salmon.

TRT (Technical Recovery Team) – Teams of scientists assembled to deal with the scientific aspects of salmon recovery planning under the ESA. In general, a separate TRT was formed in each geographic area (Recovery Domain) and was responsible for all listed ESUs in that area.

VSP (Viable Salmonid Populations) – An organizing framework for approaching risk assessments for salmon populations and ESUs. The VSP framework describes the characteristics of healthy salmon populations and assesses viability using four factors: abundance (A), growth rate/productivity (P), spatial structure (SS), and diversity (D).

Table 10–2. Number of ESUs of each species of Pacific salmon that are listed as threatened (T) or endangered (E) under the ESA, considered possible candidates for future listing (C), or not listed

Species	E	T	C	Not listed	Total	Populations[a]/ESU mean (range)
Chinook	2	7	1	7	17	19 (2–68)
Chum	–	2	–	2	4	25 (2–52)
Coho	1	3	1	2	7	22 (5–50)
Pink	–	–	–	2	2	19 (1–36)
Sockeye	1	1	–	5	7	1+ (1–3)
Steelhead	1	10	1	3	15	31 (4–81)
Totals	6	22	3	21	52	

[a] Demographically independent populations were identified according to the definition of McElhany et al. (2000) using data summarized by Gustafson et al. (2007).

(*Source:* http://www.nwr.noaa.gov/ESA-Salmon-Listings/Index.cfm; listing status as of March 2010.)

about their conservation and management will have pervasive effects on human societies. Furthermore, huge numbers ($>10^9$) of hatchery-produced salmon are released into the wild around the Pacific Rim each year (Mahnken et al. 1998); these hatchery fish help support fisheries but also interact ecologically and genetically with protected wild fish.

Developing the scientific basis for informed management and conservation of Pacific salmon also involves several layers of integration:

- integrating ecology and evolutionary biology
- integrating across diverse populations and across large geographic areas
- integrating natural sciences with social sciences, policy, and management

Most efforts aimed at salmonid conservation, and most time and money, have focused on ecological or demographic aspects of "the salmon problem" (How many fish are killed by X? How is productivity affected by Y?) (e.g., National Research Council 1996; Kareiva et al. 2000; Ruckelshaus et al. 2002). Recent research, however, has highlighted the importance of evolutionary considerations in salmon management and conservation (e.g., see the special issue of *Evolutionary Applications* [Volume 1, Issue 2, 2008] devoted to salmon evolution). In this chapter, we focus on how evolutionary processes have been considered in ESA recovery planning for Pacific salmon.

A SHORT PRIMER ON PACIFIC SALMON EVOLUTION

We begin this section with a short review of the evolutionary history of Pacific salmon and the dynamic ecosystems they inhabit, followed by a short summary of some current patterns of diversity. Understanding these patterns and processes is important to provide a context for establishing recovery goals and evaluating recovery efforts. For more detailed treatments, please refer to Groot and Margolis (1991) and Quinn (2005) (salmon ecology and life history) and to McPhail and Lindsey (1986), Hendry and Stearns (2004), Waples et al. (2008a), and references therein (salmon evolution).

By the early Miocene (15–20 million years ago), the genus *Oncorhynchus* (Pacific salmon and western trouts) had split from *Salmo* (Atlantic salmon and brown trout). Fossil evidence suggests that all the modern species of Pacific salmon had diverged by the late Miocene, which leaves approximately 6 million years for evolution within each species (Montgomery 2000). No extant genetic lineages within species are that old, however, which suggests that either 1) considerable diversity evolved but was lost during the Pleistocene; or 2) divergent lineages subsequently hybridized. It seems likely that both factors have been involved. At least four major glaciations occurred during the Pleistocene, during which time Pacific salmon were presumably restricted to ice-free refuges along the Bering Sea, in the Snake River, and to the south (McPhail & Lindsey 1986).

By approximately 5,000 years before present, after stabilization following isostatic rebound, landforms in many areas were structurally similar to those seen today (Beechie et al. 2001). These five millennia of relative stability represent roughly 1,000–3,000 salmon generations – more than enough time for evolutionary processes to have forged a dynamic equilibrium among migration, genetic drift, and natural selection. However, stability (relative to Pleistocene upheavals) does not imply constancy: salmon habitats have always been and remain dynamic across both space and time. Current natural disturbance regimes are characterized by two major types of perturbation: large but rare events (e.g., the eruption of Mt. St. Helens in 1980) and small but common events (e.g., floods or landslides). Overlaid on these disturbance regimes are atmospheric and oceanographic fluctuations that can affect large geographic areas and occur at scales of decades, centuries, or millennia. For example, Chatters et al. (1995) described post-glacial climate periods in the Columbia River basin of approximately 1,000–2,000 years duration that were alternatively "optimal," "good," or "poor" for salmon. Finney et al. (2002) identified marine productivity cycles of sockeye salmon in Alaska that spanned many centuries, and increasing evidence demonstrates that decadal-scale cycles of marine productivity (Overland et al. 2008) are also important drivers of salmon abundance.

Current patterns of diversity probably reflect a mixture of lineages from the Pleistocene and those that have evolved since (Waples et al. 2008a). Pacific salmon spawn and rear in a system of hierarchically structured stream/lake networks, and diversity within species also is typically apportioned in a hierarchical pattern. Some background level of straying by returning adults occurs naturally. Long-distance straying is rare, and most fish that stray tend to spawn in nearby populations (Quinn 2005); as a consequence, molecular genetic data generally show an isolation-by-distance pattern, with nearby populations being more similar genetically than are those that are farther apart (see Waples et al. 2001; Hendry & Stearns 2004; and references cited therein). Available evidence suggests multiple episodes of parallel evolution of Pacific salmon life-history traits. For example, many streams support both spring and fall Chinook salmon or both summer and winter steelhead, with the seasonal designation indicating the time adults enter fresh water to begin their spawning migration. These run-time differences are heritable and hence genetically based (Carlson & Seamons 2008). If the evolutionary change from fall to spring run time (or vice versa) happened only once, then we would expect that all spring Chinook populations would form one genetic lineage and all fall Chinook populations would form another lineage. But that is not the pattern found with molecular markers; instead, in most cases, spring Chinook from a given stream are more similar genetically to fall Chinook from the same stream than they are to spring Chinook from another stream (Waples et al. 2004). Other empirical data suggest a capacity for relatively rapid evolution in Pacific salmon. For example, Chinook salmon were introduced to a single New Zealand location a century ago and subsequently colonized several different rivers. Recent common garden experiments demonstrate that the populations have diverged at genetically based life-history traits typical of those found between different, closely related salmon populations in North America (Quinn et al. 2000).

Despite the notable success of some introduction efforts, the vast majority of attempts to transplant Pacific salmon within their native range have not led to sustainable populations (Withler 1982; Wood 1995). The explanation for this pattern seems to be the complex life history of the anadromous forms, which must undergo a series of precisely timed life-history transitions and lengthy migrations to complete their life cycle. A break in a single link in this chain can mean the difference between population viability and failure (Allendorf & Waples 1996).

This diverse body of empirical information can be summarized as follows:

- Pacific salmon habitats are dynamic on a variety of spatial and temporal scales.
- Each species of Pacific salmon has survived for several million years, demonstrating their capacity for resilience and evolutionary change in the face of dynamic habitat changes.
- In some cases, at least, evolutionary change can be relatively rapid, and some traits have evolved multiple times.
- On human time scales, however, Pacific salmon populations are not generally ecologically exchangeable. Their life histories can be finely tuned by adaptations to local conditions.

From these observations, we can draw some key conclusions that are relevant to recovery planning. First, if a population with distinctive life-history traits or other adaptations is lost, one might expect that a comparable population could evolve given enough time and appropriate conditions. Local populations are not generally replaceable on ecological time frames, however. Second, at any given point in time, one might expect to find a pattern of locally adapted salmon populations distributed across a landscape, and these adaptations might be relatively stable over ecological time frames. However, over evolutionary time frames (hundreds to thousands of years), gross features of salmon habitats are dynamic. Therefore, the pattern of local adaptations probably will be a shifting mosaic, with adaptation of specific populations changing as their habitats do, but the large-scale, overall pattern of a distribution of locally adapted populations being more constant over time (Waples et al. 2008a). If this picture is accurate, it suggests that a conservation program that simply attempts to preserve existing types (populations and habitats) is not likely to be sufficient in the long run; rather, a more effective strategy would be to focus on creating conditions that will promote processes that generate diversity and complexity of biological populations and the habitats on which they depend (Moritz 2002; Hilborn et al. 2003).

FEDERAL PROTECTION

By the late twentieth century, cumulative effects of human development and natural resource use had reduced many Pacific salmon populations to the point of serious conservation concern. During the 1990s, a series of comprehensive evaluations were undertaken of the status of Pacific salmon with respect to the ESA. To be protected under the ESA, an entity must meet the statutory definition of "species," which can be a taxonomic species (e.g., coho salmon), a formally recognized subspecies (apart from *O. mykiss*, there are none for Pacific salmon), or a "distinct population segment" (**DPS**). The ability to list DPSs of species that are more abundant elsewhere has been used to provide ESA protection to U.S.

populations of grizzly bear, gray wolf, and bald eagle, among others. These determinations, however, were largely made ad hoc, without a consistent framework for considering the term "distinct population segment," which has no clear biological definition and is not further explained in the ESA. Given the complex ecology and life history of Pacific salmon, a wide range of approaches might be used to group populations into distinct population segments.

To provide an operational framework for identifying salmon DPSs, the National Oceanic and Atmospheric Administration (NOAA) developed a policy (Waples 1991) based on the concept of evolutionarily significant units (**ESUs**). To be considered an ESU (and hence be potentially listable as a DPS), a salmon population (or group of populations) must meet two criteria: 1) it must be substantially isolated reproductively from other populations; and 2) it must contribute substantially to the evolutionary legacy of the species. A holistic approach, which integrates all relevant biological information, was used to identify ESUs (see also Chapter 1 by Honeycutt and colleagues). For the first criterion, general insights were provided by geographic features and tagging data that indicate fish movements, and a great deal of specific information was provided by molecular genetic data, which integrate patterns of connectivity and gene flow over evolutionary time scales. Conceptually, the second criterion focuses on adaptive differences, which are not routinely detected by molecular markers. Instead, these evaluations rely primarily on proxies for adaptation – in particular, differences in life-history traits (which might reflect local adaptations but which are also influenced by environmental factors) and ecological differences in habitats (which should promote different adaptations in populations that are sufficiently isolated). The number of listed/not listed ESUs in each species is given in Table 10–2; see Waples (1995, 2006) and Wainwright and Kope (1999) for more information about the scientific basis for the salmon listing determinations.

RECOVERY PLANNING

Several major factors – the complex patterns of biodiversity within each Pacific salmon species, the wide range of anthropogenic factors that affect them, and the long list of parties that could be substantially affected by management decisions – collectively have ensured that formal ESA recovery planning for these species is a challenging exercise. To facilitate this process, eight geographic **Recovery Domains** were established, each responsible for all listed ESUs within its boundaries (Fig. 10–1). For each domain, NOAA appointed merit-based Technical Recovery Teams (**TRT**s) to deal with scientific aspects of recovery planning and to help identify multiparty stakeholder groups with which the teams could interact on policy issues. Regular discussions between TRTs and policy groups helped the science teams focus their efforts and identified the most useful formats for presenting results and conclusions.

Although the decision to structure ESA recovery planning by geographic domains facilitated close interactions between science teams and local policy, stakeholder, and implementation groups, this also meant that up to eight separate science teams were wrestling with comparable problems and (potentially) coming

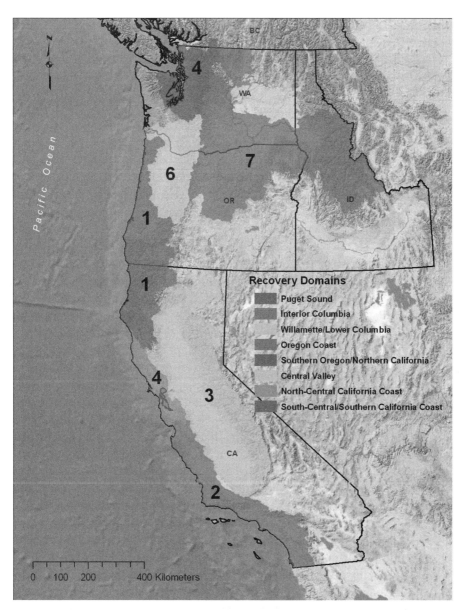

Figure 10–1: ESA recovery domains for Pacific salmon; the number of listed salmon ESUs in each domain is shown in large black numerals. *See Color Plate XI.*

to different conclusions. This situation presented a new set of challenges – how to promote creative and independent thinking without risking a chaotic mess of conflicting conclusions and recommendations. The general approach to this balancing act was to avoid enforcing conformity to any particular methodology, while at the same time expecting the different science teams to provide biologically based explanations for major differences in methods or criteria applied in different regions.

The ESA mandates that recovery plans be developed and implemented for listed species. These plans must, at a minimum, contain the following elements:

1) site-specific actions necessary for recovery; 2) objective, measurable criteria to determine when recovery has been achieved; and 3) estimates of time and costs to meet the goals. ESA recovery is defined as the point at which the species can be delisted; however, in the estimation of some, this might not be sufficient to provide all the desired benefits to society from abundant salmon populations (e.g., excess fish for tribal, commercial, or sport harvest; contributions of salmon carcasses to freshwater food webs). Therefore, each TRT discussed with its policy group the types of goals desired for that particular domain, while ensuring that (at a minimum) the recovery goals were sufficiently robust to produce sustainable salmon populations that no longer needed ESA protection.

How might such evaluations be made? To answer this question, it was necessary to identify the biological characteristics of healthy salmon populations and ESUs. The solution was development of a conceptual framework, known as viable salmonid populations (VSPs) (McElhany et al. 2000), that has served as the foundation for ESA recovery planning of Pacific salmon. Application of VSP principles involves three major steps: 1) identifying populations within ESUs; 2) assessing viability of each population; and 3) integrating across (typically many) populations to arrive at an overall assessment of viability of the ESU. In the following sections we consider these steps individually.

Population identification

Even a cursory review of the scientific literature uncovers a wide range of definitions of "population" (Andrewartha & Birch 1984; Waples & Gaggiotti 2006). Few of these definitions are objective, quantitative, or repeatable in the sense that one could give the same definition and the same data to different groups of scientists and expect to have comparable results. Therefore, there was a need for an operational definition that would provide a consistent, quantitative framework for addressing the problem of population identification. Because a major objective of this exercise was to identify units for which it is meaningful to conduct separate viability analyses, McElhany and colleagues (2000) concluded that the key criterion was demographic independence. That is, the objective was to identify units within which demographic processes are driven more by local births and deaths than by immigration. After discussing with NOAA policy makers the appropriate time scale for considering viability under the ESA, the authors arrived at the following definition (McElhany et al. 2000, p. 2–3):

> A viable salmonid population . . . is an independent population . . . that has a negligible risk of extinction due to threats from demographic variation (random or directional), local environmental variation, and genetic diversity changes (random or directional) over a 100-year time frame.
>
> . . . An independent population is any collection of one or more local breeding units whose population dynamics or extinction risk over a 100-year time period is not substantially altered by exchanges of individuals with other populations.

How much exchange of individuals can populations experience and still be demographically independent? This question would seem to be rather fundamental in ecology, but it has received surprisingly little attention. Simulations conducted by Hastings (1993) suggested that populations receiving less than approximately 10% immigration tend to have independent demographic trajectories, so 10% was used as an approximate guideline for evaluating demographic independence. Few studies, however, have directly measured natural migration rates in salmon. Therefore, it was necessary to rely on indirect evidence and geographically based proxies for immigration rates. Studies of straying by tagged hatchery fish were used to develop a general relationship between the proportion of fish that stray and the distance from the release site (Myers et al. 2006). This relationship was then used to identify a threshold distance below which demographic independence was unlikely. Geographic distances separating spawning aggregations thus became an important element in most population-definition efforts (McClure et al. 2003; Myers et al. 2006; Ruckelshaus et al. 2006).

The general absence of direct demographic evidence for independence meant that indirect genetic information was also important for population identification. However, the TRTs faced several challenges in using genetic data to draw inferences about demographically independent populations. First, because the putative threshold for demographic independence ($\sim$10% immigration rate) is high in evolutionary terms, it is difficult with genetic methods alone to distinguish migration rates that are or are not compatible with demographic independence (Waples & Gaggiotti 2006; Waples et al. 2008b). Second, although stock transfers are closely regulated today, in the past this was not the case. As a consequence, the genetic composition of many natural populations has been altered by interbreeding with non-native hatchery fish (McClure et al. 2008b; Fraser 2008), which complicates identification of historic populations. Finally, in some places, recent population bottlenecks have been severe enough to distort patterns of genetic differentiation and complicate inferences about historic population structure. For these reasons, the teams avoided the temptation to use a single, one-size-fits-all genetic metric for population identification; instead, they took a more qualitative and integrative approach that used a combination of geographic distance between spawning aggregations, physical structure of drainage basins, genetic information, demographic patterns, ecological information, and phenotypic patterns to define populations. Whenever possible, the teams identified historic but extinct populations as well as extant ones to provide a perspective on how current patterns of diversity compare to the historical template (see the next section).

One factor that was initially thought to be promising for population identification proved to be of limited practical utility. All else being equal, demographically linked populations should show more strongly correlated patterns of abundance over time than demographically independent populations (Ruckelshaus et al. 2006). However, abundance can also be strongly affected by environmental factors, and populations that share a similar environmental history can have correlated time series of abundance for reasons that have nothing to do with exchange of individuals (Moran 1953; Post & Forchhammer 2004). Furthermore, for Pacific salmon, the presence of unmarked (and unidentifiable) hatchery-origin fish on spawning grounds can also obscure natural patterns of increase or decrease. For these reasons, time–series correlations were not particularly useful for population identification, which relied on the other factors discussed earlier in text.

The following is a summary of key features of the process of identifying populations:

- The principal criterion (demographic independence) was chosen to produce units that can be the focus of separate viability analyses.
- Genetic data were instrumental in identifying populations, but a wide range of types of information was also used.
- Systematic application of these criteria to Pacific salmon ESUs produced the results shown in Table 10–2: Most ESUs include many different populations (mean of approximately twenty to thirty for all species except sockeye). The Oregon Coast ESU for coho salmon (Fig. 10–2) is typical in its geographic extent, ranging over more than 300 km of coast from Cape Blanco to the Columbia River. Twenty-one independent populations have been identified in this ESU.

Population viability

A variety of published methods are available for assessing population viability (Beissinger & McCullough 2002; Morris & Doak 2002), but most focus primarily on two demographic attributes: abundance (A) and growth rate/productivity (P). A and P are integral to the VSP framework as well, but the latter also emphasizes two additional factors (spatial structure [SS] and diversity [D]) intended to capture the importance of evolutionary processes that are often ignored in viability assessments.

Abundance and productivity

All TRTs relied to some extent on viability modeling to assess extinction risk, but details varied among teams. Fig. 10–3 outlines the conceptual framework for jointly evaluating abundance and productivity adopted by one team (McElhany et al. 2007). Implicit in this figure is the risk trade-off between A and P: A highly productive population can potentially be viable even at relatively low abundance, and a population at high abundance might not be at serious short-term risk even if it is not very productive. The dotted curves in Fig. 10–3 represent isopleths of equal risk: For each curve, the area above and to the right indicates a region of higher viability and the area lower and to the left a region of lower viability.

Assessing A/P viability for a given population involves three steps:

1. Risk isopleths are defined. Points along any given isopleth are combinations of A and P that represent equal risk of extinction over a 100-year time frame.
2. Current status of a population of interest is visualized by plotting its estimated abundance and productivity on the graph relative to the risk isopleths.
3. Current status is compared with target viability level. A discrepancy between current and target status identifies populations for which improvement is needed to achieve recovery.

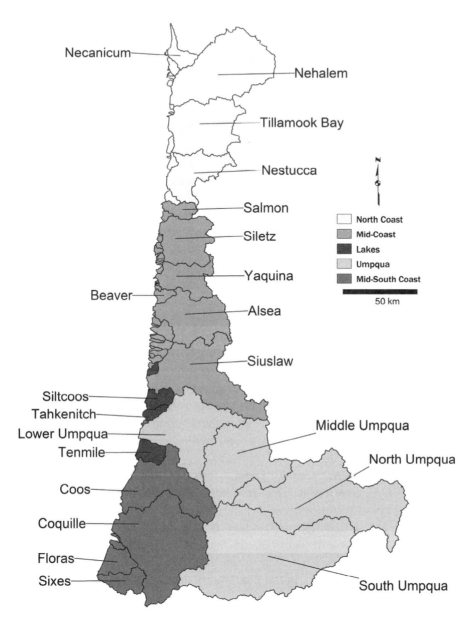

Figure 10–2: Hierarchical structuring of biodiversity within the Oregon Coast ESU for coho salmon. The twenty-one named independent populations (and numerous smaller coastal units that do not meet the independence criterion; see Lawson et al. 2007) are grouped into five biogeographic strata (indicated by shading). *Source:* Wainwright et al. (2008).

Relating abundance and productivity to extinction risk (Step 1) was accomplished using a simulation model with a stochastic recruitment function that relates spawners at time t to spawners at time $t + 1$. To estimate extinction risk for any particular set of A and P values, the model is run thousands of times to determine the fraction of runs in which abundance drops below a critical risk threshold. Because extinction probability is a function of not only the means but also the variances of A and P over time, separate plots were developed for each species, based on variance estimates averaged across multiple populations.

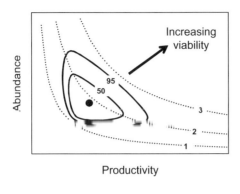

Figure 10–3: Schematic diagram depicting method for joint consideration of the VSP criteria abundance and productivity. Dotted curves are isopleths of equal extinction risk and indicate thresholds for viability levels 1, 2, and 3. For a given abundance, populations with higher productivity are more viable; for a given productivity, populations with higher abundance are more viable. Black dot and solid lines indicate point estimate and 50% and 95% confidence contours of current viability for a hypothetical population. Adapted from McElhany et al. (2007).

In Step 2, population-specific data on A and P are used to visualize the population's current status. Because the key parameters are only estimated from empirical data, a Monte Carlo approach is used to generate plausible A/P scenarios based on estimates of measurement error. The result for each population is a series of probability contours that graphically depicts the degree of uncertainty associated with point estimates of current A/P status (Fig. 10–3). The hypothetical population shown in Fig. 10–3 has relatively low abundance and relatively low productivity, and the point estimate for the A/P viability score is less than 2 (0–4 scale). The confidence contours indicate that with 95% certainty, it can be concluded that the true score is between 1 and 3.

Whereas Steps 1 and 2 are, in theory, objective and scientific, setting target recovery levels (part of Step 3) requires both technical assessments (to evaluate probability of recovery under different scenarios) and policy input (regarding risk tolerance and preferences among comparable alternatives). These topics are discussed later in the text (see "ESU viability").

Although the A/P analyses relied primarily on population dynamics, they incorporated evolutionary considerations in setting critical thresholds for reproductive failure and quasi-extinction. Based on theoretical considerations that included risks of inbreeding, loss of genetic variability, and depensation (e.g., inability to find a mate), as well as the ratio of effective to census population size, critical thresholds were identified that ranged from 50 to 300 spawners per year, depending on the size of the basin. In the simulations, if population abundance in any year dropped below this threshold, no progeny were produced. If the mean abundance was less than the critical level over any period of consecutive years that equaled a salmon generation (generally four to five years), the population was considered extinct (McElhany et al. 2007).

Most other TRTs also jointly assessed risks associated with abundance and productivity by using methods that were similar in many respects to those just described. The Oregon and Northern California Coasts TRT, however, used a somewhat different approach (Wainwright et al. 2008). Extinction risk was estimated directly using four different models, one of which (Nickelson & Lawson 1998) explicitly included small-population genetic risks. A second criterion required that productivity at low abundance be sufficient to maintain the population during periods of adverse climate conditions, and a third required that populations be above a "critical abundance" level, defined as an abundance below which small-population demographic risks were likely to become significant.

Spatial structure and diversity

The VSP parameters SS and D are closely related and, in practice, were often considered jointly. For example, in evaluations of coho salmon from the Oregon coast, three SS/D criteria (two of which focused on genetic issues) were applied at the population level (Wainwright et al. 2008):

1. Long-term harmonic mean spawner abundance must be large enough to prevent substantial loss of heterozygosity over a 100-year time frame.
2. Natural spawning by hatchery fish should be limited to minimize adverse effects on natural populations.
3. Juveniles and adults should be distributed across the historic range of the population to provide both insurance against changes in local environmental conditions and opportunities for continued local adaptation to the full variety of habitats within the population's range.

For evaluations of SS and D, a key reference point was the historical template – that is, the historical patterns of distribution and abundance, life-history traits, and the amount and quality of salmon habitats and the connectivities among them. The ESA does not mandate restoration of species to historic levels of abundance, and it might be possible for a salmon population or ESU to be viable under conditions that differ substantially from those experienced historically. However, situations in which current conditions greatly depart from the historical template can signal substantial risks to natural populations (Waples et al. 2007), so these situations were scrutinized carefully. Most teams included qualitative or quantitative metrics comparing characteristics of habitats historically and currently occupied, as well as the historical spectrum of life-history diversity. For the most part, however, the focus was not on preserving certain types but rather on conserving the processes that promote adaptation and provide resilience of a biological system to deal with future environmental challenges (Moritz 2002).

A quantitative evaluation of the types of freshwater salmon habitats that have been lost since European settlement makes explicit the close relationship between the SS and D criteria. Nearly 45% of the habitat historically occupied by anadromous salmonids in the contiguous United States is now blocked by dams or other large barriers. These losses have not been random; rather, in most geographic areas, the blocked habitats are on average wetter, higher in elevation, warmer in summer, and colder in winter than are habitats currently available to salmon (McClure et al. 2008a). Significantly, all of these habitat characteristics can strongly affect morphological and life-history traits of salmon. Wholesale losses of specific types of habitats therefore represent a major change to selective regimens experienced by salmon, with the result that certain phenotypes or life-history traits that were historically abundant are now at a selective disadvantage. Similar effects can be seen on smaller geographic scales as well. Small migration barriers (e.g., culverts and irrigation diversions) are pervasive across the western United States and primarily affect upstream areas with distinctive ecological features that support specific life-history types. Although habitat blockages on all scales are common throughout the evolutionary history of salmon and their dynamic habitats (Waples et al. 2008a), recent anthropogenic changes are creating a syndrome that is perhaps unique in the history of these species: Specific types of habitat alterations, with highly selective consequences for salmon populations, are occurring simultaneously in virtually every geographic area across the species' range.

Salmon spend a significant part of their life cycle in the ocean, which is an important determinant of growth (and, consequently, fecundity) and survival to adulthood (Pearcy 1992). Depending on the species, both the mean and variance in mortality rate can be as high or higher in the ocean as in freshwater (Bradford 1995), which indicates that the marine phase of the salmon life cycle may be as evolutionarily important as the freshwater phase. Because much less is known about the ocean ecology of salmon, risks are harder to quantify; as a consequence, the TRTs primarily evaluated the marine phase for its demographic effects on abundance and productivity. As we achieve a greater understanding of the evolutionary consequences of anthropogenic changes to natural ecosystems (e.g., size-selective harvest [Hard et al. 2008], changes to ecosystem structure and function [Myers et al. 2007; Chapter 3 by Whitham et al., 2010], climate warming and ocean acidification [Wootton et al. 2008]), assessment of these factors as they apply to the ocean ecology of salmon should be incorporated more fully into comprehensive recovery planning.

Another recent anthropogenic change with potentially widespread effects is artificial propagation, which can reduce both fitness (Araki et al. 2008) and effective population size (Ryman & Laikre 1991) of local wild populations (see reviews by Waples & Drake 2004 and Fraser 2008; also Chapter 11 by Ivy et al., 2010). As part of the within-population diversity criterion, the Interior Columbia TRT conducted an extensive evaluation of effects of artificial propagation on viability (McClure et al. 2008b). Populations that are similar to non-native hatchery stocks based on molecular genetic markers were considered to be at high risk (Cooney et al. 2007). More generally, the team developed a risk framework (Fig. 10–4) that considered 1) the fraction of natural spawners that are of hatchery origin; 2) the stock history of contributing hatchery(ies); and 3) the duration of hatchery effects. All else being equal, hatchery risks to diversity were considered greater when 1) a high fraction of natural spawners were hatchery fish; 2) non-native broodstock was used; 3) logistic constraints made it difficult to collect representative samples of broodstock; 4) hatchery practices that promote rapid domestication were used (Flagg et al. 2004; Mobrand et al. 2005); and/or 5) effects continued for a number of salmon generations. This metric can also be applied to wild fish in situations where anthropogenic changes (e.g., impediments to migration; temperature and flow alterations) have caused unnatural levels of straying.

Integrating across risk factors

Integrating across the four VSP risk factors for each population generally involved both quantitative and qualitative analyses, and methods varied among TRTs. An example of this process is that used by McElhany and colleagues (2007). For each VSP factor in each population, the Willamette–Lower Columbia River TRT assigned a risk score ranging from 0 (extinct or very low viability) to 4 (very

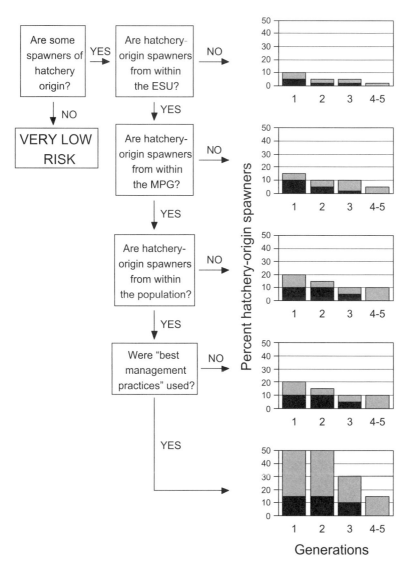

Figure 10–4: Risk factors associated with naturally spawning hatchery fish as they affect the VSP "diversity" criterion. Black indicates relatively low-risk combinations of program duration and origin and proportion of hatchery spawners; gray indicates moderate risk combinations; other combinations are high risk. MPG = major population group, a term used by this TRT to identify strata within ESUs. Modified from McClure et al. (2008b).

high viability) (Table 10–3). On this scale, a viability score of 3 would generally indicate sustainability. To reflect uncertainty in these assessments, McElhany and coworkers (2007) used expert opinion to define probability distributions for individual attribute scores. For example, the MacKenzie River population (Table 10–3) got a score of 2 (moderate viability) for SS, but this assessment only carried a "medium" level of certainty, meaning that the true status might be higher or lower. To arrive at an overall viability score (and level of certainty) for each population, McElhany and coworkers (2007) used a weighted average of means

Table 10–3. Characterization of risk and uncertainty for the four VSP factors in seven populations of Chinook salmon from the upper Willamette River

Population	VSP factor			Overall
	A/P	SS	D	
Clackamas	2 H	3 H	2 M	2 H
Molalla	0 H	2 H	1 H	1 H
North Santiam	0 H	1 H	1 M	1 M
South Santiam	0 H	1 H	1 M	1 H
Calapooia	0 H	2 H	1 M	1 H
MacKenzie	2 H	2 M	1 M	2 H
Middle Fork Willamette	0 H	0 M	1 M	0 H

Numbers are viability scores for each attribute and the population as a whole (scale is from 0 [very low viability] to 4 [very high viability]). Letters indicate the level of certainty associated with each assessment (L = low; M = medium; H = high). Based on analyses in McElhany et al. (2007).

and ranges of the individual scores. For the Chinook salmon ESU shown in Table 10–3, the overall picture is one of relatively strong certainty that the ESU is not healthy: Certainty scores for all elements were either "medium" or "high," whereas the majority of viability scores were no higher than 1 ("low") (only one element – SS in the Clackamas population – scored as high as 3 ["high"] for viability).

ESU viability

After the four VSP factors have been integrated to arrive at an overall assessment of viability of each population, a challenging question remains: How many populations have to be at what levels of viability before the ESU as a whole can be delisted? No standard solutions exist for a problem this complex, so the science teams had to develop innovative approaches. The following definition of ESU viability (Wainwright et al. 2008) provides a conceptual framework for thinking about this problem:

> A self-sustaining ESU is able to survive prolonged periods of adverse climate conditions without artificial support (including supplementation by hatchery fish) and able to maintain its genetic legacy and long-term adaptive potential. A self-sustaining ESU will be composed of diverse interconnected populations.

This definition reflects common themes expressed by all the recovery teams: Beyond simple demographic viability, a viable ESU must retain sufficient genetic legacy that it maintains adaptive potential to respond to changing physical and biotic environments. Viability thus also implies long-term stability of processes that maintain habitat diversity and other conditions necessary for full expression of life-history diversity. Salmon habitats exist in a dynamic mosaic of constantly changing local conditions (Reeves et al. 1995; Waples et al. 2008a), and it is genetic and phenotypic diversity that allows ESUs to successfully respond to these changes.

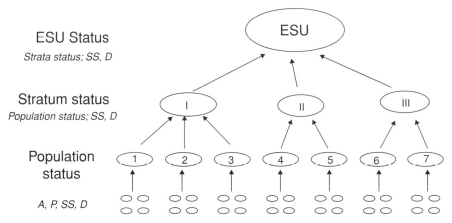

ESU Status
Strata status; SS, D

Stratum status
Population status; SS, D

Population
status

A, P, SS, D

Figure 10–5: Hierarchical framework for evaluating conservation status of Pacific salmon. Viability of each population is evaluated according to the four VSP criteria (abundance [A], productivity [P], spatial structure [SS], and diversity [D]). In most ESUs, populations are grouped into multiple strata, based on geographic, geological, ecological, life history, and/or genetic characteristics. This hypothetical example includes seven populations and three strata. The status of each stratum is assessed based on the status of the component populations, taking into consideration contributions of different populations to SS and D. Finally, status of an ESU (the smallest unit that can be listed/delisted under the ESA) is evaluated by considering the status of the component strata, again taking into consideration their contribution to SS and D at the ESU level.

Under the ESA, an endangered species is one that is "at risk of extinction *throughout all or a significant part of its range*," and a threatened species is "likely to become an endangered species in the foreseeable future *throughout all of a significant part of its range*." Delisting therefore requires a conclusion that the species is no longer threatened in either its entire range or a significant portion of its range. This curious phrase, which for decades was largely ignored in ESA listing determinations, has recently attracted a good deal of attention (Defenders of Wildlife v. Norton 2001; Vucetich et al. 2006; Waples et al. 2007). As with other key statutory language in the ESA (e.g., "distinct population segment," "likely," "foreseeable future"), the phrase "significant portion of its range" has no precise biological definition and is open to a wide range of possible interpretations. The approach used for dealing with this issue in Pacific salmon recovery planning takes into account patterns of biodiversity within each species. Most salmon ESUs contain many populations (Table 10–2) that typically are related in a hierarchical fashion; therefore, in most cases, it is possible to identify one or more intermediate levels of biological complexity between populations and ESUs (Fig. 10–5). These strata, or major population groups, reflect genetic, ecological, behavioral, and life-history patterns among populations, as well as physical features of the habitats on which salmon depend. Recognition of these intermediate levels of biodiversity provides a way to directly address the "significant portion of its range" test: If viable populations are distributed among most or all strata within an ESU, then the ESU can be considered to no longer be at risk in all *or* a significant portion of its range. Spatial structure and diversity, which were considered at the population level as described in the previous section, were thus important in evaluating whole-ESU viability as well.

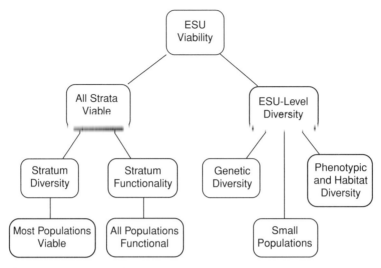

Figure 10–6: Example of integrating population criteria into ESU-wide status for the Oregon Coast ESU for coho salmon. Population-level criteria (for viability and functionality) are rolled into stratum-level criteria to determine the viability of individual strata, and the ESU is considered viable only if all strata are viable. In addition to stratum-based evaluations, some broad-scale diversity measures are best considered ESU-wide, including some genetic diversity issues, phenotypic and habitat diversity, and issues related to small (nonindependent) populations. *Source:* Modified from Wainwright et al. (2008).

Use of strata to help integrate risks across multiple populations of Oregon Coast coho salmon is illustrated in Fig. 10–6. For that ESU, the team identified five strata (Fig. 10–2) based primarily on geography, ecology, life history, and genetics. Notably, the configuration of the "Lakes" stratum reflects the recent finding that coho populations from these distinctive (but geographically disjunct) habitats are genetically similar (Ford et al. 2004). To describe characteristics of a viable ESU, the team first combined a variety of data for each population into two integrated criteria: population viability and population functionality. Viability reflects the four VSP parameters; functionality is a lower-level goal designed to ensure that even if the population is not viable, it at least has sufficient functioning habitat that could support a population with sustainable abundance. For a stratum to be viable, the team required that most of the populations in that stratum be viable and that all of the populations be functional. The stratum-viability criterion ensures that multiple viable populations exist in each stratum, thus reducing risk to the stratum from changes in local conditions and population-scale catastrophes (Good et al. 2008). The functionality criterion ensures that independent populations that are not currently viable still have sufficient habitat to contribute to future evolution of the ESU. This requirement recognizes that landscape conditions, and thus population capacity and productivity, will shift over time (Reeves et al. 1995), so populations with poor conditions at one point in time might dominate the stratum at a later time (Hilborn et al. 2003). For the ESU to be viable, the TRT required not only that all strata be viable but also that three ESU-level diversity criteria be met: broad-scale genetic diversity (interpopulation diversity, effects of selection, and effects of between-population migration), phenotypic and habitat diversity throughout the ESU, and the preservation of small (nonindependent) populations as reservoirs of potentially adaptive diversity (e.g., Wright 1948).

For most large ESUs that contain numerous populations, it is likely that multiple different scenarios could lead to overall viability and delisting, with the scenarios differing in which specific populations are required to be at which levels of viability. In general, for each ESU, the teams tried to develop a menu of delisting options. These scenarios typically had different costs, societal consequences, and probabilities of successful implementation, and these factors could be considered by managers in deciding which option(s) to pursue.

To summarize, viability analyses designed to help guide recovery of Pacific salmon have the following features:

- The overall process is guided by the concept of VSP, which provides a conceptual framework for assessing the health of populations and groups of populations, such as ESUs.
- The four VSP criteria include two (A, P) that are standard components of population-viability analyses, as well as two (SS, D) that capture evolutionary processes that are only rarely considered.
- Most ESUs include multiple geographic subregions, or strata, which represent an intermediate level of biological organization between populations and ESUs.
- All four viability criteria are considered for each population, and SS and D are also considered when integrating risks across populations to the level of strata or ESUs. Ensuring that viable populations are distributed across multiple strata helps promote long-term viability at the ESU level.
- Effectively communicating uncertainties associated with population identification and viability assessments was an explicit and important goal of the process.
- Efforts have only recently begun to formally integrate into viability analyses the consequences of evolutionary responses by salmon populations to anthropogenic changes to their environments. These evaluations will be an important complement to more traditional analyses based exclusively on demographic and ecological factors.

FUTURE PROSPECTS

Endangered species listing determinations and recovery planning for Pacific salmon represent a large and complex exercise in applied conservation biology. At numerous key steps in the process, it has been necessary to develop novel analytical approaches that integrate diverse types of information across multiple dimensions (spatial and temporal; across disciplines; across populations and species). The science teams quantified specific analyses to the extent practicable but resisted the temptation to try to reduce extremely complex evaluations (e.g., assessing whole-ESU viability) to a simple quantitative criterion or formula; instead, they have used a mix of quantitative and qualitative analyses and professional judgment. Whereas most conservation planning focuses largely on ecological issues, concerted efforts have been made to fully integrate evolutionary considerations into salmon recovery planning. This approach reflects the growing realization that many key conservation issues are eco-evolutionary in nature and therefore require joint consideration of ecological and evolutionary processes (Kinnison & Hairston 2007; see also Chapter 12 by Rhodes & Latch).

Although the examples in this chapter are drawn from Pacific salmon, most of the central themes arise more broadly in conservation planning for other taxa. A variety of methods for defining conservation units and ESUs can be found in the literature (reviewed by Fraser & Bernatchez 2001), and the choice of which method to use can be guided by the biological attributes of the taxon of interest and the specific conservation goals that one is trying to achieve. The VSP framework provides a systematic way of integrating evolutionary and ecological considerations into assessing viability and sustainability. Examples presented here only scratch the surface of the multitude of ways that these concepts might be implemented. Finally, effective conservation requires successful communication between scientists and policy makers operating at the intersection between the two disciplines. In Pacific salmon recovery planning, regular interactions between science and policy teams have helped to make this process successful, as have lengthy discussions about the most effective and useful ways for the science teams to characterize the degree of uncertainty associated with specific results and conclusions. The examples discussed here (and illustrated, e.g., in Table 10–3) might be useful in providing ideas about ways to communicate uncertainty but, in general, the best way to accomplish this will probably have to be worked out on a case-by-case basis by the parties involved, taking into consideration their respective experiences and goals.

It is widely recognized that conservation at the level of ecosystems rather than individual species is essential to address problems of long-term sustainability and that maintaining evolutionary processes is a main ingredient of an ecosystem approach to salmon conservation. Holling (1973) noted that failure to recognize and manage for the dynamic nature of ecosystems could increase rates of extirpation and extinction. Salmon ESUs were developed as dynamic meta-populations within dynamic landscapes at a variety of spatial and temporal scales, and understanding and conserving dynamic ecological processes is key to effective conservation (Rieman & Dunham 2000). Effective conservation, in turn, includes preserving the adaptive potential within ESUs to respond to changing habitats. To some degree, accomplishing this will require a balancing act: allowing sufficient adaptation to changes in ecosystems without losing the unique characteristics that define an ESU or species.

Accomplishing the latter will be challenging in landscapes increasingly dominated by anthropogenic change. Evidence is rapidly accumulating that demonstrates microevolutionary change in response to human perturbations to ecosystems (Smith & Bernatchez 2008; see also Chapter 13 by Shugart and colleagues). This change is already occurring with ESA-listed Pacific salmon. Williams and coworkers (2008) documented recent changes in a key life-history trait in threatened Snake River fall Chinook salmon, presumably due to major ecological changes associated with the federal hydropower system in the Columbia and Snake Rivers.

BOX 10: THE KERMODE BEAR: A SWIRL OF SCIENTIFIC, MANAGEMENT, AND ETHICAL VALUES IN BRITISH COLUMBIA

Kermit Ritland

Background

The Kermode bear, or "Spirit Bear," is a rare white-phase black bear (Box Fig. 10–1) that lives in the rainforests of the north coast of British Columbia (BC). Revered by the native Kitasoo and Tsimshian people, it was initially thought to be a distinct species. It is now deemed a subspecies, *Ursus americanus kermodei*, within which the black phase predominates. On certain islands off the coast of BC, the frequency of Kermode bears can be upwards of 30%, but the total number of Kermode bears is in the low hundreds. This dramatic coat-color difference has been the interest of environmentalists and ecotourists, as well as the native people. To some, the Kermode bear represents the "Great Bear Rainforest," a keystone species of ecosystem function. For others, the Kermode is a symbol of the Ice Age, yet it lives in or near forests of great economic value. The Kermode bear is clearly a model of the interactions among science, economics, and ethical values.

Case Study

Recently, environmental organizations voiced concerns that logging on islands adjacent to the preserved islands will endanger the frequency of the Kermode bear. How can molecular techniques address practical management issues as well as issues about the adaptation and evolution of this coat-color phase?

Box Figure 10–1: White-phase Kermode bear. Photo courtesy of Charlie Russell.

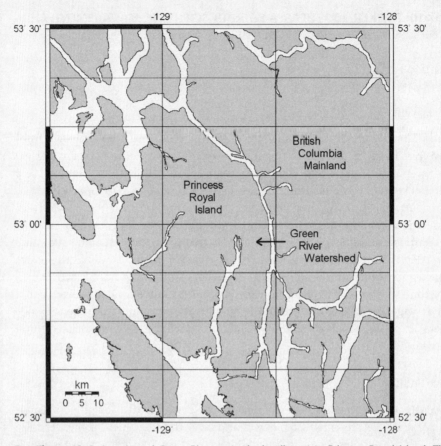

Box Figure 10–2: Location of Green River watershed, adjacent to Princess Royal Island. This island is about halfway between the north tip of Vancouver Island and the south tip of Alaska. The arrow indicates the direction of migration should logging affect Green River bear populations.

Suspicions that the color was due to a single recessive gene were confirmed in 2001 by the discovery that a single nucleotide polymorphism (SNP) at a gene termed "MC1R" showed complete correlation with white coat color when in the recessive condition (Ritland et al. 2001). MC1R is the melanocortin receptor 1 gene, and it functions near the end of a pathway that controls the production of the pigment melanin. MC1R variants (either nucleotide substitutions or deletions) have been shown to cause conspicuous color variants in many mammal species, most notably the Labrador retriever, in which a deletion near the Kermode variant causes the golden coat color (Newton et al. 2000). Also, the extinct mammoth was shown via ancient deoxyribonucleic acid (DNA) recovery to have an MC1R variant just three amino acids away from the Kermode allele (Römpler et al. 2006), and Neanderthals have a variant just nine amino acids away (Lalueza-Fox et al. 2007). In humans, more than thirty variant alleles have been identified that correlate with variation in skin and hair color (Gerstenblith et al. 2007). The exact Kermode bear variant has not been found in any other mammal, however.

What might be the impacts of logging on the Kermode bear? This question was motivated by the proposal in 2007 to selectively log portions of the Green River watershed (Box Fig. 10–2). Although this watershed does not have appreciable numbers of Kermode bears, logging can change food production for bears due to more canopy openings, and increased population size may promote migration of black bears to adjacent Kermode populations. Princess Royal Island contains the highest frequency of Kermode bears in BC, and it is separated from the Green River watershed by a narrow channel (Box Fig. 10–2) across which bears can easily swim. Because the Kermode gene is recessive, increased migration of homozygous black bears could "swamp" Kermode populations in that the influx of dominant black-phase alleles will reduce the frequency of homozygous recessive white bears (although the overall frequency of white alleles may not change over the range).

To predict the genetic consequences of this possibility, the following scenario was considered by a recent working group involving scientists from the BC Ministry of Forests and Range, and University of BC scientists. Assume that the average bear generation time is ten years and that the proposed selective logging increases bear populations by one per home range (20 km^2 for females and 300 km^2 for males), necessitating migration only for males. Migration can take place in any direction, so that approximately one in four migrants will head toward Princess Royal Island from the Green River watershed. Taking into account the area subject to logging, this translates into the potential for additional emigration to Princess Royal Island of approximately one bear every generation. Indirect evidence for historical levels of gene flow, based on Wright's F_{ST}, indicates that the numbers of migrating bears among islands is approximately two per generation (Marshall & Ritland 2002), so that migration is increased by approximately 50%.

Assuming this scenario, for an island population of 500 bears with a white-coat allele frequency of 0.33 such as found on Princess Royal Island (Ritland et al. 2001), one additional homozygous black-phase bear immigrating every generation would cause an expected 10% reduction of the white gene frequency after 52 generations or 520 years (solving for n in the equation $(499/500)^n = 0.90$). Should bear management consider 100-generation time scales, particularly if conservation budgets span only a few years? Although the Kermode bear is truly unique, should a novel SNP of conspicuous effect be worth our efforts to maintain it? The answer resides in our cultural and esthetic values.

REFERENCES

Gerstenblith MR, Goldstein AM, Fargnoli MC, Peris K, Landi MT (2007) Comprehensive evaluation of allele frequency differences of MC1R variants across populations. *Human Mutation*, **28**, 495–505.

Lalueza-Fox C, Römpler H, Caramelli D et al. (2007) A melanocortin 1 receptor allele suggests varying pigmentation among Neanderthals. *Science*, **318**, 1453–1455.

Marshall HD, Ritland K (2002) Genetic diversity and differentiation of Kermode bear populations. *Molecular Ecology*, **11**, 685–697.

Newton JM, Wilkie AL, He L et al. (2000). Melanocortin 1 receptor variation in the domestic dog. *Mammalian Genome*, **11**, 24–30.

Ritland K, Newton C, Marshall HD (2001) Inheritance and population structure of the white-phased "Kermode" black bear. *Current Biology*, **11**, 1468–1472.

Römpler H, Rohland N, Lalueza-Fox C et al. (2006) Nuclear gene indicates coat-color polymorphism in Mammoths. *Science*, **313**, 62.

Climate change looms in the future. Given uncertainties about future climate patterns, it is important that recovery actions be robust to a variety of future scenarios (Battin et al. 2007). It will not be sufficient to consider ecological factors alone, however. Climate change, and other anthropogenic changes to natural ecosystems, alter the selective regimens that salmon and other natural populations experience (e.g., Waples et al. 2008c), and these populations can be expected to evolve in response to these changing selective pressures. A key challenge in the future will be to find more effective ways to factor into conservation planning the capacity of natural populations to respond to future challenges through evolution and/or expression of adaptive phenotypic plasticity (Ghalambor et al. 2007).

Finally, after many years of anticipation, we now have the practical capability to begin to construct a bridge between genotypes and phenotypes in natural populations (Naish & Hard 2008; see also Chapter 10 by Nichols et al. and Box 10). Although molecular genetic data have played a substantial role in defining population structure and ESUs in salmon (and in many other species), the vast majority of these analyses have relied on (presumably) neutral genetic markers, which provide important information about historical levels of connectivity but little or no direct information about adaptation. It is now possible to use molecular methods to identify genes under selection, which opens up a whole new range of potential conservation applications. To take just one example, comprehensive viability analyses for species such as *O. mykiss* have to account for the genetic and ecological relationships between resident (rainbow trout) and anadromous (steelhead) forms. Scientists in several different laboratories are using a variety of molecular approaches to determine the relative importance of genetic versus environmental factors in determining life-history type and to identify the specific genes involved. Recently, Nichols and coworkers (2008) identified genomic regions in *O. mykiss* associated with morphological and physiological indices of smoltification (i.e., juvenile transition from freshwater to marine environment). Key conservation questions that this emerging type of information can potentially address include the following: If anthropogenic changes increase the cost of migration, will the genes capable of producing anadromy be lost from the population? If so, how quickly might this occur and is the process reversible? How would long-term viability of steelhead ESUs be affected by loss of the anadromous forms? Can resident forms trapped in reservoirs above dams be used to help restore self-sustaining steelhead runs?

REFERENCES

Allendorf FW, Waples RW (1996) Conservation and genetics of salmonid fishes. In: *Conservation Genetics: Case Histories from Nature* (eds. Avise JC, Hamrick JL), pp. 238–280. Chapman & Hall, New York.

Andrewartha HG, Birch LC (1984) *The Ecological Web*. University of Chicago Press.

Araki H, Berejikian BA, Ford MJ, Blouin MS (2008) Fitness of hatchery-reared salmonids in the wild. *Evolutionary Applications*, **1**, 342–355.

Battin J, Wiley MW, Ruckelshaus MH et al. (2007) Projected impacts of future climate change on salmon habitat restoration actions in a Puget Sound river. *Proceedings of the National Academy of Sciences USA*, **104**, 6720–6725.

Beamish RJ, editor (2005) *Proceedings of the 2005 NPAFC-PICES Joint Symposium on the Status of Pacific Salmon and Their Role in North Pacific Marine Ecosystems*. North Pacific Anadromous Fish Commission Bulletin No. 4, 1–337.

Beechie TJ, Collins BD, Pess GR (2001) Holocene and recent geomorphic processes, land use and salmonid habitat in two north Puget Sound river basins. In: *Geomorphic Processes and Riverine Habitat,* Water Science and Application (Vol. 4, eds. Dorava JB, Montgomery DR, Fitzpatrick F, Palcsak B), pp. 37–54. American Geophysical Union, Washington DC.

Beissinger SR, McCullough DR, editors (2002) *Population Viability Analysis.* University of Chicago Press, Chicago.

Bradford MJ (1995). Comparative review of Pacific salmon survival rates. *Canadian Journal of Fisheries and Aquatic Sciences*, **52**, 1327–1338.

Butler VL, O'Connor JE (2004) 9,000 years of salmon fishing on the Columbia River, North America. *Quaternary Research*, **62**, 1–8.

Carlson SM, Seamons TR (2008). A review of quantitative genetic components of fitness in salmonids: implications for adaptation to future change. *Evolutionary Applications*, **1**, 222–238.

Chatters JC, Butler VL, Scott MJ, Anderson DM, Neitzel DA (1995) A paleoscience approach to estimating the effects of climatic warming on salmonid fisheries of the Columbia River basin. In: *Climate Change and Northern Fish Populations* (ed. Beamish RJ), pp. 48–96. Canadian Special Publication, Fisheries and Aquatic Sciences 121.

Cooney TD, McClure MM, Carmichael R et al. (2007) Viability criteria for application to Interior Columbia Basin ESUs. Draft Technical Recovery Team document released for co-manager review. Available at http://www.nwfsc.noaa.gov/trt/trt_documents/ictrt_viability_criteria_reviewdraft_2007_complete.pdf.

Defenders of Wildlife v. *Norton* (2001) 258 F.3d 1136 (9th Cir.).

Finney BP, Gregory-Eaves I, Douglas MSV, Smol JP (2002) Fisheries productivity in the northeastern Pacific Ocean over the past 2,200 years. *Nature*, **416**, 729–733.

Flagg TA, Mahnken CVW, Iwamoto RN (2004) Conservation hatchery protocols for Pacific salmon. *American Fisheries Society Symposium*, **44**, 603–619.

Ford MJ, Teel DJ, Van Doornik DM, Kuligowski DR, Lawson PW (2004) Genetic population structure of central Oregon Coast coho salmon (*Oncorhynchus kisutch*). *Conservation Genetics*, **5**, 797–812.

Fraser DJ (2008) How well can captive breeding programs conserve biodiversity? A review of salmonids. *Evolutionary Applications*, **1**, 535–586.

Fraser DJ, Bernatchez L (2001) Adaptive evolutionary conservation: towards a unified concept for defining conservation units. *Molecular Ecology*, **10**, 2741–2752.

Ghalambor CK, McKay JK, Carroll SP, Reznick DN (2007) Adaptive versus non-adaptive phenotypic plasticity and the potential for contemporary adaptation in new environments. *Functional Ecology*, **21**, 394–407.

Good TP, Davies JR, Burke BJ, Ruckelshaus MH (2008) Incorporating catastrophic risk assessments into setting conservation goals for Pacific salmon. *Ecological Applications*, **18**, 246–257.

Groot C, Margolis I (1991) *Pacific Salmon Life Histories.* University of British Columbia Press, Vancouver.

Gustafson R, Waples RS, Myers JM et al. (2007) Pacific salmon extinctions: quantifying lost and remaining diversity. *Conservation Biology*, **21**, 1009–1020.

Hard JJ, Gross MR, Heino M et al. (2008) Evolutionary consequences of fishing and their implications for salmon. *Evolutionary Applications*, **1**, 388–408.

Hastings A (1993) Complex interactions between dispersal and dynamics: lessons from coupled logistic equations. *Ecology*, **74**, 1362–1372.

Hendry AP, Stearns SC, editors (2004) *Evolution Illuminated: Salmon and Their Relatives.* Oxford University Press, Oxford, UK.

Hilborn RJ, Quinn TP, Schindler DE, Rogers DE (2003) Biocomplexity and fisheries sustainability. *Proceedings of the National Academy of Sciences USA*, **100**, 6564–6568.

Holling CS (1973) Resilience and stability of ecological systems. *Annual Review of Ecology and Systematics*, **4**, 1–23.

Ivy JA, Lacy, RC (2010) Using molecular methods to improve the genetic management of captive breeding programs for threatened species. In: *Molecular Approaches in Natural Resource Conservation and Management* (eds: DeWoody JA, Bickham JW, Michler CH et al.), pp. 267–295. Cambridge University Press, New York.

Kareiva P, Marvier M, McClure MM (2000) Recovery and management options for spring/summer Chinook salmon in the Columbia River basin. *Science*, **290**, 977–979.

Kinnison MT, Hairston NG Jr (2007) Eco-evolutionary conservation biology: contemporary evolution and the dynamics of persistence. *Functional Ecology*, **21**, 444–454.

Lackey RT, Lach DH, Duncan SL, editors (2006) *Salmon 2100: The Future of Wild Pacific Salmon.* American Fisheries Society, Bethesda, MD.

Lawson PW, Bjorkstedt EP, Chilcote MW et al. (2007) Identification of historical populations of coho salmon (*Oncorhynchus kisutch*) in the Oregon Coast Evolutionarily Significant Unit. NOAA Technical Memorandum NMFS-NWFSC-79.

Mahnken CV, Ruggerone G, Waknitz FW, Flagg TA (1998) A historical perspective on salmonid production from Pacific Rim hatcheries. *North Pacific Anadromous Fish Commission Bulletin*, **1**, 38–53.

McClure MM, Carlson SM, Beechie TJ et al. (2008a) Evolutionary consequences of habitat loss for Pacific anadromous salmonids. *Evolutionary Applications*, **1**, 300–318.

McClure MM, Carmichael R, Cooney TD et al. (2003). Independent populations of listed Chinook salmon, sockeye salmon and steelhead Evolutionarily Significant Units in the Interior Columbia basin. Available at http://www.nwfsc.noaa.gov/trt/col/trt_pop_id.cfm.

McClure MM, Utter FM, Baldwin C et al. (2008b) Evolutionary effects of alternative artificial propagation programs: implications for the viability of endangered anadromous salmonids. *Evolutionary Applications*, **1**, 356–375.

McElhany P, Chilcote M, Myers J, Beamesderfer R (2007) Viability status of Oregon salmon and steelhead populations in the Willamette and Lower Columbia basins. NOAA-NWFSC. Seattle, WA. Available at http://www.nwfsc.noaa.gov/trt/wlc/trt_wlc_psr2007.cfm.

McElhany P, Ruckelshaus MH, Ford MJ, Wainwright TC, Bjorkstedt EP (2000) Viable salmonid populations and the recovery of evolutionarily significant units. NOAA Technical Memorandum NMFS-NWFSC 42.

McPhail JD, Lindsey CC (1986) Zoogeography of the freshwater fishes of Cascadia (the Columbia system and rivers north to the Stikine). In: *The Zoogeography of North American Freshwater Fishes* (eds. Hocutt CH, Wiley EO), pp. 615–637. John Wiley & Sons, New York.

Mobrand L, Barr J, Blankenship L et al. (2005) Hatchery reform in Washington state: principles and emerging issues. *Fisheries*, **30**, 11–23.

Montgomery DR (2000) Coevolution of the Pacific salmon and Pacific Rim topography. *Geology*, **28**, 1107–1110.

Moran PAP (1953) The statistical analysis of the Canadian lynx cycle. 2. Synchronization and meteorology. *Australian Journal of Zoology*, **1**, 291.

Moritz C (2002) Strategies to protect biological diversity and the evolutionary processes that sustain it. *Systematic Biology*, **51**, 238–254.

Morris WF, Doak DF (2002) *Quantitative Conservation Biology: Theory and Practice of Population Viability Analysis.* Sinauer Associates, Sunderland, MA.

Myers JM, Busack C, Rawding D et al. (2006) Historical population structure of Pacific salmonids in the Willamette River and lower Columbia River basins. NOAA Technical Memorandum NMFS-NWFSC-73.

Myers RA, Baum JK, Shepherd TD, Powers SP, Peterson CH (2007) Cascading effects of the loss of apex predatory sharks from a coastal ocean. *Science*, **315**, 1846–1850.

Naish KA, Hard JJ (2008) Bridging the gap between the genotype and the phenotype: linking genetic variation, selection and adaptation in fishes. *Fish and Fisheries*, **9**, 396–422.

National Research Council (1996) *Upstream: Salmon and Society in the Pacific Northwest.* The National Academies Press, Washington, DC.

Nichols KM, Felip A, Wheeler P, Thorgaard GH (2008) The genetic basis of smoltification-related traits in *Oncorhynchus mykiss*. *Genetics*, **179**, 1559–1575.

Nichols KM, Neale DB (2010) Association genetics, population genomics, and conservation: Revealing the genes underlying adaptation in natural populations of plants and animals. In: *Molecular Approaches in Natural Resource Conservation and Management* (eds: DeWoody JA, Bickham JW, Michler CH et al.), pp. 123–168. Cambridge University Press, New York.

Nickelson TE, Lawson PW (1998) Population viability of coho salmon (*Oncorhynchus kisutch*) in Oregon coastal basins: application of a habitat-based life cycle model. *Canadian Journal of Fisheries and Aquatic Sciences*, **55**, 2383–2392.

Overland J, Rodionov S, Minobe S, Bond N (2008) North Pacific regime shifts: definitions, issues and recent transitions. *Progress in Oceanography*, **77**, 92–102.

Pearcy WG (1992). *Ocean Ecology of North Pacific Salmonids.* University of Washington Press, Seattle.

Post E, Forchhammer MC (2004) Spatial synchrony of local populations has increased in association with the recent northern hemisphere climate trend. *Proceedings of the National Academy of Sciences USA*, **101**, 9286–9290.

Quinn TP (2005) *The Behavior and Ecology of Pacific Salmon and Trout.* American Fisheries Society, Bethesda, MD.

Quinn TP, Unwin MJ, Kinnison MT (2000) Evolution of temporal isolation in the wild: genetic divergence in timing of migration and breeding by introduced Chinook salmon populations. *Evolution*, **54**, 1372–1385.

Reeves GH, Benda LE, Burnett KM, Bisson PA, Sedell JR (1995) A disturbance-based ecosystem approach to maintaining and restoring freshwater habitats of evolutionarily significant units of anadromous salmonids in the Pacific Northwest. *American Fisheries Society Symposium*, **17**, 334–349.

Rieman BE, Dunham JB (2000) Metapopulations and salmonids: a synthesis of life history patterns and empirical observations. *Ecology of Freshwater Fish*, **9**, 51–64.

Ruckelshaus MH, Currens KP, Graeber WH et al. (2006) Independent populations of Chinook salmon in Puget Sound. NOAA Technical Memorandum NMFS-NWFSC-78.

Ruckelshaus MH, Levin PS, Johnson JB, Kareiva P (2002) The Pacific salmon wars: what science brings to the challenge of recovering species. *Annual Review of Ecology and Systematics*, **33**, 665–706.

Ryman N, Laikre L (1991) Effects of supportive breeding on the genetically effective population size. *Conservation Biology*, **5**, 325–329.

Schindler DE, Scheuerell MD, Moore JW et al. (2003) Pacific salmon and the ecology of coastal ecosystems. *Frontiers in Ecology and the Environment*, **1**, 31–37.

Smith TB, Bernatchez L (2008) Evolutionary change in human-altered environments. *Molecular Ecology*, **17**, 1–499.

Vucetich JA, Nelson MP, Phillips MK (2006) The normative dimension and legal meaning of endangered and recovery in the U.S. Endangered Species Act. *Conservation Biology*, **20**, 1383–1390.

Wainwright TC, Chilcote MW, Lawson PW et al. (2008) Biological recovery criteria for the Oregon Coast coho salmon evolutionarily significant unit. NOAA Technical Memorandum NMFS-NWFSC-91.

Wainwright TC, Kope RG (1999) Methods of extinction risk assessment developed for U.S. West Coast salmon. *ICES Journal of Marine Science*, **56**, 444–448.

Waples RS (1991) Pacific salmon, *Oncorhynchus spp.*, and the definition of "species" under the Endangered Species Act. *Marine Fisheries Review*, **53**(3), 11–22.

Waples RS (1995) Evolutionarily significant units and the conservation of biological diversity under the Endangered Species Act. *American Fisheries Society Symposium*, **17**, 8–27.

Waples RS (2006) Distinct population segments. In: *The Endangered Species Act at Thirty: Conserving Biodiversity in Human-Dominated Landscapes* (eds. Scott JM, Goble DD, Davis FW), pp. 127–149. Island Press, Washington, DC.

Waples RS, Adams P, Bohnsack J, Taylor BL (2007) A biological framework for evaluating whether an ESA species is threatened or endangered in a "significant portion of its range." *Conservation Biology*, **21**, 964–974.

Waples RS, Drake J (2004) Risk-benefit considerations for marine stock enhancement: a Pacific salmon perspective. In: *Stock Enhancement and Sea Ranching: Developments, Pitfalls and Opportunities* (ed. Leber KM), pp. 206–306. Blackwell, Oxford.

Waples RS, Gaggiotti O (2006) What is a population? An empirical evaluation of some genetic methods for identifying the number of gene pools and their degree of connectivity. *Molecular Ecology*, **15**, 1419–1439.

Waples RS, Gustafson RG, Weitkamp LA, et al. (2001) Characterizing diversity in Pacific salmon. *Journal of Fish Biology,* **59**(Supplement A), 1–41.

Waples RS, Pess GR, Beechie T (2008a) Evolutionary history of Pacific salmon in dynamic environments. *Evolutionary Applications,* **1**, 189–206.

Waples RS, Punt AE, Cope J (2008b) Integrating genetic data into fisheries management: how can we do it better? *Fish and Fisheries,* **9**, 423–449.

Waples RS, Teel DJ, Myers J, Marshall A (2004) Life history divergence in Chinook salmon: historic contingency and parallel evolution. *Evolution,* **58**, 386–403.

Waples RS, Zabel RW, Scheuerell MD, Sanderson BL (2008c) Evolutionary responses by native species to major anthropogenic changes to their ecosystems: Pacific salmon in the Columbia River hydropower system. *Molecular Ecology,* **17**, 84–96.

Whitham, TG, Gehring CA, Evans, LM et al. (2010) A community and ecosystem genetics approach to conservation biology and management. In: *Molecular Approaches in Natural Resource Conservation and Management* (eds: DeWoody JA, Bickham JW, Michler CH et al.), pp. 50–73. Cambridge University Press, New York.

Williams JG, Zabel RW, Waples RS, Hutchings JA, Connor WP (2008) Potential for anthropogenic disturbances to influence evolutionary change in the life history of a threatened salmonid. *Evolutionary Applications,* **1**, 271–285.

Withler FC (1982) Transplanting Pacific salmon. Canadian Tech. Rpt. Fisheries and Aquatic Sciences 1079. Dept. of Fisheries and Oceans, Vancouver, BC.

Wood CC (1995) Life history variation and population structure in sockeye salmon. *American Fisheries Society Symposium,* **17**, 195–216.

Wootton JT, Pfister CA, Forester JD (2008) Dynamic patterns and ecological impacts of declining ocean pH in a high-resolution multi-year dataset. *Proceedings of the National Academy of Sciences USA,* **105**, 18848–18853.

Wright S (1948) On the roles of directed and random changes in gene frequency in the genetics of populations. *Evolution,* **2**(4), 279–294.

11 Using molecular methods to improve the genetic management of captive breeding programs for threatened species

Jamie A. Ivy and Robert C. Lacy

Captive breeding programs are powerful tools for the conservation of our natural resources. Both animal and plant biodiversity are in global decline (International Union for Conservation of Nature [IUCN] 2008), and the trend is for levels and rates of endangerment to continue increasing (Butchart et al. 2005; see also Chapter 1 by Honeycutt and colleagues). When in-situ conservation efforts are insufficient at stopping or reversing the decline of a species, captive populations often represent the only alternative for forestalling extinction. For example, it is currently estimated that hundreds of amphibian species will soon be extinct if emergency measures are not taken to protect populations in captivity until the threats to wild populations can be halted or overcome (Gascon et al. 2007). The success of numerous captive breeding programs has been well documented. As just a few of many examples, captive breeding programs have saved the black-footed ferret (*Mustela nigripes*), Przewalski's horse (*Equus caballus przewalskii*), and the California condor (*Gymnogyps californianus*) from final extinction (after the last wild populations were extirpated) and new wild populations of the golden lion tamarin (*Leontopithecus rosalia*), Arabian oryx (*Oryx leucoryx*), and whooping crane (*Grus americana*) have been successfully re-established from captive stocks.

Conservation breeding programs aim to maintain populations that are representative of their wild counterparts, to provide a reservoir for future reintroductions and recovery efforts (see Chapter 12 by Rhodes and Latch). Thus, the genetic goals of captive population management are to minimize genetic drift, retain genetic diversity, restrict inbreeding, and limit adaptation to captivity (Lacy 1994). The foundations of most captive breeding programs are pedigree analyses, which are used to manage both the demography and genetics of captive populations (Ballou & Foose 1996). Accurate pedigrees provide information on inbreeding, the kinships among individuals, and the distributions of individual founder contributions to a population. Collectively, this information is used to produce regular breeding recommendations intended to meet demographic and genetic goals. Although pedigree analyses are quite effective, they do have limitations. The most basic limitation of pedigree-based management is that it requires complete and accurate pedigrees to be effective. Thus, the genetic and demographic management of captive populations are hampered when missing or inaccurate parentage records produce incomplete pedigrees.

When a captive pedigree is inaccurate or incomplete, molecular data have the potential to improve breeding program management. Molecular markers can

Table 11–1. Examples of studies that have incorporated molecular data into captive breeding programs

Contribution to breeding program	Species	Citation
Assessment of hybridization	Mexican gray wolf (*Canis lupus baileyi*)	Hedrick et al. 1997
	Lesser white-fronted goose (*Anser erythropus*)	Ruokonen et al. 2007
Gender identification	Old World vultures (*Gyps* species)	Reddy et al. 2007
Identification of geographic origin	Bearded vulture (*Gypaetus barbatus*)	Gautschi et al. 2003
	Galapagos tortoise (*Geochelone nigra*)	Russello et al. 2007
	Binturong (*Arctictis binturong*)	Cosson et al. 2007
Quantification of genetic differentiation	Baird's tapir (*Tapirus bairdii*)	Norton & Ashley 2004
	Chinese water deer (*Hydropotes inermis*)	Hu et al. 2007
	Scimitar-horned oryx (*Oryx dammah*)	Iyengar et al. 2007
Quantification of wild genetic diversity captured	Baird's tapir (*Tapirus bairdii*)	Norton & Ashley 2004
	Iberian wolf (*Canis lupus signatus*)	Ramirez et al. 2006
Species and subspecies identification	Chimpanzee (*Pan troglodytes*)	Ely et al. 2005
	Asian box turtles (*Cuora* species)	Spinks & Shaffer 2007

contribute to captive breeding programs in many ways, and numerous methods for incorporating molecular data into captive population management have been described (Table 11–1). The majority of widely applied methods, however, are focused on the accurate characterization of the individuals used to establish a captive stock, but they are not related to the fundamental pedigree analyses that are used to select breeding pairs for the ongoing management and maintenance of captive breeding programs. For example, although it is invaluable to know that none of the captive Mexican wolf (*Canis lupus baileyi*) lineages contain any domestic dog (*C. lupus familaris*) or coyote (*Canis latrans*) ancestry (Hedrick et al. 1997), this information informs the selection of breeding pairs only so far as to justify the inclusion of all individuals in the pool of potential breeders. Rather than focusing on previously well-described methods for using genetic data to characterize populations genetically and taxonomically, this chapter discusses the methods and prospects for incorporating molecular data into the pedigree analyses that most captive breeding programs use for ongoing genetic and demographic management.

We start the chapter by providing a brief overview of common genetic terms used in pedigree and molecular analyses. We then discuss the goals of conservation breeding programs and describe the methods typically used to genetically manage captive populations. The bulk of the chapter illustrates how molecular data can be incorporated into captive breeding programs to improve the effectiveness of pedigree-based population management. We discuss the benefits and limitations of different methods of incorporating molecular data into captive population management and highlight issues that should be considered when different methods are employed. Finally, we discuss new prospects for the continuing incorporation of molecular markers into future captive breeding programs.

GENETIC TERMS AND THEIR USE IN CAPTIVE BREEDING PROGRAMS

Pedigree and molecular analyses share many analogous terms and concepts. Although some of the terms used in both types of analyses are quite similar, there are often subtle differences between their definitions. Furthermore, the relationships between pedigree-based and molecularly based concepts are not always readily apparent.

Kinship and mean kinship

The coefficient of kinship (or consanguinity) between two individuals (f) is the probability that two alleles at a given locus, one randomly drawn from each individual, are "identical by descent" – that is, they are copies of the same piece of deoxyribonucleic acid (DNA) descended from a common ancestor (Falconer & Mackay 1996). The concept is necessarily a relative one, in that all DNA can be traced back to a common source if the pedigree is extended far enough back in time. Therefore, in practice, kinships are calculated relative to a baseline generation or source population in which all kinships are assigned to be 0, and alleles are all assumed to have independent origins. Subsequent to the base generation, if the parents of a diploid individual x are a and b, and the parents of individual y are c and d, then the kinship between individuals x and y is

$$f_{xy} = f_{yx} = \frac{1}{4}\left(f_{ac} + f_{ad} + f_{bc} + f_{bd}\right),$$

and the kinship of individual x to itself is

$$f_{xx} = \frac{1}{2}\left(1 + f_{ab}\right).$$

An individual's mean kinship ($\overline{f}$) is the average of pairwise fs between that individual and all living individuals in the population, including itself (Ballou & Lacy 1995; Lacy 1995). Under random mating, the $\overline{f}$ of an individual is the expected inbreeding coefficient of its offspring. Furthermore, an individual's $\overline{f}$ also provides a measure of its genetic representation in a population; individuals with high $\overline{f}$ have many living relatives and, thus, their alleles are well-represented, whereas individuals with low $\overline{f}$ are poorly represented because they have few living relatives.

Inbreeding coefficient

The inbreeding coefficient (F) of an individual is the probability that at a given locus, both alleles are identical by descent (Falconer & Mackay 1996). Thus, an individual's F is equal to the f between its parents. If the parents of individual x

are a and b, then the inbreeding coefficient of individual x is

$$F_x = f_{ab}.$$

Given this equality, F_x also can be substituted into the equation for calculating an individual's kinship to itself:

$$f_{xx} = \frac{1}{2}(1 + F_x).$$

Coefficient of relationship

The coefficient of relatedness (r) between two individuals is the probability that at a given locus, an allele sampled from one individual is identical by descent to at least one of the alleles at that locus in the second individual. (It should be noted that other authors have sometimes used the symbol r for the coefficient of kinship, f, as well as for other measures of relationship. We will use it here in the more consistent and precise sense of being the coefficient of relatedness as just defined.) In a noninbred, diploid population, r is equal to $2f$; however, with inbreeding, the two measures of relatedness diverge, and r becomes less than $2f$. Numerous methods for estimating r from molecular data have been proposed for a variety of markers (e.g., Queller & Goodnight 1989; Lynch & Ritland 1999; Wang 2002). The performances of microsatellite-based estimators have received the most study and have been shown to vary with the number of microsatellite loci employed, the numbers and frequency distributions of alleles at each locus, and the composition of relationship categories present in a population (Queller & Goodnight 1989; Ritland 1996; Lynch & Ritland 1999; Van de Casteele et al. 2001; Wang 2002; Milligan 2003). An individual's relationship to itself is always $r = 1$, and this measure of relatedness contains no information about the level of inbreeding. For that reason, and because r is not directly proportional to gene diversity (see next section), r is generally less useful for pedigree analyses and captive population management than is f.

Gene diversity

Gene diversity (G) is a common measure of genetic variation that can be calculated from both pedigrees and molecular data. G is defined as the heterozygosity expected in a random mating population, and it reflects both number of alleles and evenness of allele frequencies. When calculated from molecular data (e.g., microsatellites), the G of a single locus is

$$G = 1 - \sum p_i^2,$$

where p_i is the frequency of allele i, and the summation is over all alleles at a locus (Nei 1973). Multiple, single-locus estimates of G can be averaged to provide a measure of genome-wide variation. When calculated from a pedigree,

$$G = 1 - \bar{\bar{f}},$$

in which $\bar{\bar{f}}$ is the average $\bar{f}$ in the population – that is, the mean of all pair-wise kinships. An important distinction between molecular- and pedigree-based measures of G is that molecular estimates represent the heterozygosity of alleles that differ by state (i.e., in molecular structure), and pedigree-based measures represent the probability that the two alleles at a locus are not identical by descent from a common ancestor in the pedigree.

In a randomly mating population, an individual's $\bar{f}$ is the expected inbreeding coefficient of that individual's offspring. By extension, $\bar{\bar{f}}$ is the expected mean inbreeding coefficient of all offspring. Thus, $\bar{\bar{f}}$ and $\sum p_i^2$ are conceptually equivalent because they both represent the average probability that an individual is homozygous at a given genetic locus. It is important to note, however, that the meanings of "homozygous" associated with the two concepts are not identical because one refers to identity by descent (pedigree-based measure) and one refers to identity of state (molecular measure). The difference in meaning between molecular- and pedigree-based measures of G vanishes when diversity is expressed as a proportion of that in the defined baseline, or reference, population (e.g., the population founders). The proportional loss of molecular homozygosity due to accumulating kinship within a breeding population, relative to the reference population, is expected to be $\bar{\bar{f}}$. The G of the reference population in pedigree analyses is defined to be 1.0 because founders are assumed to be noninbred and to share no alleles that are identical by descent. Thus, the proportional change in heterozygosity (i.e., the decay through generations due to identity by descent) is expected to be the same whether measured by allele frequencies or pedigree-based kinships.

Gene diversity is more than just a convenient metric for quantifying the amount of genetic variation within a population. Gene diversity is proportional to the additive genetic variance in traits controlled by those loci and therefore is proportional to the expected rate of response to selection (Falconer & Mackay 1996). Thus, the loss of gene diversity is both a measure of the accumulated inbreeding that can depress fitness of individuals and a measure of the loss of the population's potential for future adaptive evolution.

Applying concepts to incomplete pedigrees

Accurate, pedigree-based calculations of kinship, inbreeding, and gene diversity are possible only when pedigrees are completely known. Many captive populations, however, have missing or questionable parentage records that create ancestry gaps that result in incomplete pedigrees. To facilitate the calculation of genetic parameters in these situations, Ballou and Lacy (1995) developed algorithms for calculating f, $\bar{f}$, and $\bar{\bar{f}}$ from only the fully known lineages within each individual's pedigree. These algorithms use the proportion of an individual's genome that can be traced to known founders (k). Individuals with completely known ancestry have a k of 1 and individuals with no known ancestry (i.e., two unknown parents) have a k of 0. For any other individual,

$$k_x = \frac{(k_a + k_b)}{2},$$

where a and b are the parents of individual x.

The kinship between individuals x and y (f'_{xy}) is the probability that an allele sampled from the known portion of y's genome is identical by descent to an allele sampled from among x's known maternal alleles, multiplied by the probability that a known allele sampled from x is maternally derived, plus the probability that the allele sampled from y is identical by descent to an allele sampled from among the known paternal alleles in x multiplied by the probability that a known allele sampled from x is paternally derived. Thus,

$$f'_{xy} = \left(f'_{my} \times \frac{k_m}{(k_m + k_p)} \right) + \left(f'_{py} \times \frac{k_p}{(k_m + k_p)} \right),$$

where m and p, respectively, refer to the dam and sire of individual x. The value of f'_{xy} is undefined if individual x has no known ancestry. The kinship of an individual to itself is the probability that two alleles sampled with replacement from the known portion of the individual's genome are both the maternal allele, plus the probability that the two sampled alleles are both the paternal allele, plus the probability that one of the sampled alleles is maternal and the other is paternal multiplied by the probability that the maternal and paternal alleles are identical by descent. Thus,

$$f'_{xx} = \left[\frac{k_m}{k_m + k_p} \right]^2 + \left[\frac{k_p}{k_m + k_p} \right]^2 + \left[2 \times \left(\frac{k_m}{k_m + k_p} \times \frac{k_p}{k_m + k} \right) \times f'_{mp} \right].$$

To calculate $\overline{f}$ and $\overline{\overline{f}}$ from incomplete pedigrees,

$$\overline{f}_x = \frac{\sum\limits_{y=1}^{N} k_y \times f'_{xy}}{\sum\limits_{y=1}^{N} k_y} \quad \text{and} \quad \overline{\overline{f}} = \frac{\sum\limits_{x=1}^{N} \sum\limits_{y=1}^{N} \left(k_x \times k_y \times f'_{xy} \right)}{\sum\limits_{x=1}^{N} \sum\limits_{y=1}^{N} \left(k_y \times k_x \right)}.$$

Although these algorithms currently provide the only option for conducting pedigree analyses on incomplete pedigrees, short of either presuming that any undocumented parents were new, unrelated founders or excluding from analysis all animals with incompletely documented ancestries, their use is suboptimal because they produce values that can be either larger or smaller than the true values that would be calculated if the pedigree was completely known. Still, it is important to note that the algorithms are unbiased as they assume that the probabilities of identity by descent are the same for alleles descended from unknown parts of the pedigree as they are for those alleles descended through traceable lineages.

AN INTRODUCTION TO CAPTIVE BREEDING PROGRAMS

The basic goal of most captive breeding programs is to maintain demographically self-sustaining populations that are genetically representative of their wild counterparts. Thus, in essence, captive breeding programs strive to prevent the evolution of captive populations away from the wild gene pool. This prevention is accomplished through careful genetic management that aims to retain genetic diversity, restrict inbreeding, and limit adaptation to captivity. Captive breeding

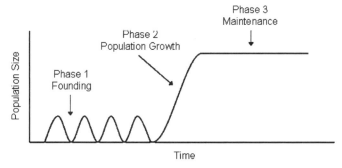

Figure 11–1. Three demographic phases experienced by a typical captive breeding program.

programs generally experience three demographic phases (Ballou & Foose 1996): founding, growth, and maintenance (Fig. 11–1).

The first phase of any captive breeding program is the founding of a captive population. Founders should genetically represent, as closely as possible, the species or population for which captive management is being initiated. Thus, the gene pool of the founders would ideally capture all possible alleles, at the same frequencies and linkage combinations observed in the wild. A common recommendation is that a captive population should have at least twenty unrelated founders, which would retain 97.5% of the gene diversity of the wild population from which they were randomly sampled (Fig. 11–2; Foose et al. 1986; Soule et al. 1986; Lacy 1994). For example, the recent population management guidelines for establishing captive assurance populations of threatened amphibians makes this explicit recommendation (Schad 2008), but then appropriately notes that usually more, and sometimes many more, than twenty individuals will need to be captured from the wild to assure that at least twenty successfully breed to establish the captive population.

Although founders are ideally unrelated, it is generally difficult if not impossible to ascertain relationships among wild-caught animals. Lacking such information, captive population founders are nearly always assumed to be unrelated. In other words, the founders are declared to be the baseline population for future kinship calculations. This means that captive breeding programs, at their best, retain the genetic variation that was present in the founders, with often considerable uncertainty regarding how well those founders represented the wild population(s) from which they came.

After captive populations are founded, the second phase of most captive breeding programs is population growth. The goal is to increase a newly founded population to a demographically self-sustaining size, while retaining both the allelic diversity and gene diversity (i.e., expected heterozygosity) captured by the founders. Captive populations should be grown as quickly as possible because slower growth generally increases the likelihood that founders will die before contributing sufficient offspring to the breeding program. The probability that a given founder allele fails to be passed to a founder's offspring, and is therefore lost from a population, is equal to $(0.5)^n$, where n is the total number of offspring produced by the founder. Thus, for a given founder allele to be retained with a 99% probability, a founder must produce at least seven offspring (Fig. 11–3).

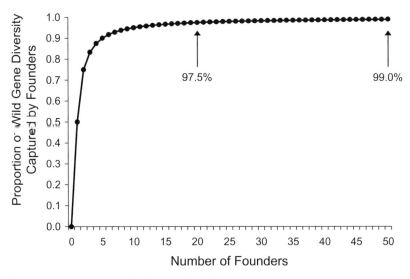

Figure 11–2. Proportion of wild gene diversity (expected heterozygosity) captured in population founders. Founders are assumed to be randomly selected from the wild population, and the proportion of gene diversity captured is calculated according to $H_f = H_w \times [1 - 1/(2N)]$, where H_f is the heterozygosity of the founders, H_w is the heterozygosity of the wild population, and N is the number of founders. It is commonly recommended that captive breeding programs secure at least twenty unrelated founders to retain approximately 97.5% of wild gene diversity.

Alleles are transmitted within linkage groups, rather than independently, however. Given the number of linkage groups in a typical vertebrate genome, a founder would need to produce at least twelve offspring to assure that all of its alleles were passed to its offspring with a 99% probability (Thompson 1995). Although the probability that a given founder's alleles will be retained increases as it produces more offspring, gene diversity will be maximized if all founders produce equal numbers of progeny. Thus, the two primary guidelines for captive population growth generally are to 1) grow the population as quickly as possible, and 2) have founders equally contribute to population growth.

The final phase of most captive breeding programs is population maintenance, during which a captive population is maintained at a carrying capacity that is usually dictated by the maximum size that can be supported by the resources (e.g., cage space) allocated by the institutions cooperating in the breeding program (Earnhardt et al. 2001). This allocation of resources is, in turn, often driven by the calculated number of breeders needed to maintain an acceptably low rate of gene diversity loss (Ballou & Foose 1996). For example, the target carrying capacity for okapis (*Okapia johnstoni*) in North American zoos is two hundred animals (Petric & Long 2008). The target was set much higher, to 480 animals (Ballou & Mickelberg 2009), for golden lion tamarins (*L. rosalia*) because the species has a shorter generation length and therefore will lose genetic diversity more rapidly over time. At the capacity stage, after the desired population size has been reached, refined genetic management becomes a priority. The best methods for retaining genetic diversity, while still limiting inbreeding, are those that minimize the average kinship in a population (Ballou & Lacy 1995; Fernandez & Toro 1999; Sonesson & Meuwissen 2001). Methods for minimizing average kinship preferentially breed

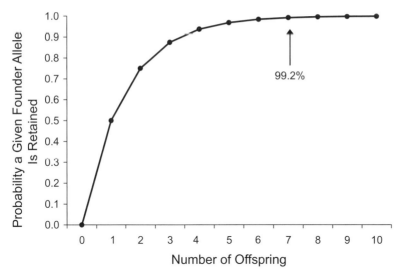

Figure 11–3. Probability that a unique founder allele is retained in a captive population as a function of the number of offspring the founder produces. For a given founder allele to be retained among the progeny with a 99% probability, a founder must produce at least seven offspring.

individuals that have the lowest $\overline{f}$, thereby equalizing founder contributions in the living population and retaining the genetic variation initially captured in the founders. By maintaining captive populations that genetically resemble their wild counterparts, captive breeding programs also aim to limit adaptation to captivity. Adaptation to captivity is a complex concept, however, because it can involve multiple genetic loci influencing multiple quantitative traits, which in turn can be subject to genotype-by-environment interactions. Furthermore, the manner in which a species may adapt to captivity is hard to predict, and the genes that may contribute to adaptation are difficult to identify.

Although the effectiveness would be difficult to evaluate, preferentially breeding those animals with the lowest $\overline{f}$ will slow adaptation to captivity. When breeding individuals are selected by $\overline{f}$, they are selected without regard for phenotype. Moreover, families with previously low survival or breeding success will be selected first for breeding, thereby working to equalize family sizes and countering, to the extent possible, natural selection for traits that might be adapted while in captivity. The equalization of family sizes is expected to halve the rate of adaptation to captivity because selection is restricted to differential reproduction among siblings within families as selection between families is removed (Haldane 1924; King 1965; Frankham & Loebel 1992; Allendorf 1993; Lande 1995; Frankham et al. 2000). Alleles that prevent survival or reproduction in the captive environment will still be removed by selection, however, and family sizes will be difficult to equalize if reproduction and/or survival in captivity is poor. Breeding pairs often produce fewer or more offspring than desired, and culling excess to equalize family sizes is not always an acceptable option for endangered species and is not an efficient use of limited resources for population maintenance. Selection of breeding pairs by mean kinship is an efficient method to equalize family sizes to the extent possible, including preferentially breeding lineages that are under-represented due to poorer-than-desired breeding in prior generations, so that through the generations, the representation of lineages is held as close to equal as possible. Although breeding-pair selection schemes that use $\bar{f}$ likely reduce adaptation to captivity, little quantitative information is available regarding the extent to which real breeding programs can approximate an equalization of family sizes, the strength of remaining within-family selection for adaptations to the captive environment, or the rate of response to such selection pressures.

METHODS FOR INCORPORATING MOLECULAR DATA INTO CAPTIVE BREEDING PROGRAMS

Resolving pedigree gaps and identifying pedigree errors

Sources of pedigree errors and types of pedigree gaps

A fundamental way that molecular data can increase the effectiveness of captive breeding programs is to improve the accuracy and completeness of pedigrees. Captive breeding programs rely on pedigree analyses (f and $\bar{f}$ calculations) to select breeding pairs that will retain genetic diversity and limit inbreeding. Thus, the true effectiveness of genetic management is dependent on the accuracies of the pedigrees on which those analyses are based. Pedigree inaccuracies can arise through human error if, for example, an offspring's sire or dam is incorrectly recorded (e.g., Tzika et al. 2009). Errors also can be introduced into a pedigree by the behavior of the animals themselves. For example, the females of some bird species are known to "dump" eggs into nests of other females (Yom-Tov 1980; Petrie & Moller 1991; Zink 2000). If multiple breeding pairs of the same species are housed together and egg dumping occurs, a female may be erroneously assumed to be the mother of all the offspring in her nest. Identifying and resolving errors improves the accuracy of kinship calculations, which increases the true effectiveness of genetic management. Furthermore, the frequency of pedigree errors is currently rarely assessed, and reports of gene diversity and inbreeding levels in captive populations usually presume that pedigrees are recorded without error. This presumption is particularly problematic because those reports are often used to assess program success, so establishing pedigree accuracy also improves our ability to evaluate the status of captive breeding programs.

In addition to improving pedigree accuracy, molecular data can be used to resolve pedigree gaps (e.g., Field et al. 1998; Zhang et al. 2005; Ivy et al. 2009). Gaps in a pedigree can arise for a variety of reasons. Before captive populations were recognized as important resources for conservation, historical parentage records were inconsistently maintained because few captive populations received regular genetic management. Thus, many contemporary captive populations have gaps deep in their pedigrees because they were initially founded during a time when accurate parentage records held little value. Captive breeding programs now strive to maintain complete pedigrees, but cases of unknown parentage still continue to arise in some situations. For example, many species are housed in groups that contain multiple males. If all males in a group are capable of siring offspring, the paternities of offspring born to the group are inherently uncertain. Pedigree gaps also can occur if a captive breeding program is composed of only a portion of the population maintained in captivity. In those situations, animals may move into and out of the managed population for a variety of reasons. If parentage records are not maintained for the portion of the population that does not belong to the captive breeding program, the ancestries of animals entering the captive breeding program from the unmanaged portion of the population are unclear. Those individuals cannot be accurately placed in the captive breeding program's pedigree because neither their parentage nor their deeper ancestries are known.

Although gaps can be distributed throughout a pedigree, they often are characterized as being either historical or contemporary (Ivy et al. 2009). Contemporary gaps occur in the recent pedigree and generally impact a limited number of individuals. For example, if a living animal with unknown parentage has not yet produced offspring, only the relationships between that individual and the rest of the population will be uncertain. Historical gaps occur deep in a pedigree and can impact a significant proportion of the living population if the animal with unknown parentage has many descendants (Fig. 11–4). For example, consider a population in which 10% of living individuals are descended from an individual with unknown parentage. Even if the sires and dams of all of the living individuals are known, the relationships between all of those individuals and the rest of the population will be uncertain due to the unknown ancestry deeper in their pedigrees. This problem is compounded as the number of historical pedigree gaps increases and higher proportions of living descendants have unknown ancestry. In a sense, all captive populations have deep historical gaps because the founders of the pedigree were assumed to be unrelated. Thus, pedigree calculations yield

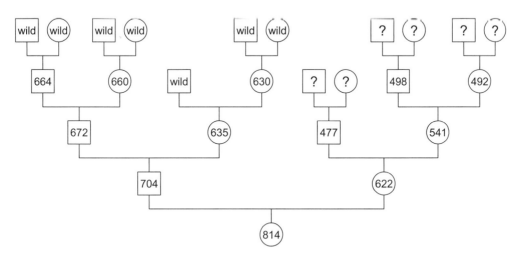

Figure 11–4. Examples of historical gaps in a captive parma wallaby (*Macropus parma*) pedigree. Squares in the pedigree represent sires, circles represent dams, individuals with "wild" parents were population founders, and individuals with "?" were born in captivity to undocumented parents. Individual 814 was living in the population as of January 1, 2007. Accurate kinships between that individual and all other individuals in the population could not be calculated because 814 had three ancestors (individuals 498, 492, and 477) with unknown parentage. Ancestor 498 had 45 additional living descendants in the population as of January 1, 2007, further hindering accurate kinship calculations for 29% of the total living population ($n = 157$).

kinship and inbreeding estimates based solely on identity by descent from common ancestors within the captive population.

Using molecular data to identify errors and resolve gaps

Numerous types of molecular markers can be used to resolve gaps and ascertain the accuracy of pedigrees. Some common markers used for parentage analyses (see DeWoody 2005 for an overview) include restriction fragment length polymorphisms (RFLPs), mitochondrial DNA (mtDNA) haplotypes, and single nucleotide polymorphisms (SNPs). Each marker has benefits and drawbacks, and each marker type has unique characteristics that must be considered before it can be effectively applied to parentage analyses. One of the most common markers used to investigate parentage is single sequence repeats (SSRs). SSRs, or microsatellites, are biparentally inherited markers that are generally assumed to be selectively neutral. They are a popular choice for parentage analyses because, when compared to other molecular markers, microsatellite data are relatively easy to collect and only a few, suitably variable loci are usually required to ascertain parentage.

Molecular data can be used to establish parentage through either exclusion or assignment. For parentage exclusion, the genetic profiles of putative parents and offspring are compared to identify incompatibilities (Fig. 11–5). A putative parent is rejected as the sire or dam of an offspring if, at a given locus, the putative parent and offspring share no alleles. The probability that a given set of molecular loci will correctly exclude a potential parent is based on both the number of alleles and allele frequencies exhibited by the assayed loci (Selvin 1980). Exclusion probabilities rise as numbers of loci increase, numbers of alleles per locus increase, and allele frequencies become more uniform. Many captive populations exhibit low genetic variation because they were founded by small numbers of individuals. This problem is often exacerbated in breeding programs for endangered species (i.e., those of greatest conservation concern) because the small, wild populations from which founders were obtained often also had low genetic variability, fewer founders were initially available to establish the captive breeding program, and it is difficult or impossible to add more founders to augment genetic variation as time goes on. Thus, on average, a greater number of molecular markers may be needed for effective parentage exclusion in captive populations than in wild populations. Although parentage exclusion can be effective at ascertaining parentage, it does have limitations. For parentage exclusion to be maximally effective, genetic profiles must be available for all potential parents. Furthermore, suites of molecular loci with low exclusion probabilities may exclude only some of all possible parents.

If parentage exclusion identifies two or more genetically plausible parents when all possible parents have been sampled, molecular data can be used to assign the most likely parent to an offspring. A number of different likelihood-based methods of parentage assignment have been proposed, but categorical allocation is usually the most appropriate for captive populations because it assigns an entire offspring to a single sire or dam (fractional allocation assigns a portion of an offspring to each potential parent; Jones & Ardren 2003). Parentage assignment can be quite effective in some situations, but there is always some uncertainty associated with assigned parentage. Parentage assignment identifies the most

ID	SEX	MICROSATELLITE LOCI												
		MP03		Pa595		Me14		Me17		Y175		G31-1		G19-1	
840	M	200	240	340	346	156	176	139	139	209	209	110	110	105	105
771	M	200	240	320	346	156	176	133	139	275	289	118	118	195	201
774	F	200	200	340	346	156	176	139	139	289	291	118	118	195	201
772	M	200	208	346	346	156	176	139	139	275	287	118	120	169	201
773	F	200	236	320	346	156	156	133	139	275	289	118	122	171	201
775	F	200	236	340	346	156	156	139	139	289	289	122	122	195	201
776	F	204	204	346	346	156	176	139	139	275	291	118	118	169	197

Figure 11–5. Example of parentage exclusion using microsatellites (data from Ivy et al. 2009). The parents of offspring 840 are 771 and 774; all other potential parents have been excluded at one or more loci. Microsatellite genotypes that are inconsistent with the parentage of 840 are highlighted in gray.

likely parent of an offspring, but any potential parent that cannot be genetically excluded has some probability of being the actual parent of the offspring. Thus, if all possible parents have been sampled, and parentage exclusion identifies two or more genetically plausible parents, captive breeding programs should attempt to increase their exclusion probabilities by adding additional marker loci before resorting to parentage assignment.

Regardless of the type of analysis employed, the effectiveness with which molecular data are able to ascertain parentage is dependent on sample availability. Genetic samples are rarely available for animals that are no longer living in a captive population because the banking of samples from all animals is not routine. Still, programs like the San Diego Zoo's "Frozen Zoo" (Benirschke 1984) provide access to many more genetic samples than were available in the past. As the number of genetic samples that are banked continues to increase, pedigree gaps and uncertain parentage will become easier to investigate. Computer simulations have suggested that for captive breeding programs that aim to minimize average kinship, recent ancestry has a greater influence than deeper ancestry on the retention of genetic diversity and the accumulation of inbreeding (Rudnick & Lacy 2008). Consequently, if a captive breeding program has a significant number of gaps spread throughout its pedigree, resolving even only the contemporary gaps can still significantly improve genetic management.

Estimating relatedness between individuals with unknown ancestry

Improving the completeness and accuracy of pedigrees can positively impact the management of captive breeding programs, but some captive pedigrees are riddled with historic gaps that cannot be directly resolved. If a significant proportion of a living population has uncertain ancestry due to irresolvable pedigree gaps, molecular data can be used to quantify relationships by estimating r for pairs of individuals. Captive breeding programs select breeding pairs based on $\overline{f}$, which are calculated from a matrix that contains the f for each possible pair of individuals living in a population (Fig. 11–6). When pedigree gaps are present, each f is calculated from only the known portions of pedigrees (Ballou & Lacy 1995). The accuracy and value of these estimates decrease, however, as less of an animal's ancestry is available for the calculations, and no estimate is possible if neither parent is known. Molecular estimates of r can be used to modify or replace f in

ID	1	2	3	4	5	$\bar{f}$
1	0.50	0.03	0.02	0.00	0.25	0.16
2	0.03	0.50	0.03	0.00	0.03	0.12
3	0.02	0.03	0.53	0.00	0.05	0.13
4	0.00	0.00	0.00	0.50	0.00	0.10
5	0.25	0.03	0.05	0.00	0.50	0.17

Figure 11–6. Kinship (f) matrix for five individuals, with resultant mean kinship values ($\bar{f}$) for each individual listed on the right. Individual identification numbers (IDs) are given across the top and down the left side of the matrix. An individual's f to itself is shaded in gray, which helps to illustrate that the matrix is symmetric. Individual 4 is not related to any other individuals in the matrix and, thus, has the lowest $\bar{f}$. Individual 3 is slightly inbred as its f to itself is greater than 0.5 (i.e., the value expected for noninbred individuals). Individual 5 has the highest average relationship to the rest of the individuals in the matrix, with $\bar{f} = 0.17$.

calculations, but genetic management is improved only if the incorporation of molecular estimates of relatedness increases the accuracy of those calculations.

Directly replacing f with an estimate of r derived from molecular data

Both r and f are measures of the ancestry shared by two individuals; thus, estimates of r could be used to replace estimates of f in the matrix used for mean kinship calculations. In the absence of inbreeding, the resulting matrix of r would be twice the matrix of f, so assessments of the proportional value of breeders would be the same. The most significant challenge to directly using molecular estimates of r in these calculations is that the estimates are notoriously inaccurate and suffer from large sampling variances (Csillery et al. 2006 and references therein). For example, unrelated individuals are expected to exhibit an r of 0.0000 in a noninbred, diploid population. When using a suite of ten microsatellite loci that each exhibit ten alleles at equal frequencies, however, the mean and standard deviation of r for 10,000 simulated pairs of unrelated individuals is 0.0002 ± 0.1069 (Fig. 11–7, a). The r distributions of full-siblings and half-siblings also overlap in this simulation (Fig. 11–7, a), further demonstrating the inaccuracy of r estimates that are based on the modest number of markers that are typically readily available for endangered species.

The accuracies of r estimators have been shown to be affected by the number of microsatellite loci employed, the numbers and frequency distributions of alleles at each locus, the composition of relationship categories present in a population, and the reference population from which allele frequencies are calculated (Queller & Goodnight 1989; Ritland 1996; Lynch & Ritland 1999; Van de Casteele et al. 2001; Wang 2002; Milligan 2003; Bink et al. 2008). Standard suites of microsatellites generally range from five to twenty loci, which are highly unlikely to produce estimates of r that are precise enough to produce calculated values of mean kinships that are sufficient to guide selection of the best breeding pairs. In the event that a suite of microsatellites is determined to produce sufficiently precise estimates of r, values of r and f still cannot be directly substituted for each other due to differences in their definitions. In a noninbred, diploid population, f can be replaced by $r/2$, but this equality fails to hold as inbreeding accumulates in a population. Additionally, whereas calculations of f from pedigrees range from

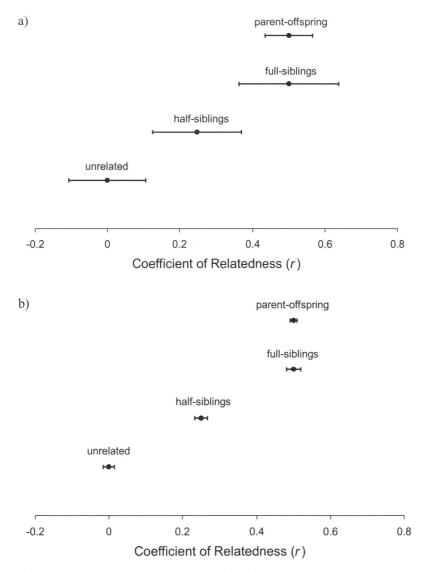

Figure 11–7. Means and standard deviations for four relationship categories, calculated from (a) 10 microsatellite loci and (b) 500 microsatellite loci. For each relationship category, 10,000 simulated pairs were used to calculate means and standard deviations. The microsatellite genotypes for all simulated individuals were drawn from loci that exhibited ten alleles at equal frequencies.

0 to 1, estimates of r based on molecular data can be negative because a pair of individuals can share fewer alleles even than what would be expected by chance if they were unrelated ($r = 0$). Yet, in the context of pedigree relationships, individuals cannot be less related than having no common ancestors. Computationally, negative values could be incorporated into $\overline{f}$ calculations, but it can be difficult to interpret negative r values in the context of f.

There is another reason to exercise caution when using molecular estimates of r to guide captive breeding programs. Conceptually, there is a subtle difference in the genetic goals being pursued by pedigree-based and molecularly based breeding strategies; pedigree-based breeding strategies (i.e., mean kinship breeding schemes) aim to keep the genetic contributions of all founders as constant as possible through subsequent generations, whereas breeding strategies that use empirical estimates of relatedness as the basis for selecting pairs preferentially breed those individuals with the most distinctive array of alleles at the sampled loci (Jones et al. 2002). In the former case, the breeding program will seek to minimize homozygosity due to common descent (autozygosity) and maintain allele frequencies across all loci as closely as possible to the frequency distribution in the wild-caught founders. In the latter case, the breeding scheme will seek to minimize homozygosity due to both 1) identity by descent, and 2) identity of state from independent origins (allozygosity), and to maintain allele frequencies at the sampled loci as evenly as possible. If large numbers of unlinked, fully neutral loci are used for the estimation of relatedness, however, then the extent of allozygosity among pair-wise comparisons should be relatively constant and estimates of r should be proportional to the true but perhaps not otherwise knowable pedigree relationships.

Estimates of r from molecular markers are often strongly influenced by the alleles that are present in only one or a few individuals, and individuals carrying rare alleles will often be accorded preference in breeding, either because of their lower estimated kinships to the population or even specifically as a means to retain rare alleles that otherwise would be at high risk of being lost from the population. It is important to note, however, that the conservation goal of captive breeding programs is rarely to preserve the specific alleles present at the sampled loci but rather to preserve diversity across the entire genome to retain, as closely as possible, the characteristics of the wild population in the captive stock. Moreover, if one or more of the sampled loci are linked to (or are themselves) loci under selection, then individuals with unique alleles might be carrying rare, deleterious alleles. Preferentially breeding these individuals will increase heterozygosity at the sampled loci, but it will do so by shifting allele frequencies away from those of the founders, possibly increasing the frequency of alleles that were under negative selection in the wild population. More work is needed on the likelihood of linkage of commonly used markers (e.g., microsatellites) to loci under selection before we can know whether rarer alleles are simply more informative about relationships, as is commonly assumed, or also sometimes reflect the consequences of past natural selection.

As microsatellite characterization technologies improve and microsatellite data become less costly to collect, it may one day be possible to use hundreds of microsatellites to calculate r for some species. Even with a suite of 500 unlinked microsatellite loci that each exhibit 10 equally frequent alleles, r estimates for a range of relatedness categories still exhibit standard deviations near 0.01, however (Fig. 11–7, b). It is currently unclear what level of accuracy is needed for r estimates to be beneficially incorporated into $\bar{f}$ calculations. Thus, additional research is necessary to identify the most likely scenarios for which estimates of r could be directly used to improve the genetic management of captive populations.

Using molecular estimates of r to identify relationship categories

When typical suites of molecular markers are employed, relatedness estimators have been shown to be sufficiently accurate for distinguishing first-order relatives from unrelated individuals (Piper & Rabenold 1992; Blouin et al. 1996; Glaubitz et al. 2003). Thus, rather than using estimates of r directly in mean kinship calculations, a more useful way of incorporating r into captive population management might be to use them for identifying degrees of relationship (Ivy et al. 2009). If pairs of individuals with little or no known ancestry could be assigned to relationship categories, inaccurate estimates of f in the matrix used for $\bar{f}$ calculations could be replaced with new values to represent specific degrees of relationship. For example, if an r estimate suggested that two individuals were first-order relatives, their f might be set to 0.25 (the theoretical value expected for that relationship).

For a given suite of microsatellites, empirically determined allele frequencies can be used to simulate pairs of individuals that represent various relationship categories. Means and standard deviations can then be calculated across hundreds or thousands of pairs to describe the expected distributions of r values representing each type of relationship (Fig. 11–7), and each pair of individuals can be assigned a degree of relationship based on the distribution into which their r falls. When the distributions of two relationship categories overlap, the midpoint between the means can be used as a cutoff for assigning a pair to one category or the other (Blouin et al. 1996). For example, consider the overlapping full-sibling and half-sibling r distributions illustrated in Fig. 11–7, a. The midpoint between the distribution means is 0.3741. Thus, for pairs that exhibit r values between the distribution means, pairs with r values greater than the cutoff would be classified as full-siblings and pairs with r values less than the cutoff would be assigned as half-siblings or to lesser degrees of relatedness.

Estimates of r theoretically can be used to discriminate among numerous relationship categories. The sampling variances associated with a given suite of molecular markers, however, limit the level of discrimination possible. As the number of potential relationship categories to be considered increases, more distribution overlap will be observed, and it will be harder to conclusively place a pair of individuals into a given relationship category. Furthermore, relationship categories of the same order often have similar means and distributions (e.g., full-siblings and parent–offspring pairs both exhibit a mean r of 0.5; Fig. 11–7). Because $\bar{f}$ calculations use f values rather than specific relationship categories, it is more important to correctly identify the order of relationship than the actual type of relationship that exists between two individuals. In other words, if r suggests that two individuals are either siblings or a parent and offspring, identifying the correct relationship category is less important than identifying that the individuals are first-order relatives with an expected r of 0.5. Standard suites of ten to twenty microsatellites should often be sufficient for distinguishing among first-order relatives, second-order relatives, and unrelated individuals with acceptable confidence, but differentiating among more distant relationship categories is likely beyond the capabilities of most molecular studies.

The suitability of a given suite of molecular markers for distinguishing among relationship categories can be evaluated by a number of methods. One option is to calculate misclassification rates for a range of relationship categories (Blouin et al. 1996). After r distributions have been generated, misclassification rates can be calculated by simulating individuals of a given relationship category and quantifying the proportion of those individuals that are incorrectly assigned to an alternate category. A second option for assessing the suitability of molecular markers for distinguishing among relationship categories is to estimate the proportion of variance explained in the marker-based relatedness by true relatedness (Van de Casteele et al. 2001). For a given relationship category, the proportion of variance explained by true relatedness can be calculated from the sampling variance associated with the distribution of r values for that category. It is currently unclear what misclassification rates and proportions of variance explained by true relatedness are needed for r estimates to be beneficially incorporated into mean kinship calculations. It is likely, however, that different levels of accuracy of estimated r will be acceptable for different pedigrees that exhibit variable amounts of unknown ancestry. For example, more accurate estimates of r may be required if little unknown ancestry is present and the pedigree-based values of f are already fairly accurate, whereas less accurate estimates of r might be acceptable if the amount of unknown ancestry in a pedigree is so large as to prohibit meaningful kinship calculations from pedigree information alone.

Molecular markers can be powerful tools for inferring relationship categories, but inconsistencies in category assignment can arise when multiple pairs of individuals are considered simultaneously. For example, it is possible for r estimates to suggest that an individual is closely related to one of a known sibling pair but unrelated to the other sibling. Although there are no mathematical reasons why these relationship category assignments cannot be incorporated into $\bar{f}$ calculations, results such as these are difficult to interpret in the context of a pedigree. Because these types of situations can arise for empirical estimates of r, it is important that the assignments of relationship categories be carefully considered both against each other and in conjunction with any ancillary pedigree information that might be available. Even if one or both parents of an individual are unknown, there may still be information available about the birth date, birth location, and possible parents. This information can be used to either support or refute relationships suggested by molecular estimates of r (Gautschi et al. 2003; Ivy et al. 2009).

Although the focus of the discussion has been on using molecular data to improve captive breeding programs for endangered species, in which the genetic goal is to minimize genetic change through generations, the methods we discuss for using molecular data also apply to programs that use artificial selection to improve aspects of performance of economic gain. Box 11 by El-Kassaby describes the use of molecular data to identify categories of kinship in a population of trees, for which direct tracking of pedigree relationships through generations would be difficult and require many years.

Using molecular data to manage breeding programs for organisms living in groups

Pedigree analyses and the genetic management of breeding programs have historically been focused on species that can be placed in pairs for breeding, so that pedigree records on parentage can be maintained and pairings can be controlled. As described earlier in this chapter, pedigree information is often incomplete

BOX 11: PEDIGREE RECONSTRUCTION: AN ALTERNATIVE TO SYSTEMATIC BREEDING

Yousry A. El-Kassaby

Problem

Forest tree breeding programs follow the classical recurrent selection scheme starting with phenotypic selection, followed by breeding and testing. Ultimately, the cycle is completed by genotypic selection of elite individuals for either starting a new breeding cycle or establishing seed orchards for the production of genetically improved seed for reforestation. Trees' long life cycle, delayed reproductive maturity, and geographically extensive and protracted testing make breeding a daunting task. Increasing programs' efficiency by either eliminating the breeding phase or simplifying testing would be of immense value. Here, we illustrate the combined use of DNA fingerprinting and pedigree reconstruction to assemble mating designs (half- and full-sibling [HS and FS] families) from natural pollination and to demonstrate the method's utility in simplifying progeny testing.

Case Study

The combined use of DNA fingerprinting and partial pedigree reconstruction is employed to assemble a mating design using 551 wind-pollinated western larch (*Larix occidentalis* Nutt) seedlings representing fourteen seed-donors from a forty-one-parent seed orchard (Funda et al. 2008). The assembled mating design encompassed 221 FS families nested within the maternal (mean: 15.8; range: 12–23) and paternal (mean: 5.4; range: 0–14) HS families (Box Fig. 11–1). This

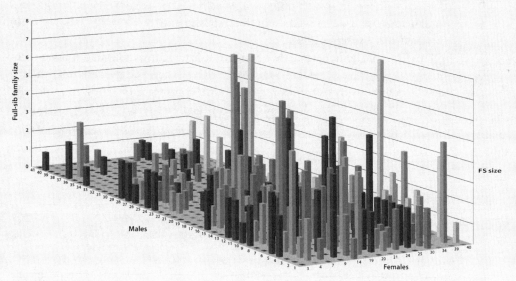

Box Figure 11–1. Three-dimensional diagram showing the formation of FS within HS families.

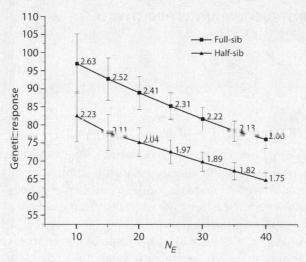

Box Figure 11–2. Rate of genetic response with its corresponding effective population size for FS and HS selection methods.

mating design, with its multiple crosses across the mating landscape, offers better genetic sampling than those produced from traditional mating designs, which restrict crossing within independent subsets of parents.

Progeny testing can be simplified if it is reduced to HS families using wind-pollinated seed and if FS families are assembled using DNA fingerprinting and pedigree reconstruction (Box Fig. 11–1). A conventional Douglas-fir (*Pseudo-tsuga menziesii* [Mirb.] Franco) FS progeny test (60 parents producing 150 FS and 50 HS families) was treated as HS, and pedigree reconstruction was implemented to assemble FS and genetic gain from the two approaches (conventional FS versus reconstructed FS from HS) was compared. The small genetic gain difference between the FS (100%) and HS (85%) testing (Box Fig. 11–2) highlights the efficiency of the proposed approach, specifically at effective population sizes similar to those used in seed-orchard establishment (Box Fig. 11–2; $> N_E = 20$) (El-Kassaby & Lstiburek 2008).

This example demonstrates the benefits of integrating molecular marker technology with advanced pedigree-reconstruction models and modern quantitative genetics methods in transforming traditional forest tree breeding to molecular and genomics-based breeding. Additionally, pedigree reconstruction has important implications in understanding the extent of genetic diversity and relatedness among members of natural and domesticated populations. This understanding is an essential component to their effective utilization and conservation.

REFERENCES

El-Kassaby YA, Lstibůrek M (2008) Breeding without breeding. *Genetics Research*, **91**, 111–120.

Funda T, Chen C, Liewlaksaneeyanawin C, Kenawy AMA, El-Kassaby YA (2008) Pedigree and mating system analyses in a western larch (*Larix occidentalis* Nutt.) experimental population. *Annals of Forest Science*, **65**, 750.

for a variety of reasons. Thus, molecular data can be a powerful tool for providing missing relationship information so that captive breeding programs can be more effective at minimizing genetic change (see Chapter 10 by Waples and colleagues). For many species that require captive breeding programs for their conservation, however, genetic management problems go beyond simple gaps in the pedigree or the uncertain ancestry of some individuals. The requirements of a species and constraints on breeding facilities often necessitate that breeding populations be maintained in large groups with multiple males and females with no possibility for controlling pairings, often little opportunity to observe parentage, and sometimes limited ability to identify and monitor individuals through their lifetimes. Examples of species that are maintained and bred within groups include antelope species that are adapted to living in herds, bats, penguins, flamingos, and other species that breed within colonies and often need the stimulation or protection of a colony before they will breed, and small species such as many frogs, insects, snails, and fishes for which it is usually impractical to manage as individuals or pairs maintained in separate enclosures. Increasingly, it is being recognized that species that are difficult or impossible to manage with traditional pedigree-based methods include many that are threatened with extinction. For example, some of the highest levels of endangerment and extinction are among the freshwater fishes, terrestrial and freshwater mollusks, coral reef inhabitants, and amphibians (Baillie et al. 2004; IUCN 2008).

Due to the difficulties inherent in trying to manage group-living species as individuals, some breeding programs rely on breeding schemes that manage movements of individuals among breeding groups, rather than on trying to control each pairing, to maximize retention of genetic diversity within and among groups. Methods such as maximum avoidance of inbreeding (Kimura & Crow 1963; Princée 1995) rotate one sex or the other through breeding groups to delay inbreeding and equalize group contributions to future generations. Such regular systems of breeding can be difficult to sustain for many species of wildlife, however, because successful reproduction is often not sufficiently reliable to prevent the failure of some groups. Still, if the survival and reproduction of groups and the transfers among groups can be carefully monitored, then the within- and between-group gene diversity, and pair-wise estimates of genetic divergence, can be used to determine the optimal transfers that retain overall genetic variation while avoiding excessive within-group inbreeding (Wang 2003, 2004).

When species must be managed as groups, molecular data can be used to monitor and enhance breeding schemes that are based on the transfer of individuals (e.g., maximum avoidance of inbreeding). The rate of gene diversity loss from groups and divergence among groups can be empirically measured to determine if breeding programs are performing as desired. This information can be particularly informative because, given the lack of control over the pattern of breeding within most groups (e.g., all the breeding could be done by one pair per group or could be spread among multiple pairs), the actual structure of genetic variation among groups could diverge substantially from that expected under theoretical models.

Molecular estimates of gene diversity also could be used in addition to or in place of models based on group histories to guide animal transfers among groups. The observed gene diversity of an overall population can be partitioned into between- (G_b) and within- (G_w) subpopulation components (Nei 1973), and the concept of kinship, defined as the probability that alleles sampled from two entities are autozygous, can be applied to between-group relationships (G_b) as well as relationships among pairs of individuals (f and r). Mathematically and conceptually, the gene diversity between entities (when scaled so that complete lack of alleles shared by common descent is assigned $G = 1$) is the same as the converse of kinship ($1 - f$) between those entities, whether the entities are individuals, breeding groups, or larger populations. Thus, in direct parallel to traditional pedigree management focused on individuals, crossing between groups with lowest kinship (highest G_b) will result in a minimization of inbreeding ($1 - G_w$), and preferentially propagating the groups that have lowest mean kinship (highest mean G_b to the array of groups) will maximally retain overall gene diversity and slow the response to natural selection by countering the between-group component of selection.

Just as the precisions of r estimates between individuals are strongly dependent on the number of loci sampled, the precision of the parallel estimates of relatedness between groups is dependent on the number of individuals (and the number of loci) sampled. Because the genotypes of multiple individuals from each group are used to estimate the relatedness between groups, those estimates will be more precise than those of relatedness between diploid individuals. Therefore, whereas the precision of estimates of genetic measures based on partial pedigree information and the usefulness of management based on pedigree calculations degrades as the level of management moves from individuals to groups, the precision of empirical estimates of population structure and the potential usefulness of molecular data to guide program management increases. Work is needed to explore the relative costs and benefits (in terms of reaching genetic goals as well as resource costs) of more intensive individual- and pedigree-based management versus coarser, group level, and empirically guided management. We are unaware of any breeding programs for group-living endangered species that are currently being guided by a molecular characterization of genetic structure each generation, but such methods may become more useful and efficient as it becomes easier, in terms of both time and money, to score large numbers of marker loci for a wider variety of species.

Identifying individuals with rare or important alleles

It has been suggested that molecular data could be used to augment captive breeding programs by characterizing genotypes at genetic loci known to be important to fitness, then managing breeding programs to maximize the retention of valuable alleles. There are two variants on this approach that are worth considering. First, we could select for animals carrying alleles believed to be especially important, such as variants at the major histocompatibility complex loci (Hedrick 2002; Hughes 1991). Second, we could measure variation at random loci, and then preferentially breed those animals that appear to carry the rarest alleles. In theory, this approach could produce a population with even more gene diversity than was present in the wild population by creating more equal allele frequencies than existed in the source population.

Although these ideas deserve more evaluation, we would caution, as have others (Vrijenhoek & Leberg 1991), that there are some potential drawbacks. First, we know only few of the many loci that might be critical to individual fitness and population viability. If we select breeding pairs on the basis of those few loci about which we do know something, we are likely to cause rapid depletion of genetic variability at other loci that may be just as important (Lacy 2000; Hedrick 2001). Because the alleles that are advantageous will depend on the environment in which the animals live, many alleles that encode adaptations important in natural environments may be neutral or even deleterious in a specific captive environment.

A strategy of preferentially breeding animals that have the rarest alleles, without trying to prejudge which alleles will be most advantageous, has perhaps more merit than attempts to select the animal with superior alleles. Even this strategy has risks, however. Initially, rare alleles may have been rare for a good reason. Selecting for them may increase frequencies of mutations that were deleterious in natural populations. We may be on safer ground if we use strategies that attempt to minimize the rate at which the populations in captivity diverge genetically from the genotypic composition of the wild populations from which they came, rather than trying to improve upon the results of the prior evolution in wild populations. If the divergence at sampled loci is indicative of genetic divergence across the genome as a whole, identifying individuals that carry unique or rare alleles at presumably neutral loci (e.g., microsatellites) also could be used as a means of selecting priority breeders that may be more likely to carry rare alleles at other loci throughout the genome. This selection of breeders is exactly the aim of incorporating molecular estimates of relatedness into pedigree analyses previously discussed, so there is little to be gained by otherwise or further targeting rare alleles for augmentation in a conservation breeding program.

Unresolved needs and unexploited opportunities

This chapter has focused on plausible methods for incorporating empirical estimates of relatedness into captive breeding programs. Areas that require further research have been mentioned throughout, but here we summarize the most critical research needs. There has been a moderate amount of work on how well genetic goals can be achieved by using various schemes for selecting priority breeding pairs (e.g., Willis 1993; Haig et al. 1994; Ballou & Lacy 1995; Willis 2001; Jones et al. 2002; Ivy et al. 2009). Much more work is needed, however, to evaluate the effectiveness of alternative methods for preserving genetic variation in captive populations, especially under various conditions of species biology, varying information availability, and rigor of management. At this time, managers of captive populations of endangered species have little guidance on which strategies to employ, particularly when it comes to incorporating molecular data into genetic management.

Incorporating molecular estimates of r *into genetic management*
The accuracies of *r* estimates need to be further explored to identify how exact *r* estimates must be to improve rather than degrade genetic management. Thus,

before estimates of *r* are incorporated into mean kinship calculations, appropriate precisions for those estimates must be defined. If estimates of *r* are to be used to place pairs of individuals into relationship categories, acceptable misclassification rates also must be identified. It is likely that different levels of *r*-estimate accuracy will be acceptable for different pedigrees, so a range of pedigrees that exhibit variable amounts of unknown ancestry should be considered when investigating these questions.

Assessing the role of selection

Almost all methods used to manage captive breeding programs were developed from genetic theory that presumes that genetic variation is neutral. The variation that is important to population persistence is, by definition, under selection (at least during some times), however. It is possible that calculations based on neutral theory provide accurate enough projections of genetic changes in captive populations, especially if the captive environment removes many of the selective pressures that sculpted genomes in the wild. Models that include plausible selective forces need to be examined, however. More needs to be known about the role of natural selection in captive populations (Frankham et al. 1986), with respect to both useful adaptation and removal of deleterious alleles (Arnold 1995; Lacy & Ballou 1998) and possibly harmful consequences for long-term population fitness and preservation of species characteristics (e.g., Frankham & Loebel 1992; Bryant & Reed 1999; Frankham et al. 2000). We need more information on the strength of natural selection in captive populations, the nature of the alleles favored or eliminated, and the rate of response under different breeding conditions.

Resources for quantitative genetic analysis

There is a largely untapped potential for captive populations of wild species to serve as a resource for studies of the genetic control and evolution of quantitative traits (Arnold 1995). Quantitative genetic partitioning of variance in traits has long been an important tool in identifying the genetic variation available for manipulation in domesticated species and in elucidating the past evolutionary forces in wild species (Falconer & Mackay 1996; Roff 1997; Lynch & Walsh 1998). Traditionally, such methods depended on specific crosses or correlations among classes of relatives, so most work has involved domesticated species or laboratory strains. It has been assumed that studies on such experimental populations that are removed from their wild ancestors by many generations of selective breeding will still reveal information about the structure of genetic variation in quantitative traits. Yet, thousands of species of wildlife are propagated in zoological parks, and some have pedigrees extending back five or more generations. Therefore, the raw material exists for quantitative genetic studies not just on guinea pigs, *Mus*, and *Drosophila*, but also on many other species representing most orders of mammals, birds, reptiles, and amphibians.

The pedigrees from most wildlife captive breeding programs are complex and, after the first generation or two, contain few cases of simple relationships (e.g., full-siblings and half-siblings) that are not also confounded by other paths of common ancestry. Therefore, many pedigrees from conservation breeding programs could provide only limited data for methods that depend on analyses of resemblance among discrete classes of relatives. With the development, however, of estimation procedures for using complex pedigrees to estimate variance components (Kruuk & Hill 2008), the pedigrees of zoo populations become a potentially vast source of data for such studies (Pelletier et al. 2009). These methods have been applied increasingly to wild populations, but the depth of pedigrees in such studies is rarely to the extent available from many captive populations. When such analyses depend on and presume accurate pedigrees, molecular genetic tools provide the means to fill in gaps, resolve uncertainties, and correct errors. As discussed earlier in the text, however, the power of molecular data to reconstruct pedigrees with confidence is often limited (Ivy et al. 2009). The mostly complete pedigrees of many captive populations may provide a more useful starting point for such work than the largely unknown pedigrees of wild populations, which often must be constructed entirely from molecular data.

"Animal models" that use matrices of relatedness estimates but do not require reconstructed pedigrees are also now available (Frentiu et al. 2008), and molecular genetic data on samples from captive populations could be a valuable resource for such work. When populations with complete pedigrees are also extensively sampled for molecular genetic characterization, they would provide the opportunity to directly compare pedigree-based and pedigree-free animal models for estimating components of genetic variance (see Chapter 6 by Nichols & Neale).

SUMMARY

The most significant limitation of using pedigree-based analyses to demographically and genetically manage captive populations is that management can be severely hindered by inaccurate or incomplete pedigrees. Molecular data can improve the genetic management of captive breeding programs when pedigree-based management is ineffective by helping to meet the standard genetic goals of retaining gene diversity and limiting the accumulation of inbreeding. The most basic use for molecular data in captive breeding programs is to resolve pedigree errors and gaps. For pedigrees with deep, unknown ancestries, empirical estimates of r also can be used to calculate kinships or to identify relationship categories for pairs of living individuals. Molecular data also could be used to guide the genetic management of group-living species through the empirical characterization of within- and between-group genetic structure, although we are unaware of any breeding programs that have yet attempted to do this. Although it is clear that there are many methods for incorporating molecular data into the management of captive breeding programs, additional research is still needed in a variety of areas to further clarify which strategies for incorporation are the most effective. Furthermore, it also should be noted that for the many species that can be maintained and propagated in tightly controlled paired breeding, the recording of parentage information, pedigree calculations, and identification of optimal breeding pairs is simple, precise, and inexpensive, so it is unlikely that it will be effective and efficient in those cases to replace traditional pedigree methods with programs guided by molecular analysis.

REFERENCES

Allendorf FW (1993) Delay of adaptation to captive breeding by equalizing family size. *Conservation Biology*, **7**, 416–419.

Arnold SJ (1995) Monitoring quantitative genetic variation and evolution in captive populations. In: *Population Management for Survival and Recovery: Analytical Methods and Strategies in Small Population Conservation* (eds. Ballou JD, Gilpin M, Foose TJ), pp. 293–317. Columbia University Press, New York.

Baillie JEM, Hilton-Taylor C, Stuart SN, editors (2004) *IUCN Red List of Threatened Species. A Global Species Assessment*. IUCN, Gland, Switzerland, and Cambridge, UK.

Ballou JD, Foose TJ (1996) Demographic and genetic management of captive populations. In: *Wild Mammals in Captivity* (eds. Kleiman DG, Lumpkin S, Allen M, Harris H, Thompson K), pp. 263–283. University of Chicago Press, Chicago.

Ballou JD, Lacy RC (1995) Identifying genetically important individuals for management of genetic diversity in pedigreed populations. In: *Population Management for Survival and Recovery: Analytical Methods and Strategies in Small Population Conservation* (eds. Ballou JD, Gilpin M, Foose TJ), pp. 76–111. Columbia University Press, New York.

Ballou JD, Mickelberg J (2009) *International Population Management Plan for Ex-situ Golden Lion Tamarins*. National Zoological Park, Washington, DC.

Benirschke K (1984) The frozen zoo concept. *Zoo Biology*, **3**, 325–328.

Bink MCAM, Anderson AD, van de Weg WE, Thompson EA (2008) Comparison of marker-based pairwise relatedness estimators on a pedigreed plant population. *Theoretical and Applied Genetics*, **117**, 843–855.

Blouin MS, Parsons M, Lacille V, Lotz S (1996) Use of microsatellite loci to classify individuals by relatedness. *Molecular Ecology*, **5**, 393–401.

Bryant EH, Reed DH (1999) Fitness decline under relaxed selection in captive populations. *Conservation Biology*, **13**, 665–669.

Butchart SHM, Stattersfield AJ, Baillie J et al. (2005) Using Red List Indices to measure progress towards the 2010 target and beyond. *Philosophical Transactions of the Royal Society of London-Series B: Biological Sciences*, **360**, 255–268.

Cosson L, Grassman LL Jr, Zubaid A et al. (2007) Genetic diversity of captive binturongs (*Arctictis binturong*, Viverridae, Carnivora): implications for conservation. *Journal of Zoology*, **271**, 386–395.

Csillery K, Johnson T, Beraldi D et al. (2006) Performance of marker-based relatedness estimators in natural populations of outbred vertebrates. *Genetics*, **173**, 2091–2101.

DeWoody JA (2005) Molecular approaches to the study of parentage, relatedness, and fitness: practical applications for wild animals. *Journal of Wildlife Management*, **69**, 1400–1418.

Earnhardt JM, Thompson SD, Marhevsky E (2001) Interactions of target population size, population parameters, and program management of viability of captive populations. *Zoo Biology*, **20**, 169–183.

Ely JJ, Dye B, Frels WI et al. (2005) Subspecies composition and founder contribution of the captive U.S. chimpanzee (*Pan troglodytes*) population. *American Journal of Primatology*, **67**, 223–241.

Falconer DS, Mackay TFC (1996) *Introduction to Quantitative Genetics*, 4th edn. Pearson/Prentice, Harlow, UK.

Fernandez J, Toro MA (1999) The use of mathematical programming to control inbreeding in selection schemes. *Journal of Animal Breeding and Genetics*, **116**, 447–466.

Field D, Chemnick L, Robbins M, Garner K, Ryder OA (1998) Paternity determination in captive lowland gorillas and orangutans and wild mountain gorillas by microsatellite analysis. *Primates*, **39**, 199–209.

Foose TJ, Lande R, Flesness NR, Rabb G, Read B (1986) Propagation plans. *Zoo Biology*, **5**, 139–146.

Frankham R, Hemmer H, Ryder OA et al. (1986) Selection in captive environments. *Zoo Biology*, **5**, 127–138.

Frankham R, Loebel DA (1992) Modeling problems in conservation genetics using captive *Drosophila* populations: rapid genetic adaptation to captivity. *Zoo Biology*, **11**, 333–342.

Frankham R, Manning H, Margan SH, Briscoe DA (2000) Does equalisation of family sizes reduce genetic adaptation to captivity? *Animal Conservation*, **3**, 357–363.

Frentiu FD, Clegg SM, Chittock J et al. (2008) Pedigree free animal models: the relatedness matrix reloaded. *Proceedings of the Royal Society of London-Series B: Biological Sciences*, **275**, 639–648.

Gascon C, Collins JP, Moore RD et al. (2007) *Amphibian Conservation Action Plan*. IUCN/SSC Amphibian Specialist Group. Gland, Switzerland.

Gautschi B, Jacob G, Negro JJ et al. (2003) Analysis of relatedness and determination of the source of founders in the captive bearded vulture, *Gypaetus barbatus*, population. *Conservation Genetics*, **4**, 479–490.

Glaubitz JC, Rhodes EO Jr, DeWoody JA (2003) Prospects for inferring pairwise relationships with single nucleotide polymorphisms. *Molecular Ecology*, **12**, 1039–1047.

Haig SM, Ballou JD, Casna NJ (1994) Identification of kin structure among Guam rail founders: a comparison of pedigrees and DNA profiles. *Molecular Ecology*, **3**, 109–119.

Haldane JBS (1924) A mathematical theory of natural and artificial selection. Part I. *Transactions of the Cambridge Philosophical Society*, **23**, 10–41.

Hedrick PW (2001) Conservation genetics: where are we now? *Trends in Ecology and Evolution*, **16**, 629–636.

Hedrick PW (2002) The importance of the major histocompatibility complex in declining populations. In: *Reproduction and Integrated Conservation Science* (eds. Wildt DE, Holt B). Cambridge University Press, Cambridge.

Hedrick PW, Miller PS, Geffen E, Wayne R (1997) Genetic evaluation of the three captive Mexican wolf lineages. *Zoo Biology*, **16**, 47–69.

Hu J, Pan H, Wan Q, Fang S (2007) Nuclear DNA microsatellite analysis of genetic diversity in captive populations of Chinese water deer. *Small Ruminant Research*, **67**, 252–256.

Hughes AL (1991) MHC polymorphism and the design of captive breeding programs. *Conservation Biology*, **5**, 249–251.

International Union for Conservation of Nature (IUCN) (2008) Available at http://www.iucnredlist.org.

Ivy JA, Miller A, Lacy RC, DeWoody JA (2009) Methods and prospects for using molecular data in captive breeding programs: an empirical example using parma wallabies (*Macropus parma*). *Journal of Heredity*, **100**, 441–454.

Iyengar A, Gilbert T, Woodfine T et al. (2007) Remnants of ancient genetic diversity preserved within captive groups of scimitar-horned oryx (*Oryx dammah*). *Molecular Ecology*, **16**, 2436–2449.

Jones AG, Ardren WR (2003) Methods of parentage analysis in natural populations. *Molecular Ecology*, **12**, 2511–2523.

Jones KL, Glenn TC, Lacy RC et al. (2002) Refining the whooping crane studbook by incorporating microsatellite DNA and leg banding analyses. *Conservation Biology*, **16**, 789–799.

Kimura M, Crow JF (1963) On the maximum avoidance of inbreeding. *Genetical Research*, **4**, 399–415.

King JL (1965) The effect of litter culling – or family planning – on the rate of natural selection. *Genetics*, **51**, 425–429.

Kruuk LEB, Hill WG (2008) Introduction. Evolutionary dynamics of wild populations: the use of long-term pedigree data. *Proceedings of the Royal Society of London-Series B: Biological Sciences*, **275**, 593–596.

Lacy RC (1994) Managing genetic diversity in captive populations of animals. In: *Restoration of Endangered Species* (eds. Bowles ML, Whelan CJ), pp. 63–89. Cambridge University Press, Cambridge.

Lacy RC (1995) Clarification of genetic terms and their use in the management of captive populations. *Zoo Biology*, **14**, 565–577.

Lacy RC (2000) Should we select genetic alleles in our conservation breeding programs? *Zoo Biology*, **19**, 279–282.

Lacy RC, Ballou JD (1998) Effectiveness of selection in reducing the genetic load in populations of *Peromyscus polionotus* during generations of inbreeding. *Evolution*, **52**, 900–909.

Lande R (1995) Breeding plans for small populations based on the dynamics of quantitative genetic variance. In: *Population Management for Survival and Recovery: Analytical Methods and Strategies in Small Population Conservation* (eds. Ballou JD, Gilpin M, Foose TJ), pp. 318–340. Columbia University Press, New York.

Lynch M, Ritland K (1999) Estimation of pairwise relatedness with molecular markers. *Genetics*, **152**, 1753–1766.

Lynch M, Walsh B (1998) *Genetic analysis of quantitative traits*. Sinauer, Sunderland, MA.

Milligan BG (2003) Maximum-likelihood estimation of relatedness. *Genetics*, **163**, 1153–1167.

Nei M (1973) Analysis of gene diversity in subdivided populations. *Proceedings of the National Academy of Sciences USA*, **70**, 3321–3323.

Norton JE, Ashley MV (2004) Genetic variability and population differentiation in captive Baird's tapirs (*Tapirus bairdii*). *Zoo Biology*, **23**, 521–531.

Pelletier F, Réale D, Watters J, Boakes EH, Garant D (2009) Value of captive populations for quantitative genetics research. *Trends in Ecology and Evolution*, **24**, 263–270.

Petric A, Long S (2008) *Population Analysis and Breeding Plan: Okapi* (Okapia johnstoni) *Species Survival Plan*. Association of Zoos and Aquariums, Silver Spring, MD.

Petrie M, Moller AP (1991) Laying eggs in others' nests: intraspecific brood parasitism in birds. *Trends in Ecology and Evolution*, **6**, 315–320.

Piper WH, Rabenold PP (1992) Use of fragment-sharing estimates from DNA fingerprinting to determine relatedness in a tropical wren. *Molecular Ecology*, **1**, 69–78.

Princée FPG (1995) Overcoming the constraints of social structure and incomplete pedigree data through low-intensity genetic management. In: *Population Management for Survival and Recovery: Analytical Methods and Strategies in Small Population Conservation* (eds. Ballou JD, Gilpin M, Foose TJ), pp. 124–154. Columbia University Press, New York.

Queller DC, Goodnight KF (1989) Estimating relatedness using genetic markers. *Evolution*, **43**, 258–275.

Ramirez O, Altet L, Ensenat C et al. (2006) Genetic assessment of the Iberian wolf *Canis lupus signatus* captive breeding program. *Conservation Genetics*, **7**, 861–878.

Reddy A, Prakash V, Shivaji S (2007) A rapid, non-invasive, PCR-based method for identification of sex of the endangered Old World vultures (white-backed and long-billed vultures) – implications for captive breeding programmes. *Current Science*, **92**, 659–662.

Ritland K (1996) Estimators for pairwise relatedness and individual inbreeding coefficients. *Genetical Research*, **67**, 175–185.

Roff DA (1997) *Evolutionary Quantitative Genetics*. Chapman & Hall, New York.

Rudnick JA, Lacy RC (2008) The impact of assumptions about founder relationships on the effectiveness of captive breeding strategies. *Conservation Genetics*, **9**, 1439–1450.

Ruokonen M, Andersson A, Tegelstrom H (2007) Using historical captive stocks in conservation: the case of the lesser white-fronted goose. *Conservation Genetics*, **8**, 197–207.

Russello MA, Hyseni C, Gibbs JP et al. (2007) Lineage identification of Galapagos tortoises in captivity worldwide. *Animal Conservation*, **10**, 304–311.

Schad K, editor (2008) *Amphibian Population Management Guidelines*. Amphibian Ark Amphibian Population Management Workshop; December 10–11, 2007; San Diego, CA. Amphibian Ark, available at www.amphibianark.org.

Selvin S (1980) Probability of nonpaternity determined by multiple allele codominant systems. *American Journal of Human Genetics*, **32**, 276–278.

Sonesson AK, Meuwissen THE (2001) Minimization of rate of inbreeding for small populations with overlapping generations. *Genetical Research*, **77**, 285–292.

Soule M, Gilpin M, Conway W, Foose T (1986) The millennium ark: how long a voyage, how many staterooms, how many passengers? *Zoo Biology*, **5**, 101–113.

Spinks PQ, Shaffer HB (2007) Conservation phylogenetics of the Asian box turtles (*Geoemydidae, Cuora*): mitochondrial introgression, numts, and inferences from multiple nuclear loci. *Conservation Genetics*, **8**, 641–657.

Thompson EA (1995) Genetic importance and genomic descent. In: *Population Management for Survival and Recovery: Analytical Methods and Strategies in Small Population Conservation* (eds. Ballou JD, Gilpin M, Foose TJ), pp. 76–111. Columbia University Press, New York.

Tzika AC, Remy C, Gibson R, Milinkovitch MC (2009) Molecular genetic analysis of a captive-breeding program: the vulnerable endemic Jamaican yellow boa. *Conservation Genetics*, **10**, 69–77.

Van de Casteele T, Galbusera P, Matthysen E (2001) A comparison of microsatellite-based pairwise relatedness estimators. *Molecular Ecology*, **10**, 1539–1549.

Vrijenhoek RC, Leberg PL (1991) Let's not throw out the baby with the bathwater: a comment on management for MHC diversity in captive populations. *Conservation Biology*, **5**, 252–254.

Wang J (2003) Maximum likelihood estimation of admixture proportions from genetic data. *Genetics*, **164**, 747–765.

Wang, J (2004) Monitoring and managing genetic variation in group breeding populations without individual pedigrees. *Conservation Genetics*, **5**, 813–825.

Wang JL (2002) An estimator for pairwise relatedness using molecular markers. *Genetics*, **160**, 1203–1215.

Willis K (1993) Use of animals with unknown ancestries in scientifically managed breeding programs. *Zoo Biology*, **12**, 161–172.

Willis K (2001) Unpedigreed populations and worst-case scenarios. *Zoo Biology*, **20**, 305–314.

Yom-Tov Y (1980) Intraspecific nest parasitism in birds. *Biological Reviews of the Cambridge Philosophical Society*, **55**, 93–108.

Zhang YP, Ryder OA, Zhao QG et al. (2005) Non-invasive giant panda paternity exclusion. *Zoo Biology*, **13**, 569–573.

Zink AG (2000) The evolution of intraspecific brood parasitism in birds and insects. *American Naturalist*, **155**, 395–405.

12 Wildlife reintroductions: The conceptual development and application of theory

Olin E. Rhodes, Jr., and Emily K. Latch

Throughout the latter part of the nineteenth century and early portion of the twentieth century, there were widespread declines in wildlife species in North America due to unregulated harvest for commercial, regulatory, and private uses as well as dramatic changes in land-use practices (Moulton & Sanderson 1999). Despite recognition of the grave situation facing many of the most popular and common wildlife species, resulting in the initiation of continent-wide conservation and management programs for both game and nongame species, many of these species were critically imperiled by the time such programs were initiated (Mackie 2000). The history of wildlife management and conservation programs in the United States is intrinsically tied to this point in time, in that many of the practices and values that are in existence today stem from the recognition that wildlife resources are not inexhaustible and that active management and protection efforts must be implemented to offset the negative impacts that our species exerts on natural resources.

One of the most effective and widely used tools employed by wildlife management and conservation organizations to recover and redistribute wildlife species within suitable habitats is the practice of species translocation (International Union for Conservation of Nature [IUCN] 1987). Indeed, species-translocation programs targeted at reintroducing or introducing wildlife species to areas of suitable or reclaimed habitat were used extensively in attempts to recover decimated species across North America beginning in the early portion of the twentieth century (Griffith et al. 1989; Fischer & Lindenmayer 2000). Early in the development of these translocation programs, most emphasis was placed on elucidating the ecology of species targeted for recovery; locating potential source populations from which to draw individuals for translocation; identifying areas of habitat suitable for population establishment; and developing viable capture, handling, and veterinary protocols for species targeted for translocation (IUCN 1998). As the logistical considerations of translocation have been optimized for growing numbers of species, the utilization of this tool has grown to encompass not only species recovery and reintroduction programs but also numerous additional applications, including population supplementation, population expansion, and nuisance-animal management (Griffith et al. 1989; Linnell et al. 1997).

The widespread utilization of species translocation as a tool to recover populations of extirpated species in combination with the expanded application of this tool in ongoing wildlife management and conservation programs has resulted in the translocation of tens of thousands of individuals representing hundreds of species throughout North America (Griffith et al. 1989; Wolf et al. 1996). Due to the primary societal focus on game species early in the twentieth century and the leading role that wildlife management agencies played in the development of species-translocation programs in North America, many of the longest running and best known examples of species translocation involve the reintroduction of game species (Maehr et al. 2001). Although numerous examples of such reintroduction programs exist, the majority of reintroduction programs involve terrestrial game species, including large herbivores (primarily Order Artiodactyla; e.g., Thomas & Toweill 1982; Maehr et al. 2001), furbearers (primarily Order Carnivora; e.g., Weckwerth & Wright 1968; Serfass et al. 1998), and upland game birds (primarily Order Galliformes; e.g., Latch et al. 2006a).

Given the timing and overall goals of many of the reintroduction programs that were conducted for game species, it is not surprising that genetic considerations rarely were taken into account as these programs were planned and implemented. Certainly, most of the biologists conducting large reintroduction programs prior to the last few decades were primarily focused on establishing species back into suitable habitats and lacked the training and technological tools that today allow wildlife managers to quickly and efficiently assess levels of genetic diversity available in potential source populations (Morrison 2002). In addition to the lack of genetic resources, many of the species that were reestablished through reintroduction programs at this point in history were severely limited in terms of potential sources from which to draw founding individuals, thus making genetic considerations (at least in the list of priorities for these programs) almost a moot issue for many biologists. The lack of suitable source populations for reintroduction of many game species created a situation that, regardless of genetic considerations, led to a series of logistical problems for reintroduction programs, almost all of which had significant genetic ramifications. For example, the limitation of source populations for many species led to widespread dispersion of individuals among geographically disparate regions, often ignoring evidence of historical and taxonomic isolation prior to the decline of the species or subspecies (Latch et al. 2006a,b). In addition, the limitations imposed by having limited source populations for many species led to the development of reintroduction practices that likely compounded existing genetic problems for many species. Examples of such practices include the use of severely limited numbers of founders for new populations and the use of serial reintroductions, a strategy that uses successful reintroductions as source populations for future reintroductions and commonly mixes individuals from multiple, successfully reintroduced populations as founders for new populations (e.g., Mock et al. 2004; Lambert et al. 2005).

Due to the probabilistic nature of population genetics theory, it seems likely that given the set of biological, demographic, and logistical circumstances associated with the reintroduction of a given species, one should be able to reconstruct the set of events and model the likelihood of the genetic outcomes of those events with some degree of statistical confidence (Sarrazin & Barbault 1996). Although this is undoubtedly true in an ideal situation, where most if not all of the information pertinent to the model is available, in practice, it often is impossible to obtain the information needed to construct such models with any degree of accuracy or precision (Lacy 1987, 2000; Mills & Smouse 1994). Many factors contribute to the difficulties encountered with the process of reconstructing the likely genetic outcomes of reintroduction programs: Some have to do with the specific parameters of the reintroduction events, and some have to do with the stochastic nature of the variables that dictate the genetic processes associated with the event (Hedrick & Miller 1992; Hedrick et al. 1996; Saccheri et al. 1998). From a logistical perspective, reconstruction of reintroduction programs can be a nightmare with a wide range of issues including a complete or partial deficit of details regarding any variables associated with reintroduction events, including variance in the age and sex ratios of the founders among reintroduction sites, variance in the numbers and relatedness of founders used and the numbers of sources represented among reintroduction sites, spatial and temporal variation in the placement of founding individuals from geographically (and presumably genetically) disparate locations, additional supplementations of individuals into previously established populations, reintroductions of individuals into previously undocumented refugial populations, mixing of putative subspecies within the same spatial context, and a general lack of information on the demography and behavior of newly founded populations during the immediate post-reintroduction period (Haig et al. 1990, 1993; Minckley 1995).

Although the problems with reconstructing and modeling genetic parameters associated with reintroduction programs mentioned earlier are daunting, they are, in fact, tractable to some extent because they do provide a discrete set of initial conditions from which to project biological processes into the future while accounting for the known sources of variance in the system (Gilpin & Soule 1986; Lacy 1987, 2000; Hedrick & Miller 1992; Haig et al. 1993; Mills & Smouse 1994). The more difficult challenge is to predict how variation associated with sampling error and stochastic changes in genetic diversity at various stages of a reintroduction program, including the initial sampling of founders and subsequent population establishment, influence the overall genetic parameters of reintroduced populations through their influence on the demographic parameters of those populations. It is this temporal sequence of events associated with species-reintroduction programs, each incurring combinations of deterministic and stochastic sources of variance in gene frequencies, that we address in this chapter. Our goal is to provide an overview of the sources of variation in gene frequencies inherent to each stage of wildlife-reintroduction programs and to associate these sources of variation with the underlying biological phenomena that contribute to the magnitude of genetic changes that occur within reintroduced populations over time. We begin by providing a brief review of the literature pertaining to genetic consequences of species-reintroduction programs, including results from both empirical and theoretical assessments of reintroduced populations. We then define and discuss the temporal sequence of events associated with reintroduction programs that we feel are critical to the maintenance or loss of genetic diversity within reintroduced populations and the individuals that comprise those populations. In addition, we provide information on a variety of biological processes that directly influence each of these events. We conclude with 1) some general recommendations for how the effects of sampling error and stochastic processes can be minimized in future reintroduction programs, and 2) suggestions for future research.

GENETIC CONSEQUENCES OF REINTRODUCTIONS

Population genetic theory predicts that most reintroductions should be accompanied by a decrease in genetic variability in the newly founded population as compared to its source(s). This decrease occurs because the founders carry only a subset of the genetic variability of their source(s) (Nei et al. 1975). Subsequent to the establishment of the new population, changes in genetic variability will be determined by those fundamental processes that govern the evolution of all populations: mutation, gene flow, natural selection, and genetic drift. Most reintroduced populations involve a small number of founders, ensuring that the effect of genetic drift is strong relative to other processes; thus, the evolution of the new population often is stochastic in nature. As long as such populations remain small, drift will continue to play a critical role in the reduction of genetic variation within them. Genetic drift also is expected to enhance divergence of newly established populations from their source populations, an effect that should be accelerated in populations that remain limited in size for many generations (Crow & Kimura 1970; Lacy 1987).

Some of the earliest empirical investigations of the genetic consequences of translocations were by George Gorman and colleagues on *Anolis* lizards in the Caribbean (Taylor & Gorman 1975; Gorman et al. 1978). The authors used electrophoretic techniques to examine levels of genetic diversity in seven introduced populations relative to their source populations. They documented reductions, relative to source populations, in the percentage of polymorphic loci for all seven introduced populations and in heterozygosity for six of seven introduced populations. The results of these studies confirmed theoretical predictions and were consistent with findings from contemporaneous research documenting reduced variability in accidentally introduced populations of *Drosophila* (Prakash et al. 1969; Prakash 1972) and land snails (Selander & Kaufman 1973). During the same time period, significant reductions in genetic diversity also were documented in bottlenecked populations of elephant seals (Bonnell & Selander 1974), wherein the bottleneck is an evolutionary process akin to reintroduction whereby a small number of individuals re-establish a population.

Following these early studies, a wealth of research has confirmed much of what was predicted concerning the genetic consequences of reintroductions; that is, reintroductions typically result in a loss of genetic diversity within newly established populations. Indeed, a large number of empirical studies now have been published that demonstrate significant reductions in genetic diversity in reintroduced populations relative to their sources (Houlden et al. 1996; Fitzsimmons et al. 1997; Broders et al. 1999; Polziehn et al. 2000; Williams et al. 2000; Hedrick et al. 2001; Latch & Rhodes 2005; Ewing et al. 2008). These empirical studies generally indicate that reductions in genetic diversity tend to be more pronounced in reintroduced populations that are established from a limited number of founders (Baker & Moeed 1987; Merila et al. 1996; Williams et al. 2003; Mock et al. 2004) or which grow slowly after establishment (Maudet et al. 2002; Williams et al. 2002; Wisely et al. 2008). Empirical data, however, also demonstrate that even when reintroductions are undertaken with large numbers of founders or experience rapid population growth after establishment (which would be expected to buffer them from losses of genetic diversity to some extent), reintroduced populations

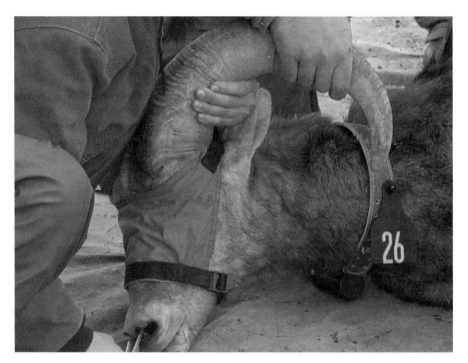

Figure 12–1: Wild captured California bighorn sheep being prepared for transport and reintroduction in Oregon. Now numbering in the thousands, the California bighorn sheep population in Oregon originated from a single transplant of twenty individuals from British Columbia, Canada. Photo courtesy of Zach Olson.

still exhibit measurable losses of genetic variation in many instances (e.g., Stockwell et al. 1996; Mock et al. 2004; Stephen et al. 2005; Witzenberger & Hochkirch 2008).

Both theoretical predictions and empirical data indicate that reductions in allelic richness are a more common (and often more severe) consequence of reintroduction than are losses in heterozygosity as rare alleles are quickly removed from newly established populations (Nei et al. 1975; Allendorf 1986; Baker & Moeed 1987; Leberg 1992; Williams et al. 2000). Similarly, entire allele lineages can be lost at rapid rates, particularly if they are underrepresented in reintroduced populations of mixed ancestry (Gompper et al. 1997; Tordoff & Redig 2001). This phenomenon is particularly common in populations founded from limited numbers of individuals, a situation that will intensify the effects of genetic drift following a reintroduction, often resulting in dramatic shifts in the allele frequency distribution of a reintroduced population relative to its source (Fitzsimmons et al. 1997; Luikart et al. 1998; Rowe et al. 1998; Rhodes et al. 2001; Olech & Perzanowski 2002; Williams et al. 2002; Mock et al. 2004), relative to other native or reintroduced populations (Baker & Moeed 1987; Leberg 1991; Rhodes et al. 1995; Perez et al. 1998; Stephen et al. 2005), or relative to theoretical expectations (Scribner & Stuwe 1994; Fitzsimmons et al. 1997; Figure 12–1). Depending on the numbers and relatedness of founders used and the magnitude of the stochastic processes influencing reintroduced populations, significant genetic divergence from source populations can occur rapidly, over fewer than five generations in some situations (Rowe et al. 1998; Cardoso et al. 2009). At one-fourth the effective population size of the nuclear genome, the mitochondrial genome in particular is at risk during reintroduction events as genetic drift may act rapidly to change haplotypic frequencies in newly reintroduced populations (Stephen et al. 2005).

Not all empirical data support theoretical predictions regarding the loss of genetic diversity in reintroduced populations. For instance, several studies have failed to demonstrate any measurable decrease in genetic variation in reintroduced populations relative to their sources (Merila et al. 1996; Forbes & Boyd 1997; Vernesi et al. 2003; Drauch & Rhodes 2007; Hicks et al. 2007; Eldridge & Killebrew 2008; LeGouar et al. 2008; Vonholdt et al. 2008) or relative to other remnant populations (Rowe et al. 1998; Larson et al. 2002; DeYoung et al. 2003; Williams et al. 2003). Such outcomes often have been attributed to specific attributes of the reintroduction, such as the use of large founding population sizes, mixtures of sources, or multiyear supplementation efforts. Inbreeding avoidance also has been invoked as a mechanism whereby genetic variation could be maintained within reintroduced populations well beyond the time frame predicted by traditional population genetics models (Vonholdt et al. 2008).

In addition, some investigations have demonstrated that although genetic divergence clearly has occurred between reintroduced populations and their sources, the genetic signature (or legacy) of the reintroduction event may persist for many generations (Leberg et al. 1994; Leberg & Ellsworth 1999; Latch & Rhodes 2005). Reasons suggested for the retention of a clear, multigenerational genetic signature from source populations within reintroduced populations include large numbers of founders and/or rapid population expansion (Figure 12–2).

Gene flow among networks of reintroduced populations or into reintroduced populations by naturally recolonizing individuals also has been documented (DeYoung et al. 2003; Latch & Rhodes 2005; Hicks et al. 2007). Provided that such gene-flow events do not represent unfit matings (e.g., hybridization between species or subspecies; Latch et al. 2006a,b), these external genetic contributions can buffer reintroduced populations from losses of genetic diversity. For example, Vernesi and colleagues (2003) noted an atypically high level of genetic variation in a reintroduced population of wild boars. Further investigation revealed that the genetic variation in this population represented a mixture of lineages from both reintroduced and naturally recolonizing individuals. Despite the potential ameliorating effects of gene flow on losses of genetic diversity within reintroduced populations, dispersal events resulting in emigration of individuals also may exacerbate losses of genetic variation by reducing the effective number of breeders. Most species exhibit increased dispersal and limited site fidelity following a translocation event (e.g., Linnell et al. 1997; Tuberville et al. 2005), and this behavior has been correlated with postrelease mortality (Miller et al. 1999).

Perhaps the most challenging aspect of elucidating the processes that contribute to the conservation or loss of genetic diversity associated with reintroduction programs is the fact that the long-term establishment of such populations is the result of a series of interdependent sampling events distributed through time.

Figure 12–2: Eastern subspecies of the wild turkey being reintroduced in northern Indiana. At one time reduced to approximately 30,000 individuals, reintroduction efforts on a massive scale have brought wild turkey populations back to a population size exceeding 5 million in the United States. Photo courtesy of Lee Humberg.

Although the fundamental processes that contribute to changes in gene frequencies during each sampling event are well characterized from a theoretical perspective (e.g., genetic drift, founder effect, bottleneck, inbreeding, and selection), the interactions of these processes within the context of these sampling events are difficult to generalize. This same problem has been recognized within captive breeding programs focused on reintroducing threatened and endangered species to the wild (see Chapter 11 by Ivy & Lacy). For example, Lacy (1994) outlined three distinct phases associated with captive breeding programs during which genetic sampling processes are critical to the maintenance of genetic diversity within populations: the founder phase, the growth phase, and the capacity phase.

Whereas the distinctive nature and intensive management options associated with captive populations make them quite different from reintroductions involving the translocation of wild individuals, the recognition that disparate elements of the reestablishment process can interact over time to influence levels of genetic diversity in reintroduced populations is critical to our efforts to understand how management decisions can dictate genetic outcomes within reintroduction programs. Like captive breeding programs, reintroductions of wild species also incorporate three distinct sampling periods, although the nature of these three phases are fundamentally different in these two management activities (see Chapter 11). For the purposes of this chapter, we define these three phases as 1) the founding event phase (the founder event resulting from the initial sampling of individuals from the source population); 2) the population establishment phase (the bottleneck caused by the initial population establishment); and 3) the population growth phase (the stochastic and deterministic factors influencing long-term population growth).

The study of the temporal components of the wildlife-reintroduction process has been largely absent from studies of the genetic consequences of reintroductions yet likely is the key to reconciling the different outcomes observed in empirical studies. Certainly, these phases of the reintroduction process are difficult to separate, in part because they are largely dependent on one another in both space and time. By breaking down reintroduction into these distinct temporal components and investigating the relative contributions of deterministic and stochastic processes to changes in gene diversity at each phase, we can begin to characterize the relative importance of the demographic and evolutionary parameters that influence the overall reintroduction event. Ultimately, this approach to the study of the genetic consequences of reintroduction programs should yield improvements in our ability to predict the outcomes of competing reintroduction strategies and, thus, to design and implement significantly more efficient reintroduction programs.

THREE MAJOR SAMPLING PERIODS THAT DETERMINE GENETIC EFFECTS

Founding event phase

Species-reintroduction programs begin with an initial sampling event during which founders for newly created populations are trapped and transferred to unoccupied habitats. The efficiency of this initial sampling event, from the perspective of the capture and transfer of genetic information, varies as a function of numerous factors associated with the logistical and biological attributes of the actual capture occurrence. One of the first sources of variation in the extent of genetic information that is captured during a reintroduction event is the number of and levels of genetic diversity in the populations from which founding individuals were selected. Certainly, the genetic characteristics of each potential source population and the level of genetic differentiation exhibited among source populations that are used in a reintroduction program greatly influence the potential quantity of genetic diversity that can be transferred to any newly established population, in the form of diversity sequestered both within and among individuals.

It is interesting that the method and timing of capture of individuals from source populations also can directly influence the levels of genetic diversity sampled from those populations. For example, capture techniques that target individuals when they are nonrandomly distributed either temporally or spatially, relative to biological factors like sex, age, or degree of relatedness, can significantly affect the correlations of genes among individuals in the sample and thus the representativeness of the sample in terms of genetic variation available within the source population (Latch & Rhodes 2006). Capture techniques that take advantage of spatial or temporal associations of individuals stemming from social or family structure are not independent, random samples of the genetic variation present within a source population and thus can bias samples of genetic diversity represented in founding individuals used for reintroductions.

Another important consideration associated with the initial sampling event(s) of any reintroduction program is the number of individuals used to found new populations. Ultimately, the maximum extent of genetic diversity that can be captured at any gene locus is $2N$, where N is the number of founding individuals. Constraints on the numbers of founding individuals that are used to initiate new populations often are encountered and can stem from limitations of an economic and/or logistical nature. For example, the costs of translocating individuals, when weighed against the number of available sites targeted to receive reintroduced individuals, may result in the movement of only a few individuals to many locations rather than the movement of a large number of individuals to one or a few sites. Alternatively, the availability of individuals for capture and transplant to new habitats may be a limiting factor, thus resulting in a simple logistical restriction on the numbers of individuals that can be used to populate each new reintroduction site. Capture myopathy, resulting in the death of individuals during transport or immediately subsequent to release, is another aspect of the initial sampling event of reintroduction programs that may decrease numbers of founders below desirable levels. Capture myopathy likely is elevated in reintroduced populations as a result of increased stress associated with the handling and transport of animals (McArthur et al. 1986; Letty et al. 2000).

The sex and age composition of founding cohorts also can introduce constraints on our ability to sample genetic diversity during reintroductions. The implications of sex ratio, for example, are quite critical in terms of the potential for transfer of genetic diversity in founders to subsequent generations (Ewens 1982; Caballero 1994). The number of females used to initiate a population dictates the potential number of mitochondrial lineages that can be present in the population as well as the potential numbers of offspring that can be generated subsequent to the reintroduction event. From a genetic perspective, it is desirable that founding females not be from the same maternal lineages (to allow for the maximization of mitochondrial types that are transferred into newly established populations) and, in an optimal situation, females would be pregnant at the time of the translocation to allow for the capture of additional genetic diversity from the source population (e.g., extending N beyond just the number of individuals relocated) that otherwise might not be represented in the founding individuals.

The number of males used to found new populations also plays a vital role in the efficiency with which genetic diversity is sampled and transferred to reintroduced populations. For example, numbers of male founders often are small (as compared to females) in reintroduction programs involving polygynous species, whereas sex ratios are more evenly distributed in reintroductions of species known to be monogamous. Although such an approach may seem to be biologically realistic, at least with regard to the mating tactics employed by individual species, little is known about the actual effects of translocation events on the mating structure of most species immediately subsequent to translocation events. From a genetic perspective, it is advantageous to maximize the numbers of males contributing genes in the population as quickly as possible to facilitate the capture of genetic diversity from founding individuals. In the absence of active selection of more genetically variable mates by females, polygynous species in particular may be vulnerable to severe reductions in the effective number of founders if only a few founding males have the opportunity to pass along their genes to future generations.

Age structure also may play a role in the transfer of genetic variation in a reintroduced population. For example, adult-founded populations, in which reproduction occurs immediately after release, may be more affected by mutation accumulation than populations founded by juveniles. This has been hypothesized to occur due to a delay in attainment of reproductive maturity by juveniles, allowing time for selection to purge deleterious recessive alleles, although this phenomenon depends on the numbers of individuals released, generation time, carrying capacity of the habitat, and so forth (Kleiman et al. 1991; Robert et al. 2004b). For species that exhibit high variance in reproductive success as a function of age (e.g., older, socially dominant males securing a vast majority of matings or older females with higher offspring survival), the release of adults may be a more logistically successful strategy (Sigg et al. 2005). Alternatively, the idea that equalizing the age structure of individuals in reintroduction events might lower the variance in reproductive contributions among males over time, particularly for long-lived species, is an especially important concept to consider when planning reintroduction programs.

Population establishment phase

A variety of factors can affect the maintenance and transfer of genetic variation during the initial generations subsequent to a reintroduction event. Newly founded populations are extremely vulnerable to a suite of behavioral, environmental, adaptive, and stochastic processes during the initial postrelease portion of population establishment. The cumulative impact of these processes can reduce genetic diversity within reintroduced populations to levels far below those that might be expected based on the initial event (Williams et al. 2002). In particular, processes that most significantly contribute to rates of loss of genetic diversity during the population establishment phase of reintroductions are those that influence population growth rates and variance in reproductive contributions among individuals.

Most certainly, there is a large stochastic component to the loss of genetic diversity during the initial generations of population establishment associated with any reintroduction. Mortality of founding individuals as they encounter new predators, new diseases, new habitat features, and so forth often is observed within newly reintroduced populations. In turn, variance in reproductive contributions among founding individuals due to stochasticity in breeding opportunities, mortality of offspring, and physiological attributes of individuals also is a prominent feature of newly reintroduced populations. The impact of these stochastic processes on species-specific rates of loss of genetic diversity within reintroduced populations is confounded by a number of factors, such as generation time, numbers of offspring, life span, and age at first reproduction, all of which directly influence the number of opportunities each founder may have for their genes to become incorporated into the population.

From a deterministic perspective, selection coefficients within reintroduced populations also may change in response to new environmental pressures. Thus, natural selection can increase variance in survival and/or reproductive success among individuals in newly established populations, thereby eroding genetic variation more quickly than would be expected if all individuals contributed equally to the next generation (see Box 12 by Worthen et al). Captive breeding programs exemplify the importance of minimizing reproductive variance to the conservation of genetic diversity and use this principle to maximize retention of genetic variability by equalizing the contribution of founder individuals (Ballou & Lacy 1995). In a reintroduced population, where we have no control over mating, natural selection can readily result in a loss of genetic diversity by increasing the variance in reproductive contributions among individuals. The genetic ramifications of natural selection can be particularly influential in small populations, where each founder contains a large proportion of the total genetic variation available for incorporation into future generations.

Environmental factors also play a critical role in determining the initial rates of growth for newly founded populations, influencing factors such as survival, reproductive potential, dispersal, degree of intra- and interspecific competition, and individual health. Clearly, the quality and distribution of essential resources within the environment can limit the reproductive success of individuals and thus the potential growth rates of reintroduced populations. In turn, the distribution of habitat elements within the environment can influence the dispersion of individuals, creating opportunities for intra- and interspecific competition, potentially limiting the survival and/or reproductive success of individuals due to competition for critical resources such as food and cover (see Chapter 9 by Leberg and colleagues). Another aspect of habitat quality, measured both as a function of the availability of resources that are critical to the survival and reproduction of individuals as well as of the actual physical space available for individuals to occupy, is its influence on the movement behavior of individuals, most specifically the probability of dispersal. Dispersal of individuals from newly established populations often can lead to the death of those individuals due to increased risks of mortality and reduced chance of finding habitats that are suitable for occupation (Miller et al. 1999). Clearly, the dispersal of founders or their offspring during the initial generations of a reintroduction can lead to loss of genetic potential within the population. When appropriate, reintroduction programs often combat the hazards associated with dispersal of founders by conducting what are commonly referred to as "soft" releases (Davis 1983; Bright & Morris 1994; Carbyn et al. 1994). In these situations, the founding individuals are kept within enclosures in their new habitats for a period of acclimation, the length of which is dependent on the movement capabilities of the species being

BOX 12: GENETIC RAMIFICATIONS OF RESTORATION
OF BLIGHT-RESISTANT AMERICAN CHESTNUT

Lisa Worthen, Charles H. Michler, and Keith E. Woeste

Reintroduction of a forest tree species presents some challenges unlike those encountered by wildlife ecologists. In the case of chestnut, for example, breeders seek to incorporate specific genetic changes (genes for resistance to chestnut blight) into the genome of the organism that will be reintroduced. These changes will be necessary but not sufficient to assure the survival of the reintroduced individuals. Although the goal of plant introduction is like that of animal introduction – the re-establishment of a healthy, adapted, and self-sustaining population – it remains to be seen if this goal can be achieved in habitats that are permanently altered by the presence of an invasive, exotic disease. Molecular tools offer conservation geneticists the opportunity to monitor the ways in which exotic genes for resistance to chestnut blight will interact with the genome of chestnut over space and time. The dynamic coevolution of both host and pathogen will determine the success of the reintroduction.

Trees are long-lived, perennial organisms, often with large populations and continuous distributions; they present a persistent source of inoculum and opportunity for infection. Thus, trees are particularly susceptible to exotic pests and epiphytotic disease. The most common method for managing plant species threatened by new pests or diseases is backcross breeding. Resistance genes are transferred from a donor species into germplasm of a recipient species through a process of repetitive intermating and selection among the progeny for resistant individuals, which are then deployed to replace or mate with the remaining native population. The American chestnut (*Castanea dentata*) provides a compelling example of management of a tree species in the face of chestnut blight, a catastrophic disease that followed the introduction of the fungus *Cryphonectria parasitica* probably in 1904 (Merkel 1906).

All *Castanea* species are diploid ($2N = 2X = 24$), monoecious, primarily wind pollinated, and self-incompatible (McKay 1942). They freely hybridize with congeners and generally bear three heavy seeds per cupule, although offspring from some combinations suffer from low vigor or male sterility (Jaynes 1975). Because *C. parasitica* usually only kills the aboveground parts of an infected American chestnut, researchers used populations of root-crown collar sprouts to show that American chestnuts maintained high levels of genetic variation, with most variation occurring within populations (Kubisiak & Roberds 2003).

Efforts to breed blight resistance into American chestnuts began in the 1940s, when extirpation of the species was believed imminent, but the program was abandoned when commercial phenotypes were not quickly recovered and blight resistance in F1 hybrids proved insufficient (Clapper 1952; Jaynes 1974). The American Chestnut Foundation (TACF), a nonprofit organization, took up the monumental task and began breeding native American chestnuts to a BC1 [(Chinese × American) × American] chestnut tree from the early U.S. Department of Agriculture breeding program that showed unusually good form and blight resistance (Diller & Clapper 1969). Soon after its inception, TACF organized

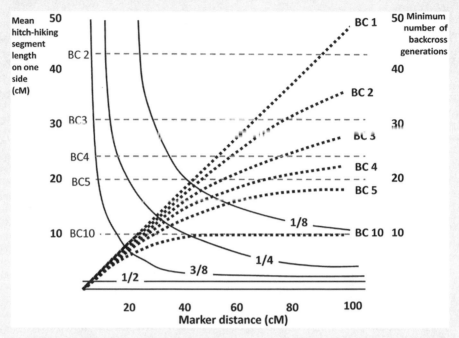

Box Figure 12–1: Expected mean length (cM) of intact donor chromosome (hitch-hiking) on one side of an introgressed locus without marker-assisted selection (gray) or with selection for a recipient genome marker at the indicated distance (cM) for backcross 1–10 (BC1–10; black dotted lines). Number of backcross generations required to reduce donor segment length by the indicated proportion (black solid lines) if a recipient genome marker at the indicated distance (cM) is used to assist selection. Figure adapted from Hospital (2001).

into state chapters that sought surviving American chestnut trees for intermating with resistant crosses with the goal of introducing locally adapted genotypes into each state's breeding population (Burnham et al. 1986).

An important issue for reintroduction genetics associated with catastrophic disease outbreaks is the extent to which strong selection (a selective sweep in which the genes for resistance to *Cryphonectria* become fixed) and linkage drag will result in the retention of large sections of the genome of Chinese chestnut within future American chestnut populations. Population geneticists call this effect "hitch-hiking." Although the length of a hitch-hiking chromosomal segment is expected to be large in early backcross generations (Box Fig. 12–1; Frisch & Melchinger 2001), by screening segregating populations for markers from the recipient genome, the proportion of donor genome retained can be minimized (Hospital 2001). In the near term, the size of the chromosomal segment expected to hitch-hike into American chestnut depends primarily on the selection pressure (s, the relative advantage an allele from Chinese chestnut that confers resistance to *C. parasitica* over the alleles found in American chestnut) and the recombination rate (r), such that hitch-hiking is efficient when $r < s$ (Maynard Smith & Haigh 1974). The extent to which genomic regions from Chinese chestnut (*C. mollissima*, the donor species) will be retained in American chestnut over the long term is difficult to predict. The models for linkage disequilibrium during hitch-hiking are complex, even for single, beneficial alleles (Eriksson et al. 2008), and the breeding program for American chestnut includes the introgression of at least three resistance genes, potentially with incomplete dominance and epistatic interactions. It remains to be seen if gene flow from reintroduced populations of American chestnut will rescue native American chestnut or initiate a separate and evolutionarily distinct lineage.

REFERENCES

Burnham CR, Rutter PA, French DW (1986) Breeding blight-resistant chestnuts. In: *Plant Breeding Reviews* (Vol. 4), pp. 347–397. AVI Publishing Co., Westport, CT.

Clapper RB (1952) Relative blight resistance of some chestnut species and hybrids. *Journal of Forestry*, **50**, 453–455.

Diller JD, Clapper RB (1969) Asiatic and hybrid chestnut trees in the eastern United States. *Journal of Forestry*, **67**, 328–331.

Eriksson A, Fernstrom P, Mehlig B, Sagitov S (2008) An accurate model for genetic hitch-hiking. *Genetics*, **178**, 439–451.

Frisch M, Melchinger AE (2001) Marker-assisted backcrossing for simultaneous introgression of two genes. *Crop Science*, **41**, 1716–1725.

Hospital F (2001) Size of donor chromosome segments around introgressed loci and reduction of linkage drag in marker-assisted backcross programs. *Genetics*, **158**, 1363–1379.

Jaynes RA (1974) *Genetics of chestnut*. U.S. Forest Service, USDA, Washington, DC.

Jaynes RA (1975) Chestnut. In: *Advances in Fruit Breeding* (eds. Janick J, Moore J), pp. 490–503. Purdue University Press, West Lafayette, IN.

Kubisiak TL, Roberds JH (2003) Genetic variation in natural populations of American chestnut. *Journal of the American Chestnut Foundation*, **16**, 42–48.

Maynard Smith J, Haigh J (1974) The hitch-hiking effect of a favorable gene. *Genetical Research*, **23**, 23–35.

McKay JW (1942) Self-sterility in the Chinese chestnut (*Castanea mollissima*). *Proceedings of the American Society of Horticultural Science*, **41**, 156–160.

Merkel HW (1906) A deadly fungus on the American chestnut. *Annual Report of the New York Zoological Society*, **10**, 97–103.

reintroduced. Soft releases have been demonstrated to reduce initial "prospecting" behavior immediately after the reintroduction event (Davis 1983; Bright & Morris 1994).

In addition to the influence of movement behavior on the retention of genetic diversity within newly established populations, a variety of other species-specific behavioral attributes can directly influence the rate of loss of genetic diversity within reintroduced populations. For example, the mating tactics (e.g., polygyny, monogamy, and assortative mating) of some species naturally introduce a large degree of reproductive variance among individuals comprising a newly founded population (Sigg et al. 2005). Polygyny, in particular, is a mating tactic that can severely reduce mating opportunities among males and introduce a large variance in male reproductive contributions to the next generation. Although variance in mating opportunities is often age-related, with older males thought to have more reproductive success than younger males, a variety of outcomes is commonly observed that can ameliorate this variance in reproductive contributions within and among generations. For instance, even in highly structured mating systems with clear age-related reproductive variance, younger males often will have some reproductive opportunities; of course, as individuals age, their reproductive opportunities are expected to increase. Thus, time can serve as a means to equalize reproductive contributions among males over rather than within generations in long-lived species. Unfortunately, the initial few generations subsequent to a reintroduction are critical to the establishment of the genetic foundation of the new population; polygyny can work against a species if many of the founding males do not live long enough or are not allowed to contribute their genes to subsequent generations. Thus, it is particularly important to include adequate numbers of adult male founders in reintroductions involving species that demonstrate strong patterns of age-related variance in male reproductive success (Sigg et al. 2005). Although we do not discuss each individual instance in detail, it should be noted that these same types of considerations also must be taken into account with species that exhibit other reproductive strategies such as monogamy, polyandry, and polygnyandry.

Likewise, social structure can play a large role in the transfer of genetic variation from one generation to the next (Sugg et al. 1996). Social hierarchies leading to nonrandom associations of individuals, resulting in assortative mating of any type, can influence rates of conservation or loss of genetic information within newly created populations. For example, social structure that promotes positive assortative mating may significantly reduce the reproductive contributions of dissimilar pairs of founding individuals while enhancing the contributions of more similar pairs. The effects of such a social system on newly founded populations, which often are composed of individuals thrown together from a variety of social groups, may result in the loss of considerable amounts of genetic diversity from the population in the initial generations. Alternatively, a social structure that promotes negative assortative mating may result in numerous individuals being excluded from mating opportunities if founders exhibit high levels of relatedness, even leading to the forced dispersal of some individuals from the core population. From a spatial perspective, social structures that are associated with various forms of territoriality also may limit breeding opportunities among founding individuals and lead to the loss of genetic information from the population during the initial generations. In contrast to the scenarios just mentioned, a breakdown of social structure leading to hybridization between founding individuals and individuals of other subspecies or species also may have a devastating effect on the genetic composition of a newly created population, ultimately leading to the loss of significant portions of the genetic diversity of the founding individuals and introgression of foreign genetic material into the core population (Storfer 1999).

Population growth phase

Subsequent to the initial few generations, the length of time over which a reintroduced population expands and reaches a local or regional equilibrium is a critical component in determining the overall loss of genetic diversity associated with population establishment. In other words, the length of the bottleneck associated with population growth over time has a strong influence on the proportion of the genetic diversity that is retained within a reintroduced population. There are a number of factors that contribute to the extent and length of bottlenecks in reintroduced populations: biological, environmental, behavioral, and spatial constraints on population growth. The interaction of these factors determines the rate of growth or recovery of population stability within reintroduced populations and, ultimately, the amount of genetic diversity that is retained within these populations over time.

Biological constraints on population growth are related to the ecology of individual species and thus vary widely among reintroduced species. Certainly, the longevity, time to sexual maturation, and generation time of a species are critical to its ability to pass on genetic information to future generations. In turn, the opportunity to sample the genomes of the members of a population depends greatly on numbers of offspring per individual and the survival of those offspring to sexual maturity. For example, short-lived species that mature quickly, exhibit fast generation times, and have large numbers of surviving offspring would be expected to capture and transfer genetic information quite efficiently under ideal conditions – thus allowing the species to rapidly recover in size with little additional loss of genetic diversity (e.g., Sigg et al. 2005; Witzenberger & Hochkirch 2008). Under less than ideal conditions – for instance, when offspring mortality is high – the efficiency of sampling of the genomes of individuals within a reintroduced population is significantly lowered and, due to the short life span of the adult population, accelerated losses of genetic diversity from the population and increased levels of inbreeding would be expected to occur (Jamieson et al. 2007).

Alternatively, in long-lived species that mature slowly and have long generation times and low numbers of offspring, sampling of the genomes of individuals in a population can be highly inefficient and depends on the ability of mature individuals to survive and reproduce over a long enough time span to capture a significant portion of their genetic diversity in subsequent generations (e.g., Drauch & Rhodes 2007). Obviously, the potential for overlapping generations within long-lived species is beneficial to the capture of genetic diversity from individuals, but it depends profoundly on the survival rates of both adults and their offspring. When mortality rates are high for any life-history stage of such species, the rate of loss of genetic diversity and increases in levels of inbreeding in the population can be substantial over time.

Environmental constraints on population growth rates include any external factors that significantly restrict the growth of reintroduced populations and will result in the loss of genetic diversity over time. In particular, newly reintroduced populations must contend with issues pertaining to habitat quality, which in turn dictate the capacity of the available habitat to sustain any population at a given size. Diseases and parasites also may limit the growth of reintroduced populations due to either infection transferred into the population by its founders or the exposure of the reintroduced population to new diseases and parasites to which a species has little or no innate defenses (e.g., Thorne & Williams 1988). Additional factors, such as competition for critical resources or predation, also may play significant roles in limiting population growth in reintroduced populations, especially in situations where populations have been re-established into marginal habitats.

Behavioral constraints, such as those discussed in the previous section on initial population establishment, also may have profound consequences for population growth rates and, thus, for the conservation or loss of genetic diversity over time. Mating systems, social structure, territoriality, and dispersal all contribute to the potential for interactions among individuals and to the ability of populations to overcome the effects of genetic drift. Any behavioral mechanisms that restrict breeding opportunities among individuals or that limit the rate of population growth can accelerate the stochastic loss of genetic diversity over time, especially within a population that was established from a limited pool of founding individuals. Ultimately, the same behavioral mechanisms that proximally influence the loss of genetic diversity in a newly founded population also can ultimately contribute to the gradual loss of genetic diversity from that same population over subsequent generations.

Spatial constraints on population growth rates and the conservation of genetic diversity have both physical and biological aspects. For example, some species may be reintroduced to islands of habitat that have a limited carrying capacity, due to either the actual properties of the habitat (e.g., islands, high altitude habitats) or the influence of outside forces (e.g., poaching, urbanization). In such cases, the actual asymptotic size of a population may be a function of the physical properties of its surroundings and thus may limit population growth in a manner that is detrimental to the maintenance of genetic diversity within the population. Alternatively, reintroduced populations may be one component within a network of spatially proximate populations, the persistence of which is dependent on the connectivity among populations. Although in ideal conditions such a scenario would be expected to slow the rate of loss of genetic diversity within and among populations due to dispersal and gene flow (Polans & Allard 1989; Latch & Rhodes 2005), the loss of populations within the matrix can lead to the collapse of the system and the isolation of various components within the network. Reintroduced populations in particular are vulnerable to the collapse of the larger network of populations due to the fact that the network itself often is composed of a series of reintroduced populations that have been reestablished over a period of years. Both the failure of some of the reintroduced populations in the network and the likelihood that many of the reintroduced populations share much of the same genetic diversity can lead to situations where populations lose genetic diversity over time even in the face of genetic rescue from neighboring populations. Supplemental or progressive releases may help to alleviate some of the losses of genetic variability caused by metapopulation processes (Robert et al. 2004a; Drauch et al. 2008), particularly if founders are ill-adapted to the environment or otherwise prone to failure.

All of the factors discussed in this section can influence the growth rate of a reintroduced population, which, in turn, dictates the length of the population bottleneck that is experienced by the population. In general, the shorter the length of the bottleneck, the more genetic diversity is retained within the reintroduced population as the opportunities for deterministic and stochastic forces to reduce genetic diversity are limited. Of course, population recovery is a continuous variable, and the time to asymptotic population size for reintroduced populations varies widely both within and between species (Bouzat et al.

2009). The temporal component of the population-recovery process, which is directly tied to the number of generations that a population experiences between the initial reintroduction and the failure or stabilization of the population, is governed by many of the biological, environmental, behavioral, and spatial constraints discussed earlier in text and often dictates the level of genetic diversity that a reintroduced population will retain (Jamieson et al. 2007; Witzenberger & Hochkirch 2008).

It is interesting that although the length of the bottleneck alone is likely a good predictor of the loss of genetic diversity that will be experienced by a reintroduced population during the population growth phase, it is not always a good predictor of the fate of the population. For example, a population that recovers quickly to an asymptotic population size may ultimately suffer from inbreeding depression. Rapid growth of a reintroduced population results in the conservation of much of the genetic diversity in the population, including deleterious recessive alleles that contribute to the lethal equivalent load of the population. In such a scenario, the population may experience a decrease in effective population size, be susceptible to the negative effects of inbreeding, and ultimately fail despite the fact that much of the genetic diversity contained in the population subsequent to the reintroduction was retained (although see Jamieson et al. 2007). Alternatively, a population that survives a prolonged bottleneck may have gradually paid the cost of inbreeding and purged itself of the majority of its detrimental genetic burden (although empirical evidence for this phenomenon is limited; Leberg & Firman 2008). Although the population may have lost a substantial portion of its overall genetic diversity, its individuals are able to withstand inbreeding and the population persists, despite the overall loss of variation. Obviously, the scenarios listed earlier in the text represent only two of a multitude of possible outcomes related to the length of the bottleneck associated with a reintroduction, but they clearly demonstrate the complexities associated with this stage of the reintroduction process.

CONCLUSION AND RECOMMENDATIONS

The primary impediment to understanding the genetic effects of reintroductions of wildlife species is not demonstrating that change in gene frequencies has or has not occurred over time, it is understanding the magnitude to which each of the three phases of the reintroduction process has contributed to the conservation or loss of genetic diversity within such populations. Of course, because of our interest in the long-term viability of reintroduced populations, we are most often concerned when there has been a loss of diversity within such populations; and there are a plethora of examples demonstrating losses of genetic diversity in reintroduced populations documented in the literature. Although documented cases of the loss of genetic diversity in reintroduced populations are empirically important and of direct interest to the wildlife managers of these populations, studies demonstrating genetic losses in reintroduced populations do not, by themselves, advance our understanding of the processes that lead to these losses.

What then, are the scientific contributions that have the best chance of informing our approaches to reintroduction programs for wildlife species? We return here to the idea that the solid theoretical foundation of population genetics provides us with a diversity of opportunities to develop testable predictions concerning the genetic consequences of species reintroductions. Using our understanding of the ecology of species and the probabilistic nature of genetic processes, we can devise and implement experimental reintroduction programs that test discrete hypotheses concerning the factors that contribute most significantly to the conservation or loss of genetic diversity within reintroduced populations during their various phases of establishment (see Chapter 10 by Waples and colleagues). Such experimental approaches to the elucidation of genetic processes occurring within reintroduction programs, using theoretical predictions evaluated with empirical data, provide a powerful means of hypothesis testing that can be incorporated directly into the framework of existing or proposed reintroduction events (Sarrazin & Barbault 1996). Alternatively, the development of more refined, flexible, analytical tools and simulation models that allow researchers to reconstruct existing reintroduction programs and evaluate competing scenarios that would produce observed empirical outcomes would be useful in addressing this problem and would allow us to take better advantage of the large quantities of genetic data already collected for a number of reintroduced species.

Undeniably, we need a clearer understanding of how the various phases of population establishment contribute to the conservation or loss of genetic diversity in reintroduced populations. It also is clear that adequate post-translocation monitoring of reintroduced populations is critical to any attempt to obtain this understanding. Without such monitoring it is unlikely that we will progress significantly in our efforts to improve the genetic outcomes associated with reintroduction programs involving wildlife species. In addition, there will be many more contributions to the literature that document the loss of genetic diversity in reintroduced populations, but without consideration of the genetic effects of each of the three phases, it will be difficult to identify plausible explanations for exactly how those losses occurred. With careful consideration of genetic effects at each phase, it is likely that we can improve the genetic success of reintroduction programs, at least for some species, with proper planning and strategies that incorporate mechanisms to ameliorate losses of genetic diversity across all three phases of wildlife reintroduction.

REFERENCES

Allendorf FW (1986) Genetic drift and the loss of alleles versus heterozygosity. *Zoo Biology*, 5, 181–190.

Baker AJ, Moeed A (1987) Rapid genetic differentiation and founder effect in colonizing populations of common mynah (*Acridotheres tristis*). *Evolution*, 41, 525–538.

Ballou JD, Lacy RC (1995) Identifying genetically important individuals for management of genetic diversity in pedigreed populations. In: *Population Management for Survival and Recovery: Analytical Methods and Strategies in Small Population Conservation* (eds. Ballou JD, Gilpin M, Foose TJ), pp. 76–111. Columbia University Press, New York.

Bonnell ML, Selander RK (1974) Elephant seals: genetic variation and near extinction. *Science*, 184, 908–909.

Bouzat JL, Johnson JA, Toepfer JE et al. (2009) Beyond the beneficial effects of translocations as an effective tool for the genetic restoration of isolated populations. *Conservation Genetics*, **10**, 191–201.

Bright PW, Morris PA (1994) Animal translocation for conservation: performances of dormice in relation to release method, origin, and season. *Journal of Applied Ecology*, **31**, 699–708.

Broders HG, Mahoney SP, Montevecchi WA, Davidson WS (1999) Population genetic structure and the effect of founder events on the genetic variability of moose, *Alces alces*, in Canada. *Molecular Ecology*, **8**, 1309–1315.

Caballero A (1994) Developments in the prediction of effective population size. *Heredity*, **73**, 657–679.

Carbyn LN, Armbruster HJ, Mamo C (1994) The swift fox reintroduction program in Canada from 1983 to 1992. In: *Restoration of Endangered Species: Conceptual Issues, Planning, and Implementation* (eds. Bowles ML, Whelan CJ), pp. 247–269. Cambridge University Press, Cambridge.

Cardoso M, Eldridge MDB, Oakwood M et al. (2009) Effects of founder events on the genetic variation of translocated island populations: implications for conservation management of the northern quoll. *Conservation Genetics*, **10**, 1719–1733.

Crow JF, Kimura M (1970). *An Introduction to Population Genetics Theory*. Harper and Row, New York.

Davis MH (1983) Post-release movements of introduced marten. *Journal of Wildlife Management*, **47**, 59–66.

DeYoung RW, Demarais S, Honeycutt RL et al. (2003) Genetic consequences of white-tailed deer (*Odocoileus virginianus*) restoration in Mississippi. *Molecular Ecology*, **12**, 3237–3252.

Drauch AM, Fisher BE, Latch EK, Fike JA, Rhodes OE (2008) Evaluation of a remnant lake sturgeon population's utility as a source for reintroductions in the Ohio River system. *Conservation Genetics*, **9**, 1195–1209.

Drauch AM, Rhodes OE (2007) Genetic evaluation of the lake sturgeon reintroduction program in the Mississippi and Missouri Rivers. *North American Journal of Fisheries Management*, **27**, 434–442.

Eldridge WH, Killebrew K (2008) Genetic diversity over multiple generations of supplementation: an example from Chinook salmon using microsatellite and demographic data. *Conservation Genetics*, **9**, 13–28.

Ewens WJ (1982) On the concept of effective population size. *Theoretical Population Biology*, **21**, 373–378.

Ewing SR, Nager RG, Nicoll MAC et al. (2008) Inbreeding and loss of genetic variation in a reintroduced population of Mauritius kestrel. *Conservation Biology*, **22**, 395–404.

Fischer J, Lindenmayer DB (2000) An assessment of the published results of animal relocations. *Biological Conservation*, **96**, 1–11.

Fitzsimmons NN, Buskirk SW, Smith MH (1997) Genetic changes in reintroduced Rocky Mountain bighorn sheep populations. *Journal of Wildlife Management*, **61**, 863–872.

Forbes SH, Boyd DK (1997) Genetic structure and migration in native and reintroduced Rocky Mountain wolf populations. *Conservation Biology*, **11**, 1226–1234.

Gilpin ME, Soule ME (1986) Minimum viable populations: the processes of species extinction. In: *Conservation Biology: The Science of Scarcity and Diversity* (ed. Soule ME), pp. 13–34. Sinauer, Sunderland, MA.

Gompper ME, Stacey PB, Berger J (1997) Conservation implications of the natural loss of lineages in wild mammals and birds. *Conservation Biology*, **11**, 857–867.

Gorman GC, Kim YJ, Yang SY (1978) The genetics of colonization: loss of variability among introduced populations of Anolis lizards (*Reptilia, Lacertilia, Iguanidae*). *Journal of Herpetology*, **12**, 47–51.

Griffith B, Scott JM, Carpenter JW, Reed C (1989) Translocation as a species conservation tool: status and strategy. *Science*, **245**, 477–480.

Haig SM, Ballou JD, Derrickson SR (1990) Management options for preserving genetic diversity: reintroduction of Guam rails to the wild. *Conservation Biology*, **4**, 290–300.

Haig SM, Belthoff JR, Allen DH (1993) Population viability analysis for a small population of red-cockaded woodpeckers and an evaluation of enhancement strategies. *Conservation Biology*, 7, 289–301.

Hedrick PW, Gutierrez-Espeleta GA, Lee RN (2001) Founder effect in an island population of bighorn sheep. *Molecular Ecology*, 10, 851–857.

Hedrick PW, Lacy RC, Allendorf FW, Soule ME (1996) Directions in conservation biology: comments on Caughley. *Conservation Biology*, 10, 1312–1320.

Hedrick PW, Miller PS (1992) Conservation genetics: techniques and fundamentals. *Ecological Applications*, 2, 30–46.

Hicks JF, Rachlow JL, Rhodes OE, Williams CL, Waits LP (2007) Reintroduction and genetic structure: Rocky Mountain elk in Yellowstone and the western states. *Journal of Mammalogy*, 88, 129–138.

Houlden BA, England PR, Taylor AC, Greville WD, Sherwin WB (1996) Low genetic variability of the koala *Phascolarctos cinereus* in south-eastern Australia following a severe population bottleneck. *Molecular Ecology*, 5, 269–281.

International Union for Conservation of Nature (IUCN) (1987) *Translocation of living organisms: introductions, reintroductions, and restocking.* IUCN Position Statement, Gland, Switzerland: IUCN.

International Union for Conservation of Nature (IUCN) (1998) *Guidelines for Reintroductions.* Prepared by the IUCN/SSC Re-introduction Specialist Group, Gland, Switzerland.

Jamieson IG, Tracy LN, Fletcher D, Armstrong DP (2007) Moderate inbreeding depression in a reintroduced population of North Island robins. *Animal Conservation*, 10, 95–102.

Kleiman DG, Beck B, Dietz JM, Dietz A (1991) Costs of a reintroduction and criteria for success: accounting and accountability in the golden lion tamarin conservation program. *Symposia of the Zoological Society of London*, 62, 125–142.

Lacy RC (1987) Loss of genetic diversity from managed populations: interacting effects of drift, mutation, immigration, selection, and population subdivision. *Conservation Biology*, 1, 143–158.

Lacy RC (1994) Managing genetic diversity in captive populations of animals. In: *Restoration and Recovery of Endangered Plants and Animals* (eds. Bowles ML, Whelan CJ), pp. 63–89. Cambridge University Press, Cambridge.

Lacy RC (2000) Considering threats to the viability of small populations using individual-based models. *Ecological Bulletins*, 48, 39–51.

Lambert DM, King T, Shepherd LD et al. (2005) Serial population bottlenecks and genetic variation: translocated populations of the New Zealand saddleback (*Philesturnus carunculatus rufusater*). *Conservation Genetics*, 6, 1–14.

Larson S, Jameson R, Bodkin J, Staedler M, Bentzen P (2002) Microsatellite DNA and mitochondrial DNA variation in remnant and translocated sea otter (*Enhydra lutris*) populations. *Journal of Mammalogy*, 83, 893–906.

Latch EK, Applegate RD, Rhodes OE (2006a) Genetic composition of wild turkeys in Kansas following decades of translocations. *Journal of Wildlife Management*, 70, 1698–1703.

Latch EK, Harveson LA, King JS, Hobson MD, Rhodes OE (2006b) Assessing hybridization in wildlife populations using molecular markers: a case study in wild turkeys. *Journal of Wildlife Management*, 70, 485–492.

Latch EK, Rhodes OE (2005) The effects of gene flow and population isolation on the genetic structure of reintroduced wild turkey populations: are genetic signatures of source populations retained? *Conservation Genetics*, 6, 981–997.

Latch EK, Rhodes OE (2006) Evidence for seasonal variation in local genetic structure in the wild turkey. *Animal Conservation*, 9, 308–315.

Leberg PL (1991) Effects of bottlenecks on genetic divergence in populations of the wild turkey. *Conservation Biology*, 5, 522–530.

Leberg PL (1992) Effects of population bottlenecks on genetic diversity as measured by allozymes electrophoresis. *Evolution*, 46, 477–494.

Leberg PL, Ellsworth DL (1999) Further evaluation of the genetic consequences of translocations on southeastern whitetailed deer populations. *Journal of Wildlife Management*, 63, 327–334.

Leberg PL, Firman BD (2008) Role of inbreeding depression and purging in captive breeding and restoration programs. *Molecular Ecology*, **17**, 334–343.

Leberg PL, Stangel PW, Hillestad HO, Marchinton RL, Smith MH (1994) Genetic structure of reintroduced wild turkey and white-tailed deer populations. *Journal of Wildlife Management*, **58**, 698–711.

LeGouar P, Rigal F, Boisselier-Dubayle MC et al. (2008) Genetic variation in a network of natural and reintroduced populations of Griffon vulture (*Gyps fulvus*) in Europe. *Conservation Genetics*, **9**, 349–359.

Letty J, Marchandeau S, Clobert J, Aubineau J (2000) Improving translocation success: an experimental study of anti-stress treatment and release method for wild rabbits. *Animal Conservation*, **3**, 211–219.

Linnell JDC, Aanes R, Swenson JE, Odden J, Smith ME (1997) Translocation of carnivores as a method for managing problem animals: a review. *Biodiversity Conservation*, **6**, 1245–1257.

Luikhart G, Allendorf FW, Cornuet JM, Sherwin WB (1998) Distortion of allele frequency distributions provides a test for recent population bottlenecks. *Journal of Heredity*, **89**, 238–247.

Mackie RJ (2000) History of management of large mammals in North America. In: *Ecology and Management of Large Mammals in North America* (eds. Demarais S, Krausman PR), pp. 292–320. Prentice Hall, Upper Saddle River, NJ.

Maehr DS, Noss RF, Larkin JL (2001) *Large Mammal Restoration: Ecological and Sociological Challenges in the 21st Century*. Island Press, Washington, DC.

Maudet C, Miller C, Bassano B et al. (2002) Microsatellite DNA and recent statistical methods in wildlife conservation management: applications in Alpine ibex (*Capra ibex* [*ibex*]). *Molecular Ecology*, **11**, 421–436.

McArthur RA, Geist V, Johnston RH (1986) Cardiac responses of bighorn sheep to trapping and radio instrumentation. *Canadian Journal of Zoology*, **64**, 1197–1200.

Merila J, Bjorklund M, Baker AJ (1996) The successful founder: genetics of introduced *Carduelis chloris* (greenfinch) populations in New Zealand. *Heredity*, **77**, 410–422.

Miller B, Ralls K, Reading RP, Scott JM, Estes J (1999) Biological and technical considerations of carnivore translocation: a review. *Animal Conservation*, **2**, 59–68.

Mills LS, Smouse PE (1994) Demographic consequences of inbreeding in remnant populations. *The American Naturalist*, **144**, 412–431.

Minckley WL (1995) Translocation as a tool for conserving imperiled fishes: experiences in western United States. *Biological Conservation*, **72**, 297–309.

Mock KE, Latch EK, Rhodes OE (2004) Assessing losses of genetic diversity due to translocation: long-term case histories in Merriam's turkey (*Meleagris gallopavo merriami*). *Conservation Genetics*, **5**, 631–645.

Morrison ML (2002) *Wildlife Restoration: Techniques for Habitat Analysis and Animal Monitoring*. Island Press, Washington, DC.

Moulton MP, Sanderson J (1999) *Wildlife Issues in a Changing World, 2nd edn*. CRC Press, Boca Raton.

Nei M, Maruyama T, Chakraborty R (1975) The bottleneck effect and genetic variability in populations. *Evolution*, **29**, 1–10.

Olech W, Perzanowski K (2002) A genetic background for reintroduction program of the European bison (*Bison bonasus*) in the Carpathians. *Biological Conservation*, **108**, 221–228.

Perez T, Albornoz J, Nores C, Dominguez A (1998) Evaluation of genetic variability in introduced populations of red deer (*Cervus elaphus*) using DNA fingerprinting. *Hereditas*, **129**, 85–89.

Polans NO, Allard RW (1989) An experimental evaluation of the recovery potential of ryegrass populations from genetic stress resulting from restriction of population size. *Evolution*, **43**, 1320–1324.

Polziehn RO, Hamr J, Mallory FF, Strobeck C (2000) Microsatellite analysis of North American wapiti (*Cervus elaphus*) populations. *Molecular Ecology*, **9**, 1561–1576.

Prakash S (1972) Origin of reproductive isolation in the absence of apparent genic differentiation in a geographic isolate of *Drosophila pseudoobscura*. *Genetics*, **72**, 143–155.

Prakash S, Lewontin RC, Hubby JL (1969) A molecular approach to the study of genic heterozygosity in natural populations. IV. Patterns of genic variation in central, marginal, and isolated populations of *Drosophila pseudoobscura. Genetics,* **61,** 841–858.

Rhodes OE, Buford DJ, Miller M, Lutz RS (1995) Genetic structure of reintroduced Rio Grande turkeys in Kansas. *Journal of Wildlife Management,* **59,** 771–775.

Rhodes OE, Reat EP, Heffelfinger JR, deVos JC (2001) Analysis of reintroduced pronghorn populations in Arizona using mitochondrial DNA markers. *Proceedings of the 19th Biennial Pronghorn Antelope Workshop,* **19,** 45–54.

Robert A, Sarrazin F, Couvet D (2004a) Influence of the rate of immigration on the fitness of restored populations. *Conservation Genetics,* **5,** 673–682.

Robert A, Sarrazin F, Couvet D, Legendre S (2004b) Releasing adults versus young in reintroductions: interactions between demography and genetics. *Conservation Biology,* **18,** 1078–1087.

Rowe GT, Beebee JC, Burke T (1998) Phylogeography of the natterjack toad *Bufo calamita* in Britain: genetic differentiation of native and translocated populations. *Molecular Ecology,* **7,** 751–760.

Saccheri I, Kuussaari M, Kankare M et al. (1998) Inbreeding and extinction in a butterfly metapopulation. *Nature,* **392,** 491–494.

Sarrazin F, Barbault R (1996) Reintroduction: challenges and lessons for basic ecology. *Trends in Ecology and Evolution,* **11,** 474–478.

Scribner KT, Stuwe M (1994) Genetic relationships among alpine *ibex Capra ibex* populations re-established from a common ancestral source. *Biological Conservation,* **69,** 137–143.

Selander RK, Kaufman DW (1973) Self-fertilization and genetic population structure in a colonizing land snail. *Proceedings of the National Academy of Sciences USA,* **70,** 186–190.

Serfass TL, Brooks RP, Novak JM, Johns PE, Rhodes OE (1998) Genetic variation among populations of river otters in North America: considerations for reintroduction projects. *Journal of Mammalogy,* **79,** 736–746.

Sigg DP, Goldizen AW, Pople AR (2005) The importance of mating system in translocation programs: reproductive success of released male bridled nailtail wallabies. *Biological Conservation,* **123,** 289–300.

Stephen CL, Whittaker DG, Gillis D, Cox LL, Rhodes OE (2005) Genetic consequences of reintroductions: an example from Oregon pronghorn antelope (*Antilocapra americana*). *Journal of Wildlife Management,* **69,** 1463–1474.

Stockwell CA, Mulvey M, Vinyard GL (1996) Translocations and the preservation of allelic diversity. *Conservation Biology,* **10,** 1133–1141.

Storfer A (1999) Gene flow and endangered species translocations: a topic revisited. *Biological Conservation,* **87,** 173–180.

Sugg DW, Chesser RK, Dobson FS, Hogland JL (1996) Population genetics meets behavioral ecology. *Trends in Ecology and Evolution,* **11,** 338–342.

Taylor CE, Gorman GC (1975) Population genetics of a "colonizing" lizard: natural selection for allozyme morphs in *Anolis grahami. Heredity,* **35,** 241–247.

Thomas JW, Toweill DE (1982) *Elk of North America: Ecology and Management.* Stackpole Books, Mechanicsburg, PA.

Thorne ET, Williams ES (1988) Disease and endangered species: the black-footed ferret as a recent example. *Conservation Biology,* **2,** 66–74.

Tordoff HB, Redig PT (2001) Role of genetic background in the success of reintroduced peregrine falcons. *Conservation Biology,* **15,** 528–532.

Tuberville TD, Clark EE, Buhlmann KA, Gibbons JW (2005) Translocation as a conservation tool: site fidelity and movement of repatriated gopher tortoises (*Gopherus polyphemus*). *Animal Conservation,* **8,** 349–358.

Vernesi C, Crestanello B, Pecchioli E et al. (2003) The genetic impact of demographic decline and reintroduction in the wild boar (*Sus scrofa*): a microsatellite analysis. *Molecular Ecology,* **12,** 585–595.

Vonholdt BM, Stahler DR, Smith DW et al. (2008) The genealogy and genetic viability of reintroduced Yellowstone grey wolves. *Molecular Ecology,* **17,** 252–274.

Weckwerth RP, Wright PL (1968) Results of transplanting fishers in Montana. *The Journal of Wildlife Management,* **32,** 977–980.

Williams CL, Lundrigan B, Rhodes OE (2003) Analysis of microsatellite variation in tule elk. *Journal of Wildlife Management*, **68**, 109–119.

Williams CL, Serfass TL, Cogan R, Rhodes OE (2002) Microsatellite variation in the reintroduced Pennsylvania elk herd. *Molecular Ecology*, **11**, 1299–1310.

Williams RN, Rhodes OE, Serfass TL (2000) Assessment of genetic variance among source and reintroduced fisher populations. *Journal of Mammalogy*, **81**, 895–907.

Wisely SM, Santymire RM, Livieri TM, Mueting SA, Howard J (2008) Genotypic and phenotypic consequences of reintroduction history in the black-footed ferret (*Mustela nigripes*). *Conservation Genetics*, **9**, 389–399.

Witzenberger KA, Hochkirch A (2008) Genetic consequences of animal translocations: a case study using the field cricket, *Gryllus campestris* L. *Biological Conservation*, **141**, 3059–3068.

Wolf CM, Griffith B, Reed C, Temple SA (1996) Avian and mammalian translocations: update and reanalysis of 1987 survey data. *Conservation Biology*, **10**, 1142–1154.

13 Evolutionary toxicology

Lee R. Shugart, Chris W. Theodorakis, and John W. Bickham

It is clear that scientific studies that investigate the influence of environmental pollutants on evolutionary processes are applicable to all species. The discussion here, however, deals with organisms other than humans, and it recognizes, of course, that distinctions, at the molecular level, between toxic effects to humans and to other species often are not discretely distinguishable. The important distinction to be made here is that studies of the effects of environmental contaminants on human health are ultimately concerned with the health of individuals, whereas studies that deal with natural populations of wildlife or other organisms focus on the well-being of populations.

Evolutionary toxicology is the study of the effects of pollutants on the genetics of natural populations (Bickham & Smolen 1994; Bickham et al. 2000; Matson et al. 2006). The emergence of evolutionary toxicology as a new field of scientific investigation was noted in the early 1990s at the Napa Conference on Genetic and Molecular Ecotoxicology held at Yountville, California, in October 1993 (Bickham & Smolen 1994). The stated goal of this new endeavor was to identify environmental pollutants that influence evolutionary processes and to quantify the extent of their effects. The procedures and techniques of modern molecular biology to population genetics were invoked to accomplish this goal.

Evolution includes the change in inherited traits of a population from one generation to the next. Natural factors that drive genetic diversity are gene flow, mutation, natural selection, bottlenecks, and genetic drift (Belfiore & Anderson 2001; Staton et al. 2001). Mutations to genes are changes in the nucleotide sequence of an organism's genetic material that can produce new or altered traits in individuals. These changes, in turn, can produce disease, illness, or susceptibility in numerous ways. Gene flow as a result of interactions with immigrating individuals into a population can cause a variety of effects, including increased genetic diversity, homogenization of genetic patterns across populations, and even hybridization among species. Mutations may be spontaneous or induced by exposure to genotoxins (i.e., ultraviolet and high-energy radiation or certain chemicals). Both processes serve to increase genetic diversity through the introduction of new alleles into a population. Reduced genetic variability can be

We thank Ms. Windy Rose, Dr. Susan Anderson, Ms. Jessica Leet, and Dr. Sean Richards for their permission to include material from their recently published reviews (Rose & Anderson 2005; Leet & Richards, in press).

caused by bottlenecks, which are drastic reductions in a population's size due to natural or manmade events, combined with genetic drift, which produces the random elimination of alleles from a small population.

Chemical toxicants and radiation have the potential to affect genetic systems in two general ways. First, somatic effects result from the direct interaction between a mutagen and the DNA of an organism. These interactions can occur in any cell in the body, including somatic and reproductive tissues, and there is a wide range of consequences for the health and reproductive success of the organism. Environmental pollutants that directly change the integrity of an organism's DNA with subsequent adverse effects are designated genotoxicants. Moreover, genotoxicants can be either natural or man-made mutagens that an organism encounters in its environment. Conversely, environmental pollutants that do not directly interact with the DNA are nongenotoxicants.

Second, population genetic effects can be caused by genotoxicants or nongenotoxicants that have a sufficiently profound effect on survival and reproduction to create a measurable impact on genetic patterns. This latter class of environmental effects results from impaired reproductive success that leads to selection or other indirect changes in population genetic structure or diversity (Rose & Anderson 2005). These effects include the reduction of genetic diversity that results from population bottlenecks with subsequent genetic drift that typically eliminates rare or low-frequency alleles. Reduction in genetic diversity also results from selection at survivorship loci that allows certain individuals to live and successfully reproduce in the contaminated environment with the concomitant loss of the alleles that allow successful reproduction in pristine sites. An increase in genetic diversity results when pollutants induce genetic mutations that create new alleles found only at the contaminated sites. Finally, contaminated environments might also create ecological sinks that alter the patterns of gene-flow. These patterns represent emergent effects and are not necessarily predictable from a knowledge of the toxicity mechanisms of the contaminants. In other words, genotoxicants like polycyclic aromatic hydrocarbons (PAHs), ionizing radiation, and certain heavy metals can have population effects similar to those of nongenotoxicants like many organochlorines and other classes of compounds that act as endocrine disruptors. Notwithstanding the fact that the actions of these compounds at the molecular level are entirely different, their emergent population effects might be similar.

DISRUPTION OF GENE STRUCTURE BY ENVIRONMENTAL POLLUTANTS

Human knowledge about toxic effects of chemicals dates back to early history and the use of plant and animal extracts in hunting (Egyptian papyrus roll ~1,500 BC). A milestone in the study of adverse effects of chemicals on living organisms was the observation by Paracelsus in the sixteenth century that there is a link between the dose of a poison and its biological effect. The enormous industrial and economic development that occurred after World War II resulted in an unprecedented release of untreated anthropogenic chemicals into the environment. Nevertheless, at that time, the idea that chemicals present in the environment might be dangerous to human health was essentially nonexistent. In the 1960s, however, our thinking that chemicals posed no adverse environmental risks began to change (Carson 1962). In the 1970s, toxicological investigations expanded to the study of pollutant effects on populations and entire ecosystems with the emergence of ecotoxicology as a prominent new scientific discipline (Newman 1998; Walker et al. 2006). Today, the detrimental health effects of many of the chemicals found in the environment are well documented, and the biological mechanisms responsible for their toxicity are under extensive investigation (Yu 2005).

Biomarkers

The term *biomarker* is currently defined as a biological response expressed at or below the organism level of biological organization and which can be related to "the exposure to" or "the effect of" chemicals in the exposed organisms (McCarthy & Shugart 1990; Peakall & Shugart 1993).

Because the health of the organism can be affected on exposure to a chemical, toxicological research often focused on those biological measures that might be indicative of toxicity (Depledge et al. 1993). The rationale that the identification and use of sufficiently sensitive biological responses of toxicity (biomarkers) may be useful tools of toxicity assessment hinged on an important paradigm of ecotoxicology: "organisms exposed to chemicals may demonstrate a state of toxicity which is related to the capacity of the organism to withstand the toxic stress." Toxicity is initiated upon exposure with the interaction of environmental chemicals and cellular biomolecules. The potential for the initial toxic effect to cascade through the organism starting at the biochemical level and progressing to the organism level will depend on the resiliency of the individual. Subsequently, deleterious effects may be observed at the population level of biological organization. Thus, biomarkers are viewed as measurable biological responses in an organism exposed to stressors present in its environment and, when appropriately applied, certain cause–effect relationships can be inferred. In some instances, they are predictors of the adverse consequences of that exposure.

Several key workshops, symposia, and conferences were held during the late 1980s and through the 1990s to identify and articulate many of the important concepts associated with biomarkers (Table 13–1). From a historical viewpoint, these various venues were instrumental in establishing the framework, as currently practiced by the scientific community, for the implementation of biomarkers to evaluate environmental toxicity.

Genetic ecotoxicology

Background
Genetic ecotoxicology is the study of the effects of substances or agents (e.g., chemicals and radiation) on the genetic material of natural populations, and subsequent related population- and community-level responses (Anderson et al. 1994). It is a complex discipline that integrates knowledge from diverse fields

Table 13–1. Important workshops, symposia, and conferences on the development, application, and evaluation of biological markers in environmental health

Venue	Topic	Reference
ACS Symposium 1990, Los Angeles, CA, Fall 1988	Biomarkers of environmental contamination	McCarthy & Shugart 1990
8th Pellston Workshop, Keystone, CO, July 23–28, 1989	Biomarkers: Biochemical, physiological, and histological markers of anthropogenic stress	Huggett et al. 1992
NATO Advanced Research Workshop, Netherlands Institute of Sea Research, Texel, The Netherlands, May 12–17, 1991	Biomarkers: Research and application in the assessment of environmental health	Peakall & Shugart 1993
International Workshop, Certosa di Pontignano, Siena, Italy, May 25–27, 1992	Nondestructive biomarkers in vertebrates	Fossi & Leonzio 1994
NATO Advanced Research Workshop, Luso, Portugal, June 1–5, 1992	Use of biomarkers in assessing health and environmental impacts of chemical pollutants	Travis 1993
National Institute of Environmental Health Sciences (NIEHS) Conference, Yountville, CA, October 12–15, 1993	Napa conference on genetic and molecular ecotoxicology	Anderson et al. 1994
NIEHS Study Group, Spring and Fall, 1994	Ecotoxicity and human health: A biological approach to environmental remediation	de Serres & Bloom 1995
North Atlantic Treaty Organization (NATO) Advanced Research Workshop, Cieszyn, Poland, September 21–25, 1997	Biomarkers: A pragmatic basis for remediation of severe pollution in Eastern Europe	Peakall et al. 1999

such as genetic toxicology, ecology, molecular biology, and population genetics (Forbes 1998; Newman and Jagoe 1996).

The basic scientific principles representative of genetic ecotoxicology emerged from observations made in early studies in radiobiology (Rose & Anderson 2005). In the 1950s, radiation exposure was shown to alter the development, growth, and reproduction of mammals, fishes, and invertebrates. Growth retardation, suppression of cell division, and modified cell differentiation were detected in radiation-exposed organisms. Radiation exposure was also linked to gonad sterility as well as reduced fecundity, hatching success, and fertilization success in both vertebrate and invertebrate species. By the 1960s, congenital and developmental abnormalities that occurred in animals exposed to radiation were shown to be associated with chromosome damage and mutations. Similar pathological responses were noted upon exposure to certain chemicals (genotoxicants).

As a subfield of ecotoxicology, genetic ecotoxicology can trace its origin to the search for biological responses (biomarkers) that were indicators of genotoxicity (McCarthy & Shugart 1990; Peakall & Shugart 1993). Scientists were acutely aware that changes in certain biological responses (biomarkers) of organisms could be attributed to the toxic effects of environmental pollutants and exploited these observations in their effort to document environmental health issues (Stein et al. 1992). Specific genetic damage, which was shown to be initiated by the interaction of a genotoxicant with the DNA molecule, became a biomarker for studies in genetic ecotoxicology (Varanasi et al. 1981; Shugart et al. 1992; Shugart & Theodorakis 1998). Early application (e.g., Martineau et al. 1988) of DNA damage as a biomarker for exposure to environmental genotoxicants was greatly facilitated by the availability of sensitive and specific analytical methods that had been developed for human health studies (Rahn et al. 1982).

Significance of early studies with DNA adducts

DNA adduct is the term used to describe the moiety that results from the covalent bonding of a chemical to the DNA molecule. DNA adducts can occur in living organisms upon exposure to chemicals in their environment. The base guanine is by far the most prevalent target, although adducts have been reported for all bases. DNA adduct levels, measured at any point in time, reflect tissue-specific rates of adduct formation and removal, which depend on chemical activation, DNA repair, adduct instability, and tissue turnover. Adduct formation in humans appears to be indicative of molecular dosimetry and suggestive of increased human cancer risk. Nevertheless, this relationship has been defined for only a few carcinogens and remains a compelling challenge for future investigations. The chemical structure of most DNA adduct moieties identified in the early 1990s was characterized by chemically specific techniques using mass, fluorescence, and nuclear magnetic resonance spectrometry (Hemminki et al. 1994). Such information was invaluable in determining the stereo- and region-selectivity of enzymatic reactions involving chemical activation.

The early studies on DNA-adduct detection and identification were part of a broad investigation of chemical carcinogenesis in humans. As early as the late 1940s, it was known that certain chemicals, upon ingestion by the test animal, would bind to cellular macromolecules (Miller & Miller 1947). Two decades later, more detailed studies showed that PAHs could bind to mouse-skin DNA (Brookes & Lawley 1964; Boshman & Heidelberger 1967). Subsequent investigations in animals exposed to potential carcinogenic chemicals focused on two main research areas: 1) metabolism, and 2) reaction with cellular macromolecules. It was soon realized that these two areas were linked because these hydrocarbons undergo metabolic activation within cells to intermediates that react covalently with nucleic acids (Grover & Sims 1968; Brookes & Heidelberger 1969; Gelboin 1969). It should be noted that PAHs are ubiquitous in nature and, for experimental design, many can be obtained in pure form. Some individual PAHs (i.e., dibenz[a,h]anthracene and benzo[a]pyrene [BaP]) are known to cause cancer in experimental animals. For these reasons, and especially because the cellular mechanisms for DNA-adduct formation for all chemicals tested were shown to be similar, this class of chemicals, in particular BaP, became the "gold standard" for DNA adduct studies with respect to metabolic activation, covalent binding to DNA, and detection methodologies (Phillips 1990).

Investigations concerned with the metabolic activation of chemicals showed that the metabolism of chemicals is similar in most organisms and occurs via a cellular detoxication/toxication system in steps: Phase I (biotransformation) and Phase II (conjugation) reactions. The Phase I system involves oxidation by various monooxygenase reactions including epoxidation, hydroxylation, and dealkylation and is catalyzed by the cytochrome P-450 monooxygenase- or mixed-function oxidase (MFO) enzyme system, which are iron-containing hemoproteins. The terminal component of this system, cytochrome P-450, exists in multiple forms. The resulting products of Phase I metabolism may be converted to dihydrodiols and/or conjugated (Phase II) with glutathione, glucuronic acid, or sulfate. An important feature of the MFO enzyme system is that the activities and concentrations of specific isoenzymes of cytochrome P-450 can be induced by exposure to xenobiotics. Paradoxically, during the course of metabolism, certain xenobiotics may form reactive electrophilic intermediates, which are at higher energy contents than the parent compounds and thus have the potential to interact covalently with nucleophilic sites in DNA, ribonucleic acid (RNA), and protein. The metabolism and disposition of PAHs in aquatic organisms (in particular, fish) were well understood in the 1980s (Stegeman 1981; Varanasi et al. 1981, 1989).

Sensitive and specific methods for DNA-adduct measurements were in place by the early 1990s (Phillips 1990). The most frequently used methods included immunoassays and immunohistochemistry (Poirier 1981; Kriek et al. 1984) using adduct-specific antisera, electrochemical detection (Yamamoto & Ames 1987), fluorescence (Rahn et al. 1982; Jeffrey 1991), mass spectrometry (Fedtke & Swenberg 1991), and ^{32}P-postlabeling (Randerath et al. 1981). These methodologies, which were developed primarily for human and experimental animal investigations, were applied to studies with environmental species (Varanasi et al. 1986; Shugart 1998).

Future studies in genetic ecotoxicology

The field of genetic ecotoxicology has expanded significantly in the past several years to include investigations that focus on toxic exposure and subsequent population-level effects with evolutionary implications. The integration of population genetics with genetic ecotoxicology can provide a useful approach for evaluating the long-term and high-order effects of environmental genotoxic pollutants (Würgler & Kramers 1992; Shugart et al. 2003). Biomarkers of genotoxicity are measures of individual responses to environmental pollutants that will improve assessment of organismal fitness (Bickham & Smolen 1994; Theodorakis & Shugart 1999; Theodorakis & Wirgin 2002; Matson et al. 2006; Theodorakis et al. 2006). Thus, to differentiate between evolutionary toxicology and genetic ecotoxicology, it should be noted that the former is essentially the field of science that describes how organisms adapt to polluted environments, whereas the latter describes the genotoxic effects of environmental pollutants from the molecular level to the population-genetic level (Leet and Richards in press).

Population-level consequences

Change in gene frequency is at the heart of adaptation, speciation, and evolution. Bickham and colleagues (2000) and Rose and Anderson (2005) provide

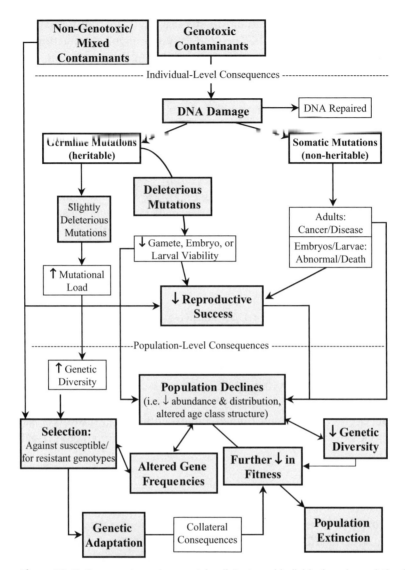

Figure 13–1: Exposure to environmental pollutants and individual- and population-level consequences. By permission: Rose WL, Anderson S I (2005) Genetic ecotoxicology. In: *Encyclopedia of Toxicology*, 2nd edn. (ed. Wexler P), pp. 126–132. Elsevier Ltd., Oxford, UK.

thorough discussions of the complex processes and pathways by which environmental pollutants can interfere with the genetic makeup of individuals and have consequences at the population level.

Figure 13–1 (taken from Rose & Anderson 2005) details a model that relates pollutant (usually a toxic chemical) exposure to population-level consequences. The model incorporates the potential for a pollutant to act as a genotoxicant or a nongenotoxicant.

Genotoxicants act by disrupting gene structure at the cellular and molecular levels either by directly causing DNA damage or via cellular mechanisms, including induction of oxidative stress, inhibition of DNA repair (which eventually can result in DNA damage), and chromosomal breaks or rearrangements. DNA damage not repaired or repaired incorrectly can lead to adverse health effects, or stress, on the individual (Shugart 1996, 1998). These effects are depicted in Figure 13–1 as individual-level consequences to genotoxicant exposure and can be responsible for population-level consequences such as reproductive impairment or high mortality rates. Population-level consequences, in turn, can cause demographic effects that result in reduction of genetic variability in populations or the alteration of gene-flow patterns. Thus, although a genotoxicant initially exerts its effects at the molecular level, it has the potential to initiate a cascade of responses that may influence evolutionary processes, including adaptation and extinction.

Nongenotoxic pollutants are toxic substances that do not alter the genetic material of the exposed organism (Colborn 1994). They may impair reproductive success or lead to selection and indirect changes in population structure (Figure 13–1). A frequently cited example of a nongenotoxic pollutant is dichlorodiphenyltrichloroethane (DDT), a synthetic chemical used as an agricultural insecticide. This chemical and its breakdown products are toxic to embryos and disrupt calcium absorption, resulting in eggshell thinning in birds. Population reduction in several exposed species of birds was due to DDT exposure. Many environmental pollutants, including DDT, are classified as endocrine disruptors, which are exogenous substances that act like hormones in the endocrine system and disrupt the physiologic function of endogenous hormones. Endocrine disruptors are linked to several different adverse biological effects that can affect survival in animals (Colborn et al. 1993). One such way in which environmental pollutants, including endocrine disruptors, can affect survival is by altering gene expression. Epigenetics is the study of the mechanisms of temporal and spatial control of gene activity during the development of complex organisms. Currently, the term is used to describe those processes that establish heritable states of gene expression without altering the DNA sequence of the organism. These processes result in gene expression that is stable between cell division and sometimes between generations. The latter epigenetic phenomenon is referred to as "transgenerational inheritance" (Anway et al. 2005).

Thus, both genotoxic and nongenotoxic pollutants have the capability of impacting population genetic processes by a variety of mechanisms and pathways. The complexities of molecular, cellular, physiological, and reproductive processes that are involved in genetic ecotoxicology belie the relative simplicity of higher-order effects expressed at the population-genetic or evolutionary level. Notwithstanding the fact that recent advances in ecotoxicology have included the discovery of broad new classes of toxic effects (i.e., endocrine disruptors) and the fact that human society now produces chemicals that have never been produced by natural processes and to which all organisms are naïve, we know of no population-genetic or evolutionary processes that are unique to evolutionary toxicology. Thus, the methods and conceptual basis for studies in evolutionary toxicology can be derived directly from evolutionary biology and conservation biology (Bickham & Smolen 1994).

Basically, population genetic responses to xenobiotic stress fall into four categories: alterations in relative amount of genetic diversity; alterations in allele or genotype frequencies as a result of contaminant-induced selection; alterations in gene flow and the genetic relationships among populations; and alterations in allele or genotype frequencies as a result of altered mutation rate (Theodorakis & Wirgin 2002, van Straalen & Timmermans 2002). Reductions in genetic diversity may occur via population crashes or bottlenecks, reduced reproductive output or recruitment (both from within and outside the population), or alterations in the relative birth and death rates. Anthropogenic reductions in genetic diversity are of concern because reduced genetic diversity may affect the growth, evolutionary plasticity, sustainability, and probability of extinction of populations. Also, reduced genetic diversity may lead to increased inbreeding and subsequent fixation of deleterious mutations. Furthermore, it has recently been argued that population genetic diversity may influence higher-level processes such as community structure and ecosystem function (Medina et al. 2007), so that pollutant-induced modifications in genetic diversity may affect these parameters as well (the effects of genetic diversity on community structure has been termed "community genetics" [Whitman et al. 2006]). In addition, polluted habitats are often highly modified, and habitat alteration/destruction and pollution may have additive or synergistic effects on population genetic diversity. For example, it has been found that individuals with lower genetic diversity are more susceptible to the effects of pollution than are individuals with greater genetic diversity (Nowak et al. 2007; Prus-Gowacki et al. 2006). Thus, habitat disturbance or resource exploitation affects on genetic diversity (Hoffmann & Daborn 2007; DiBattista 2008) could exacerbate the effects of environmental pollution on impacted populations. Although the "genetic erosion hypothesis" (van Straalen & Timmermans 2002) holds that pollution should reduce the level of genetic diversity, some studies have found that pollution may increase genetic diversity (Theodorakis and Shugart 1997; Baker et al. 2001; Cohen 2002; Matson et al. 2006; Theodorakis et al. 2006; Tsyusko et al. 2006). Possible explanations for the increase in genetic diversity include altered patterns of dispersal, increased rate of mutation or genomic rearrangements, or diversifying selection (Theodorakis and Shugart. 1997; Baker et al. 2001; Cohen 2002; Theodorakis et al. 2006; Tsyusko et al. 2006).

Alteration of gene or allele frequencies due to contaminant-induced selection may occur if individuals with certain genotypes are more susceptible than other individuals to contaminant exposure. Contaminant-induced selection may also affect the genetic diversity and evolutionary plasticity of populations, at least for loci that are under selective pressure and those that are linked to them. The ecological consequences and significance of genetic adaptations to environmental pollutants are discussed in more detail in Theodorakis and Shugart (1998).

Allele frequencies and genetic relationships may also be affected by pollution-induced changes in gene flow. Because gene flow is often important for maintaining sustainability of populations, particularly in fragmented habitats, alterations of gene flow may affect population persistence or sustainability (Theodorakis & Wirgin 2002). Patterns of genetic relatedness among populations may also be a consequence of alterations of dispersal or occurrence of extinction–recolonization or source–sink dynamics. Thus, patterns of interpopulation genetic diversity may be used as an indication that these events are occurring (Theodorakis & Wirgin 2002). Finally, environmental contamination may serve as a dispersal barrier and lead to decreased gene flow and increased genetic isolation of contaminated populations. Such a situation has been found for plants living in heavy metal–contaminated soils (Mengoni et al. 2001).

Increased mutation rates may also affect genetic diversity of populations. Such mutations lead to increased genetic load in the population, which may affect average fitness of the population. An increased mutation rate is particularly important for small populations, in which increased genetic load may lead to decreased fitness, which would then lead to reduced population size, resulting in increased inbreeding and fixation of deleterious alleles, which would further reduce fitness, and so forth. Thus, such populations may spiral toward extinction in a process known as "mutational meltdown" (Gabriel & Bürger 1994).

Finally, population genetic responses can make important contributions to ecological risk assessments because alterations in population genetic structure or diversity may be sensitive or early-warning indicators of other effects, such as alterations of dispersal and recruitment, population growth or dynamics, loss of species, and changes in community structure (Cronin & Bickham 1998; Theodorakis & Shugart 1998). The study of the effects of pollution on genetic diversity and changes in population genetic structure due to bottlenecks, selection, or mutation is the focus of "evolutionary toxicology" (Bickham & Smolen 1994).

METHODS AND TECHNIQUES APPLICABLE TO EVOLUTIONARY TOXICOLOGY

Overview

There are numerous methods and techniques available to detect structural damage to DNA and other individual-level consequences that result upon exposure to environmental pollutants (Shugart et al. 1992; Shugart 1996, 1998). Genomic technologies are rapidly advancing and provide powerful tools to evaluate subsequent genetic consequences at the population level. The incorporation of these sophisticated technologies to measure change to genetic material extends our knowledge of contaminant effects beyond traditional parameters, such as reproductive and phenotypic effects to population-genetic and evolutionary effects (D'Surney et al. 2001; Rose & Anderson 2005; Leet & Richards, in press).

With respect to genotoxicants, application of appropriate methods and techniques of genetic ecotoxicology can document the cellular progression of genotoxic stress (Shugart 1996) from pollutant exposure that leads to damage of the organism's genetic apparatus, to change in gene expression and other types of irreversible somatic cellular damage like chromosomal damage. Connecting measurable, individual-level effects to eventual population-level consequences (Figure 13–1) is an important aim of evolutionary toxicology. Costs and effort will restrict experimental design, and these frequently result in the use of several methods of evaluation to support findings and gain the knowledge to draw conclusions. This restricted experimental design encourages a weight-of-evidence approach to establish a cause–effect relationship between events at the molecular level and consequent changes at higher levels of biological organization (Walker et al. 2006).

There is a variety of genomic technologies and methods that can be used to generate molecular markers to evaluate a population's genetic structure. All have specific advantages and disadvantages that must be considered when developing a research strategy and drawing conclusions (D'Surney et al. 2001; Rose & Anderson 2005; Leet & Richards, in press). A brief overview is given for several techniques currently used to evaluate genetic diversity. Three methods selected evaluate polymorphism either by analyzing for protein polymorphism (i.e., several alternate forms [alleles] of proteins that relate directly to gene expression) using allozymes or DNA polymorphism (i.e., sequence differences at a particular site in the DNA usually as a result of nucleotide differences or variable numbers of repeated nucleotides) using randomly amplified polymorphic DNA (RAPD) or microsatellite methods. The final method uses microarray technology to address changes in gene expression that occur throughout the genome.

Allozymes

Allozymes are enzymes that vary in their electrophoretic mobility and are indicative of different alleles of single genetic loci. Allozyme genotyping can be an informative and rapid way to evaluate genetic diversity (Bickham et al. 2000). This technique relies on electrophoretic mobility in buffers at specific pH. Genetic diversity is apparent only if amino-acid substitutions lead to change in the net charge of the protein. Because these loci code for proteins with specific functions, ecotoxicological studies using these markers often find diversity that is associated with contaminant-induced selection (Theodorakis & Wirgin 2002). Because they are coding loci, however, selective constraints limit the amount of genetic diversity that can be maintained in populations. Thus, the technique has limited value for evaluating short-term and finer-scale genetic variation. Although they are considered an "older" technology, allozymes have been used in recent evolutionary toxicology studies (Laroche et al. 2002; Mulvey et al. 2002; Roark et al. 2005).

More sensitive proteomic technologies for evaluating protein expression throughout the genome are currently being developed (Bradley et al. 2002; Shrader et al. 2003), and these technologies may lead to identification of candidate loci for studies of selection in polluted populations. For example, Mosquera and colleagues (2003) used two-dimensional gel electrophoresis to examine genetic diversity in Mediterranean mussels (*Mytilus galloprovincialis*). Thirty-three polymorphic loci were found, based on electrophoretic mobility of alleles. Rather than relying on electrophoretic motilities, future studies could use high-throughput proteomic sequencing technologies, such as mass spectrometric techniques, to identify alleles based on amino-acid sequence or mass spectrum. Genetic analyses using amino-acid sequences would be more powerful than those based on identification of electrophoretic variants.

Microsatellites

Microsatellites are noncoding regions of DNA composed of arrays of repeating motifs of short nucleotide patterns. Differences in the number of repeat motifs in an array define microsatellite polymorphisms. Primers (less than twenty-five nucleotides in length) to genomic DNA containing microsatellite loci are developed. Polymerase chain reaction (PCR) amplification produces base-pair products that are separated by gel electrophoresis. The microsatellite allele bands are analyzed using statistical programs to evaluate population-genetic structure. Genetic diversity is assessed on the basis of heterozygosity, allelic richness, and gene differentiation. Microsatellites are rapidly evolving loci that often display high amounts of genetic variation. Thus, microsatellite analysis is conducive to many situations, including field studies, where it may not be possible to collect large numbers of individuals as well as in situations where fine-scale discrimination of recently diverged populations are needed (Leet et al. 2007). If large numbers of microsatellite markers are available, they can also be used for "population genomic" techniques to identify markers linked to functional loci. Finally, because microsatellites are noncoding loci, they are not subject to selective constraints, so mutations are not quickly removed from the population. Because of their high mutability, microsatellites can be used to detect recent mutations. Microsatellite markers have been recently used to examine pollutant-mediated effects on genetic diversity in fish (Whitehead et al. 2003), mice (Berckmoes et al. 2005), mollusks (Piñeira et al. 2008), and plants (Tsyuko et al. 2006). They have also been used to examine mutation rates in swallows (Ellegren et al. 1997).

Dominant anonymous DNA markers

These types of markers typically produce "DNA fingerprint"–like banding patterns on agarose or polyacrylamide gels. Genetic diversity is then quantified on the basis of band presence or absence. Because "homozygous present" bands are indistinguishable from "heterozygous present/absent" bands, they are referred to as "dominant markers" (Theodorakis & Shugart 1998). The main advantage of such techniques is that they analyze large numbers of loci simultaneously and so often reveal high levels of genetic diversity. Two types of dominant markers that have been used in ecotoxicological studies are RAPD and amplified fragment length polymorphisms (AFLP). The following paragraph focuses on PCR-based analysis. Restriction fragment length polymorphism (RFLP)-based analysis of minisatellites has been used in a limited number of studies (e.g., Yauk et al. 2000) and is reviewed elsewhere (Theodorakis & Wirgin 2002).

Various methods have been developed to assay anonymous genetic markers that use PCR to amplify a large number of bands from which patterns or associations can be determined. The best of these is the AFLP method, which uses genomic DNA digested with various restriction enzymes, which are then ligated to short pieces of DNA of known sequence (adaptors). PCR primers that are complementary to the adaptors and the ends of the restriction fragment are then used to amplify these DNA fragments. PCR primers with three random bases at the 3' end are used to reduce the number of amplified bands into a manageable number. These DNA fragments are then separated by polyacrylamide gel electrophoresis and identified by silver staining, stained with fluorescent DNA-binding dyes, or labeled with fluorescent probes. The process results in a banding pattern with typically 20–100 bands per reaction. The RAPD technique also employs PCR amplification of DNA sequences but uses a random set of short primers to generate typically ten to thirty bands per reaction. Problems of reproducibility stem from different PCR conditions resulting in different band profiles and the fact that different laboratories typically produce different profiles. The AFLP technique is more labor-intensive than the related RAPD technique. The reproducibility of AFLP bands is greater, however; thus, this method is recommended. A third technique is inter-simple sequence repeat (ISSR) polymorphisms, which use microsatellite-based primers to amplify DNA sequences between closely spaced microsatellite loci (Pradeep et al. 2002).

The investigations of Theodorakis and Shugart (1999) and Theodorakis and colleagues (2006) demonstrate the utility of anonymous genetic markers as an effective tool to evaluate a population's genetic structure in environmental species. In particular, anonymous genetic markers are useful to screen for patterns among populations with different exposure histories and for identifying markers of selection. The latter is described in more detail in Box 13. The main advantage of the approach is that such markers have applicability to a wide variety of species. The main disadvantage is that rigorous standardization and quality controls must be adhered to in order to ensure reproducibility. Anonymous genetic markers have been used to examine population genetics in contaminated environments for plants (Muller et al. 2004), invertebrates (Ross et al. 2002; De Wolf et al. 2004; Piñeira et al. 2008; Gardeström et al. 2008; Martins et al. 2009), and fish (McMillan et al. 2006; Theodorakis et al. 2006; Williams & Oleksiak 2008).

Organellar DNA

Certain organelles – mitochondria and chloroplasts – contain their own DNA, which can be used for genetic analyses. Mitochondrial DNA (mtDNA) is a closed, circular molecule that exists in multiple copies within individual cells. It is almost always maternally inherited, does not undergo recombination, and contains thirteen protein-encoding genes, two ribosomal RNAs (rRNAs), and twenty-two transfer RNAs (tRNAs) with little intergenic space (Theodorakis & Wirgin 2002). Because mtDNA consists of highly conserved regions interspersed with regions of moderate-to-high genetic variability, it lends itself quite well to PCR-based DNA sequence analysis. The mitochondrial genome evolves at a much higher rate than genomic DNA and so provides a much finer-scale resolution than do genomic coding sequences (Theodorakis & Wirgin 2002). Because of its clonal nature of inheritance, mtDNA is also more sensitive to the effects of population bottlenecks than genomic DNA is. Because mtDNA is readily amenable to DNA sequencing, it can provide a large amount of genetic information, and it can be used in phylogeographic analysis. Recent evolutionary toxicology studies that have employed mtDNA include studies on rodents (Matson et al. 2000; Baker et al. 2001; Theodorakis et al. 2001), frogs (Matson et al. 2006), and invertebrates (Kim et al. 2003; Rocha-Olivares et al. 2004; Chung et al. 2008). Although, theoretically, all copies of mtDNA in a cell (or an organism) should be genetically identical clones, mutations in mtDNA may also lead to a condition known as heteroplasmy, a condition in which there are different mtDNA clones in a cell with different DNA sequences. Heteroplasmy can be used as a marker of increased mutation rates (Wickliffe et al. 2002; Matson et al. 2006), but it also may complicate analyses using mtDNA in population genetic studies.

Chloroplast DNA can also be used in genetic studies of plants in contaminated environments (Mengoni et al. 2001). Many of the same characteristics for mtDNA mentioned earlier in text can also be ascribed to chloroplast DNA. In addition, chloroplast DNA in plants may contain microsatellite sequences (Mengoni et al. 2001; Provan et al. 2001), which combine characteristics of both organelle DNA and nuclear microsatellites.

DNA sequencing and single nucleotide polymorphisms

Analyses of genomic DNA sequences employ some of the advantages of allozymes and mtDNA. The loci are usually of known function, so assessing significance of alterations in DNA sequence may be straightforward. Also, DNA sequences provide a lot of genetic information per locus and are amenable to phylogenetic analyses. Because of the diploid nature of these loci, however, DNA sequencing is not as straightforward. This difficulty can be overcome by using techniques such as single-strand conformational polymorphism to rapidly identify heterozygotes and homozygotes (Grompe 1993). Also, the amount of genetic diversity may be limited by functional and selective constraints. This limitation may be alleviated by examining genetic diversity in noncoding regions, such as introns, but not without sacrificing information on the functional significance of such variation. Analyses employing nuclear coding loci often use either loci with known significance to toxicological responses or fitness or loci with high variability. For example, Tanguy and colleagues (2002) examined genetic variation in the metallothionein gene in oyster (*Crassostrea gigas*) populations exposed to metals. Besides finding evidence of pollution-mediated effects on genetic diversity, tolerance to metals was also dependent on genotype in laboratory experiments and in the field (Tanguy et al. 2002).

In some situations, nuclear coding loci may be both highly variable and of functional significance. One example would be major histocompatibility complex (MHC) genes, which have a high amount of genetic variability. In fact, erosion of genetic variability in these genes has been associated with loss of immune function and increased parasite load (Allen 2008). These genes have been used in ecotoxicological studies in rodents (Pfau et al. 2001) and fish (Cohen et al. 2006).

Another tactic for using genomic loci is to use single nucleotide polymorphisms (SNPs). SNPs are point mutations that result in single base-pair divergence among DNA sequences (Brumfield et al. 2003). SNPs can be present in both coding and noncoding regions of the genome and thus can be used for analyses requiring neutral (e.g., estimates of migration or demographic parameters) and non-neutral markers (e.g., identifying signatures of selection). Because use of SNPs allows the simultaneous analysis of multiple loci and because of the fact that they are the most widespread class of sequence variation, they can provide much genetic information (Brumfield et al. 2003; Schlötterer 2004). SNPs can be identified by a wide variety of methods that are applicable to almost any species, so they can be readily applied to studies of evolutionary toxicology. SNPs are commonly analyzed using conventional methods such as DNA sequencing, allele-specific PCR, and single strand conformation polymorphism (SSCP; Brumfield et al. 2003), but alternative high-throughput methods such as GeneChip assays, mass spectrometric analysis, and pyrosequencing can also be used (Syvänen 2001; see also Chapter 4 by DeWoody and colleagues). Although SNPs hold much promise for evolutionary toxicology, they have not yet been used for this purpose.

DNA microarrays

The incorporation of transcriptomics (the analysis of global gene expression) to evaluate changes in DNA sequences and expressed messenger RNA (mRNA) has greatly improved the detection of altered gene expression in organisms exposed to environmental pollutants (Snape et al. 2004). Nucleic acid microarray technology now enables the simultaneous screening of thousands of genes for differential expression among individuals and populations (Aardema & MacGregor 2002; Watanabe et al. 2007). Gene chips are prepared by printing on glass microscope slides PCR-amplified complementary DNA (cDNA) sequences for particular genes. Relative abundance of gene sequences or transcribed sequences can be compared between samples. Wullschleger and Weston from the Environmental Sciences Division of the Oak Ridge National Laboratory discuss in Box 13 a multistep procedure that incorporates analytical advances from the biomedical community with microarrays to determine up- and down-regulated genes for plants exposed to an environmental stress. Two potential applications of microsatellites in population genetic studies include 1) identification of candidate loci that might be under selection pressure from environmental contaminants, and 2) use as a tool for high-throughput genetic analysis.

Identification of candidate loci using microarrays involves identifying which loci are potentially responsible for tolerance mechanisms and determining if these loci are genetically variable a posteriori. There are two potential strategies that could be employed. First, gene-expression patterns in exposed and nonexposed individuals (either field collected or exposed in the laboratory) can be compared. Second, tolerant and resistant individuals can be obtained – either from polluted and nonpolluted sites or as a result of laboratory selection studies – and exposed to a pollutant. The gene-expression patterns in tolerant individuals could then be compared to those in sensitive individuals. In either case, differentially expressed genes could be cloned and sequenced for use in population genetic studies. Because use of microarrays requires known sequences for cDNA probes, such analyses may not be available for nonmodel organisms. Further information on using microarrays and other techniques for identifying candidate loci involved in adaptation to environmental variables can be found in Hoffmann and Willi (2008).

Microarrays can also be used as a high-throughput platform for determining DNA sequences (Chakravarti 1999). In this case, different alleles are immobilized on microarrays, and the array is hybridized to the genomic DNA of interest. Under stringent hybridization conditions, only those genomic DNA sequences that are completely homologous to the probes on the array will hybridize and provide a signal. By using this approach, genetic diversity in thousands of different loci may be monitored simultaneously (Chakravarti 1999). This technique, however, has yet to be widely applied in studies of population genetics, let alone evolutionary toxicology.

BOX 13: MICROARRAYS AND MOLECULAR PHENOTYPES

Stan D. Wullschleger and David J. Weston

Toxicogenomics is an emerging field of investigation that promises to add new insights to understanding how organisms respond to chemical contaminants or other stressors in aquatic and terrestrial environments (Snell et al. 2003; Klaper & Thomas 2004). Although the challenges likely to be encountered in incorporating genomics into ecotoxicological studies are many (Snape et al. 2004), a range of gene expression and profiling methods are available to help address the complex interactions of a species with its environment. Microarrays are such a technology (Gibson 2002) and provide a platform from which the genome-wide response of an organism to a given toxicant can be related to underlying genes and gene networks. Few groups, however, have tapped the full potential of microarrays especially as they relate to accelerating the discovery of toxicant pathways for biota in aquatic and terrestrial ecosystems.

Scientists from Oak Ridge National Laboratory are tackling this challenge by coupling microarrays with analytical advances from the biomedical community (Lamb et al. 2006). In a multistep procedure (Box Figure 13–1), microarrays are used to determine the up- and down-regulation of genes for plants exposed to a stressor (Step 1). The state of that organism is determined by scanning the observed fingerprint or phenotype against a compendium database that includes many expression signatures derived for plants exposed to a range of stressors. A novel, weighted gene coexpression network approach (Horvath et al. 2006)

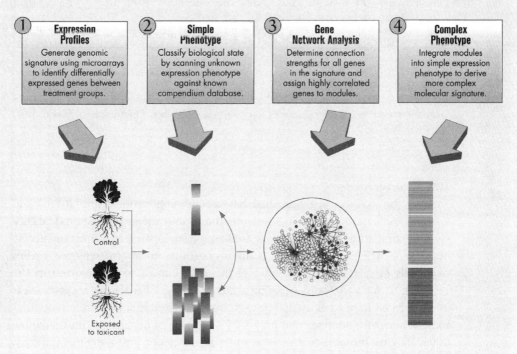

Box Figure 13–1: Steps involved in creating a genomic signature using gene expression data from microarrays.

is then used to determine signaling networks and core genes underlying the expression phenotype (Step 3). This information is integrated with network properties to create a more complex and informative molecular signature (Step 4).

Weston and colleagues (2008) argue that such an approach, if successful and further verified, will allow the biologist to classify the molecular phenotype of an organism subjected to a given stressor and then link that to the genes and signaling pathways that govern the response. Microarrays used in this manner could provide a unique approach or early-warning indicator to the response of aquatic and terrestrial organisms to single- and perhaps multiple-chemical stressors in any environment. Although still early in the development phase, this signature-based approach coupled with network analysis could prove useful as a tool to interrogate the mechanisms that underlie complex biological responses to environmental stressors. It is likely that this concept will be applied in ecological genomics, stress physiology, ecotoxicology, and evolutionary biology. Thus, the use of genomic signatures or molecular phenotypes, as determined by microarray analysis, could serve as a new tool for scientific discovery.

REFERENCES

Gibson G (2002) Microarrays in ecology and evolution: a preview. *Molecular Ecology*, 11, 17–24.

Horvath S, Zhang B, Carlson M et al. (2006) Analysis of oncogenic signaling networks in Glioblastoma identifies ASPM as a molecular target. *Proceedings of the National Academy of Sciences USA*, **103**, 17402–17407.

Klaper R, Thomas MA (2004) At the crossroads of genomics and ecology: the promise of a canary on a chip. *BioScience*, **54**, 403–412.

Lamb J, Crawford ED, Peck D et al. (2006) The connectivity map: using gene-expression signatures to connect small molecules, genes, and disease. *Science*, **313**, 1929–1935.

Snape JR, Maund SJ, Pickford DB, Hutchinson TH (2004) Ecotoxicogenomics: the challenge of integrating genomics into aquatic and terrestrial ecotoxicology. *Aquatic Toxicology*, **67**, 143–154.

Snell TW, Brogdon SE, Morgan MB (2003) Gene expression profiling in ecotoxicology. *Ecotoxicology*, **12**, 475–483.

Weston DJ, Gunter LE, Rogers A, Wullschleger SD (2008) Connecting genes, coexpression modules, and molecular signatures to environmental stress phenotypes in plants. *BMC Systems Biology*, **2**, 16.

Which marker to use

Each of the markers just described has specific applications, limitations, and advantages (Schlötterer 2004). Therefore, the specific marker that should be chosen for population or evolutionary toxicology studies depends on the specific application. For example, if one is interested in analyzing for selective sweeps or signals of selection, then either coding loci known to be involved in the response to such toxicants or using a large number of markers (e.g., AFLPS) is advisable (Whitehead et al. 2003). Codominant markers such as microsatellites may be better for analyses that are based on allele frequencies (Whitehead et al. 2003). Highly mutable loci such as microsatellites and minisatellites (Ellegren et al. 1997; Yauk et al. 2000) might be used in both situations. Also, recent research has found that RAPD markers may provide evidence of somatic mutations or other genetic rearrangements (De Wolf et al. 2004). Thus, RAPD markers may be well suited to examining effects of contaminants in general, but it raises an additional caution about interpreting differences in RAPD marker patterns on the basis of traditional population genetic theory alone. Also, even though allozyme analysis is often considered to be an "obsolete" technology, it consistently uncovers patterns of selection that might not be seen using other markers (Mulvey et al. 2002). In general, it should be kept in mind that any marker has not only its own advantages but its own pitfalls as well, and none are free from the possibility of genotyping errors. Such pitfalls should be thoroughly understood when interpreting data, and judicious methods of identifying and correcting for genotyping errors should also be employed (Bonin et al. 2004; Pompanon et al. 2005).

Assessing causality

Demonstrating population genetic effects of pollution may be problematic for a number of reasons. Such effects may occur as a result of alterations of population growth and dynamics due to acute or chronic toxicity, alterations in behavior, changes in community dynamics (e.g., competition, predatory/prey), or changes in productivity or trophic structure of ecosystems. Thus, changes in population genetic structure and diversity may be emergent properties of perturbed systems rather than direct effects of the toxicants themselves, so determination of the contributing factors to such perturbations may be difficult. One difficulty may be associated with the demography and dynamics of the populations themselves. For example, large effective population sizes or higher dispersal rates may mask population genetic effects of pollution (Pfau et al. 2001; McMillan et al. 2006). Several studies have also found a lack of genetic divergence among contaminated and reference populations, even when it was experimentally demonstrated that organisms from contaminated sites were more tolerant of pollution and that this tolerance was heritable (Muller et al. 2004; McMillan et al. 2006). An explanation for this finding could be 1) rapid evolution of tolerance without reduction of population size, or 2) a small number of loci involved in tolerance relative to the number of loci examined (Muller et al. 2004; McMillan et al. 2006). Conversely, it may appear as though there are contaminant-induced population genetic alterations when none are present due simply to stochastic fluctuations in genetic or demographic parameters. Also, there are many natural deterministic factors that may affect population genetic diversity (e.g., habitat structure, biotic or abiotic variables), irrespective of xenobiotic contamination. Consequently, separating natural from anthropogenic effectors of genetic diversity is an important consideration for population genetic monitoring. In addition, any two populations would be expected to differ in terms of genetic diversity and gene frequencies due to differences in evolutionary history and gene flow. Therefore, merely comparing a contaminated population with a reference population may not provide meaningful information about effects of contamination on population genetic structure.

A more robust approach would be to use a weight-of-evidence approach. The "criteria for assessing causality" outlined by Adams (2005) may be used but, because of the distinctive nature of evolutionary toxicology, some of these criteria may be directly applicable to determining population genetic effects of pollution, whereas others may require some modification. These seven causality criteria include 1) strength of association, 2) consistency of association, 3) specificity of association, 4) time order or temporality, 5) biological gradient, 6) experimental evidence, and 7) plausibility. It should be noted that it is often not possible to meet all of these criteria, but the strength of the causality argument increases with increasing number of criteria that are met. A description of how population genetic analyses can fit these criteria is given in the following section, and additional discussion can be found in Theodorakis and Shugart (1998), Theodorakis and Wirgin (2002), Baker and colleagues (2001), and Theodorakis (2003).

Strength of association

This criterion deals with whether the cause and effect coincide (Adams 2005) or if the endpoint in question is sensitive to pollutant stress. Because any two populations are expected to be genetically divergent, even if they reside in identical environments (due to different evolutionary histories), sampling from multiple reference sites and, if possible, multiple contaminated sites is required. Then, one way of demonstrating that cause and effect coincide or that genetic diversity is sensitive to pollutant stress would be to determine if the difference between the contaminated site(s) and any reference site is greater than what would be expected between any two reference sites. For example, a necessary criterion for establishing that pollution affects population genetic structure would be to determine if the amount of genetic diversity or frequency of a particular allele were lower at multiple polluted sites than at multiple reference sites. Examination of pollutant effects in multiple populations would minimize bias due to population-specific responses. Notable examples include studies that determined genetic variation in the estuarine fish *Fundulus heteroclitus* using allozyme (Mulvey et al. 2002; Roark et al. 2005) and AFLP analyses (McMillan et al. 2006).

Additionally, effects of contamination on population genetic structure could be inferred by determining if the observed patterns are different from what would be expected on the basis of evolutionary theory (e.g., expected genetic relationships based on inferred patterns of gene flow, common ancestry, or geographic distance and distribution). Therefore, careful choice of geographic locations of reference sites could allow testing to see if patterns of genetic diversity and relatedness among reference and polluted sites conform to those predicted by evolutionary theory. For example, Ross and colleagues (2002) examined population genetic diversity in crustaceans, and they chose reference sites that were both north and south of the contaminated site. This choice assured that differences among sites were not due to latitudinal gradients.

When determining if pollution affects population genetic structure, it is also advisable to pay close attention to possible genetic subdivisions or cryptic community diversity. For example, Woodward and colleagues (1996) examined heterozygosity in benthic chironomids and found that differences in heterozygosity between polluted and reference sites were due to a heterogeneous distribution of allele frequencies within the population rather than pollution ("Wahlund effect"). Such an increase in genetic subdivision, however, could also be an effect of anthropogenic disturbance due to physical habitat disturbance or heterogeneously distributed contaminant concentrations. As an even more extreme example of how genetic differences can obfuscate pollution-induced effects, Rocha-Olivares and colleagues (2004) examined genetic diversity in marine harpacticoid copepods (a type of benthic crustacean). They found that pollution-induced changes in population genetic structure were due to the presence of two cryptic species (species that are superficially indistinguishable). In the polluted sites, one of the species was absent, which may appear as a change in the genetic diversity if it is assumed that the two species are part of the same intraspecific population.

In addition, differences in genetic diversity or genotype frequencies between contaminated and reference populations may be due to random factors (e.g., genetic drift, founder effects, stochastic disturbances) or neutral genetic variability (not affected by selection). To test for random effects, bootstrapping procedures (multiple subsampling with replacement) or other mathematical simulations could be used to test if the observed patterns of genetic diversity are statistically different from those generated by random sampling of genotypes. For example, Williams and Oleksiak (2008) used AFLPs to study population genetic structure in an estuarine fish, *F. heroclitis*. They calculated F_{ST} values for each of the AFLP bands and compared these values to simulated distributions of F_{ST} values based on neutrality theory. They identified some markers that were different from expectations based on neutral theory and suggested that these "outliers" may be markers of loci that are under selection by the contaminants. To control for effects due to population history or environmental variables not related to pollution, they examined three separate populations affected by similar contaminants. A slightly different approach was applied by Timmermans and colleagues (2007). They used the ratio of silent to nonsilent mutations in DNA sequences to test the hypothesis of neutral evolution in the coding and promoter regions of the metallothionein gene in metal-exposed populations of the springtail *Orchesella cincta* (a type of soil arthropod). Cohen (2002) used a similar approach to examine MHC loci in polychlorinated biphenyl (PCB)- and PAH-contaminated populations of *F. heroclitus* and found evidence of diversifying selection (resulting in an increased genetic diversity) in contaminated sites. Although rejection of the hypothesis of neutral evolution or nonstochastic distribution of genetic variation is a necessary line of evidence for inferring pollution-induced selection as a causal factor, it is not sufficient in and of itself to unequivocally demonstrate selection. The rejection of the hypothesis can be an important contribution, however, to multiple lines of evidence establishing strength of association and weight of evidence.

Consistency of association

Although demonstration of strength of association in population genetic studies is a necessary criterion for establishing causality, it is not sufficient alone. Another criterion would be demonstration of consistency of association: In other words, are the effects corroborated by other investigators and/or at other places or times, or are similar effects seen in other species from the same locations? For population genetic studies, consistency of association may also apply to the consistent patterns seen when using multiple genetic techniques (e.g., concomitant responses for mtDNA, microsatellites, and allozymes) or multiple loci (e.g., multiple mitochondrial or microsatellite markers, multiple allozyme loci) and multiple types of analyses. For example, Theodorakis and Shugart (1997) and Theodorakis and colleagues (1998) found that RAPD markers that were elevated in frequency in radionuclide-contaminated populations of western mosquitofish (*Gambusia affinis*) in Tennessee were also elevated in frequency in radionuclide-contaminated populations of eastern mosquitofish (*G. holbrooki*) in South Carolina, and Southern blot experiments suggested that the DNA sequences of these loci were homologous. Tsyusko and colleagues (2006) also examined radionuclide-contaminated populations of two species of cattails (*Typha* spp.) and found similar increases in the amount of genetic diversity in both species. They cautioned, however, that their results also could have arisen due to factors such as mode of reproduction (sexual versus asexual) or to patterns of gene flow mediated by natural or anthropogenic factors not related to contamination. In another study, Whitehead and colleagues (2003) examined population structure of the Sacramento sucker (*Catostomus occidentalis*) using both AFLP and microsatellites and found concordant patterns among both markers. In another study, Piñeira and colleagues (2008), using a combination of AFLPs, microsatellites, and fitness-related markers, examined effects of an oil spill on the genetic diversity in mollusks.

It should also be noted that in some cases, different markers may not give the same answer due to the nature of the markers themselves. The same answers may be due to an advantage in evolutionary genetic studies (Whitehead et al. 2003). For example, rapidly evolving and conserved markers may be used to differentiate between patterns of recent genetic change and evolutionary history. Also, because mtDNA is maternally inherited and nuclear loci are biparentally inherited, comparisons of mitochondrial and nuclear genetic diversity may give an indication of sex-biased dispersal (Theodorakis et al. 2001).

Specificity of association

This criterion addresses the process of determining if the effect is diagnostic of exposure (Adams 2005) or, in other words, distinguishing stressor effects from environmental variability. The significance for population genetic studies is that there are environmental factors other than pollution that act as selective pressures or influence genetic diversity via modulation of dispersal, recruitment, population size, growth, and/or subdivision. One method of distinguishing between these natural and anthropogenic effectors would be to determine if patterns of population genetic diversity are concordant with indicators of exposure and effect, both at the individual and population level. Indicators of exposure include chemical body burden data and biomarkers of exposure, and indicators of effect include population- or community-level parameters (e.g., population declines), gross injury (e.g., malformations, fish or bird kills), or biomarkers of effect. For example, Keane and colleagues (2005) found that in populations of dandelion (*Taraxacum officinale*), tissue concentrations of heavy metals and particulate matter were correlated with levels of genetic diversity. Additionally, Benton and colleagues (2002) found that population genetic diversity was correlated to average level of DNA damage in snail (*Pleurocera canaliculatum*) populations. In another study, Maes and colleagues (2005) found that genetic diversity was decreased in polluted populations of eels (*Anguilla anguilla L.*) and that individual levels of heterozygosity were negatively correlated to metal body burdens. Additional discussion of the integration of genetic biomarkers and genetic diversity in ecotoxicological studies can be found in Theodorakis (2001). Also, if contaminant-induced selection is suspected of contributing to patterns of frequencies of alleles or other genetic markers (i.e., individuals with a particular allele have a selective advantage in contaminated habitats), then the fitness components or biomarkers of deleterious effects should be genotype-dependent. For example, Laroche and colleagues (2002) found that in the flounder populations mentioned earlier in text, individual levels of DNA and fitness in individual fish were correlated to genotype and heterozygosity in a manner consistent with population genetic structuring. Furthermore, if contaminant-induced selection on particular alleles or genotypes is contributing to the observed differences in population genetic structure, then the association between biomarker response or fitness and genotype should be seen in contaminated populations but not in uncontaminated populations (such an association in the absence of contamination would indicate that this is a general phenomenon and not related to contamination). For example, Theodorakis and Shugart (1997) found that levels of DNA damage and fecundity were dependent on genotype in radionuclide-contaminated populations of *G. affinis*, but there was no such association in the noncontaminated populations.

Additionally, the experimental design should be such that the reference sites are as similar as possible to the study site and/or the environmental conditions of the reference sites bracket those of the study site. For example, if the contaminated site is a small stream with a silt substrate, then similar streams could be chosen for reference sites. There would still be some environmental variation even between the most similar of sites, so in another tact, the reference sites could be chosen to represent a distribution of environmental conditions. It could then be determined if the patterns of genetic diversity of contaminated sites fall within or outside of this distribution. These analyses should also examine the relationship between geographic distance or spatial structuring and genetic distance or population structure using tests such as Mantel tests or spatial autocorrelation. Such tests can be used to differentiate between genetic relationships mediated by contaminants and those due to geographic distance or distribution. For example, Mulvey and colleagues (2002) found that there was no correlation between geographic distance and genetic distance among populations of mummichog, but Mantel tests showed that there was a correlation between genetic distance and levels of contaminants in the sediments. Natural biogeography should also be taken into account when trying to determine if there are any effects of pollution on population genetics. For instance, in the Whitehead and colleagues (2003) study mentioned earlier in text, it was found that biogeographical factors such as location within a drainage, location of the drainage basin, and watershed geography described the data better than historical pesticide exposure did. Using analysis of molecular variance (AMOVA), they found that watershed geography, rather than contamination history, accounted for much of the partitioning of genetic variation among sites. Similarly, Tsyusko and colleagues (2006) found that latitude, geography, and watershed location had a small but significant impact on microsatellite diversity in cattails near Chernobyl, Ukraine. To examine such relationships in the future, one promising approach may be to incorporate landscape genetic analyses into evolutionary toxicological studies. Landscape genetics analyzes spatial genetic data without the requirement of identifying discrete populations in advance (Manel et al. 2003). As the name implies, it involves spatially explicit interpretations of patterns of genetic variation and gene flow in the context of landscape structure (Manel et al. 2003; Storfer et al. 2006). This could involve integrating spatial analyses and Geographic Information System (GIS) and remote sensing technologies into population genetic and ecotoxicological studies (Storfer et al. 2006).

Examination of the environmental effectors of population genetic structure could also contribute to differentiation between natural and contaminant effects on populations. For example, multivariate techniques could be used to discern which environmental variables contribute the most to genetic differences among reference sites, and then it could be determined if the difference between contaminated and reference sites is different than what would be expected based on these confounding variables. Or, environmental contamination could be included as one of the environmental variables, and multivariate statistics could be used to determine if contamination was one of the major contributors to genetic differences between populations. For example, Timmermans and colleagues (2007), Janssens and coworkers (2007, 2008), and Laroche and colleagues (2002) used multivariate techniques such as principle component analysis and redundancy analysis to determine correlations among contaminant concentrations, fitness parameters, and genotype. These techniques can also be used to test the relationship between genotype and other environmental variables that might affect genetic diversity irrespective of the level of contamination.

Finally, to determine specificity of association, population genetic analyses might benefit from integration with population demographic studies, which might include mark-recapture or grid-trapping studies to obtain estimates of parameters such as population size, sex ratios, age structure, and patterns of dispersal and movement (Matson et al. 2000; Baker et al. 2001). Such studies may also be used to differentiate between alternative hypotheses used to explain differences in genetic diversity, such as changes in mutation or migration rates (Matson et al. 2000; Baker et al. 2001).

Time order or temporality

This criterion implies that the assumed stressor precedes the observed effect and that the pollution effect must diminish when the stressor is remediated. Because it may take many generations for population genetic effects to become manifest, demonstration of this criterion may be difficult for long-lived species. Reversal of effects of pollution is also dependent on gene flow. For species with shorter generation times and higher levels of gene flow, reversal of population genetic effects may be relatively rapid. Reversal would be particularly true for selection-mediated changes in population genetic structure if individuals with resistant genotypes were at a selective disadvantage in noncontaminated environments. In addition, in some cases, it may be possible to monitor differences in population genetic structure before and after a specific event – for example, a chemical or oil spill (Piñeira et al. 2008) or construction of a factory or power plant. Monitoring would be possible only if prior information on genetic diversity is known. If not, then genetic changes after the event can be monitored over time and compared to temporal patterns of nonimpacted populations.

Population genetic effects are distinctive ecotoxicological end points in that the history of the population is reflected in current patterns of genetic diversity. The patterns may contribute to determining whether the stressors preceded population genetic effects. For one thing, different measurements of genetic variability may be able to distinguish between recent and historical effects. For example, number of alleles is more sensitive to recent events than is average heterozygosity. There are newer tests of recent population bottlenecks, and some of them rely on calculating the observed levels of heterozygosity and comparing them to those expected under models that assume mutation/drift equilibrium (Berckmoes et al. 2005). Also, certain markers (e.g., microsatellites) may evolve more quickly than others (e.g., allozymes) or may be more sensitive to the effects of bottlenecks (e.g., mtDNA). Differences between polluted and reference sites being revealed by number of alleles or haplotypes for microsatellites and mtDNA, but not for average heterozygosity, haplotype diversity, or by allozyme analysis, would suggest that the differentiation between contaminated and reference sites has occurred relatively recently.

Phylogenetic relationships may also be a way of discriminating between recent and historical influences on genetic diversity (Avise 1998). The discipline of phylogenetics examines evolutionary relationships between populations or between alleles/haplotypes, which are usually represented visually by a phylogenetic tree. Phylogenetic relationships between populations would be important in discerning patterns of gene flow, as well as in experimental design. For example, in choosing reference sites, it would be best not to develop a sampling scheme where a contaminated site is in one clade (i.e., a branch on a phylogenetic tree representing a group of evolutionarily related individuals) and all reference sites are in a different clade. Otherwise, differences between contaminated and reference sites may be due to evolutionary history rather than contamination. It would be better to either select reference sites that are in the same clade as the contaminated site or choose reference and contaminated sites that are homogeneously distributed among clades. Also, the topology of the tree may provide information as to the relative age of alleles or haplotypes: The younger haplotypes or alleles should be distributed among the terminal branches, whereas the older ones would be more deeply "rooted" in the tree, and this information can be used to infer mutational history. The topology of the dendrogram may also give information about demographic parameters; for example, a tree with long terminal branches may indicate recent population expansion or growth or the direction of dispersal among populations (Nielsen & Beaumont 2009).

A sub-branch of phylogenetics is termed "phylogeography," which integrates evolutionary relationships among alleles or haplotypes with their biogeographical distributions. This integration might provide information as to the effects

of geography on genetic relationships among populations, which can be used
to determine if the pollution has caused the effects. There are statistical phylogenetic approaches: For example, a technique known as nested clade analysis
(Templeton 2008) can be used to distinguish between effectors of gene flow and
evolutionary history on population genetic structure. This technique has been
used by Kim and colleagues (2003), for example, to infer that heavy-metal pollution has restricted gene flow. There has been some recent controversy over
this technique, however: some have argued that the analysis is inherently and
fatally flawed (Petit 2008), whereas others have defended the reproducibility of
the analysis given proper experimental design and rigor of analysis (Garrick et al.
2008; Templeton 2008). Nonetheless, there are also other statistical tests that
can be used for analysis of phylogenetics, and these tests can be applied in a
geographic context (Knowles 2009; Knowles & Maddison 2002; Templeton 2004;
2009; Nielsen & Beaumont 2009).

Biological gradient

This criterion asserts that a gradient in pollution should correspond to a gradient
in genetic diversity and genotype frequencies. Data from biological gradients can
be gathered by sampling sites that are various distances from a known source
of contamination or by sampling various locations that are known to have differing levels of contamination. In one such study, Yauk and colleagues (2000)
used microsatellites to calculate mutation rates in herring gulls, and they found
that the mutation rate decreased with increasing distance from the source. In
addition, Berckmoes and colleagues (2005) found that genetic distance among
populations of mice increased with increasing distance from a metal ore smelter,
but the experimental design precluded distinguishing between effects of contamination and other factors. DeWolf and colleagues (2004) also found that
population genetic diversity varied in concordance with a gradient in metals in
a marine gastropod. We should be cautioned, however, that there may be other
gradients that affect genetic diversity (e.g., temperature, soil chemistry) that may
covary independent of pollution exposure. For instance, genetic diversity may
increase in a downstream trend in creeks and rivers regardless of contamination.
Because contaminant concentrations increase downstream of a discharge pipe
(due to dilution, sequestration, and degradation of contaminants), there may be
a correlation between genetic diversity and contaminant concentrations even
though there is no mechanistic connection. In addition, wastewater effluents
may alter the environment in ways that have nothing to do with contamination
(e.g., by increasing water flow in a small stream), and this alteration may indirectly affect genetic diversity. With judicious experimental design and careful
analysis of data, however, such effects can be factored into the interpretation of
the data.

Experimental evidence

Experimental evidence for evolutionary toxicological effects may include laboratory exposures, in situ caging studies, or microcosms/mesocosms. To establish
evidence of causality, similar responses should be seen for field and laboratory
exposed populations. For example, Gardeström and coworkers (2008) exposed

laboratory populations of copepod crustaceans to copper-spiked sediment. Using AFLP markers, they found that exposed populations not only showed reduced genetic diversity but also increased interpopulation differentiation (increased F_{ST} values). They did not find, however, concordant decreases in abundance. They therefore concluded that the observed effects on diversity and F_{ST} were due to selection for resistant genotypes, rather than to genetic bottlenecks or drift (Gardeström et al. 2008). This hypothesis could be further tested by using coalescence-based simulations of null distributions of F_{ST} values discussed earlier in text (Williams & Oleksiak 2008), by using breeding studies to examine heritability of resistance, or by determining if fitness components were genotype-dependent.

In general, genotype dependence of fitness components (e.g., survival, reproduction) can be demonstrated in experimental studies. For example, it has been found that survival in mercury-exposed mosquitofish (*Gambusia holbrooki*) in the laboratory was correlated with allozyme genotype and that this correlation was relatable to population genetic patterns in contaminated populations in the field (Heagler et al. 1993). In subsequent work, Tatara and coworkers (2002) found an increased frequency of the resistant genotypes in microcosm populations of mercury-exposed mosquitofish that was stable over multiple generations. Ross and colleagues (2002) examined genetic diversity in metal-contaminated populations of isopods and prawns and found contamination-associated patterns of genetic diversity in the isopods but not the prawns. Experimental laboratory exposures found that mortality in metal-exposed prawns was not genotype-dependent, but it was for metal-exposed isopods. Another experiment focusing on microcrustaceans was reported by Martins and coworkers (2009). They collected *Daphnia longispina* from acid mine drainage–contaminated and reference populations and cultured them in the laboratory. They then exposed the laboratory populations to metals and recorded average time to death (TTD). Using the Mantel test, they found statistically significant correlations between genetic distance and differences in TTD among populations. They also found that TTD was associated with genotype within populations. All of the studies just mentioned are significant because they provide supporting evidence for patterns of genetic diversity seen in the field.

An alternative experiment would be to artificially select for tolerant and resistant phenotypes and then determine if differences in genotype correspond to differences seen between contaminated and reference sites in the field. Such an approach was attempted by Roark and coworkers (2005) to determine if laboratory selection of *F. heteroclitus* exposed to PCBs would result in shifts in allele frequencies similar to those seen in PCB-contaminated field populations, but the results were equivocal. In a related approach, Theodorakis and Shugart (1997) examined *G. affinis* from two populations: a radionuclide-contaminated population (Pond 3513) that was founded by intentional introduction of individuals from a reference population (Crystal Springs) in 1977 and that reference population. They collected individuals from Crystal Springs and caged them in Pond 3513. Not only did they find that survival and levels of DNA damage were genotype-dependent in the caged individuals, they also found that the genetic distance between the survivors and the indigenous Pond 3513 population was smaller than that between the survivors and the original Crystal Springs population. Crystal Springs fish caged in a noncontaminated pond, however, were still more similar to the original Crystal Springs population than to the population in the pond in which they were caged.

Plausibility

This criterion requires a biologically plausible mechanism whereby the stressors can induce the effects. In an evolutionary toxicological context, this would require that contaminants exist (or have existed in the past) at levels great enough to affect fitness components (growth, survival, reproductive success, embryo/larval development), recruitment, mutation rates, or gene flow. Studies that demonstrate that a contaminant causes effects on population dynamics or demographic parameters could also be used to fulfill this criterion. If there is a hypothesis that the data indicate that there is selection for certain alleles or genotypes, there must be a mechanism for such selection that is consistent with previous findings or theoretical principles. For example, Roark and colleagues (2001) found that allozyme heterozygosity was correlated with levels of hypoxia in mosquitofish in an urban river. Because these allozymes were involved in aerobic respiration, and such loci have been associated with allele-specific differences in respiration (Tatara et al. 2001), a mechanism of adaptation to hypoxia is plausible.

In another notable example, Timmermans and coworkers (2007) and Janssens and colleagues (2007, 2008) examined the DNA sequence of the metallothionein coding and promoter regions in the springtail *O. cincta*. They found that allele frequencies and patterns of genetic diversity differed between metal-contaminated and reference populations. They found one promoter allele in particular, allele "pmtD2" that was elevated in contaminated sites, and multivariate analyses indicated that the frequency of this allele was most strongly associated with environmental cadmium concentrations. Tajima's D test, which is based on the ratio of silent to nonsilent mutations, rejected the hypothesis of neutral variation at this locus. In subsequent studies, Janssens and coworkers (2007) cloned the various promoter alleles into luciferase expression vectors and examined the levels of cadmium-induced expression in vitro. They found that the expression levels mediated by the pmtD2 allele were much higher than those mediated by many other alleles. These findings are consistent with results that showed that heavy-metal tolerance in invertebrates is associated with levels of metallothionein expression and that levels of cadmium-induced metallothionein expression is an additive genetic trait (i.e., subject to selection; reviewed in Timmermans et al. 2007).

The previous studies focused on noncoding *cis*-regulatory elements of genes, but genetic variation may occur in the coding portion of the genes as well. In such an instance, the plausibility criterion would require demonstration that genetic variation results in nonsilent substitutions in amino acid sequence, that such modifications occur within the functional regions of the molecule, and that the variation has ramifications for health or fitness parameters. For example, Cohen (2002) and Cohen and coworkers (2006) found that genetic diversity of MHC alleles in *F. heteroclitus* was apparently affected by environmental contamination. They found evidence of increased genetic diversity, possibly due to diversifying selection. These findings were also concordant with a contaminant-induced alteration in parasite communities (a possible selection pressure) and with the finding that MHC diversity was correlated to parasite load (a possible link to fitness). DNA sequencing of these alleles found that many of the variable locations were in functional regions of the protein, such as antigen-binding sites (Cohen et al. 2006). Continuing work is pursuing evidence of relative immune function among the genotypes (Cohen et al. 2006).

CASE STUDIES

Azerbaijan

Since 1995, studies have been conducted on the effects of pollution on fish and wildlife populations in Azerbaijan, with special interest in the city of Sumgayit. Azerbaijan sits in the midst of the hot spot of Caucuses biodiversity, which is one of the twenty-five richest and most threatened areas on the planet (Myers et al. 2000). The country faces difficult challenges as it develops its post–Soviet era economy. In particular, impacts resulting from poor environmental safety practices during the Soviet era are severe in Baku, Sumgayit, and in many oil fields around the country. The situation in Sumgayit is particularly grave, in that contamination of the environment from multiple industrial sources and the wastewater treatment plant is extensive and threatens the health of city residents. In fact, Sumgayit was recently recognized as one of the world's ten most contaminated cities by the Blacksmith Institute. To make matters worse, many refugees from the occupied areas of western Azerbaijan now live in Sumgayit, including in its industrial zone. These people are particularly at risk due to their living conditions and daily activities.

Wildlife in and around Sumgayit show the effects of chemical exposures. Biomarker studies were conducted of turtles and frogs from the contaminated wetlands (Figs. 13–2 and 13–3) scattered around the city compared to animals collected at more pristine sites located in the mountains and elsewhere around the country (Matson et al. 2005a,b). Within Sumgayit are several sites with high levels of pollutants of various kinds. Mercury is a widespread contaminant in Sumgayit. It has been found at high levels in ponds along the Caspian Sea coast such as the one adjacent to the wastewater treatment plant, as well as near the chlor-alkali plant (the site where an estimated 1,566 tons of mercury have been spilled; Bickham et al. 2003). PAHs are commonly found in oil or produced as by-products of combustion. Some PAHs are highly toxic and, like mercury, are mutagenic. PAHs are high in certain ponds, including the ones near the wastewater treatment plant, as are other potentially mutagenic chemicals like PCBs and many organochlorine pesticides like lindane and DDT. Residues of these chemicals have been found in sediments and in the tissues of turtles and frogs (Swartz et al. 2003; Matson et al. 2005a,b). The presence of such a complex mixture of

Figure 13–2: Two panoramic views of the wetlands adjacent to the industrial wastewater treatment plant in Sumgayit, Azerbaijan (photos courtesy of Cole Matson). In the top panorama (a), the Caspian Sea can be seen in the background.

Figure 13–3: The ponds adjacent to the industrial wastewater treatment plant in Sumgayit, Azerbaijan, are home to a variety of fish and wildlife species that are the focus of ecotoxicological studies (photos courtesy of Cole Matson). (A) shows two species of freshwater turtles including the Caspian turtle (*Mauremys caspica*, closest in the foreground) and European pond turtles (*Emys orbicularis*, the remaining turtles). (B) shows a dice snake (*Natrix tessellata*) consuming a mosquitofish (*Gambusia* sp.). The mosquitofish is an invasive introduced to Europe from the United States. (C) shows the marsh frog (*Rana ridibunda*).

contaminants in the environment and in animal tissues is cause for great concern but, as discussed earlier in text, it does not by itself confirm damage to the wildlife populations.

Biomarker analyses were conducted to determine if the chemical exposures are affecting wildlife. These studies were designed to detect genetic damage in somatic cells of turtles and frogs and, if such damage were present, to identify which contaminant or contaminants were the cause of the damage. To do this, two methods were used to detect genetic damage in blood cells. The micronucleus assay is commonly used as a biomarker of genetic damage; our collaborator, Dr. Grigoriy Palatnikov of the Karaev Institute of Physiology in Baku, has examined blood smears from hundreds of turtles, frogs, and other animals during the course of these studies. Micronuclei form in blood cells as a result of chromosomal breaks or aneuploidy produced by exposures to mutagens. We also use a procedure called flow cytometry (FCM), which is a more automated analysis. Cell-to-cell variation in DNA content results from chromosomal breaks and rearrangements caused by exposure to chemical contaminants (Bickham et al. 1992, 1994). Both the micronucleus and FCM tests are capable of detecting genetic damage in wildlife and fish and have proven to be sensitive biomarkers to the effects of a variety of chemical mutagens and radioactivity (Bickham 1990, 1994). Increases in the frequencies of micronuclei or in the coefficient of variation of cellular DNA content are indicators of genotoxic exposure and effect.

Studies on turtles and frogs showed that micronuclei and FCM tests revealed genetic damage in blood cells from animals taken at the wastewater treatment plant ponds, as well as ponds adjacent to the chlor-alkali plant, compared to matched reference sites. Correlation analyses showed mercury and PAHs to be the likely causes of the observed genetic damage (Swartz et al. 2003; Matson et al. 2005a,b).

Population genetic analyses of frogs were conducted using nucleotide sequences of the mtDNA control region (Matson et al. 2006). Specifically, nucleotide sequences were examined from frogs from Sumgayit, Ali Bairamly, and Alti Agach. The latter two sites served as reference or pristine sites because frogs from those areas did not show evidence of genotoxic damage using FCM or micronuclei.

The mtDNA is a small piece of DNA located outside the cell nucleus. It shows strict maternal inheritance, and typically an individual has only one form of mtDNA. Using the mtDNA sequences, it was shown that frogs from Sumgayit have lower genetic diversity, including both haplotype and nucleotide diversity, than do frogs from the reference areas. A closer examination of the patterns of distribution of the different forms of mtDNA revealed that gene flow, or the effective migration of female frogs from one site to another and subsequent reproduction, was predominantly into the Sumgayit region. Therefore, Sumgayit has acted as an ecological sink, which is indicative of a long-term problem with successful reproduction. That is, excess reproductive output from surrounding pristine areas provides an input of migrants into Sumgayit, and the resident frogs in Sumgayit have relatively poor reproductive success or survival. Thus, the genetics data revealed an important ecological effect of the chemical exposures.

In addition, evidence was found of an increased mutation rate of the mtDNA at the most contaminated site in Sumgayit. The ponds adjacent to the wastewater treatment plant had the highest levels of haplotype and nucleotide diversity among the sites examined within Sumgayit. Moreover, a few individuals were observed that possessed two forms of mtDNA. These "heteroplasmic" frogs show the initial stages of new mutations, and it is assumed that in a few generations the new mutations will become "fixed" in their offspring. The observation of heteroplasmy in rare in nature. Matson and colleagues (2006) examined a total of 207 marsh frogs. Of that total, only seven frogs were observed to be heteroplasmic. Of these, six were from the wastewater treatment plant ponds and one was from an adjacent area. Frogs from other ponds in Sumgayit or from the reference areas showed no evidence of heteroplasmy.

New mutations initially occur in the heteroplasmic state but become fixed due to a bottleneck in the population of mitochondria that occurs during oogenesis in each generation (Pakendorf & Stoneking 2005). Thus, heteroplasmy is typically an ephemeral characteristic and is not expected to last long within a population. It is usually the association of a common haplotype with a rare haplotype; the rare haplotype is inferred to be the new mutation. New mutations are expected to be rare (depending on the population size), to have limited distributions because they have not existed for sufficient time to be broadly distributed, and potentially to exist in the heteroplasmic state. In studies of evolutionary toxicology, it is desirable to differentiate among naturally occurring and pollution-induced mutations. Although a new mutation can never be directly observed to result from a contaminant, the co-occurrence of new mutations with contaminants is strong evidence of cause and effect. In the case of the marsh frogs of Sumgayit, it is precisely the occurrence of heteroplasmy, unique rare alleles, and the subsequent higher diversity estimates for the most contaminated sites that supports the conclusion that these are contaminant-induced mutations.

Considering all of the findings just summarized, it is apparent that wildlife from the Sumgayit industrial zone is experiencing genetic damage both at the somatic level (i.e., cells from the body of the animal) as well as at the population level. In this series of studies, the presence of potentially genotoxic and nongenotoxic chemicals was documented in sediment samples and tissue samples of frogs and turtles. Biomarker studies confirmed somatic chromosomal damage, and correlation analyses indicated that it was likely due to mercury and PAHs, both of which are known genotoxins. On a regional scale, population genetic studies showed that effective female migration primarily into Sumgayit is potentially the result of reduced reproductive output and evidence of an ecological sink. Overall reduced genetic variability for the Sumgayit region is interpreted as the result of an historic bottleneck with subsequent diversity loss due to genetic drift that likely occurred as a result of the construction of a large number of chemical plants beginning in the 1940s. At a finer geographic scale, the presence of heteroplasmic individuals and higher diversity estimates at the most contaminated site within Sumgayit is evidence of induced mutations and thus an increase in the mutation rate. This study illustrates the necessary connections to be made among exposure, biomarker effects, and emergent population genetic effects that are at the core of evolutionary toxicology.

Pigeon River

The Pigeon River is located in North Carolina and Tennessee and drains into the Tennessee River. Since 1908, the Pigeon River has been severely impacted by paper-mill discharges. Since 1988, biomonitoring studies have found that red-breast sunfish (*Lepomis auritis*) have exhibited a reduction in fitness parameters, health indices, and population density. In addition, there has been evidence that fish populations at the contaminated sites may have been ecological sinks and may have been largely sustained by immigration from neighboring nonimpacted streams. The flow of water in this river has been diverted in one section so that the Pigeon River kilometer (PRK) 42 site no longer receives contaminated flow from upstream. Hence, the biomarker and bioindicator analyses (molecular, biochemical, physiological, and population- and community-level end points) measured at these sites indicated that the degree of contaminant impact decreased in the order of PRK 89 > PRK 103 (upstream of the mill) > PRK 27 > PRK 42 Little River > Little Pigeon River. Thus, the biomarker data suggest (at least partial) recovery at site PRK 42 subsequent to diversion of the main river channel. Effects of this contamination on population genetic structure were not examined, however. To this end, DNA was analyzed using RAPD. Genetic analyses included genetic distance between populations, levels of gene flow among populations, and genetic relatedness among the individual fish. It was found that the level of genetic diversity was higher in the Pigeon River populations than in the reference populations. Patterns of genetic distances among populations may have been impacted by the pulp-mill contaminants because they could not be completely explained by drainage patterns. Phylogeographic analysis, maximum likelihood (MLE) analysis, and assignment tests indicated a higher immigrant/emigrant ratio in the contaminated sites, which suggests that the contaminated populations may be more "sinklike." This finding is consistent with the hypothesis that pulp-mill contamination lowers the population density of the affected redbreast sunfish populations because populations in which the density is much lower than the carrying capacity would produce fewer emigrants, and any immigrants would be more likely to become established than in high-density, nonimpacted populations. A sinklike population may experience higher genetic diversity if it receives genetic input from many different populations (thus farther from migration–drift equilibrium than other populations). Finally, a cladistic analysis using neighbor-joining and minimum-spanning trees, and MLE analysis, suggested an elevated mutation rate in the more impacted sites. This finding is consistent with previous research that found an elevated level of DNA strand breakage in the liver and a higher mutagenic potential (using the Ames assay) in muscle extracts in these fish. Thus, the higher genetic diversity in the Pigeon River populations may have been affected by altered gene flow and mutational processes as a result of pulp-mill effluent discharge.

This study illustrates the ways in which multiple lines of evidence can be used to infer causality between chemical exposure and population genetic effects. Regarding the strength of association criterion, this study used multiple reference sites and multiple contaminated sites. The consistency-of-association criterion was met because similar results were obtained with different analyses. Also, the specificity-of-association criterion was met because the patterns of genetic diversity could not be explained by geography alone. The time-order/temporality criterion was met when a change in the river channel diverted contaminated water away from a previously contaminated stretch of river. The patterns of genetic diversity in this site were more similar to the reference site than to the contaminated sites in the same river, indicating possible recovery. In addition, the biological gradient criterion was met because sites closer to the paper mill were more divergent – in terms of pollution or genetic diversity – from reference sites than were the downstream sites. Thus, a gradient of biological responses ranging from biochemical biomarker to community-level effects was observed. The plausibility criterion was met because previous studies showed that the levels of contaminants in this river system were high enough to affect fitness components and population demography, and extracts from these sunfish were found to be mutagenic in vitro (Theodorakis et al. 2006).

STATISTICAL METHODS FOR EVALUATING POPULATION GENETIC STRUCTURE

There are a number of ways in which genetic data can be analyzed. Many early studies used traditional analyses, such as analyses of genetic diversity, population fixation indices, or Mantel tests. A number of approaches have been developed in the last ten to fifteen years, however, that have been getting increased usage in recent years. These include AMOVA, multivariate analyses, coalescence-based approaches, MLE analyses, and Bayesian analyses.

Although it was initially published in 1992 (Excoffier et al. 1992), the AMOVA was not widely applied in evolutionary toxicology until after the turn of the twenty-first century. This analysis is analogous to the parametric analysis of variance tests (Excoffier et al. 1992). AMOVA can test the proportion of overall genetic variation attributable to variation within populations, between populations, and between groups of populations in a region. It can also be used to determine the relative partitioning of genetic variability among groups (e.g., contaminated versus reference sites), between populations within groups, or within populations. In evolutionary toxicology studies, this analysis has been applied to rodents (Theodorakis et al. 2001; Berckmoes et al. 2005), fish (Whitehead et al. 2003; Theodorakis et al. 2006), springtails (Timmermans et al. 2007), and *Daphnia* (Martins et al. 2009).

Another type of analysis that is receiving more attention is multivariate analysis of genetic data. Techniques such as nonmetric multidimensional scaling principal component analysis and redundancy analysis have recently been used in evolutionary toxicology studies (Laroche et al. 2002; Theodorakis et al. 2006; Timmermans et al. 2007). Such analyses can be used to examine genetic relationships among populations without forcing them into a bifurcated tree and to examine relationships between allele frequencies and contaminant concentrations or other environmental variables.

Recently, the use of coalescent theory has also gained increased attention in population genetic studies of contaminated sites (Rosenberg & Nordborg 2002).

Coalescence is a stochastic model that uses simulated genealogies to make inferences regarding the evolutionary processes that shape observed patterns of polymorphism and genetic diversity, both within and between populations. Coalescent models are not used to construct phylogenies but rather estimate parameters
such as recombination, migration, selection, and so forth that may give rise to
the genealogical patterns. The coalescent is commonly used as a simulation tool
to test hypotheses, for example, by simulating a distribution of data sets under
a null hypothesis (e.g., no mutation, selection, migration) and determining how
frequently the observed patterns of polymorphism in an actual data set are simulated. Coalescence theory is also the basis for "population genomic" analyses
(Luikart et al. 2003). This method was used with *F. heroclitus* from Superfund
sites in the eastern United States (McMillan et al. 2006; Williams & Oleksiak
2008). Analysis of multiple loci, such as AFLPs or large numbers of microsatellite
loci, is used to identify loci that show interpopulation relationships that significantly deviate from those expected based on neutral evolution (Luikart et al.
2003). Finally, the simulated distribution of genealogies can be used to calculate
confidence intervals for various population parameters, such as heterozygosity
and migration rates. The coalescent is often used as a basis for MLE and Bayesian
analyses.

MLE analyses are those that use a fixed set of data and underlying probability
model to choose the values of the model parameters that make the data "more
likely" than any other values (Holder & Lewis 2003). MLE can, for example, be
used to estimate population parameters such as mutation and migration events,
given a set of genetic data and a particular model (i.e., an equation that describes
a particular distribution; Beerli & Felsenstein 2001). MLE will then randomly
choose values for the model parameters (e.g., "constants" of an equation) until
values are found that produce a model that best fits the data. MLE can also be used
to construct phylogenies or "trees." In this case, all possible trees are constructed,
and the likelihood that each tree explains the data is calculated. These likelihoods
are based on the chance that different genotypes would have a common ancestor. This chance can be calculated on the basis of the number of different bases
in the two sequences, the rate of mutation from one base to another, and the
type of mutation required to produce two different sequences (e.g., transitions
from one purine to another are more likely than transversion mutations between
purines and pyrimidines). As one would expect, MLE procedures can be quite
computationally intensive and, even with advanced computer programs, such
calculations may take days or even weeks to calculate. In evolutionary toxicology
studies, a Markov Chain Monte Carlo MLE approach was used to estimate asymmetric patterns of gene flow between contaminated and reference populations of
fish (Theodorakis et al. 2001).

Bayesian analysis is similar to MLE analysis in that it employs a likelihood
function, which in Bayesian analyses is a conditional distribution that stipulates
the probability for the observed data given a particular value of the parameters
of the model (Beaumont & Rannala 2004). This analysis is different from an
MLE analysis, in which the outcome is a probability distribution for the data
given a fixed set of parameters. Bayesian analyses may be considered to be fully
probabilistic because not only are the data treated as a random variable, but the

parameters of the model are also treated as random variables. An appealing feature of Bayesian analysis is that it incorporates previous background information or estimations of known parameters (called "priors") into the model specifications. If detailed estimates of this prior information are not available, partial or incomplete information can be used. A Bayesian analysis will use the probability distribution of the data and the distribution of the priors to calculate the posterior distribution. The posterior distribution can be used to make inferences about the parameters, such as a point estimate of the parameter and its 95% confidence interval. Bayesian analysis can also be used in the analysis of genetic structure of a population, to calculate traditional F_{ST} values, to infer changes in population size, to detect the genetic signatures of selection, and for phylogenetic analysis (Holder & Lewis 2003; Beaumont & Rannala 2004). Recent evolutionary toxicology studies that have used Bayesian analysis include laboratory exposures as well as field studies (Gardeström et al. 2008). Bayesian analyses can also be used to calculate assignment tests (Beaumont & Rannala 2004).

Assignment tests are analyses that use genetic information to establish the likelihood that any particular individual would be found in any given population (Manel et al. 2005). Every individual is assigned to a population in which it has the greatest likelihood of occurrence, given its genotype. Individuals that are incorrectly assigned (i.e., sampled from one population but "assigned" to another) may be immigrants from a different sampled population. This test can also identify individuals that may be from a population other than one that was sampled, if the likelihood of that individual occurring in the sampled population is low. For example, Theodorakis and colleagues (2001, 2006) used assignment tests to examine source–sink dynamics in fish and rodent populations exposed to contaminants. Assignment tests can be based on MLE, Bayesian, or multivariate methods (Manel et al. 2005).

Finally, it should be mentioned that the analyses outlined are not foolproof, and that each type of analysis has its limitations and pitfalls, as well as advantages. Thus, just because differences are found between contaminated and reference populations, it should not automatically be assumed that these differences are due to pollution. Due diligence should be exercised to take into account possible alternative explanations, confounding factors, multiple etiologies, and possible errors or misinterpretations of the results. For example, if MLE, Bayesian analysis, or assignment tests indicate that gene flow occurs between contaminated and reference populations, examination of geographic parameters, dispersal abilities, and life-history traits of the species in question should be used to ensure that such dispersal is likely or even possible. In addition, alternative explanations, such as lack of significant genetic subdivision among populations, a high frequency of shared genotypes among sites, or an asymmetric distribution of private alleles, should also be investigated. Also, most of the analyses rely on specific assumptions – such as mutation–drift or migration–drift equilibrium – that might not be met in contaminated populations. For multiple-loci markers, such as AFLPs or other dominant loci, assumptions of independence among loci may be violated (Luikart et al. 2003). Thus, proper attentiveness to potential pitfalls and limitations of each analysis should be used to avoid inappropriate interpretation or overinterpretation of data.

SUMMARY

Studies in the field of evolutionary toxicology focus on changes to the genetic and evolutionary processes of natural populations that occur from exposure to environmental pollution. Research indicates that integration of population genetics with genetic ecotoxicology provides a useful approach for evaluating postulated long-term and higher-order effects. Detecting and quantifying the influence that environmental pollutants exert on genetic diversity and fitness continue to be the major research challenges facing investigators in this field. Techniques and methodologies of molecular biology are currently being applied to address this challenge. Even though many factors contribute to the success or failure of natural populations, the science of evolutionary toxicology is at a stage of maturity sufficient to begin to delineate the contributions from environmental pollution. Nevertheless, to successfully bridge the gap in our knowledge between exposure to the individual and subsequent population-level consequences, new approaches are needed to help us understand the fundamental biological mechanisms at play.

REFERENCES

Aardema MJ, MacGregor JT (2002) Toxicology and genetic toxicology in the new era of "Toxicogenomics": impact of "-Omics" technologies. *Mutation Research.* **499**, 13–25.

Adams SM (2005) Assessing cause and effect of multiple stressors on marine systems. *Marine Pollution Bulletin*, **51**, 649–657.

Allen DL (2008) Making sense (and antisense) of myosin heavy chain gene expression. *American Journal of Physiology – Regulatory, Integrative and Comparative Physiology*, **295**, R206–R207.

Anderson S, Sadinski W, Shugart L et al. (1994) Genetic and molecular ecotoxicology: a research framework. *Environmental Health Perspectives*, **102**, 13–17.

Anway MD, Cupp AS, Uzumcu M, Skinner MK (2005). Epigenetic transgenerational actions of endocrine disruptors and male fertility. *Science*, **308**, 1466–1469.

Avise JC (1998). The history and purview of phylogeography: a personal reflection. *Molecular Ecology*, **7**, 371–379.

Baker R, Bickham A, Bondarkov M et al. (2001). Consequences of polluted environments on population structure: the bank vole (*clethrionomys glareolus*) at Chornobyl. *Ecotoxicology*, **10**, 211–216.

Beaumont MA, Rannala B (2004). The Bayesian revolution in genetics. *Nature Reviews Genetics*, **5**, 251–261.

Beerli P, Felsenstein J (2001). Maximum likelihood estimation of a migration matrix and effective population sizes in a subpopulation by using a coalescent approach. *Proceedings of the National Academy of Sciences USA*, **98**, 4563–4568.

Belfiore NM, Anderson SL (2001). Effects of contaminants on genetic patterns in aquatic organisms: a review. *Mutation Research*, **489**, 97–122.

Benton MJ, Malott ML, Trybula J, Dean DM, Guttman SI (2002). Genetic effects of mercury contamination on aquatic snail populations: allozyme genotypes and DNA strand breakage. *Environmental Toxicology and Chemistry*, **21**, 584–589.

Berckmoes V, Scheirs J, Jordaens K et al. (2005). Effects of environmental pollution on microsatellite DNA diversity in wood mouse (*apodemus sylvaticus*) populations. *Environmental Toxicology and Chemistry*, **24**, 2898–2907.

Bickham JW (1990) Flow cytometry as a technique to monitor the effects of environmental genotoxins on wildlife populations. In: *In Situ Evaluations of Biological Hazards of Environmental Pollutants* (eds. Sandhu SS, Lower WR, de Serres FJ, Suk WA, Tice RR), pp. 97–108. Plenum, New York.

Bickham JW (1994) Genotoxic responses in blood detected by cytogenetic and cytometric assays. In: *Nondestructive Biomarkers in Vertebrates* (eds. Fossi MC, Leonzio C), pp. 141–152. Lewis Publishers, Boca Raton, FL.

Bickham JW, Matson CW, Islamzadey A, et al. (2003). The unknown environmental tragedy in Sumgayit, Azerbaijan, *Ecotoxicology, 12*, 507–510.

Bickham JW, Sandhu S, Herbert PDN, Chikhi L, Athwal R (2000). Effects of chemical contaminants on genetic diversity in natural populations: implications for biomonitoring and ecotoxicology. *Mutation Research, 463*, 33–51.

Bickham JW, Sawin VL, Burton DW, McBee K (1992). Flow cytometric analysis of the effects of triethylenemelamine on somatic and testicular tissues of the rat. *Cytometry, 13*, 368–373.

Bickham JW, Sawin VL, Smolen MJ, Derr JN (1994). Further flow cytometric studies of the effects of triethylenemelamine on somatic and testicular tissues of the rat. *Cytometry, 15*, 222–229.

Bickham JW, Smolen MJ (1994). Somatic and heritable effects of environmental genotoxins and the emergence of evolutionary toxicology. *Environmental Health Perspectives, 102*, 25–28.

Bonin A, Bellemain E, Bronken Eidesen P et al. (2004). How to track and assess genotyping errors in population genetics studies. *Molecular Ecology, 13*, 3261–3273.

Boshman LM, Heidelberger C (1967) Binding of tritium-labeled polycyclic hydrocarbons to DNA of mouse skin. *Cancer Research, 27*, 1678–1682.

Bradley BP, Shrader EA, Kimmel DG, Meiller JC (2002) Protein expression signatures: application of proteomics. *Marine Environmental Research, 54*, 373–377.

Brookes P, Heidelberger C (1969) Isolation and degradation of DNA from cells treated with tritium-labeled 7,12-dimethylbenz[a]pyrene: studies on the nature of the binding of this carcinogen to DNA. *Cancer Research, 29*, 157–161.

Brookes P, Lawley PD (1964) Evidence for the binding of polynuclear aromatic hydrocarbons to the nucleic acids of mouse skin: relation between carcinogenic power of hydrocarbons and their binding to deoxyribonucleic acid. *Nature, 202*, 781–790.

Brumfield RT, Beerli P, Nickerson DA, Edwards SV (2003) The utility of single nucleotide polymorphisms in inferences of population history. *Trends in Ecology & Evolution, 18*, 249–256.

Carson R (1962) *Silent Spring. The New Yorker* magazine, June 16, June 23, and June 30.

Chakravarti A (1999) Population genetics – making sense out of sequence. *Nature Genetics, 21*(Suppl 1), 56–60.

Chung PP, Hyne RV, Mann RM, Ballard JWO (2008) Genetic and life-history trait variation of the amphipod *melita plumulosa* from polluted and unpolluted waterways in eastern Australia. *Science of the Total Environment, 403*, 222–229.

Cohen S (2002) Strong positive selection and habitat-specific amino acid substitution patterns in MHC from an estuarine fish under intense pollution stress. *Molecular Biology and Evolution, 19*, 1870–1880.

Cohen S, Tirindelli J, Gomez-Chiarri M, Nacci D (2006) Functional implications of major histocompatibility (MH) variation using estuarine fish populations. *Integrative and Comparative Biology, 46*, 1016.

Colborn T (1994) The wildlife/human connection: modernizing risk decisions. *Environmental Health Perspectives, 102*, 55–59.

Colborn T, vom Saal FS, Soto AM (1993) Developmental effects of endocrine-disrupting chemicals in wildlife and humans. *Environmental Health Perspectives, 101*, 378–384.

Cronin MW, Bickham JW (1998) A population genetic analysis of the potential for a crude oil spill to induce heritable mutations and impact natural populations. *Ecotoxicology, 7*, 259–278.

D'Surney SJ, Shugart LR, Theodorakis CW (2001) Genetic markers and genotyping methodologies: an overview. *Ecotoxicology, 10*, 201–204.

De Wolf H, Blust R, Backeljau T (2004) The population genetic structure of *littorina littorea* (mollusca: Gastropoda) along a pollution gradient in the Scheldt estuary (the Netherlands) using RAPD analysis. *Science of the Total Environment, 325*, 59–69.

Depledge MH, Amaral-Mendes JJ, Daniel B et al. (1993) The conceptual basis of the biomarker approach. In: *Biomarkers: Research and Application in the Assessment of Environmental Health* (eds. Peakall DB, Shugart LR), pp. 15–29. NATO ASI Series H: Vol. 68. Springer-Verlag, Heidelberg.

de Serres FJ, Bloom AD, editors (1995) *Ecotoxicity and Human Health*. Lewis Publishers, Boca Raton, FL.

DiBattista JD (2008) Patterns of genetic variation in anthropogenically impacted populations. *Conservation Genetics*, **9**, 141–156.

Ellegren H, Lindgren G, Primmer CR, Møller AP (1997) Fitness loss and germline mutations in barn swallows breeding in Chernobyl. *Nature*. **389**, 593–596.

Excoffier L, Smouse P, Quattro J (1992) Analysis of molecular variance inferred from metric distances among DNA haplotypes: application to human mitochondrial DNA restriction data. *Genetics*, **131**, 479–491.

Fedtke N, Swenberg JA (1991) Quantitative analysis of DNA adducts: the potential for mass spectrometric techniques. In: *Molecular Dosimetry and Human Cancer: Analytical, Epidemiological and Social Considerations* (eds. Groopman JD, Skipper PL), pp. 171–188. CRC Press, Boca Raton, FL.

Forbes VE, editor (1998) *Genetics and Ecotoxicology*. Taylor & Francis, Philadelphia, PA.

Fossi MC, Leonzio C, editors (1994) *Nondestructive Biomarkers in Vertebrates*. CRC Press, Boca Raton, FL.

Gabriel W, Bürger R. (1994) Extinction risk by mutational meltdown: synergistic effects between population regulation and genetic drift. In: V Loeschcke, J Tomuik, and SK Jain (Eds.) *Conservation Genetics*, pp 69–84, Birkhäuser, Basel.

Gardeström J, Dahl U, Kotsalainen O et al. (2008) Evidence of population genetic effects of long-term exposure to contaminated sediments – a multi-endpoint study with copepods. *Aquatic Toxicology*, **86**, 426–436.

Garrick R, Dyer R, Beheregaray L, Sunnucks P (2008) Babies and bathwater: a comment on the premature obituary for nested clade phylogeographical analysis. *Molecular Ecology*, **17**, 1401–1403.

Gelboin HV (1969) A microsome-dependent binding of benzo[a]pyrene to DNA. *Cancer Research*, **29**, 1272–1279.

Grompe M (1993) The rapid detection of unknown mutations in nucleic acids. *Nature Genetics*, **5**, 111–117.

Grover PL, Sims P (1968) Enzyme-catalyzed reactions of polycyclic hydrocarbons with deoxyribonucleic acid and protein in vitro. *Biochemical Journal*, **110**, 159–165.

Heagler MG, Newman MC, Mulvey M, Dixon PM (1993) Allozyme genotype in mosquitofish, *Gambusia holbrooki*, during mercury exposure: Temporal stability, concentration effects and field verification. *Environmental Toxicology and Chemistry*, **12**, 385–395.

Hemminki K, Dipple A, Shuker DEG et al. (1994) *DNA Adducts: Identification and Biological Significance*. IACR Scientific Publications No. 125, IARC, Lyon.

Hoffmann AA, Daborn PJ (2007) Towards genetic markers in animal populations as biomonitors for human-induced environmental change. *Ecology Letters*, **10**, 63–76.

Hoffmann AA, Willi Y (2008) Detecting genetic responses to environmental change. *Nature Reviews Genetics*, **9**, 421–432.

Holder M, Lewis PO (2003) Phylogeny estimation: traditional and Bayesian approaches. *Nature Reviews Genetics*, **4**, 275–284.

Huggett RJ, Kimerle RA, Mehrle M Jr, Bergman HL, editors (1992) *Biomarkers: Biochemical, Physiological, and Histological Markers of Anthropogenic Stress*. Lewis Publishers, Boca Raton, FL.

Janssens TKS, Lopéz RR, Mariën J et al. (2008) Comparative population analysis of metallothionein promoter alleles suggests stress-induced microevolution in the field. *Environmental Science and Technology*, **42**, 3873–3878.

Janssens TK, Marien J, Cenijn P et al. (2007) Recombinational micro-evolution of functionally different metallothionein promoter alleles from *Orchesella cincta*. *BMC Evolutionary Biology*, **7**, 88.

Jeffrey AM (1991) Application of fluorescence to analysis of genotoxicity. In: *Molecular Dosimetry and Human Cancer: Analytical, Epidemiological and Social Considerations* (eds. Groopman JD, Skipper PL), pp. 249–261. CRC Press, Boca Raton, FL.

Keane B, Collier MH, Rogstad SH (2005) Pollution and genetic structure of North American populations of the common dandelion (*Taraxacum officinale*). *Environmental Monitoring and Assessment*, **105**, 341–357.

Kim SJ, Rodriguez-Lanetty M, Suh JH, Song JI (2003) Emergent effects of heavy metal pollution at a population level: *Littorina brevicula* a study case. *Marine Pollution Bulletin*, **46**, 74–80.

Knowles L (2009) Statistical phylogeography: The use of coalescent approaches to infer evolutionary history. *Annual Review of Ecology, Evolution, and Systematics*, **40**, 593–612.

Knowles LL, Maddison WP (2002) Statistical phylogeography. *Molecular Ecology*, **11**, 2623–2635.

Kriek E, Den Engelse L, Scherer E, Westra JG (1984) Formation of DNA modification by chemical carcinogens. Identification, localization and quantification. *Biochimica et Biophysica Acta*, **738**, 181–201.

Laroche J, Quiniou L, Juhel G, Auffret M, Moraga D (2002) Genetic and physiological responses of flounder (*Platichthys flesus*) populations to chemical contamination in estuaries. *Environmental Toxicology and Chemistry*, **21**, 2705–2712.

Leet JK, Kovach MJ, Shaw J, Richards SM (2007). Comparison of Genetic Diversity in Mouse Populations Historically Exposed to Polycyclic Aromatic Hydrocarbons. *Abstracts of the 2007 Society of Environmental Toxicology and Chemistry*. November 11–15, 2007. Milwaukee, Wisconsin.

Leet J, Richards S (in press) Genetic ecotoxicology. In: *Ecotoxicology Research Development* (ed. Santos, EB). Nova Science Publication, Inc.

Luikart G, England PR, Tallmon D, Jordan S, Taberlet P (2003) The power and promise of population genomics: from genotyping to genome typing. *Nature Reviews Genetics*, **4**, 981–994.

Maes G, Raeymaekers J, Pampoulie C et al. (2005). The catadromous European eel *Anguilla anguilla* (L.) as a model for freshwater evolutionary ecotoxicology: relationship between heavy metal bioaccumulation, condition and genetic variability. *Aquatic Toxicology*, **73**, 99–114.

Manel S, Gaggiotti OE, Waples RS (2005) Assignment methods: matching biological questions with appropriate techniques. *Trends in Ecology & Evolution*, **20**, 136–142.

Manel S, Schwartz MK, Luikart G, Taberlet P (2003) Landscape genetics: combining landscape ecology and population genetics. *Trends in Ecology & Evolution*, **18**, 189–197.

Martineau D, Lagace A, Beland P et al. (1988) Pathology of stranded beluga whales (*Delphinapterus leucas*) from the St. Lawrence Estuary, Quebec, Canada. *Journal of Comparative Pathology*, **98**, 287–311.

Martins N, Bollinger C, Harper RM, Ribeiro R (2009) Effects of acid mine drainage on the genetic diversity and structure of a natural population of *Daphnia longispina*. *Aquatic Toxicology*, **92**, 104–112.

Matson CW, Lambert MM, McDonald TJ et al. (2006) Evolutionary toxicology: population-level effects of chronic contaminant exposure on the marsh frogs (*Rana ridibunda*) of Azerbaijan. *Environmental Health Perspectives*, **114**, 547–552.

Matson CW, Palatnikov G, Islamzadeh A et al. (2005a) Chromosomal damage in two species of aquatic turtles (*Emys orbicularis* and *Mauremys caspica*) inhabiting contaminated sites in Azerbaijan. *Ecotoxicology*, **14**, 1–13.

Matson CW, Palatnikov G, McDonald TJ et al. (2005b) Patterns of genotoxicity and contaminant exposure: evidence of genomic instability in the marsh frogs (*Rana ridibunda*) of Sumgayit, Azerbaijan. *Environmental Toxicology and Chemistry*, **24**, 2055–2064.

Matson CW, Rodgers BE, Chesser RK, Baker RJ (2000) Genetic diversity of *Clethrionomys glareolus* populations from highly contaminated sites in the Chornobyl region, Ukraine. *Environmental Toxicology and Chemistry*, **19**, 2130–2135.

McCarthy JF, Shugart LR, editors (1990) *Biological Markers of Environmental Contamination*. Lewis Publishers, Inc., Boca Raton, FL.

McMillan AM, Bagley MJ, Jackson SA, Nacci DE (2006) Genetic diversity and structure of an estuarine fish (*Fundulus heteroclitus*) indigenous to sites associated with a highly contaminated urban harbor. *Ecotoxicology*, **15**, 539–548.

Medina MH, Correa JA, Barata C (2007) Micro-evolution due to pollution: possible consequences for ecosystem responses to toxic stress. *Chemosphere*, **67**, 2105–2114.

Mengoni A, Barabesi C, Gonnelli C et al. (2001) Genetic diversity of heavy metal–tolerant populations in *Silene paradoxa* L. (*Caryophyllaceae*): a chloroplast microsatellite analysis. *Molecular Ecology*, **10**, 1909–1916.

Miller EC, Miller JA (1947) The presence and significance of bound aminoazo dyes in the livers of rats fed *p*-dimethylaminoazobenzene. *Cancer Research*, **7**, 469–480.

Mosquera E, López J, Alvarez G (2003) Genetic variability of the marine mussel *Mytilus galloprovincialis* assessed using two-dimensional electrophoresis. *Heredity*, **90**, 432–442.

Muller L, Lambaerts M, Vangronsveld J, Colpaert J (2004) AFLP-based assessment of the effects of environmental heavy metal pollution on the genetic structure of pioneer populations of *Suillus luteus*. *New Phytologist*, 297–303.

Mulvey M, Newman MC, Vogelbein W, Unger MA (2002) Genetic structure of *Fundulus heteroclitus* from PAH-contaminated and neighboring sites in the Elizabeth and York rivers. *Aquatic Toxicology*, **61**, 195–209.

Myers N, Mittermeier RA, Mittermeier CG, Da-Fonseca GAB, Kent J (2000) Biodiversity hotspots for conservation priorities. *Nature*, **403**, 853–858.

Newman MC (1998) *Ecotoxicology*. Ann Arbor Press, Chelsea, MI.

Newman MC, Jagoe CH, editors (1996) *Ecotoxicology – A Hierarchical Treatment*. Lewis Publishers, Boca Raton, FL.

Nielsen R, Beaumont MA (2009) Statistical inferences in phylogeography. *Molecular Ecology*, **18**, 1034–1047.

Nowak C, Jost D, Vogt C et al. (2007) Consequences of inbreeding and reduced genetic variation on tolerance to cadmium stress in the midge *Chironomus riparius*. *Aquatic Toxicology*, **85**, 278–284.

Pakendorf B, Stoneking M (2005) Mitochondrial DNA and human evolution. *Annual Review of Genomics and Human Genetics*, **6**, 165–183.

Peakall DB, Shugart LR, editors (1993) *Biomarkers: Research and Application in the Assessment of Environmental Health*. NATO ASI Series H: Vol. **68**. Springer-Verlag, Heidelberg.

Peakall DB, Walker CH, Migula P, editors (1999) *Biomarkers: A Pragmatic Basis for Remediation of Severe Pollution in Eastern Europe*. NATO Science Series: 2. Environmental Security: Vol. **54**. Kluwer Academic Publishers, London.

Petit RJ (2008) The coup de grace for the nested clade phylogeographic analysis? *Molecular Ecology*, **17**, 516–518.

Pfau RS, McBee K, Van Den Bussche RA (2001) Genetic diversity of the major histocompatibility complex of cotton rats (*Sigmodon hispidus*) inhabiting an oil refinery complex. *Environmental Toxicology and Chemistry*, **20**, 2224–2228.

Phillips DH (1990) Modern methods of DNA adduct determination. In: *Handbook of Experimental Pharmacology* (eds. Cooper CS, Grover PL), pp. 503–546. Springer-Verlag, Heidelberg, Germany.

Piñeira J, Quesada H, Rolán-Alvarez E, Caballero A (2008) Genetic impact of the Prestige oil spill in wild populations of a poor dispersal marine snail from intertidal rocky shores. *Marine Pollution Bulletin*, **56**, 270–281.

Poirier MC (1981) Antibodies to carcinogen-DNA adduct. *Journal of the National Cancer Institute*, **67**, 515–519.

Pompanon F, Bonin A, Bellemain E, Taberlet P (2005) Genotyping errors: causes, consequences and solutions. *Nature Reviews Genetics*, **6**, 847.

Pradeep R, Sarla N, Siddiq EA (2002) Inter simple sequence repeat (ISSR) polymorphism and its application in plant breeding. *Euphytica*, **128**, 9–17.

Provan J, Powell W, Hollingsworth PM (2001) Chloroplast microsatellites: new tools for studies in plant ecology and evolution. *Trends in Ecology & Evolution*, **16**, 142–147.

Prus-Gowacki W, Chudzińska E, Wojnicka-Pótorak A, Kozacki L, Fagiewicz K (2006) Effects of heavy metal pollution on genetic variation and cytological disturbances in the *Pinus sylvestris* L. population. *Journal of Applied Genetics*, **47**, 99–108.

Rahn R, Chang S, Holland JM, Shugart LR (1982) A fluorometric-HPLC assay for quantitating the binding of benzo[a]pyrene metabolites to DNA. *Biochemical and Biophysical Research Communications*, **109**, 262–269.

Randerath K, Reddy MV, Gupta RC (1981) 32P-post-labeling test for DNA damage. *Proceedings of the National Academy of Sciences*, USA, **78**, 6126–6129.

Roark, SA, Andrews JF, Guttman SI (2001) Population genetic structure of the western mosquitofish, *Gambusia affinis*, in a highly channelized portion of the San Antonio River in San Antonio, TX. *Ecotoxicology*, **10**, 223–227.

Roark SA, Nacci D, Coiro L, Champlin D, Guttman SI (2005) Population genetic structure of a nonmigratory estuarine fish (*Fundulus heteroclitus*) across a strong gradient of polychlorinated biphenyl contamination. *Environmental Toxicology and Chemistry*, **24**, 717–725.

Rocha-Olivares A, Fleeger JW, Foltz DW (2004) Differential tolerance among cryptic species: a potential cause of pollutant-related reductions in genetic diversity. *Environmental Toxicology and Chemistry*, **23**, 2132–2137.

Rose WL, Anderson SI (2005) Genetic ecotoxicology. In: *Encyclopedia of Toxicology*, 2nd edn. (ed. P. Wexler), pp. 126–132. Elsevier Ltd., Oxford, UK.

Rosenberg NA, Nordborg M (2002) Genealogical trees, coalescent theory and the analysis of genetic polymorphisms. *Nature Reviews Genetics*, **3**, 380–390.

Ross K, Cooper N, Bidwell JR, Elder J (2002) Genetic diversity and metal tolerance of two marine species: a comparison between populations from contaminated and reference sites. *Marine Pollution Bulletin*, **44**, 671–679.

Schlötterer C (2004) The evolution of molecular markers – just a matter of fashion? *Nature Reviews Genetics*, **5**, 63–69.

Shrader EA, Henry TR, Greeley MS Jr, Bradley BP (2003) Proteomics in Zebrafish exposed to endocrine disrupting chemicals. *Ecotoxicology*, **12**, 485–488.

Shugart LR (1996) Molecular markers to toxic agents. In: *Ecotoxicology – A Hierarchical Treatment* (eds. Newman MC, Jagoe CH), pp. 135–161. Lewis Publishers, Boca Raton, FL.

Shugart LR (1998) Chapter 8: Structural damage to DNA in response to toxicant exposure. In: *Genetics and Ecotoxicology* (ed. Forbes VE), pp. 151–168. Taylor & Francis, Philadelphia, PA.

Shugart LR, Bickham J, Jackim E et al. (1992). DNA alterations. In: *Biomarkers: Biochemical, Physiological, and Histological Markers of Anthropogenic Stress* (eds. Huggett RJ, Kimerle RA, Mehrle PM, Bergman HL), pp. 125–153. Lewis Publishers, Boca Raton, FL.

Shugart LR, Theodorakis CW (1998) New trends in biological monitoring: application of biomarkers to genetic ecotoxicology. *Biotherapy*, **11**, 119–127.

Shugart LR, Theodorakis CW, Bickham AM, Bickham JW (2003) Genetic effects of contaminant exposure and potential impacts on animal populations. In: *Handbook of Ecotoxicology*, 2nd edn. (eds. Hoffman DJ, Rattner BA, Burton GA Jr, Cairns J Jr), pp. 1129–1147. Lewis Publishers, Boca Raton, FL.

Snape JR, Maund SJ, Pickfor DB, Hutchinson TH (2004) Ecotoxicogenomics: the challenge of integrating genomics into aquatic and terrestrial ecotoxicology, *Aquatic Toxicology*, **67**, 143–154.

Staton JI, Schizas NV, Chandler GT, Coull BC, Quanttro JM (2001) Ecotoxicology and population genetics: the emergence of "phylogeographic and evolutionary ecotoxicology." *Ecotoxicology*, **10**, 217–222.

Stegeman JJ (1981) Polynuclear aromatic hydrocarbons and their metabolism in the marine environment. In: *Polycyclic Hydrocarbons and Cancer* (Vol. 3; eds. Gelboin HV, Ts'o POP), pp. 1–25. Academic Press, New York.

Stein JE, Collier TK, Reichert WL et al. (1992) Bioindicators of contaminant exposure and sub-lethal effects: studies with benthic fish in Puget Sound, Washington. *Environmental Toxicology and Chemistry*, **11**, 701–714.

Storfer A, Murphy M, Evans J et al. (2006) Putting the "landscape" in landscape genetics. *Heredity*, **98**, 128–142.

Swartz CD, Donnelly KC, Islamzadey A et al. (2003) Chemical contaminants and their effects in fish and wildlife from the industrial zone of Sumgayit, Republic of Azerbaijan. *Ecotoxicology*, **12**, 509–521.

Syvänen AC (2001) Accessing genetic variation: genotyping single nucleotide polymorphisms. *Nature Reviews Genetics*, **2**, 930–942.

Tanguy A, Boutet I, Bonhomme F, Boudry P, Moraga D (2002) Polymorphism of metallothionein genes in the pacific oyster *Crassostrea gigas* as a biomarker of response to metal exposure. *Biomarkers*, **7**, 439–450.

Tatara CP, Mulvey M, Newman MC (2002) Genetic and demographic responses of mercury-exposed mosquitofish (*Gambusia holbrooki*) populations: temporal stability and reproductive components of fitness. *Environmental Toxicology and Chemistry*, **21**, 2191–2197.

Tatara CP, Newman MC, Mulvey M (2001) Effect of mercury and gpi-2 genotype on standard metabolic rate of eastern mosquitofish (*Gambusia holbrooki*). *Environmental Toxicology and Chemistry*, **20**, 782–786.

Templeton AR (2004) Statistical phylogeography: methods of evaluating and minimizing inference errors. *Molecular Ecology*, **13**, 789–809.

Templeton AR (2008) Nested clade analysis: an extensively validated method for strong phylogeographic inference. *Molecular Ecology*, **17**, 1877–1880.

Templeton AR (2009) Statistical hypothesis testing in intraspecific phylogeography: nested clade phylogeographical analysis vs. approximate Bayesian computation. *Molecular Ecology*, **18**, 319–331.

Theodorakis CW (2001) Integration of genotoxic and population genetic endpoints in biomonitoring and risk assessment. *Ecotoxicology*, **10**, 245–256.

Theodorakis CW (2003) Establishing causality between population genetic alterations and environmental contamination in aquatic organisms. *Human and Ecological Risk Assessment*, **9**, 37–58.

Theodorakis CW, Bickham JW, Lamb T, Medica PA, Lyne TB (2001) Integration of genotoxicity and population genetic analyses in kangaroo rats (*Dipodomys merriami*) exposed to radionuclide contamination at the Nevada test site, USA. *Environmental Toxicology and Chemistry*, **20**, 317–326.

Theodorakis CW, Lee KL, Adams SM, Law CB (2006) Evidence of altered gene flow, mutation rate, and genetic diversity in redbreast sunfish from a pulp-mill–contaminated river. *Environmental Science Technology*, **40**, 337–386.

Theodorakis CW, Shugart LR (1997) Genetic ecotoxicology II: population genetic structure in mosquitofish exposed in situ to radionuclides. *Ecotoxicology*, **6**, 335–354.

Theodorakis CW, Shugart LR (1998) Genetic ecotoxicology III: the relationship between DNA strand breaks and genotype in mosquito fish exposed to radiation. *Ecotoxicology*, **7**, 227–235.

Theodorakis CW, Shugart LR (1999) Natural selection in contaminated habitats: a case study using RAPD genotypes. In: *Genetics and Ecotoxicology* (ed. Forbes VE), pp. 123–150. Taylor & Francis, New York.

Theodorakis CW, Wirgin I (2002) Genetic responses as population-level biomarkers of stress in aquatic ecosystems. In: *Biological Indicators of Aquatic Ecosystem Health* (ed. Adams SM), pp. 147–186. American Fisheries Society, New York.

Timmermans M, Ellers J, Van Straalen N (2007) Allelic diversity of metallothionein in *Orchesella cincta* (L.): traces of natural selection by environmental pollution. *Heredity*, **98**, 311–319.

Travis CC, editor (1993) *Use of Biomarkers in Assessing Health and Environmental Impacts of Chemical Pollutants* (Vol. 250). NATO ASI Series, Series A: Life Sciences. Plenum Press, London.

Tsyusko OV, Smith MH, Oleksyk TK, Goryanaya J, Glenn TC (2006) Genetics of cattails in radioactively contaminated areas around Chornobyl. *Molecular Ecology*, **15**, 2611–2625.

van Straalen N, Timmermans M (2002) Genetic variation in toxicant-stressed populations: an evaluation of the "genetic erosion" hypothesis. *Human and Ecological Risk Assessment*, **8**, 983–1002.

Varanasi U, Nishimoto M, Reichert WL, Eberhart B (1986) Comparative metabolism of benzo[a]pyrene and covalent binding to hepatic DNA in English sole, starry flounder and rat. *Cancer Research*, **46**, 3817–3824.

Varanasi U, Stein JE, Hom T (1981) Covalent binding of benzo[a]pyrene to DNA in fish liver. *Biochemical and Biophysical Research Communications*, **103**, 780–789.

Varanasi U, Stein JE, Nishimoto M (1989) Biotransformation and disposition of polycyclic aromatic hydrocarbons (PAH) in fish. In: *Metabolism of Polycyclic Aromatic Hydrocarbons in the Aquatic Environment* (ed. Varanasi U), pp. 93–184. CRC Press, Boca Raton, FL.

Walker CH, Hopkin SP, Sibly RM, Peakall DB, editors (2006) *Principles of Ecotoxicology*, 3rd edn. Taylor & Francis, Boca Raton, FL.

Watanabe H, Takahachi E, Nakamura Y et al. (2007) Development of *Daphnia magna* DNA microarray for evaluating the toxicity of environmental chemicals. *Environmental Toxicology and Chemistry*, **26**, 669–676.

Whitehead A, Anderson SL, Kuivila KM, Roach JL, May B (2003) Genetic variation among interconnected populations of *Catostomus occidentalis*: implications for distinguishing impacts of contaminants from biogeographical structuring. *Molecular Ecology*, **12**, 2817–2833.

Whitham TG, Bailey JK, Schweitzer JA, Shuster SM, Bangert RK, LeRoy CJ et al. (2006). A framework for community and ecosystem genetics. *Nat Rev Genet* **7**: 510–523.

Wickliffe JK, Chesser RK, Rodgers BE, Baker RJ (2002) Assessing the genotoxicity of chronic environmental irradiation by using mitochondrial DNA heteroplasmy in the bank vole (*Clethrionomys glareolus*) at Chernobyl, Ukraine. *Environmental Toxicology and Chemistry*, **21**, 1249–1254.

Williams LM, Oleksiak MF (2008) Signatures of selection in natural populations adapted to chronic pollution. *BMC Evolutionary Biology*, **8**, 282.

Woodward LA, Mulvey M, Newman MC (1996) Mercury contamination and population level responses in chironomids: Can allozyme polymorphism indicate exposure? *Environmental Toxicology and Chemistry* **15**, 1309–1316.

Würgler FE, Kramers PG (1992) Environmental effects of endotoxins (eco-genotoxicology). *Mutagenesis*, **7**, 321–327.

Yamamoto Y, Ames BN (1987) Detection of lipid hydroperoxides and hydrogen peroxide at picomole levels by HPLC and isoluminol chemiluminescence assay. *Free Radical Biology & Medicine*, **3**, 359–361.

Yauk CL, Fox GA, McCarry BE, Quinn JS (2000). Induced minisatellite germline mutations in herring gulls (*Larus argentatus*) living near steel mills. *Mutation Research* **45**, 211–218.

Yu MH (2005) *Environmental Toxicology, 2nd edn. Biological and Health Effects of Pollutants*. CRC Press, Boca Raton, FL.

Index

conservation, of species (*cont.*)
    ESU viability and, 240, 254–257, 258
    future applications of, 257–262
    integration strategies for, 241
    methodologies for, 247–248
    molecular approaches to, 262
    population identification in, 246–248
    population viability and, 248–254
    Recovery Domains in, 244–245
    risk factor integration in, 252–254
    spatial structure and diversity
      assessments in, 251–252
    terms for, 240
    TRTs in, 244
    VSP and, 246, 254, 257, 258
  with pedigree reconstruction, 285–286
  phylogenetics and, 13–17
    delimiting species and, 14–15
    PSC and, 14
  pollen and seed movement and, with
    landscape fragmentation, 206–207
    management strategies for, 207
  promotion of, from hybridization,
    182–184
  with wildlife reintroduction, 296–314
    development of, 296
    founding event phase of, 303–305
    genetic consequences of, 299–303
    population establishment phase of,
      305–310
    population growth phase of, 310–313
    population theory and, 297–298, 299
    variation predictions for, 298
conspecific sperm precedence (CSP),
    181–182
crustaceans. *See* rusty crayfish, hybridization
    among, zone dynamics as factor in
CSP. *See* conspecific sperm precedence

Darwin, Charles, 1
  geological study by, 1
DDT. *See* dichlorodiphenyltrichloroethane
deoxyribonucleic acid (DNA)
  bar coding for, 18–19
  evolutionary toxicology as influence on,
    evidence of, 321
    adduct studies in, 324–325
    in anonymous markers, 331–332
    detection methods for, 331–337
    marker selection criteria for, 336–337
    in MHC, 333
    with microarrays, 334
    in organelles, 332–333
    with sequencing, 333–334
    in SNPs, 333–334
  fingerprinting from, in conservation
    management, 74, 75
  RAPD and, 90
  sexing assays with, 74–77
deserts, climate change in, community
    genetics and, 66
  drought-adaptive genotypes in, 66

dichlorodiphenyltrichloroethane (DDT), 327
Dinizia excelsa, pollen movement for,
    202–203
discovery of species. *See* species discovery,
    rates of
disease resistance
  in American chestnut trees, 308–309
  in GM crops, 36
"distinct population segments" (DPS),
    243–244
DNA. *See* deoxyribonucleic acid
DNA adduct studies, 324–325
  measurement methods in, 325
  phases of, 324–325
Douglas-firs, association genetics in, 148–151
  mapping studies for, 149–151
  population genomics in, 149
  QTL mapping for, 149
DPS. *See* "distinct population segments"
dune restoration, 214–215
  Ammophila breviligulata and, 214–215
  molecular and phenotypic data for,
    215

E. cyclocarpum, pollen movement for,
    204–206
  mean parameters for, 205
  study sites for, 204–205
ecosystem genetics, 50
  case studies for, 51
  climate change and, 66–68
    in deserts, 66
    in mountain forests, 66
    prediction models for, 67–68
  conservation and management in, 52–53
    population analysis for, 52–53
    three-way interactions in, 52
  for foundation species, 50
    as community drivers, 50
    definition of, 55–56
    dependent communities influenced by,
      51–52
    heritable phenotypes in, 56–57
  for GEOs, 61–62, 66
    ecological consequences of, 62, 65–66
    ecosystem phenotypes in, 62
    fitness of, 62
    as foundation species, 61–62
    native species hybridization by, 62
    nontarget phenotypes in, 62
  heritable phenotypes, 53–59
    AFLP molecular markers for, 56
    for arthropods, 54–55
    for birds, 55
    conservation consequences for, 56–59
    in foundation species, 56–57
    for insects, 55
    Mantel tests for, 56
    for microbes, 55
    with species-area relationships, 58–59
    species differentiation from, 57–58
    support for similar genotypes, 56